Silicon Nitride Bioceramics

B. Sonny Bal • Bryan J. McEntire
Giuseppe Pezzotti
Editors

Silicon Nitride Bioceramics

Springer

Editors
B. Sonny Bal
Orthopaedic Surgeon and Attorney
Bal Consulting LLC
Columbia, MO, USA

Bryan J. McEntire
Emeritus Scientific Officer
SINTX Technologies, Inc.
Salt Lake City, UT, USA

Giuseppe Pezzotti
Professor, Ceramic Physics Laboratory
Kyoto Institute of Technology
Sakyo-ku, Matsugasaki, Kyoto, Japan

ISBN 978-3-031-67046-6 ISBN 978-3-031-67047-3 (eBook)
https://doi.org/10.1007/978-3-031-67047-3

This Springer imprint is published by the registered company Springer Nature Switzerland AG
The registered company address is: Gewerbestrasse 11, 6330 Cham, Switzerland

Foreword

Silicon Nitride Bioceramics

Today, silicon nitride is one of the brightest shining stars in the field of bioceramics: besides its load-bearing qualities, it has an exceptionally interesting bioactivity, demonstrating not only osteoconductivity but also both antibacterial and antiviral effects.

In this book, you will learn more about this material, going all the way from the basics to the forefront of fundamental research, and including its clinical application. Outstanding experts in the field will guide you throughout, including academics like Profs. Giuseppe Pezzotti, Ichiro Nishimura, Robert Heimann, and Stuart Hampshire, as well as Drs. Sonny Bal and Bryan McEntire. They have been instrumental in bringing silicon nitride to the medical device market and thereby to medical clinics and operating rooms for the benefit of patients. While silicon nitride's clinical implementation is already a success for some applications, there is still a lot of research to be done for others.

At Uppsala University, myself, Prof. Håkan Engqvist and colleagues have, for over a decade, tried to contribute to the story of silicon nitride with knowledge on how to develop this material into functional coatings for implants, starting as a local project that later grew into a European collaboration on Life-Long Joints, followed by a European training network on materials and implants for the spine, Nu-Spine, in which I had the pleasure to work with Dr. Bryan McEntire. Our fellow researchers within that network were able to enjoy his collaborative nature, and he was always ready to share his expertise. It is therefore no surprise that Dr. McEntire has been a key driving force behind delivering this knowledge to you.

Wishing you an enjoyable journey on the road to learning about the vast possibilities of silicon nitride and especially the science behind it!

Division of Biomedical Engineering,
Department of Materials Science and
Engineering, Uppsala University
Uppsala, Sweden
30 June 2024

Cecilia Persson

Preface

When Terwilliger and Lange performed their initial sintering studies on silicon nitride (Si_3N_4) in the early 1970s,[1] no one ever imagined that this unique ceramic would someday become a sought-after biomaterial. The interest in Si_3N_4 then was its potential as high-temperature heat-engine components in automotive and aerospace applications. While this hope was never fully realized, credit has to be given to the governments of Europe, the USA, and Japan for sponsoring significant research programs that substantially improved Si_3N_4's physical, thermal, mechanical, and surface chemical properties. Alongside this preeminent undertaking, several smaller research groups quietly began to investigate the biological properties of Si_3N_4. Surprisingly, in vitro and in vivo small animal studies demonstrated its excellent biocompatibility. This led to a seminal human clinical trial conducted in Australia in 1986 that used Si_3N_4 as a spinal arthrodesis device. Uniquely, Si_3N_4 was the first synthetic material to ever be used for this purpose, and a 30+ year follow-up of surviving patients continued to show positive outcomes.[2] During the past two decades, interest in Si_3N_4 as a biomaterial has continued to grow. Not only has its excellent osteoconductivity been validated, but it has also now been shown to be antipathogenic. While remaining bio-friendly to mammalian cells, it is lethal to most nosocomial bacteria, fungi, and viruses (including SARS-CoV-2)[3]. Today, Si_3N_4 stands out as one of the few ceramics with established applications across diverse fields, including industry, transportation, aerospace, and increasingly, biomedicine.

---

[1]Terwilliger, G.R.; Lange, F.F. Hot-Pressing Behavior of Si_3N_4. *J. Am. Ceram. Soc.* **1974**, *57*, 25–29; and Terwilliger, G.R.; Lange, F.F. Pressureless Sintering of Si_3N_4. *J. Mater. Sci.* **1975**, *10*, 1169–1174, doi:10.1007/BF00541399.

[2]Mobbs, R.J.; Rao, P.J.; Phan, K.; Hardcastle, P.; Choy, W.J.; McCartney, E.R.; Druitt, R.K.; Mouatt, C.A.L.; Sorrell, C.C. Anterior Lumbar Interbody Fusion Using Reaction Bonded Silicon Nitride Implants: Long Term Case Series of the First Synthetic ALIF Spacer Implanted in Humans. World Neurosurg. 2018, 120, 256–264.

[3]Pezzotti, G. Silicon Nitride: A Bioceramic with a Gift. ACS Appl. Mater. Interfaces 2019, 11, 26619–26636.

This compilation stems from the unique and multifaceted history of Si_3N_4. While acknowledging its prominent past, this book primarily focuses on Si_3N_4's remarkable future potential as a biomaterial. It serves as the first comprehensive work to explore its biomedical properties and applications in-depth, including detailed analyses of the groundbreaking research behind its osteoconductivity and antipathogenic characteristics. The book is structured to provide readers with a foundational understanding of biomaterials. Subsequent chapters delve into Si_3N_4's structure, bulk properties, and current industrial processing methods. Dedicated chapters explore the intricacies of Si_3N_4's surface chemistry, osteoconductivity, and antipathogenicity. The book then investigates Si_3N_4 composites and coatings, demonstrating that these materials exhibit properties as favorable as bulk Si_3N_4. Two specific applications in orthopedics are subsequently highlighted: spinal arthrodesis and total joint arthroplasty. A dedicated chapter meticulously compares Si_3N_4 with zirconia toughened alumina (ZTA), the current biomedical ceramic market leader. The book concludes by examining the emerging applications of Si_3N_4 in dentistry and prospects for its use in various other biomedical devices.

This book wouldn't be possible without the dedication and expertise of its esteemed contributors. We extend our sincere gratitude to each author and co-author, all world-renowned scientists, and medical practitioners, who have generously shared their invaluable knowledge and insights on Si_3N_4. Additionally, we recognize the vast foundation of research that underpins this field. Our deepest appreciation goes to the countless scientists, engineers, and technicians whose tireless efforts in researching and publishing their findings on Si_3N_4 have paved the way for the advancements explored in this volume. We stand in awe of your collective contribution to our understanding of this remarkable material.

Columbia, MO, USA — B. Sonny Bal
Salt Lake City, UT, USA — Bryan J. McEntire
Kyoto, Japan — Giuseppe Pezzotti

Dedication and Acknowledgments

This book is dedicated to the late Bill Allen MD, who served as the Chief of Orthopaedic Surgery at the University of Missouri, and to Delbert Day, Ph.D., Curators Professor Emeritus at Missouri University of Science & Technology, and founder of Mo-Sci Corporation, in Rolla, Missouri. These mentors encouraged me to pursue research in ceramic materials, a venture that led to my meeting with now-retired Professor Mohamed Rahaman, with whom I had a lasting collaboration at the Rolla campus. Together, we did work on bioactive glasses, and later, on silicon nitride ceramics. That work, in turn, led to my joining SINTX Technologies after I retired from a surgical career. During my leadership of SINTX Technologies, I was fortunate to work with Professor Giuseppe Pezzotti from the Kyoto Institute of Technology, and Dr. Bryan McEntire, Emeritus Scientific Officer of SINTX Technologies. Together, we had an exciting journey researching silicon nitride, making several discoveries along the way that ultimately led to this book. I hope that future scientists, engineers, and investors will take our work further. A very special thanks to all the authors who made this book possible, with their tireless dedication and hard work. On a personal note, I am very blessed and fortunate to have a wonderful and patient family, especially my wife Dana, who has made all the good things happen in my life.

B. Sonny Bal, MD, MBA, JD, Ph.D.

The foundational knowledge and skills acquired during my education at the University of Utah, particularly under the guidance of Professors Henry Eyring, Ivan B. Cutler, Ron Gordon, and Anil Virar, were instrumental in my ability to contribute to this book. Their instruction fostered a deep interest in physical chemistry, particularly the field of ceramics, for which I remain deeply grateful. Additionally, several individuals significantly impacted my understanding of silicon nitride and its unique properties. David Richerson and Dr. Carr Lane Quackenbush were pivotal in demonstrating the potential applications of this seemingly brittle material. I'm deeply grateful for the opportunity to work with and for Dr. Ashok

Khandkar. His entrepreneurial leadership led to the first commercial introduction of silicon nitride as a biomaterial. I am also profoundly indebted to Professor Giuseppe Pezzotti for his mentorship and collaborative spirit. His exceptional intuition, scientific understanding, and unwavering dedication have significantly contributed to our current knowledge of silicon nitride as a biomaterial. Professor Pezzotti's pioneering work in this field is currently recognized by the scientific community and it will undoubtedly continue to be acknowledged by future generations of researchers. Lastly, I acknowledge the invaluable support from my wife, Darla, whose support was essential in compiling this volume.

Bryan J. McEntire, MBA, Ph.D.

I am grateful to my collaborators, administrative staff, and students for their invaluable contributions to my professional and personal development. I would also like to acknowledge the formative influence of my education in Italy, which exposed me to Cartesian rationalism, and my experiences in Japan, which instilled in me the value of diligence and perseverance that are part of Confucian Philosophy. Finally, this work is dedicated to Iori, a young samurai, who is my constant source of inspiration.

Giuseppe Pezzotti, Ph.D.

Contents

About the Editors

B. Sonny Bal, MD, MBA, JD, Ph.D., is an orthopedic surgeon and attorney with extensive experience in joint replacement surgery and the law. He has served as a Professor of Orthopaedic Surgery at the University of Missouri in Columbia, Missouri, and most recently, as CEO and President of SINTX Technologies, a ceramic materials development and manufacturing company in Salt Lake City, UT, USA. He resides in Columbia, MO, USA, with his family.

Bryan J. McEntire, MBA, Ph.D., is Emeritus Scientific Officer for SINTX Technologies Inc, a position he held from 2012 until retiring in 2021. He was Vice President of Manufacturing from 2004 and Vice President of Research from 2006. Dr. McEntire has more than 45 years of experience in research, development, and production of advanced ceramics, including positions at Ceramatec (Plant Manager, Salt Lake City, UT, USA), Norton/TRW Ceramics (Vice President, Northboro, MA, USA), Saint-Gobain Advanced Ceramics (General Manager, E. Granby, CT, USA), and Applied Materials Corporation (Senior Director, Santa Clara, CA, USA). Dr. McEntire is the author or co-author of over 150 peer-reviewed publications, an inventor or co-inventor of 15+ patents, and a Fellow of the American Ceramic Society. He resides in Salt Lake City, UT, USA, with his family.

Giuseppe Pezzotti, Ph.D., is a Professor and the leader of the Ceramic Physics Laboratory of the Kyoto Institute of Technology since 2000; and was Vice President for International Affairs and Collaborative research in Health Science between 2017 and 2024. He holds five doctoral degrees in Materials Science, Quantum Chemistry, Orthopedics, Immunology, and Food Chemistry, all obtained in Japan during the past 37 years. He was one of the first foreign nationals to obtain a tenured full professor position in a Japanese Government University and the only Western-born trustee Vice President in a National Japanese University. He is an Adjunct or Visiting Professor at numerous other Japanese, Italian, and US institutions of higher education. Professor Pezzotti has published 770 scientific papers, 1 book as a single author, 15 chapters, and holds 14 patents. He is a Fellow of the Academy of Science (Bologna Institute) and was awarded the prestigious Prize for Scientific Research of the Ministry of Education of Japan in 2017. In 2021, he received the title of Knight of the Order of the Star of Italy. He resides in Kyoto, Japan, with his family.

Bioceramics: An Overview

Robert B. Heimann

1 Introduction

Bioceramics are a subset of biomaterials that can be defined as biocompatible ceramic materials that are used to replace, repair, or reconstruct tissues. Bioceramics are being developed, optimized, tested, certified, and deployed to interact with biological systems. These materials aim to replace a part or a function of the human body in physiological, safe, reliable, economical, and esthetically acceptable ways [66]. Recent developments suggest that bioceramics are destined to make a large impact on modern medicine as parts of prosthetic devices used in orthopedic, dental, cardiovascular, ophthalmological, and reconstructive surgeries, tissue engineering scaffolds, and drug and gene delivery.

To achieve optimum biomedical functionality and long-term biocompatibility, any foreign material implanted in the human body must abide by a set of properties that will assure no deleterious interaction with living tissue. One of the most important performance requirements of bioceramics is their *biocompatibility*. Biocompatibility is neither a single specific material property nor a single biological phenomenon but is meant to describe a collection of properties and processes that involve different but interdependent interactions between an inorganic material and living tissue. In addition, the term encompasses not only biological safety as stipulated by the ISO 10993 [36–38] and ASTM F981-04 [3] standards but also the mechanical and chemical characteristics, the design, sterilization procedures, and even the packaging of a medical material or device. In a broader context, biocompatibility does not just mean the absence of cytotoxicity but provides for the requirement that a material performs safely in a given application. This also means that different applications of a particular material will require different conditions.

R. B. Heimann (✉)
Technische Universität Bergakademie Freiberg, Freiberg, Germany
e-mail: robert.heimann@ocean-gate.de

B. S. Bal et al. (eds.), *Silicon Nitride Bioceramics*,
https://doi.org/10.1007/978-3-031-67047-3_1

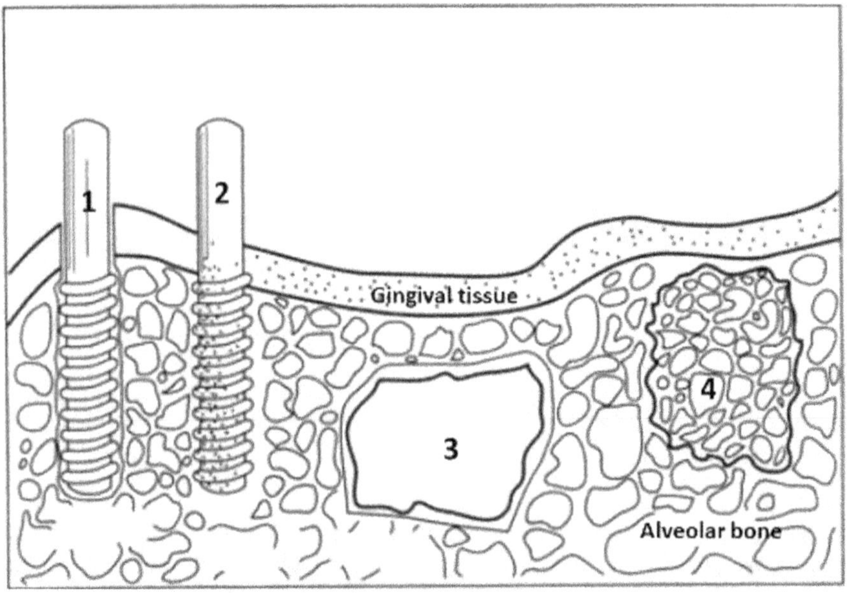

Fig. 1 Classification of bioceramics according to their biological behavior. (1) bioinert single crystal alumina dental implant, (2) bioactive hydroxylapatite coating on a metallic dental implant, (3) surface active bioglass®, and (4) bioresorbable ß-tricalcium phosphate cavity filling. (Modified after Heness and Ben-Nissan and used with permission under Creative Commons License, 2004 [34])

As a consequence, a material, be it a metal, a ceramic, or a polymer that is biocompatible or otherwise suitable in one specific medical application or environment may not be useful in others.

Understanding the various mechanisms with which bioceramic materials interact with living tissue requires knowledge of the nature of the tissue at the interface of an implant. Connective tissues forming the supporting and connecting structures of the body include collagenous, elastic, and reticular fibers, adipose tissue, cartilage, and bone. Bioceramic materials interact chiefly with bone. If a ceramic material is introduced into the body, it inevitably triggers an inflammatory tissue response [60] since no foreign material placed in contact with living tissue known today is completely biocompatible. The only substances conforming totally to the native biocompatibility requirements of living tissue are those fabricated by the body itself. Any other material is recognized by the body as foreign and thus, will trigger some type of host-tissue response. Figure 1 shows the four general types of response allowing different means of attachment of implants to the musculoskeletal system.

The two major types of bioceramics in clinical use today are *bioinert structural ceramics* (e.g., alumina, partially stabilized zirconia, and carbon), and *bioactive ceramics* (e.g., osseoconductive calcium orthophosphates and surface-active bioglasses). Bioinert ceramics are mechanically strong. They are wear- and corrosion-resistant materials with excellent tribological properties and have high structural integrity and reliability in service. Hence, they are used in load-bearing applications such as femoral heads in hip endoprosthetic devices, spacers in spine surgery, parts in knee arthroplasty, and polycrystalline and single crystal forms, such as tooth root implants. In contrast, bioactive ceramics and coatings are composed of hydroxylapatite (HAp), resorbable tri- and tetracalcium phosphates, calcium carbonates, as well as surface-active bioglasses. They are used for their osseoconductive and, in combination with bone growth-supporting proteins, osseoinductive properties, but they are not particularly known for their structural integrity [30]. Silicon nitride (Si_3N_4) might be considered a unique and perhaps ideal ceramic that bridges a gap between bioinert and bioactive materials. It can serve a structural role (e.g., spinal fusion implants and prosthodontic devices) and therefore is often classified with alumina and zirconia as a bioinert material. Yet it possesses both osseoconductive and osseoinductive properties similar to phosphate, carbonate, and bioglass ceramics, and can therefore also be considered bioactive [28, 29].

Osseoconductivity refers to the ability of a material to support the in-growth of bone cells, blood capillaries, and perivascular tissue into the operation-induced gap between the implant body and the existing (cortical) bone bed. *Osseoinductivity* refers to the ability to transform undifferentiated mesenchymal precursor stem cells into osseoprogenitor cells that precede endochondral ossification. The process relies crucially on the osseoinductive action of non-collagenous proteins (NCPs).

Pure and ion-substituted osseoconductive hydroxylapatites are predominantly used as coatings for endoprosthetic hip and dental root implants. Bioresorbable tricalcium phosphates are used as bone cavity fillers and bone restoration materials, and tetracalcium phosphate is an ingredient in bone and dental cement. Currently, significant effort is being expended to develop ways to integrate bioactive ceramics with several resilient polymers to form hybrid biological-inorganic nanocomposites capable of mimicking the salient properties of natural biological materials [27, 41].

2 Intrinsic Properties of Bioceramics

2.1 Bioinert Ceramics

Bioinert ceramics excel due to their high compressive strength, moderate tensile and flexural strengths, and their excellent corrosion, scratch, and wear resistance. These properties have earned alumina, zirconia, titania, silicon nitride, and diamond-like carbon important roles in designing, manufacturing, and clinical deployment of various load-bearing implants [40]. However, the typical low fracture toughness of these ceramic materials and their limited ability to integrate with soft and hard tissues are restricting factors affecting their clinical performance and utilization. As bioceramics in most applications are integrated with mechanically strong and resilient metallic materials, their intrinsic properties must be carefully assessed, and their mutual interaction must be considered to avoid antagonistic cross effects. Figure 2

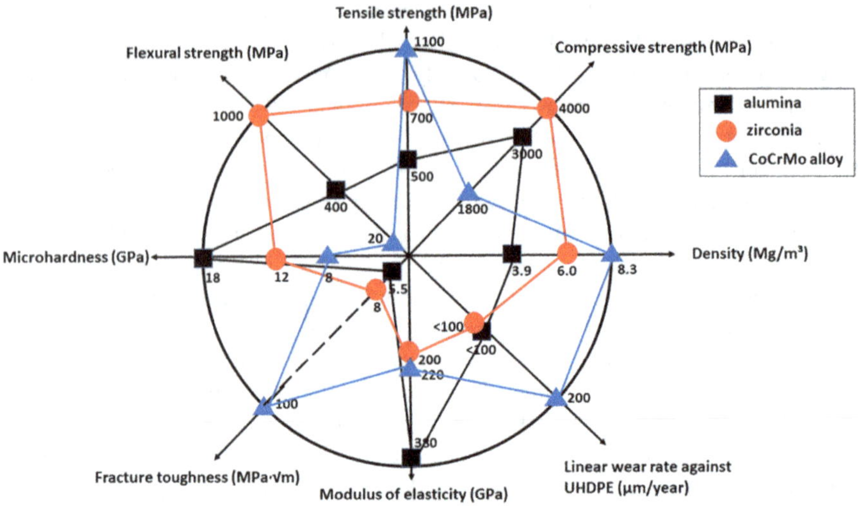

Fig. 2 Mechanical performance diagram of bioinert alumina and zirconia ceramics compared to CoCrMo alloy used as resilient implant material [27]. (© Used with permission from Walter de Gruyter and Company)

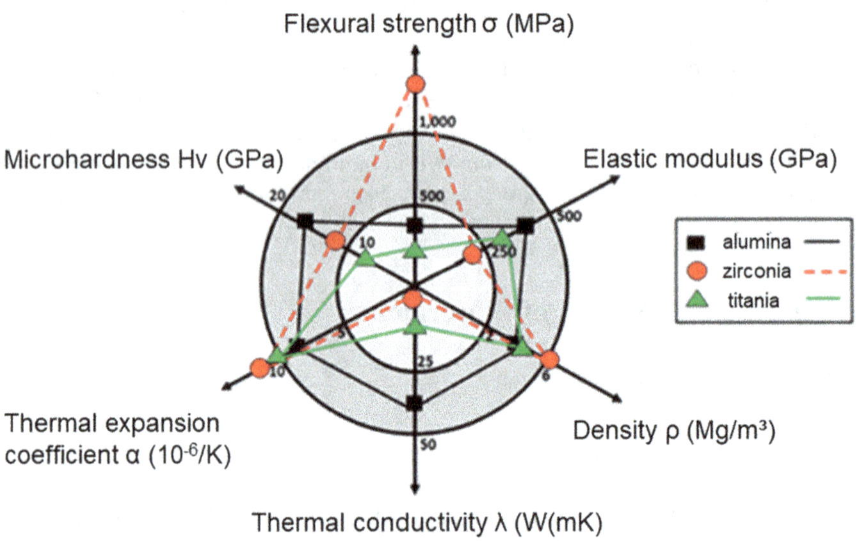

Fig. 3 Mechanical, thermal, and elastic performance diagram of bioinert alumina, zirconia, and titania ceramics

shows the mechanical performance diagram of alumina (black squares) and zirconia (red dots) compared to CoCrMo alloy (blue triangles).

Figure 3 confirms that titania, although a biotolerant bioceramic, is inferior to alumina and zirconia in its mechanical, thermal, and elastic performance. However,

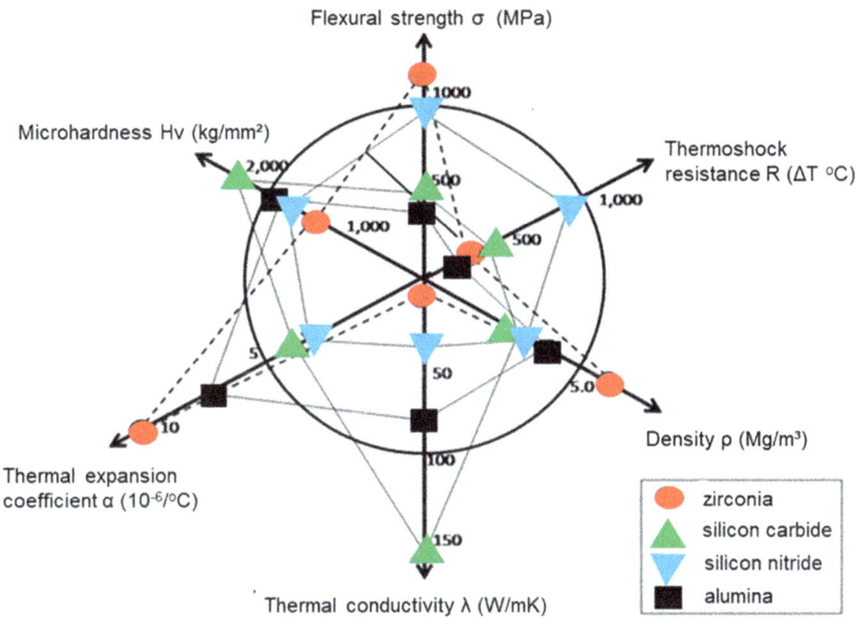

Fig. 4 Mechanical and thermal performance diagram of four typical advanced structural ceramics: zirconia, silicon carbide, silicon nitride, and alumina. Owing to the dependency of the properties on processing conditions, the values given are only tentative [25]. (© Used with permission from John Wiley & Sons)

it may provide a suitable interlayer bond to anchor a hydroxylapatite coating securely to a Ti6Al4V implant surface [23].

Figure 4 compares several mechanical and thermal properties of the oxide ceramics alumina and zirconia to those of the non-oxide ceramics silicon carbide and silicon nitride.

Since the microstructure of ceramic materials is highly dependent on their processing parameters (maximum temperature of sintering, duration of sintering, thermal ramping, purity, and grain size distribution of precursor powders, porosity, presence of sintering additives, and others), the resulting mechanical, thermal, and elastic properties of the final biomedical device vary widely [48] as shown in Tables 1 and 2. In addition, the intrinsic properties of bioceramic materials depend to a high degree on the types of pre-sintering consolidation by cold uniaxial, isostatic, or hot isostatic pressing [25].

One of the purported advantages of bioinert ceramic structural materials is the observation that they do not release potentially cytotoxic constituents into the human body that would interfere with normal metabolic functions. However, on the downside, they do not show positive interaction with living tissue such as promotion of cell growth, proliferation, and spreading. Instead, the body usually responds to these bioinert materials by forming a non-adherent fibrous capsule of connective tissue around the implant that in the case of bone remodeling manifests itself by a

Table 1 Selected property ranges of structural oxide ceramics for medical application

Property	Units	Alumina	Zirconia	Titania	CoCrMo[a]
Density	Mg/m^3	3.7–3.9	5.6–6.1	3.9–4.1	8.2–8.4
Compressive strength	MPa	2000–4000	1200–5000	800–1000	1600–2000
Tensile strength	MPa	300–630	120–700	250–300	1000–1340
Flexural strength	MPa	280–420	900–1400	30–100	~20
Fracture toughness	$MPa{\cdot}m^{1/2}$	3–5	7–10	2–3	~100
Elastic modulus	GPa	350–400	200–220	230–290	200–220
Shear modulus	GPa	140–160	50–90	90–110	87
Microhardness	GPa	15–20	12–13	8–10	8–10
Thermal expansion coefficient	$10^{-6}/K$	7.5–8.5	10–12	8–12	13
Thermal conductivity	W/m·K	30–40	2.5–3.0	5–11	13

Heimann [25, 27]
[a]ASTM F1537, ASTM F799

Table 2 Selected property ranges of structural non-oxide ceramics for medical application

Property	Units	RBSN	SSN	Titanium nitride[a]	Diamond
Density	Mg/m^3	2.7	3.2	5.21	3.52
Flexural strength	MPa	150–350	500–1000		
Fracture toughness	$MPa{\cdot}m^{1/2}$	1.5–3.0	5–8	2–4	~2
Elastic modulus	GPa	120–220	300–330	200–650	900–1200
Microhardness	GPa	17–30	40	4–20	>150
Thermal expansion	$10^{-6}/K$	3.2	3.2	9.5	0.8
Thermal conductivity	W/m·K	10–30	15–50	20–40	>2200

Heimann [25], Bailey et al. [4]
Note: *RBSN* reaction-bonded silicon nitride, *SSN* sintered silicon nitride
[a]Bailey et al. [4]

shape-mediated *contact osteogenesis*. Consequently, only compressive forces are transmitted across the bone-bioceramic interface ("bony on-growth"). This also means that the resistance of bioinert ceramics to tensile and shear forces as well as their fracture toughness are limited and must be carefully considered when designing biomedical structures and devices. As will be shown below, deposition of a thin bioceramic hydroxylapatite coating on the surface of a metallic implant will prevent the formation of this connective tissue capsule and provide a high level of osseoconductivity by supporting the ingrowth of bone cells (i.e., osseointegration). Today, osseointegration is seen as a foreign body reaction (FBR) whereby interfacial bone is formed as a defense reaction to shield the foreign material implant from the tissue [2].

2.1.1 Alumina and Alumina Composites

Alumina (corundum, sapphire, α-Al_2O_3) is the workhorse of the structural ceramics industry. It excels by its very high hardness (Vickers microhardness over 20 GPa), high abrasion resistance, and chemical inertness against most acids and alkalis. Due to its exceptional chemical stability and mechanical strength, alumina exhibits superior performance in diverse aggressive environments. These applications encompass the mining and chemical industries, metal manufacturing and processing, as well as demanding fields like ceramic armor, aerospace engineering, and biomedical applications. Mechanically strong, abrasion-resistant, and thoroughly biocompatible corundum single crystals have been successfully used as dental root implants [1].

However, these advantageous properties are partially offset by its rather low tensile and flexural strengths and fracture toughness. While these properties may affect the wear performance of alumina in biomedical devices, the decision to include alumina in bioengineering designs must be judiciously assessed to avoid catastrophic failure in service under the harsh conditions prevailing in the human body. Extremely pure and fine-grained alumina used in femoral heads for hip endoprostheses shows noticeably higher strength, fracture toughness, and hardness compared to common industrial high alumina ceramics (Table 3). Such pure, fine-grained polycrystalline alumina materials have been used for over 50 years [10] for

Table 3 ISO 6474 norm requirements and properties of clinically utilized alumina-based ceramics

Property	Units	High alumina	ISO 6474-1: 2010	ISO 6474-2: 2012	BIOLOX® *forte*	BIOLOX® *delta*
Al_2O_3 content	Vol.%	>99.7	>99.5	>60	>99.8	~82
ZrO_2 content	Vol.%			<30		~17
SiO_2 + Na_2O	Vol.%	<0.02	0.1			
Cr_2O_3	Vol.%					0.5
SrO	Vol.%					0.5
Density	Mg/m^3	3.98	>3.90	>3.94	3.98	4.37
Average grain size (Al_2O_3)	μm	3.6	<7	<4.5	1.7	0.6
Average grain size (ZrO_2)	μm					0.3
4-point bending strength	MPa	280–420	400–500	>450	630	1380
Fracture toughness	MPa·√m	3–4	4–6	–	3.2	6.5
Modulus of elasticity	GPa	350–400	380–420	–	410	360
Vickers hardness	GPa	24	>20		20	19
Wear rate per 10^6 cycles	mm^3				8.3	0.6

Heimann [27]

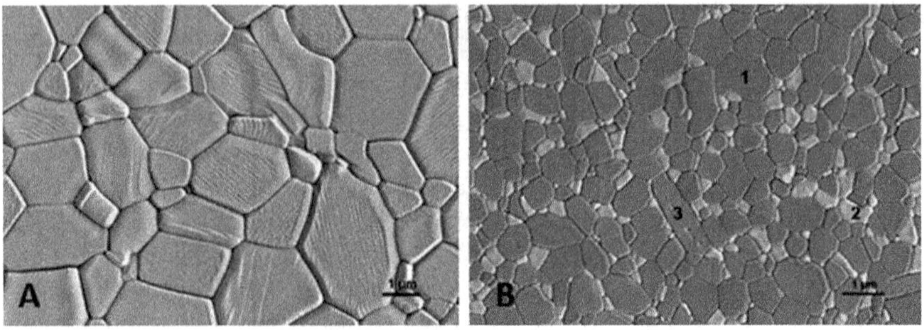

Fig. 5 (**a**) Microstructure of thermally etched polycrystalline BIOLOX®*forte*. (**b**) Microstructure of thermally etched composite BIOLOX®*delta*: (1) Alumina matrix grain (dark grey), (2) Zirconia reinforcing grain (light grey), (3) Strontium hexaluminate platelet. (© Used with permission from CeramTec GmbH, Plochingen, Germany)

femoral heads. Today, there exists a large variety of clinical options to combine femoral heads and acetabular cups [6].

The ISO 6474 standard specifies the characteristics of biocompatible and biostable ceramic bone substitute materials based on high-purity alumina for use as bone spacers, bone replacements, and components of orthopedic joint prostheses. High-strength alumina ceramics for medical use can be generated by grain-boundary engineering. Suppression of grain growth at the high sintering temperature required for optimum densification is achieved by adding small amounts of magnesium oxide in the range of a few tenths of a percent. Reaction of magnesium oxide along the grain boundaries of alumina will result in a thin surface layer consisting of spinel ($MgAl_2O_4$) that acts as a barrier against the grain boundary movement that is normally associated with the process of recrystallization. Large grains within a matrix of much smaller grains will act as stress concentration centers and thus, considerably reduce the fracture toughness of the ceramic material. Grain-boundary engineering will effectively suppress recrystallization and result in a microstructure typical of third-generation BIOLOX®*forte* material marketed by CeramTec GmbH in Germany (Fig. 5a).

The ISO 6474-2:2012 standard relates to high-purity alumina matrix materials reinforced with zirconia to form a zirconia-toughened alumina (ZTA) compound. This results in a so-called duplex structure with high fracture strength and toughness. Since Y-PSZ has a lower elastic modulus (around 210 GPa) compared to alumina (~400 GPa; Table 2), cracks of critical size tend to move toward the zirconia particle aggregates as shown schematically in Fig. 6. Dissipation of the crack energy happens by forcing tetragonal t-ZrO_2 particles to transform into monoclinic m-ZrO_2. The resulting compressive stress that is generated tends to counteract crack movement, accounting for the extraordinary strength of this modern bioceramic material known as BIOLOX®delta.

BIOLOX®*delta* is composed of high-purity alumina (~82 vol.%) with additions of about 17 vol.% tetragonal Y-stabilized zirconia, 0.5 vol.% chromium oxide (Cr_2O_3) as well as 0.5 vol.% strontium oxide (SrO). During sintering at temperatures

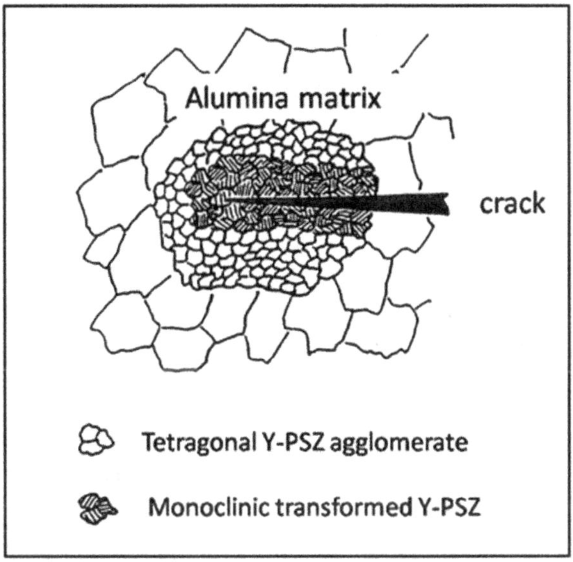

Fig. 6 Schematic rendering of the reinforcement mechanism of an alumina-zirconia duplex ZTA ceramics with agglomerated Y-PSZ particles [25]. (© Used with permission from Walter de Gruyter and Company)

as high as 1800 °C, the zirconia-toughened alumina (ZTA) matrix incorporates chromium oxide in a solid solution coloring it pink, and tiny platelets of strontium hexaluminate ($SrAl_{12}O_{19}$) form. They act as potent crack arresters (Fig. 5). Consequently, the fracture toughness of BIOLOX®*delta* is twice as high as BIOLOX®*forte* and its wear rate is dramatically reduced (Table 3).

2.1.2 Zirconia

Zirconia has found various applications in biomedical devices, most importantly as a hard and tough structural material for femoral heads in hip endoprostheses and as a material for restorative dentistry, including inlays, onlays, and tooth veneers. Since stabilized zirconia shows substantially larger fracture toughness compared to alumina, it is advantageously applied in prosthetic devices [12].

Zirconia exists in several polymorphic crystallographic structures: monoclinic, tetragonal, and cubic. On heating, the monoclinic *m*-ZrO_2 phase is stable up to 1170 °C. Beyond this temperature, it transforms by a so-called martensitic (diffusionless) shear mechanism to the tetragonal *t*-ZrO_2 form that is stable up to 2370 °C. On further heating, cubic *c*-ZrO_2 forms. It is stable up to its melting point of 2700 °C. Since the transformation results in a volume change of about 5 vol.%, in pure (i.e., unalloyed monoclinic zirconia), the elastic limit and yield strength will be exceeded on cooling. Cracks form leading eventually to the complete structural disintegration of monolithic zirconia bodies. However, alloying of zirconia with oxides such as yttria, calcia, magnesia, ceria, or scandia delays this deleterious phase

Table 4 Characteristic properties of Y-stabilized zirconia and ZIOLOX® (CeramTec GmbH)

Property	Units	Range of values of Y-stabilized zirconia	ZIOLOX®
Density	Mg/m^3	6.05–6.09	6.09
Zr content	%	95–97	~95
Y content	%	3–6	4.5–5.5
Average grain size	μm	0.2–0.4	<0.5
Vickers hardness	GPa	12–13	12.5
Modulus of elasticity	GPa	150–210	210
Compressive strength	MPa	>2000	
Tensile strength	MPa	>650	
Flexural strength	MPa	900–1300	>900
Fracture toughness	$MPa{\cdot}m^{1/2}$	7–10	~8
Thermal conductivity	W/m·K	2.5–3.0	

Heimann [27]

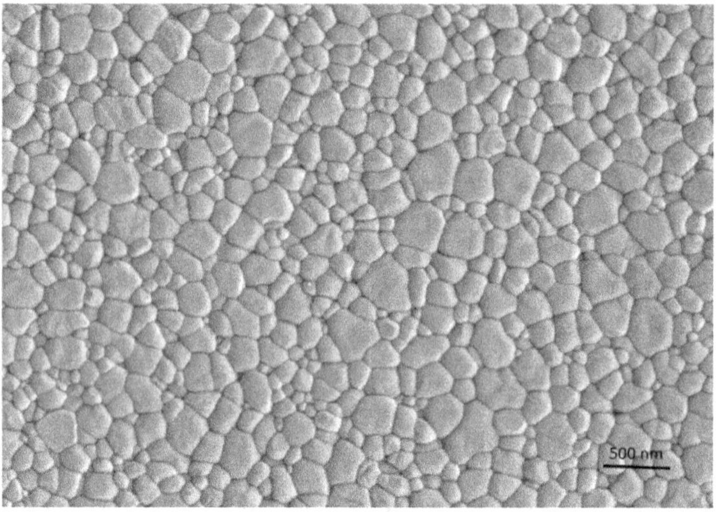

Fig. 7 Microstructure of ZIOLOX®. (© Used with permission from CeramTec GmbH, Plochingen, Germany)

transformation by stabilizing the tetragonal structure. This process is at the very heart of the important mechanism of *transformation toughening* of zirconia ceramics.

Biomedical applications of Y-stabilized zirconia as tough ceramics for femoral heads or dental restoration material require maximization of the mechanical properties, in particular flexural strength and fracture toughness. This requires careful adjustment of the relative amounts of ZrO_2-phases. Table 4 shows selected properties of commercially available Y-TZP (ISO 13356:2016) and ZIOLOX® material formerly marketed by CeramTec GmbH. Figure 7 shows the uniform microstructure of ZIOLOX® with an average grain size of less than 500 nm.

Despite the advantageous mechanical properties of Y-stabilized zirconia, concern has been raised about its performance as a femoral ball for hip endoprostheses. In 1996, the British Medical Device Agency (MDA) published reports that zirconia femoral heads suffered surface degradation by roughening during autoclaving owing to their hydrothermal instability [17]. Consequently, substantially increased friction and mechanical wear rates were apparent. In addition, cracks were detected in femoral heads sold under the commercial name Prozyr® by Saint Gobain Ceramiques Desmarqueste, caused by the fact that the inner core of the ball was not as dense as required. Hence, cracks originating near the ball and neck boundary quickly propagated into the inner, less compact core. In response to these serious flaws, the use of Y-TZP in joint arthroplasty has been largely abandoned in many countries today.

In contrast, zirconia is presently copiously used in all kinds of dental devices including implants for dental roots, inlays, onlays, and color-matched tooth veneers [47]. Indeed, zirconia has emerged as a versatile dental bioceramic due to its excellent esthetic outcome in terms of color, opacity, and shine that perfectly mimics the appearance of natural teeth. In addition, the mechanical strength of Y-stabilized zirconia makes it an ideal material to withstand the masticatory stresses dental prostheses are subjected to and the positive interaction with tissue decreases peri-implant inflammatory reactions.

2.1.3 Carbon

Carbon is the element on which all life is based on Earth. Two allotropic modifications of carbon play important roles in biomaterials engineering today, pyrolytic graphite and diamond-like carbon (DLC), with nanostructured graphene and carbon nanotubes (CNTs) waiting in the wing destined to be developed for therapeutic and diagnostic ("theragnostic") application [57].

Pyrolytic graphite belongs to the family of turbostratic carbons and is structurally closely related to graphite [63]. It consists of covalently bonded stacks of hexagonal nets of carbon atoms with sp^2-hybridization that are rotationally disordered in the crystallographic c-direction, (i.e., perpendicular to the layer stacks). Whereas in ordered graphite, the layers can easily slide over each other, making it a solid lubricant, in turbostratically disordered graphite, wrinkles or distortions exist within the layers that resist sliding, giving pyrolytic turbostratic graphite improved stabilization and durability compared to ordered graphite. Currently, low-temperature isotropic pyrolytic graphitic carbon (LTIC) with its turbostratic structure is the material of choice to fabricate artificial heart valves. However, the material is brittle and its hemocompatibility is still insufficient for prolonged clinical use. As a result, the risk of thrombosis forces patients to frequently take anti-coagulation medication for the remainder of their lives. To suppress thrombogenicity requires the development of a material with better surface properties and improved hemocompatibility. Considerable research is underway to achieve this goal.

Diamond is the ultrahard, wear- and corrosion-resistant, tribological superior, and electrically insulating crystalline allotrope of carbon with the highest thermal conductivity of all known materials. However, the most ubiquitous carbon allotrope for biomedical applications is diamond-like carbon (DLC) which, in contrast to fully sp^3-hybridized diamond, is an amorphous carbon material with mixed sp^3/sp^2-bonding. Depending on the preparation conditions, a large number of different types of DLC films exist distinguished by the absence or presence of hydrogen in its amorphous matrix [8]. DLC coatings are being developed as alternate bearing surfaces for endoprosthetic hip and knee implants, designed to combat wear in articulating implant devices [14]. DLC coatings on Ti6Al4V alloy substrates have shown superior wear resistance, not in the least owing to their strong adherence. In addition, immersion and electrochemical tests have revealed improved stability of DLC coatings on Ti6Al4V substrates compared to 316 L stainless steel and CoCrMo alloy substrates [44].

However, several studies have shown that DLC coatings on articulating joints may fail in service owing to partial or complete delamination some years after implantation. This delayed delamination is thought to be caused by crevice corrosion of an adhesion-promoting metal carbide or Si interlayer. Delamination can also occur by a slowly advancing crack in a thin carbidic interlayer due to stress corrosion cracking. Nevertheless, when stable coating adhesion can be achieved, DLC-coated articulating implants show essentially no wear of the coating up to 100 million articulations in a hip simulator, corresponding to about 100 years of in vivo service [22].

2.1.4 Silicon Nitride

Silicon nitride is an emerging bioceramic material with unique properties destined to become a clinically relevant part of the biomedical engineering toolbox for application in a variety of medical fields [28, 29]. These applications are based on high compressive and flexural strengths, high fracture toughness, low friction coefficient, high wear, and corrosion resistance, improved medical imaging ability as a radiolucent material, and enhanced biocompatibility, osseointegration, and antimicrobial activity against bacteria and viruses. Silicon nitride was found to be a "killer ceramic" for Gram-negative bacteria likely based on chemically driven mechanisms, related to its pH-dependent surface chemistry [52, 53] since in an alkaline pH environment, the release of ammonium ions causes lysis of bacterial proteins. Hence, silicon nitride takes on a hybrid function between bioinert and bioactive ceramics, bridging the conceptual gap that separates these functionally different bioceramic materials. Synthesis, properties, and present and future medical applications of silicon nitride are described in detail in subsequent chapters.

2.1.5 Ultrahard Transition Metal Nitride

Wear debris particles released from articulating metallic bearing interfaces of prosthetic implants may cause metal hypersensitivity and local tissue necrosis, whereas particles released from a polymeric acetabular cup are known to promote aseptic implant loosening and osteolysis, aptly referred to as "particle disease" [18]. Wear-reducing bioceramic coatings of implants potentially increase scratch resistance, enhance wettability, and reduce wear and ion release of articulating surfaces.

Commercially available ultrahard coatings for implants include titanium nitride (TiN, [21]), titanium niobium nitride (TiNbN, [15]), titanium carbonitride (TiCN, [58]), silicon nitride [20], zirconium nitride (ZrN, [56]), chromium nitride (CrN, [16]), and tantalum nitride (TaN, [5]). These coatings are still at the early research stage and were found to be non-cytotoxic and able to reduce wear of articulating parts of hip and knee prostheses. However, coating adhesion is of concern, as poor adhesion causes partial or full delamination and excessive wear. Shtansky et al. [62] provided a comprehensive account of the failure modes of such thin films and coatings.

TiN plasmonic nanoparticles have been shown to provide dipolar resonances in the range required for phototherapeutic applications to combat cancer. When excited with 800 nm laser light, these nanoparticles provide a broadband resonance peak [45]. In advanced photothermal cancer therapy, plasmonic light-activated TiN nanoparticles may be infused into the blood of a patient and tend to concentrate inside a tumor. When an external laser light beam of the same frequency as that of the surface plasmon is shone into the tumor mass, the TiN nanoparticles will heat up by resonance, selectively killing the cancerous cells without affecting surrounding healthy tissue.

2.1.6 Dental Porcelains

Dental porcelains are used to replace dental superstructures, including crowns, bridges, inlays, onlays, tooth veneers, and artificial teeth. Increasing demand for esthetic restorations and the encouraging performance of all-ceramic restorations in permanent dentition have led to significant advances in dental ceramics. Although dental glass-ceramics achieve highly esthetic results today, their relatively low strength and brittle nature prohibit their predictable use in high-stress applications unless supported by a high-strength substrate.

Dental porcelains encompass feldspathic multiphase ceramic material of the ternary system SiO_2-Al_2O_3-K_2O/Na_2O. *High-fusing dental porcelain* is typically composed of 75 mass.% SiO_2, 15 mass.% Al_2O_3, and 10 mass.% K_2O. This composition is positioned in the two-phase region mullite + liquid of the ternary phase diagram. Compositional fluctuation, for example, by evaporation of alkali oxides during sintering may move the system into the three-phase region mullite + tridymite (or cristobalite) + liquid. *Low-fusing dental porcelain* consisting typically of 60 mass.% SiO_2, 10 mass.% Al_2O_3, and 30 mass.% K_2O contains leucite ($KAlSi_2O_6$) as a crystalline component embedded in an aluminum silicate glassy phase. Tetragonal leucite is stable at ambient conditions but undergoes a martensitic phase transformation to a cubic phase at 625 °C. This transformation toughens the ceramic matrix in the same way as martensitic transformation toughens zirconia (see above).

2.2 *Bioactive Ceramics*

In contrast to bioinert ceramics discussed in Sect. 2.1, bioactive ceramics (i.e. osseoconductive materials) show positive interactions with living tissue that may even include differentiation of immature stem cells towards osteocytes (i.e. osseoinductive behavior). These ceramics exhibit chemical bonding to the bone along the interface triggered by the adsorption of bone growth-mediating proteins at the surface of a bioactive ceramic within the first few seconds after implantation. Hence, the presence of bioactive ceramics will support biochemically-mediated strong *bonding osteogenesis* that allows, in addition to compressive forces, to some degree tensile and shear forces to be transmitted through the interface ("bony in-growth"). Typical bioactive ceramic materials are calcium orthophosphates including hydroxylapatite, tri- and tetracalcium phosphates, calcium carbonates, silica, and bioglasses.

2.2.1 Hydroxylapatite

The reason hydroxylapatite is copiously applied to biomedical devices (e.g., coatings of the stem of hip endoprostheses, densified monolithic implants for dental root replacement, suitable material for filling bone cavities, and bone growth-supporting 3D-scaffolds) is grounded on its chemical and structural similarity to native osseous tissue, (i.e., bone-like apatite). It is generally believed that the bioactivity of calcium phosphates is strongly associated with the formation of hydroxylcarbonate apatite (HCA) [43]. This non-stoichiometric bone-like apatite consists of calcium- and hydroxyl-deficient carbonated hydroxylapatite in which some $[PO_4]^{3-}$ lattice sites are replaced by $[HPO_4]^{2-}$ ions to yield $Ca_{10-x}[(HPO_4)_x(PO_4)_{6-x}](OH,O,Cl,F,CO_3,\square)_{2-x}\cdot nH_2O$ ($0 < x < 1$; $0 < n < 2.5$) [46].

In addition to its outstanding biological performance, hydroxylapatite can be inexpensively produced, is biologically safe, and can relatively easily be certified for clinical use. Beyond this, the incorporation of metabolically important ions such as magnesium, strontium, silicon, copper, zinc, or cobalt into calcium phosphate-based bone graft materials is destined to provide an economic and feasible solution for bone defect healing.

Hydroxylapatite is predominantly used as a coating for metallic implants in orthopedic and dental applications. The preferred method of deposition is thermal

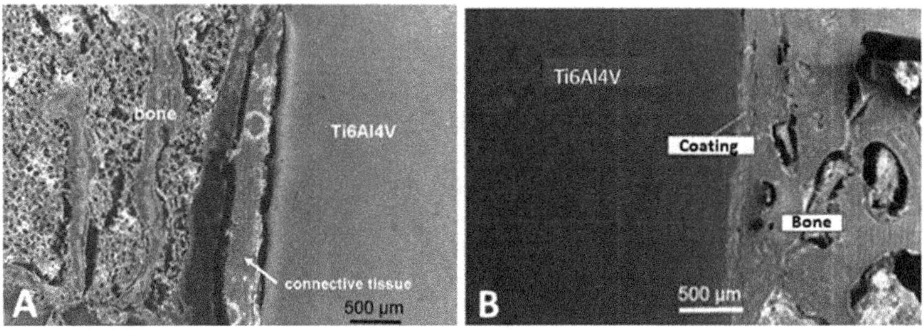

Fig. 8 (**a**) Ti6Al4V implant positioned in the lateral condyle of a canine femur without coating, showing the formation of a layer of connective tissue separating the implant from the bone. (**b**) Ti6Al4V implant positioned in the lateral condyle of a canine femur with a titania bond coat/HAp coating assembly [39]. (© Used with permission from John Wiley & Sons)

spraying, approved by the FDA and many other national medical regulation authorities. Plasma spraying provides a fast, economical, well-controlled, and well-researched technology to deposit coatings over about 50 μm thickness [24]. However, the extremely high temperature of the plasma jet at up to 15,000 K leads to fast thermal dehydroxylation, followed by thermal decomposition of hydroxylapatite [28, 29]. This has to be factored in when designing plasma-sprayed hydroxylapatite coatings. Hydroxylapatite coatings provide endoprosthetic implants with osseoconductive properties [30]. In particular, such coatings suppress the formation of a fibrous connective tissue capsule separating a metallic implant from the bone and thus, preventing solid osseointegration. Figure 8a shows a fibrous capsule of connective tissue that is absent in Fig. 8b.

2.2.2 Other Calcium Orthophosphates

State-of-the-art materials for rapidly resorbable calcium orthophosphate ceramics are tricalcium phosphates, either as the low-temperature modification (β-TCP) or one of its high-temperature modifications (α-TCP, α'-TCP). Phase-pure ß-tricalcium phosphate is being used either in the form of sintered ceramic monolithic shapes (for example, Cerasorb®; KSI tricalcium phosphate®; chrOs™; BioResorb®) or as coatings for implants. Fast resorbable alkali-bearing calcium orthophosphates (CAOPs) are supplied as granular materials, but increasingly also as porous spongiosa-like bodies or pastes [42]. Porosity in monolithic ceramic bodies can be introduced by bloating agents, organic spacers, or by utilization of naturally occurring cellular structures such as sponges.

Biphasic calcium phosphates (BCP) are composed of different ratios of stable hydroxylapatite and more soluble β-tricalcium phosphate (β-TCP). BCPs have significant advantages over other calcium phosphate ceramics due to their controlled bioactivity and resorption kinetics that guarantee the stability of the biomaterial while efficiently promoting bone in-growth [19].

Transition metal-substituted calcium hexaorthophosphates of the general formula $CaTi_{4-x}Zr_x(PO_4)_6$ with NaSiCON structure were found to provide dense, well-adhering coatings with excellent biocompatibility and osseoconductivity [31] as well as comparatively high thermal stability and low solubility in vivo compared to hydroxylapatite. Such compounds attain substantial ionic conductivity when doped with highly mobile ions such as sodium or lithium. These mobile ions can move in response to an outside electric field within the cavities of the crystalline structure. This suggests interesting options to design so-called fourth-generation biomaterials that utilize bioelectric properties for theragnostic purposes [49] or electrically stimulated bone growth [26].

2.2.3 Calcium Carbonates

The marine environment is rich in mineralizing organisms with porous structures, some of which are currently being used as bone graft materials [7]. Their exoskeletons are built up by calcium carbonate phases such as calcite, aragonite, or magnesian calcite. Organism-templated biomineralization provides biological pathways potentially able to emulate nature's physiological mechanisms for biomedical purposes [41, 68]. Biomimetic strategies have led to the investigation of naturally occurring porous structures as useful templates for bone growth, in particular cancellous bone.

2.2.4 Silica

Marine sponges of the phylum *Porifera* can develop skeletons composed of unique rigid multiphase SiO_2- or $CaCO_3$-based composites. The beneficial properties of the three different phases present, (i.e., amorphous silica, biopolymers such as collagen or chitin, and crystalline aragonite), extend the biomimetic potential of marine sponges for material scientists toward the bioinspired design of novel multiphase composite materials for several biomedical applications [41, 50]. While the functional role of silica–collagen composites in bone development is still enigmatic and unclear in many aspects, recent research indicates that they may trigger precipitation of calcium phosphates. The complex hierarchical mesoscale structures of sponge or diatomaceous biosilica combine hard and stiff inorganic components with soft and tough organic tissues to produce composite materials with extraordinary fracture resistance.

2.2.5 Bioglasses

Bioactive glass compositions are predominately found in the system $CaO\text{-}Na_2O\text{-}P_2O_5\text{-}SiO_2$. The first of the series of bioglasses introduced to the biomedical industry was 45S5 Bioglass® with a composition of 45% SiO_2, 24.5% CaO,

24.5% NaO_2, and 6% P_2O_5 by mass. The application of bioactive glasses for medical use is a relatively recent accomplishment. Hence, there have been comparatively few but generally successful clinical applications of these materials, revealing osseoconductive, antibacterial, and antifungal properties. Today, medical uses of bioglasses include a wide range of applications [33]. These applications include dental implants, alveolar ridge augmentation, cervical and lumbar interbody fusion, spinal fusion for adolescent idiopathic scoliosis, bone graft substitutes, bone tissue engineering scaffolds, coatings for surgical screws and wires, cochlear implants, and others.

A common feature of bioactive glasses is the time-dependent kinetic modification of their surfaces during implantation on contact with extracellular fluid (ECF). While bioglasses are generally non-resorbable, they release sodium and calcium ions that trigger a cascade of biological reactions and eventually nucleate a thin layer of biological hydroxylcarbonate apatite (HCA). This layer provides a bonding interface with bony tissue that is mechanically so strong that in many cases, the interfacial strength of adhesion exceeds the cohesive strength of the implant, or the bone tissue bonded to it [32].

2.2.6 Other Calcium-Based Bioceramics

Calcium titanate (perovskite, $CaTiO_3$) likely supports osseointegration and may be needed for enhanced implant efficacy [64]. The bonding of a plasma-sprayed hydroxylapatite coating layer to a titanium substrate is thought to be mediated by a thin reaction film of calcium titanate. However, experimental evidence of such a reaction layer is scant. Its visualization by electron microscopy even at high magni fication is precluded by its exiguity owing to the very short diffusion paths of Ca^{2+} and Ti^{4+} ions, respectively, that render any potential reaction zone extremely thin. Enhanced apatite nucleation by calcium titanate surfaces is possibly controlled by the presence of an epitaxial structural relation between the (022) lattice plane of calcium titanate and the (001) lattice plane of hydroxylapatite [65]. Some research has been performed aimed at utilizing sphene (calcium titanium silicate, $CaTiSiO_5$) as a biocompatible osseoconductive coating on titanium-based implants [9].

Calcium silicate phases such as larnite (ß-Ca_2SiO_4) and wollastonite (α-$Ca_3[Si_3O_9]$) have been investigated for their application as ceramic biomaterials since calcium silicate glasses and ceramics were found to bond easily to living bone, and to form, by a biomimetic process, an apatitic surface layer when exposed to simulated body fluid [11]. CaO-SiO_2 glasses were shown to form a tight bond with bone mediated by a thin apatite layer formed adjacent to the bone tissue. The cyclic silicate trimer $[Si_3O_9]^{6-}$ group of pseudowollastonite (ß-$Ca_3[Si_3O_9]$) possibly acts as an active site for heterogeneous, stereo-chemically promoted nucleation of hydroxylapatite in simulated body fluid.

Calcium magnesium silicates such as diopside ($CaMgSi_2O_6$), merwinite ($Ca_3MgSi_2O_8$), and monticellite ($CaMgSiO_4$) have been suggested and experimentally verified as bioceramics for medical application, showing sufficient mechanical stability, adhesion strength, and biocompatibility when applied to metallic implant surfaces [54]. Diopside has shown significant potential to form an apatitic layer during incubation in SBF [59]. Nanostructured merwinite coatings not only improved the corrosion resistance of the implant substrate, but also enhanced the bioactivity, mechanical stability, and cytocompatibility of AZ91 Mg alloy [55]. The sorosilicate baghdadite ($Ca_3Zr[(Si_2O_7)O_2]$) has been suggested as a suitable biocompatible coating for Ti-based implants [35]. Other potentially advantageous silicate-based bioceramics such as melilite-type hardystonite ($Ca_2Zn[Si_2O_7]$) and åkermanite ($Ca_2Mg[Si_2O_7]$) are under scrutiny, owing to their ability to inhibit bone resorption and promote osteoblast differentiation, osteocalcin secretion, and enhanced alkaline phosphatase activity.

3 Designing with Bioceramics

Designing with bioceramic materials involves optimization of their mechanical, elastic, and thermal properties as well as their biological functionality by tailoring appropriate processing conditions and surface structures.

3.1 Grain Boundary Engineering

High-strength alumina ceramics for medical use as femoral heads for hip endoprostheses can be obtained by grain-boundary engineering. Densification of polycrystalline ceramic materials requires the application of high temperatures where recrystallization promotes the formation of larger grains. However, large grains within a matrix of much smaller grains act as stress concentration centers, and thus, they considerably reduce the fracture toughness of the ceramic. Consequently, grain growth must be suppressed by adding small amounts of magnesium oxide in the range of a few tenths of a percent. In the grain boundary region, the free surface energy is increased so that MgO tends to concentrate along the grain boundaries. Its reaction with alumina results in a thin surface layer of spinel ($MgAl_2O_4$) at the grain boundary. This phase acts as a barrier to the grain growth that is normally associated with the process of recrystallization.

3.2 Manufacturing of Femoral Balls

Femoral balls of hip endoprostheses made from bioinert alumina or zirconia have to sustain high mechanical stresses, resorption/corrosion by aggressive body fluid, and abrasive wear over the lifetime of the implant in the human body of 20–25 years.

Table 5 Important mechanical and functional properties of alumina femoral balls (after [67])

Material property	Prerequisite for
High hardness	Wear resistance over many years
High wear resistance	Low risk of particle-induced osteolysis
No plastic deformation under load	Retains implant dimension
No elastic deformation under load	Retains implant dimension
Fine-grained microstructure	Excellent surface finish, low coefficient of friction
Dense (near-zero porosity)	Mechanical stability
High flexural strength	High load-bearing capability
High compressive strength	High fracture strength
Good fatigue resistance	Improved reliability over many years
High Weibull modulus	Reliable product
Extreme scratch resistance	Product integrity
Extreme corrosion resistance	Strong biocompatibility
High chemical purity	Bioinert behavior
Electrical insulator	No galvanic or fretting corrosion

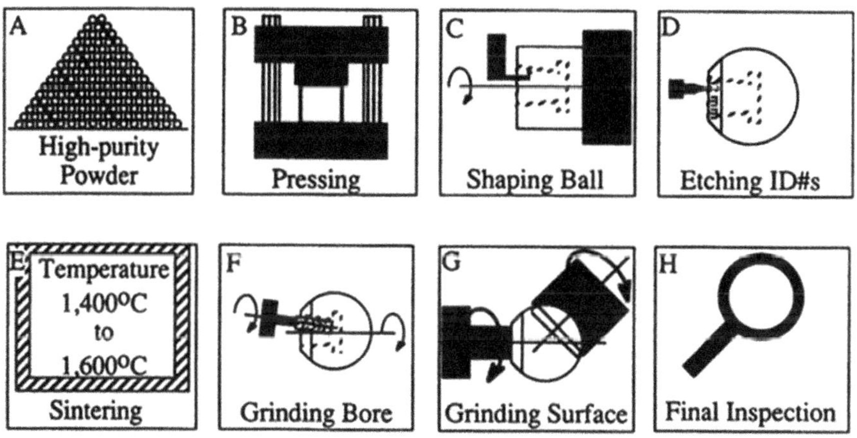

Fig. 9 Schematic of the manufacturing process of ceramic femoral balls [13]. (© Used with permission from Walter de Gruyter and Company)

Some important properties of alumina femoral joint heads required for long-term performance are listed in Table 5 [67].

Figure 9 shows schematically the manufacturing process of ceramic femoral ball heads [13]. High-purity alumina powder (a) is consolidated by cold uniaxial pressing (b) and subsequently turned on a CNC-controlled lathe to form the inside taper and the outside spherical surface (c). After laser-etched engraving for identification (d), the ceramic green body is densified by sintering at temperatures up to 1600 °C in an oxidizing or inert gas atmosphere (e). Finishing by grinding with diamond tools to a very low roughness value of <1 μm is performed to achieve a low coefficient of friction (f and g). Final inspection (h) will assure an extremely high degree of

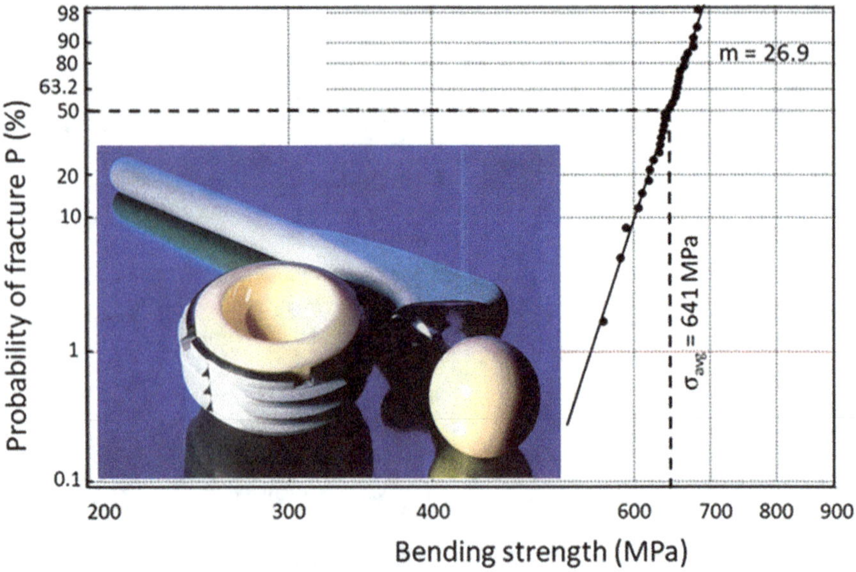

Fig. 10 Weibull fracture probability plot of BIOLOX®*forte* bars subjected to a four-point bending test. At the mean strength value of 641 MPa, there is a probability that 50% of the investigated specimens will fail. The high Weibull modulus $m = 26.9$ attests to the high engineering reliability of the material [27]. (© Courtesy of CeramTec GmbH, Plochingen, Germany. Inset: BIOLOX®*forte* high-end hip endoprosthesis with an alumina femoral ball (right) articulating against an alumina acetabular cup. © Image courtesy Professor Gerd Willmann †, CeramTec GmbH, Plochingen, Germany)

reliability required for the long-term survival of the ball in the aggressive body environment. This high reliability is confirmed by a low stochastic Weibull probability of failure (Fig. 10).

A cumulative Weibull plot can assess the functional reliability of the ceramic medical implants. The stochastic Weibull ln-ln distribution is frequently used in survival analysis of alumina femoral heads. The probability P_f of failure during a three- or four-point flexural test is given by $P_f = 1 - \exp.[-(\sigma\text{-}\sigma_\theta)]^m$, whereby σ is the applied stress, σ_θ is the failure stress, and m is the so-called Weibull modulus. Linearization yields $\ln(1\text{-}P_f) = -(\sigma\text{-}\sigma_\theta)^m$ and finally $\ln[\ln(1/(1\text{-}P_f))] = m\cdot\ln\sigma - m\cdot\ln\sigma_\theta$. Then, m is the slope of the linear plot of $x = \ln\sigma$ vs $y = \ln[\ln(1/(1\text{-}P_f))]$.

The mechanical integrity of alumina bioceramics has led to long-term survival predictions for these materials when subjected to subcritical flexural stress. Figure 10 shows a cumulative Weibull plot as representative of the failure probability of typical femoral heads made from high-purity alumina such as BIOLOX®*forte*. The very low data scatter and the high average flexural strength of 641 MPa (d_{50}) attest to extraordinary reliability as indicated by a high value of m, (i.e., the slope of the straight line of the ln-ln plot).

3.3 *Designing Biological Functionality*

Different routes of designing favorable biological functionality of bioceramics are being taken to improve the ingrowth and osseogenic differentiation of bone cells. Since in biological apatite, Ca^{2+} ions are partially substituted by metabolically important cations such as Na^{+}, Mg^{2+}, Sr^{2+}, K^{+}, and some trace elements such as Pb^{2+}, Ba^{2+}, Zn^{2+}, and Fe^{2+}, modifying hydroxylapatite by adding these substitutional ions confer enhanced biological properties on osseoconductive coatings. In addition, the replacement of PO_4^{3-} anions with CO_3^{2-}, SiO_4^{4-}, and SO_4^{2-} as well as OH^{-} with Cl^{-}, F^{-}, O^{2-}, and CO_3^{2-} contributes to a host of biochemical pathways in which bone matter is involved [61]. Even small substitutions may have very significant effects on thermal stability, solubility, osteoclastic and osteoblastic response in vitro, and degradation and bone regeneration in vivo. A host of less-common ions of the lanthanide and transition metal groups of the PSE have been incorporated into osseoconductive bioglasses to improve their biological and physical properties [51].

Biological functionality of hydroxylapatite coatings can also be engineered by micro-patterning [69] or advanced techniques to modify and structure the surface of coatings aiming to enhance roughness and fractality that control to a large degree the adhesion, proliferation, and spreading of osseogenic cells, preconditions for successful and lasting osseointegration.

4 Concluding Remarks

Novel medical materials options beyond those discussed above are expected to sprout from the extremely wide range of available bioceramic materials [27]. However, although the scope and potential of practical application of such materials are broad, examples of implementation of actual and tangible clinical applications with near-term commercial horizons are rather rare because of many antagonistic issues that still need to be resolved. At a time when stringent safety and standardization requirements, validation regulations, an increasing number of litigation cases, and compatibility issues are important factors counteracting the growth of the bioceramics market, this situation has severe repercussions on return of investment expectations of companies engaged in producing devices and solutions based on bioceramic materials.

Whereas novel applications of bioceramic materials in tissue engineering, biosensing, and bioimaging have already been deployed or are close to commercialization, other more ambitious applications are much farther off including bioelectronic fourth-generation biomaterials [49], photonic devices, or biologically inspired computers based on artificial neural networks (ANNs) [41]. Here lies a potential socio-economic gold mine to be explored in the near future.

References

1. Akagawa, Y., Hashimoto, M., Kondo, N. et al. (1986). Initial bone-implant interfaces of submergible and supramergible endosseous single-crystal sapphire implants. J. Prosth. Dent., 55(1), 96–100.
2. Albrektsson, T., Chrcanovic, B., Jacobsson, M. and Wennerberg, A. (2017). Osseointegration of implants—a biological and clinical overview. JSM Dent. Surg., 2(3), 1022 (6 pages).
3. ASTM F981-04. (n.d.). Standard practice for assessment of compatibility of biomaterials for surgical implants with respect to effect of materials on muscle and bone. ASTM International, West Conshohocken, PA, 19428-2959, USA.
4. Bailey, E., Ray, N.M.T., Hector, A.L. et al. (2011). Mechanical properties of titanium nitride nanocomposites produced by chemical synthesis followed by high-P,T treatment. Materials, 4(10), 1747–1762.
5. Balagna, C., Faga, M.G. and Spriano, S. (2012). Tantalum-based multilayer coating on cobalt alloys in total hip and knee replacement, Mater. Sci. Eng. C. 32, 887–895.
6. Ben-Nissan, B., Choi, A.H. and Cordingley, R. (2008). Alumina ceramics. In: *Bioceramics and their Clinical Applications.* Woodhead Publ. Ser. Biomater., pp. 223–242.
7. Ben-Nissan, B., Choi, A.H. and Green, D.W. (2019). Marine derived biomaterials for bone regeneration and tissue engineering: Learning from Nature. In: Choi, A., Ben-Nissan, B. (eds). *Marine-Derived Biomaterials for Tissue Engineering Applications.* Springer Series in Biomater. Sci. Eng., 14, 51–78.
8. Bewilogua, K. and Hofmann, D. (2014). History of diamond-like carbon films—from first experiments to worldwide applications. Surf. Coat. Technol., 242, 214–225.
9. Biasetto, L., Elsayed, H., Bonollo, F. and Colombo, P. (2016). Polymer-derived sphene biocoatings on cp-Ti substrates for orthopedic and dental implants. Surf. Coat. Technol., 301, 140–147.
10. Boutin, P. (1972). L'arthroplastie total de la hanche par prothèse en alumine frittée. Revue. Chir. Orthop., 58, 229–246.
11. Buga, C., Hunyadi, M., Gácsi, Z. et al. (2019). Calcium silicate layer on titanium fabricated by electrospray deposition. Mater. Sci. Eng. C, 98, 401–408.
12. Chevalier, J. and Gremillard, L. (2017). Zirconia as a biomaterial. In: Reference Module in Materials Science and Materials Engineering. Comprehensive Biomaterials II. Vol. 1, 122–144.
13. Clarke, I.C. and Willmann, G. (1994). Structural ceramics in orthopedics. In: Cameron, H.U. (ed.). *Bone Implant Interface*. St. Louis, Baltimore, Boston, Chicago, London, Madrid, Philadelphia, Sydney, Toronto: Mosby. pp. 203–252.
14. Dalibón, E.L., Escalada, L., Simison, S., Forsich, C., Heim, D. and Brühl, S.P. (2017). Mechanical and corrosion behavior of thick and soft DLC coatings. Surf. Coat. Technol., 312, 101–109.
15. D'Ambrosi, R., Nuara, A., Mariani, I. et al. (2021). Titanium niobium nitride mobile-bearing unicompartmental knee arthroplasty results in good to excellent clinical and radiographic outcomes in metal allergy patients with medial knee osteoarthritis. J. Arthropl., 36(1), 140–147.
16. De Villiers, D., Traynor, A., Collins, S.N. et al. (2015). Chromium nitride coating for large diameter metal-on-polyethylene hip bearings under extreme adverse hip simulator conditions, Wear. 328–329, 363–368.
17. Deville, S., Gremillard, L., Chevalier, J. and Fantozzi, G. (2017). A critical comparison of methods for the determination of the aging sensitivity in biomedical grade yttria-stabilized zirconia. https://arxiv.org/pdf/1710.04449. (Accessed 4 Oct 2021).
18. Dumbleton, J.H., Manley, M.T. and Edidin, A.A. (2002). A literature review of the association between wear rate and osteolysis in total hip arthroplasty, J. Arthroplasty. 17, 649–661.
19. Ebrahimi, M., Botelho, M.G. and Dorozhkin, S.V. (2017). Biphasic calcium phosphates bioceramics (HA/TCP): concept, physicochemical properties and the impact of standardization of study protocols in biomaterials research. Mater. Sci. Eng. C, 71, 1293–1312.

20. Filho, L., Schmidt, S., Leifer, S. et al. (2019). Towards functional silicon nitride coatings for joint replacements. Coatings. 9, 1–10.
21. Gallegos-Cantú, S., Hernandez-Rodriguez, M.A.L., Garcia-Sanchez, E. et al. (2015). Tribological study of TiN monolayer and TiN/CrN (multilayer and superlattice) on Co–Cr alloy, Wear. 330–331, 439–447.
22. Hauert, R., Thorwarth, K. and Thorwarth, G. (2013). An overview on diamond-like carbon coatings in medical applications. Surf. Coat. Technol., 233, 119–130.
23. Heimann, R.B. (2007). Novel approaches towards design and biofunctionality of plasma-sprayed osteoconductive calcium phosphate coatings on biomedical implants: The concept of bond coats. In: Pannone, P.J. (ed.) *Trends in Biomaterials Research*. pp. 1–80. Nova Science Publishers, Inc., New York.
24. Heimann, R.B. (2008). *Plasma Spray Coating. Principles and Applications*. 2nd ed. Weinheim: Wiley-VCH.
25. Heimann, R.B. (2010). *Classic and Advanced Ceramics. From Fundamentals to Applications*. Wiley-VCH Verlag GmbH & Co. KGaA, Weinheim, Germany.
26. Heimann, R.B. (2017). Calcium (Ti, Zr) hexaorthophosphate bioceramics for electrically stimulated biomedical implant devices: A position paper. Amer. Mineral., 102, 2170–2179.
27. Heimann, R.B. (ed.). (2020). *Materials for Medical Application*. Walter de Gruyter GmbH, Berlin, Boston.
28. Heimann, R.B. (2021a). Functional plasma-sprayed hydroxylapatite coatings for medical application: Clinical performance requirements and key property enhancement. J. Vac. Sci. Technol. A, 39(5), 050801.
29. Heimann, R.B. (2021b). Silicon nitride, a close to ideal ceramic material for medical application. Ceramics, 4, 208–223.
30. Heimann, R.B. and Lehmann, H.D. (2015). Bioceramic Coatings for Medical Implants. Trends and Techniques. Weinheim: Wiley-VCH.
31. Heimann, R.B., Schürmann, N. and Müller, R.T. (2004). In vitro and in vivo performance of Ti6Al4V implants with a plasma-sprayed osteoconductive hydroxylapatite-bioinert titania bond coat "duplex" system: An experimental study in sheep. J. Mater. Sci. Mater. Med., 15(9), 1045–1052.
32. Hench, L.L. (1991). Bioceramics: From concept to clinic. J. Am. Ceram. Soc., 74(7), 1487–1510.
33. Hench, L.L. (2013). Chronology of bioactive glass development and clinical applications. New J. Glass Ceram., 3(2), Article ID: 30885, 7 pages.
34. Heness, G., Ben-Nissan, B. (2004). Innovative bioceramics. Mater. Forum., 27, 104–114. Inst. Mater. Eng. Australasia Ltd.
35. Huang, L., Ji, H., Liang, Y., Xie, Y. and Zheng, X. (2011). Bone replacing material of baghdadite coated-titanium alloy, useful for inducing formation of bone-like apatite in simulated body fluid, comprises titanium and its alloy as matrix and coating is deposited on matrix by plasma spray coating. Chinese patent CN102049065-A.
36. ISO 10993 (2003). Biological evaluation of medical devices. Part 8: Selection and qualification of reference materials for biological tests.
37. ISO 10993 (2010). Biological evaluation of medical devices. Part 10: Tests for irritation and skin sensitization
38. ISO 13356 (2016). Ceramic materials based on yttria-stabilized tetragonal zirconia (Y-TZP)
39. Itiravivong, P., Promasa, A., Laiprasert, T., Techapomgworachai, T., Kuptnitazsaikul, S., Thanakit, V. and Heimann, R.B. (2003). Comparison of tissue reaction and osteointegration of metal implants between hydroxyapatite/Ti alloy coat: an animal experimental study. J. Med. Assoc. Thailand, 86(3), S422–S430.
40. Ivanova, E.P., Bazaka, K. and Crawford, R.J. (2014). Bioinert ceramic biomaterials: advanced applications. In: Ivanova, E.P., Bazaka, K. and Crawford, R.J. (eds). *New Functional Biomaterials for Medicine and Healthcare*. pp. 173–186. Woodhead Publishing Series in Biomaterials. Woodhead Publishing Limited.

41. Jelinek, R. (2013). *Biomimetics. A molecular perspective.* Berlin, Boston: De Gruyter.
42. Knabe C., Adel-Khattab, D. and Müller-Mai, C. (2020). Biological performance of biomaterials and biomedical devices. In: Heimann, R.B. (ed.). Materials for Medical Application. pp. 369-385.Walter de Gruyter GmbH, Berlin, Boston.
43. LeGeros, R.Z. and LeGeros, J.P. (1984). Phosphate minerals in human tissues. In: Nriagu, J.O. and Moore, P.B (eds). *Phosphate Minerals.* Springer: Berlin, Heidelberg, pp. 351–385.
44. Li, J., Li, Z.L., Tu, J.P., Jin, G., Li, L.L., Wang, K.T. and Wang, H.R. (2019). In vitro and in vivo investigations of a-C/a-C:Ti nanomultilayer coated Ti6Al4V alloy as artificial femoral head. Mater. Sci. Eng. C, 99, 816–826.
45. Li, W., Guler, U., Kinsey, N. et al. (2014). Refractory plasmonics with titanium nitride: Broadband metamaterial absorber. Adv. Mater., 26(47), 7959–1765.
46. Liu, C., Huang, Y., Shen, W. and Cui, J. (2001). Kinetics of hydroxyapatite precipitation at pH 10 and 11. Biomaterials, 22, 301–306.
47. Madfa, A.A., Al-Sanabani, F.A., Al-Qusami, N.H. et al., (2014). Use of zirconia in dentistry: An Overview. The Open Biomater. J., 5, 1–9.
48. Navarro, M., Michiardi, A., Castaño, O. and Planell, J.A. (2008). Biomaterials in orthopedics. J. R. Soc. Interface, 5, 1137–1158.
49. Ning, C.Y., Zhou, L. and Tan, G.X. (2015). Fourth generation biomedical materials. Materials Today, 19(1), 2–3.
50. Niu, L., Jiao, K., Ryou, H. et al. (2013). Multiphase intrafibrillar mineralization of collagen. Angew. Chem. (Intern. Ed.), 52, 5762–5766.
51. Pantulap, U., Arango-Ospina, M. and Boccaccini, A.R. (2022). Bioactive glasses incorporating less-common ions to improve biological and physical properties. J. Mater. Sci. Mater. Med., 33,3. https://doi.org/10.1007/s101856-021-06626-3.
52. Pezzotti, G., Bock, R.M., McEntire, B.J. et al. (2016). Silicon nitride bioceramic induce chemically driven lysis in *Porphyromonas gingivalis*. Langmuir, 32(12), 3024–3035.
53. Pezzotti, G., Bock, R.M., McEntire, B.J. *et al* (2018). In vitro antibacterial activity of oxide and non-oxide bioceramics for arthroplastic devices: I. In situ time-lapse Raman spectroscopy, R. Soc. Chem. 143, 3708–3721.
54. Pouroutzidou, G.K., Theodorou, G.S., Kontonasaki, E. et al. (2019). Effect of ethanol/TEOS ratios and amount of ammonia on the properties of copper-doped calcium silicate nanoceramics. J. Mater. Sci. Mater. Med., 30:98.
55. Razawi, M., Fathi, M., Savabi, O. et al. (2014). Nanostructured merwinite bioceramic coating on Mg alloy deposited by electrophoretic deposition. Ceram. Intern. A, 40(7), 9473–9484.
56. Reyna, A.L.P., Fritz, B., Schwiesau, J. et al. (2018). Metal ion release barrier function and biotribological evaluation of a zirconium nitride multilayer coated knee implant under highly demanding activities wear simulation. J. Biomech., 79, 88–96.
57. Rifai, A., Pirogova, E. and Fox, K. (2019). Diamond, carbon nanotubes and graphene for biomedical applications. In: Narayan, R. (ed.). *Encyclopedia of Biomedical Engineering*. Elsevier, Inc.
58. Sáenz de Viteri, V., Barandika, G., Bayón, R. et al. (2015). Development of Ti-C-N coatings with improved tribological behavior and antibacterial properties, J. Mech. Behav. Biomed. Mater. 55, 75–86.
59. Sahahinejad, E. and Vahedifard, R. (2017). Deposition of nano-diopside coatings on metallic biomaterials to stimulate apatite-forming ability. Mater. Design, 123, 120–127.
60. Schlosser, M. (2020). In vivo evaluation of inflammatory and immunological reactions. In: Heimann, R.B. (ed.). *Materials for Medical Application*. pp. 396–422. Walter de Gruyter GmbH, Berlin, Boston.
61. Shepherd, J.H., Shepherd, D.V. and Best, S.M. (2012). Substituted hydroxyapatites for bone repair. J. Mater. Sci. Mater. Med., 23(10), 2335–2347.
62. Shtansky, D.V., Levashov, E.A. and Sukhorukova, I.V. (2015). Multifunctional bioactive nanostructured films. In: Mucalo, M. (ed.) *Hydroxyapatite (Hap) for Biomedical Applications*. Woodhead Publ. Ser. Biomater., pp. 159–188.

63. Toth, P. (2021). Nanostructure quantification of turbostratic carbon by HRTEM image analysis: State of the art, biases, sensitivity and best practices. Carbon, 178, 688–707.
64. Webster, T.J., Ergun, C., Doremus, R.H. and Lanford, W.A. (2003). Increased osteoblast adhesion on titanium-coated hydroxyapatite that form $CaTiO_3$. J. Biomed. Mater. Res. Part A, 67A(3), 976–980.
65. Wei, D., Zhou, Y., Jia, D. and Wang, Y. (2007). Structure of calcium titanate/titania bioceramic composite coatings on titanium alloy and apatite deposition on their surfaces in a simulated body fluid. Surf. Coat. Technol., 201, 8715–9722.
66. Williams, D.F. (1990). Biocompatibility: an overview. In: D.F. Williams (ed.), *Concise Encyclopedia of Medical & Dental Materials*, Pergamon Press, Oxford.
67. Willmann, G. (1995). Ceramic hip joint heads made from zirconia and alumina—a comparison. In: G. Fishman, A. Clare and L. Hench (eds), *Bioceramics. Materials and Application*, Ceramic Transactions 48, p. 83, The American Ceramic Soc, Westerville.
68. Wolf, S.E., Schüßler, M., Böhm, C.F. and Demmert, B. (2019). Frontiers in bio-inspired mineralization: Addressing mimesis of four-dimensional, hierarchical, and nonclassical growth characteristics of biominerals. In: Bezarra, U.T., Ferreira, H.S. and Barbosa, N.P. (eds). *Bio-Inspired Materials*. Frontiers in Biomaterials, 6, 160–176. Singapore: Bentham Science Publishers Pte. Ltd.
69. Zhao, C., Xia, L.G., Zhai, D. et al. (2015). Designing ordered micropatterned hydroxyapatite bioceramics to promote the growth and osteogenic differentiation of bone marrow stromal cells. J. Mater. Chem. B, 3, 968–976.

Structure and Bulk Properties of Silicon Nitride

Stuart Hampshire

1 Introduction

Silicon nitride is one of the major structural engineering ceramics that has been developed over many years of intensive research. It possesses high flexural strength, good fracture resistance, high hardness, excellent wear resistance, and good creep resistance at high temperatures where metals cannot be used. These properties make it a unique material for applications such as cutting tools, bearings, turbochargers, engine valves, and other wear-resistant parts used under aggressive conditions. Excellent in-depth reviews have been published by Riley [1] and Petzow and Herrmann [2].

Silicon nitride is a covalent, crystalline compound, with normal stoichiometric composition of Si_3N_4, that was first synthesized by Sainte-Claire Deville and Wöhler [3]. It only came to prominence nearly a century later when it was developed as a refractory material by the Carborundum Company through carbothermal reduction of SiO_2 in the presence of nitrogen. Collins and Gerby [4] reported its use as thermocouple tubes, crucibles for molten metals, and also rocket nozzles, formed by nitriding silicon powder compacts. Interest began to grow for potential use in applications such as engines. Parr et al. [5] published a comprehensive review of its physical and chemical properties and structure, outlining the technology that they had developed and is now known as *Reaction-Bonded Silicon Nitride* (RBSN). This would be the *major* processing route for silicon nitride ceramics for at least the next decade. The challenge was to develop strong and thermally shock-resistant ceramics that would enable their use in high-temperature structural applications such as gas turbines.

S. Hampshire (✉)
The Bernal Institute, University of Limerick, Limerick, Ireland
e-mail: stuart.hampshire@ul.ie

B. S. Bal et al. (eds.), *Silicon Nitride Bioceramics*,
https://doi.org/10.1007/978-3-031-67047-3_2

Nitridation of silicon was also the process for producing silicon nitride powder but initial attempts to sinter Si_3N_4 powder compacts were unsuccessful. Its high energy covalent chemical bonds result in low self-diffusivity such that atomic species only become sufficiently mobile for densification to commence at temperatures >1850 °C where decomposition occurs. Deeley et al. [6] investigated alternative approaches and succeeded in achieving reasonably dense materials by hot-pressing with various oxide sintering additives. A fully dense ceramic was produced by hot-pressing silicon nitride containing MgO at 1850 °C under a uniaxial pressure of 23 MPa. Flexural strength was found to be substantially higher than that of RBSN [7].

In the 1970s, a major research program was initiated in the USA to produce a ceramic gas turbine [8], which was mirrored by similar programs in Europe and Japan. This more intensive research period resulted in a better understanding of the role of sintering additives in the densification and microstructural development of silicon nitride-based ceramics and their relationships with mechanical and thermal properties [9]. Improvements in powder manufacture, forming techniques, and alternative firing processes led to a new "family" of ceramics including RBSN, hot-pressed silicon nitride (HPSN), sintered silicon nitride (SSN), sintered reaction-bonded silicon nitride (SRBSN), gas-pressure sintered silicon nitride (GPSSN), hot isostatically pressed silicon nitride (HIPSN), spark plasma sintered silicon nitride (SPSSN), and silicon nitride "alloys" or solid solutions termed SiAlONs [10], based on their elemental components. These fabrication and firing processes are covered in a later chapter.

2 Atomic Bonding and Crystal Structure

The atomic bonding in silicon nitride is predominantly directionally covalent because of the necessity of N forming three bonds to Si as shown in Fig. 1a. The three Si atoms are arranged at the vertices of a planar triangle with bond angles of 120° each. Bonding around the nitrogen atoms is therefore considered as sp^2, in which the s, p_x, and p_y orbitals hybridize to form three bonds to silicon atoms, with the p_z orbitals non-bonding. The bonding leads to a basic structural framework of $Si\text{-}N_4$ tetrahedra as shown in Fig. 1b with Si-N bond lengths of approximately 1.6 Å. These are slightly distorted and joined by sharing nitrogen corners so that each nitrogen is common to three tetrahedra.

Silicon nitride exists in three known crystallographic modifications, α and β [11], both with hexagonal symmetry, and a cubic spinel phase, $\gamma\text{-}Si_3N_4$ [12], produced only at high pressure. In the cubic modification, Si occupies both the tetrahedral A site and the octahedral B site in the spinel structure. The β-phase crystallizes in the hexagonal space group $P6_3$ and its formula unit cell is Si_6N_8. It is composed of 8-membered puckered rings of alternating Si and N atoms, which join in turn to form layers perpendicular to the *c*-axis with a stacking sequence ABAB as shown in Fig. 2a [13] and forming long continuous channels in the c-direction.

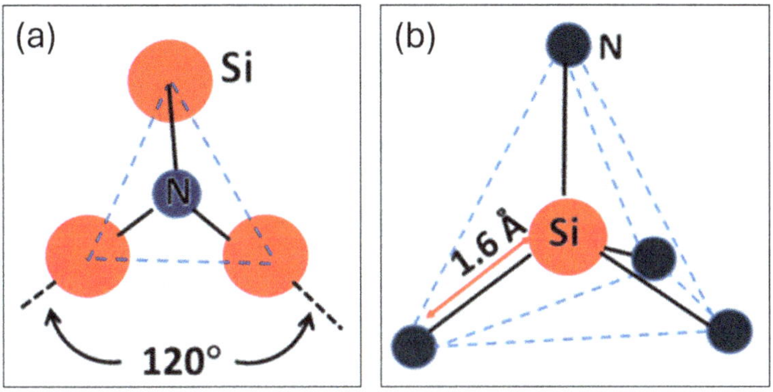

Fig. 1 Atomic arrangements in silicon nitride: (**a**) 3 Si atoms around each N atom at the vertices of a planar triangle, (**b**) Si-N_4 tetrahedral structural unit

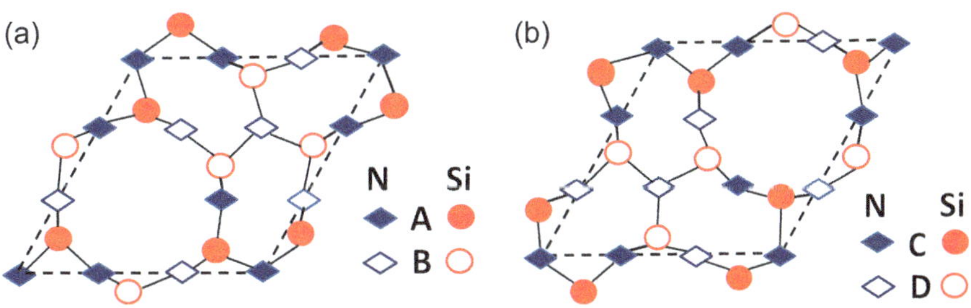

Fig. 2 (**a**) AB Si-N layers in β-Si_6N_8: stacking is ABAB; (**b**) CD Si-N layers in α-$Si_{12}N_{16}$: stacking is ABCD

Table 1 Cell dimensions and theoretical densities of α and β silicon nitrides

α-silicon nitride cell dimensions	a-axis: 0.775–0.777 nm c-axis: 0.5614–0.569 nm c/a ~ 0.70
Theoretical density (g.cm^{-3})	3.168–3.188
β-silicon nitride cell dimensions	a-axis: 0.759–0.761 nm c-axis: 0.271–0.292 nm c/a ~ 0.37
Theoretical density (g.cm^{-3})	3.19–3.202

The α-phase crystallizes in the hexagonal space group P31c. The c-dimension of α is approximately twice that of β, as shown in Table 1, and thus its formula unit cell is $Si_{12}N_{16}$. Where the layers of atoms in the β-phase are linked along the [001] direction in the sequence ABAB, the α-structure has the sequence ABCDABCD. The CD layer, shown in Fig. 2b [13], is like the AB layer except that it is rotated by 180° on the c-axis. The long continuous channels in the c-direction of the β-phase are therefore closed off to create two large interstices in the unit cell at ⅓, ⅔, ⅜, and ⅔,

⅓, and ⅞, which can accommodate other ions. Further crystal structure determinations and investigations of the Si-O-N system [14] showed that α is a defect structure with up to 1 in 30 nitrogen atoms replaced by oxygen with corresponding Si vacancies. As shown in Table 1, the unit-cell dimensions for twenty-six different α-silicon nitrides [15] vary widely. As the same precise method was used by the same investigator, this variation is well outside any experimental error and is thought to be due to the substitution of oxygen for nitrogen and reduced valency Si^{3+} present in the interstitial sites to give a composition: ${Si^{3+}}_{0.3}\,{Si^{4+}}_{11.6}\,{O^{2-}}_{0.5}\,{N^{3-}}_{15.5}$.

α-Si_3N_4 transforms to β-Si_3N_4 at temperatures above ~1400 °C but the reverse transformation has never been reported. Thus, the α- and β-phases are not considered to be low- and high-temperature forms. The α → β transformation may be classified as a reconstructive transformation of 2nd coordination, a process that only occurs when the transforming specie is in contact with a solvent. The greater solubility of the less stable α-Si_3N_4 drives it into solution where it breaks down into its simpler SiN_4 tetrahedra that are reconstructed as the less soluble, more stable β-Si_3N_4 upon precipitation from solution.

Both α and β silicon nitrides can form solid solutions with Al_2O_3 and are known as SiAlONs. β-SiAlON is formed when oxygen (O^{2-}) replaces nitrogen (N^{3-}) in the β-Si_6N_8 structure and silicon (Si^{4+}) is replaced by aluminum (Al^{3+}) to maintain overall charge neutrality [16, 17]. The general composition is:

$$Si_{6\text{-}z}Al_zO_zN_{8\text{-}z}$$

retaining the 6:8 metal-to-non-metal ratio of ß-Si_6N_8, with z values in the range 0–4.2.

Al^{3+} can also partially replace Si^{4+} in the α-$Si_{12}N_{16}$ unit cell if, at the same time, charge compensation is effected by accommodation of other ions, M = Li^+, Ca^{2+}, Y^{3+}, or rare earth lanthanide ions (RE^{3+}), in the two interstitial sites. The general composition [13] is:

$$M_xSi_{12\text{-}(m+n)}Al_{(m+n)}O_nN_{(16\text{-}n)}$$

$x = m/v$ where v is the valence of the M cation, and x should be a value between 0.3 and 2 (the maximum value because of 2 interstitial sites per unit cell). α-SiAlONs are more N-rich compositionally than β-SiAlONs.

3 Porous Silicon Nitride and Microstructure

As outlined above [5], Reaction-bonded Silicon Nitride, RBSN, was the first form of the ceramic to be fabricated from its constituent elements of Si and N [18]:

$$3\ Si_{(s)} + 2\ N_{2(g)} \rightarrow Si_3N_{4(s)};\quad \Delta G^n \approx -723 + 0.315T\ \text{kJmol}^{-1} \qquad (1)$$

RBSN is normally manufactured from silicon powder compacts. These may be fabricated as billets, or in the shape of the component being produced, by isostatic pressing. Nitriding is carried out under N_2 in the temperature range 1200–1450 °C with a slow heating rate to avoid problems associated with the exothermic nature of Eq. (1). The true density of silicon is 2.330 g.cm^{-3} and that of silicon nitride is 3.187 g.cm^{-3} so a volume expansion of 21.7% occurs during nitride formation. Classical sintering, and thus shrinkage, do not happen during the process; however, considerable internal rearrangement occurs as the silicon nitride grows within the pre-existing pores of the Si powder compact and, therefore, the original dimensions remain virtually unchanged during nitriding (linear shrinkage <0.1%). This allows complex shapes to be machined to final dimensions, usually after an initial annealing step under argon at 1100–1200 °C. This results in neck formation between particles because of surface diffusion and this process strengthens the Si compact for the machining. On completion of nitridation, close dimensional tolerances are maintained, and the product requires little or no subsequent machining, thus lending the process very effectively to mass production.

During nitriding below the melting point of silicon, thin films of nitride initially form on the Si particle surfaces, but because of the compressive strain in these films, microcracks occur exposing the underlying silicon, which allows the reaction to continue [18]. In the early stages of nitridation, a background mat of mainly α-Si_3N_4 forms, and the pores of the Si compact become filled with needle-like crystals. Small areas of Si are also converted to grains of β-Si_3N_4. A major part of the nitriding process occurs over 50–70 h.

Pure Si melts at ~1413 °C and the nitridation reaction above this temperature occurs according to:

$$3\,Si_{(l)} + 2\,N_{2(g)} \rightarrow \beta{-}Si_3N_{4(s)};\ \ \Delta G^n \sim -874 + 0.405T\ \text{kJmol}^{-1} \qquad (2)$$

The β-Si_3N_4 crystallites grow into the molten Si held within the matrix forming a microstructure of discrete β-Si_3N_4 grains surrounded by a dense mat of α-Si_3N_4 crystals and porosity in the range 15–20 vol.%, with ~80 vol.% pores with dimensions less than 0.1 μm [18]. It is recognized that the steady development of an α-Si_3N_4 whisker mat is responsible for increases in mechanical strength during reaction-bonding. Various impurities, such as Fe, Ca, and oxygen, affect the course of the reaction, the formation of liquid Si, and the extent of β-Si_3N_4 formation.

4 Dense Silicon Nitride and Microstructure

4.1 Sintered Silicon Nitride (SSN)

As outlined above, dense silicon nitride was initially fabricated by hot-pressing silicon nitride powder containing oxide sintering additives at 1850 °C [7]. It was not understood at that stage that each particle of silicon nitride powder was covered in a surface layer of silica (SiO_2) and that during sintering, the additives react with the silica and some of the nitride to form an oxynitride liquid phase, which promotes densification. It was subsequently found that silicon nitride powder compacts containing mixed yttria and alumina additives, sintered in a nitrogen atmosphere (0.1 MPa) at 1750–1900 °C, allowed densification without resorting to hot-pressing. The overall reaction may be summarized by [19, 20]:

$$\alpha\text{-}Si_3N_4 + SiO_2 + Y_2O_3 + Al_2O_3 \rightarrow \alpha\text{-}Si_3N_4 + \text{Y-Si-Al-O-N liquid phase}$$
$$\rightarrow \beta\text{-}Si_3N_4 + \text{Y-Si-Al-O-N secondary phase (usually amorphous)} \quad (3)$$

The α-structure dissolves in the liquid phase at the sintering temperature and transforms to β-Si_3N_4 through a solution-diffusion-precipitation process. Prismatic hexagonal β-Si_3N_4 crystals are precipitated from the liquid and grow along their c-axes to form an interlocking microstructure of high aspect ratio β grains. Upon cooling, the liquid forms an intergranular Y-Si-Al-O-N glass, as shown in the scanning electron micrograph of Fig. 3.

The kinetics of pressureless sintering and α → ß transformation were studied by Hampshire and Jack [19] for silicon nitride with either MgO or Y_2O_3 as a single additive. The three stages of liquid phase sintering were identified as follows and are shown in the schematic shrinkage–time plot of Fig. 4.

1. ***Particle Rearrangement*** during which the initial liquid is formed. The total amount of shrinkage depends on the volume of liquid and the rate depends on the viscosity; this stage is also the incubation period for the α → ß transformation.

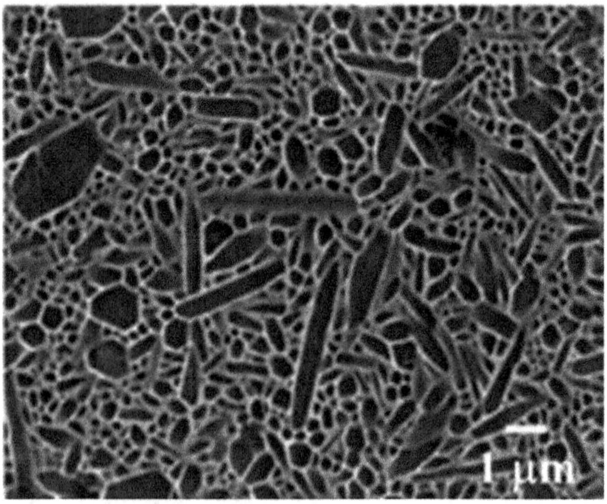

Fig. 3 Scanning electron micrograph (SEM) showing β-Si_3N_4 grains (dark crystals) surrounded by Y-Si-Al-O-N glass phase (light) (Original additive composition: 6 wt.% Y_2O_3/ 2 wt.% Al_2O_3)

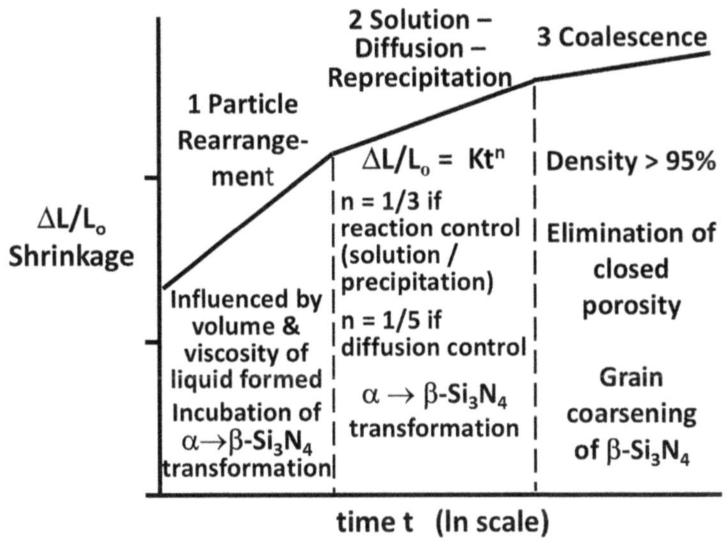

Fig. 4 Schematic ln-shrinkage—ln-time plot for liquid phase sintering of silicon nitride ceramics. (After Hampshire and Jack [19])

2. ***Solution-diffusion-reprecipitation***, where shrinkage is given by:

$$\Delta V/V_o = Kt^n \tag{4}$$

where K is a constant, t is time, and $n = 1/3$ if solution into or precipitation from the liquid is rate controlling, as was found for the MgO additive, and $n = 1/5$ if diffusion through the liquid is rate-controlling, as was found for the Y_2O_3 additive, where diffusion through a more viscous oxynitride liquid is much slower and full densification without pressure is not achieved [19]; the α → ß transformation occurs during this stage and is more rapid with Y_2O_3.

3. ***Coalescence*** or final elimination of closed porosity at relative densities >95%, during which further growth of the elongated β grains occurs but this depends critically on grain boundary liquid/glass chemistry; final density is >99%. Grain growth during this stage is considered to occur via diffusion-controlled Ostwald ripening [21].

The types and amounts of additives used for sintering determine the nature and quantity of the resulting grain boundary phase. As the grain boundary composition changes, the aspect ratios (length to diameter) of the β grains vary. Grain coarsening also occurs as sintering time or temperature is increased. The glass phase is found at triple points (TP) or as intergranular films (IGF) between β-Si_3N_4 grains as shown in Fig. 5.

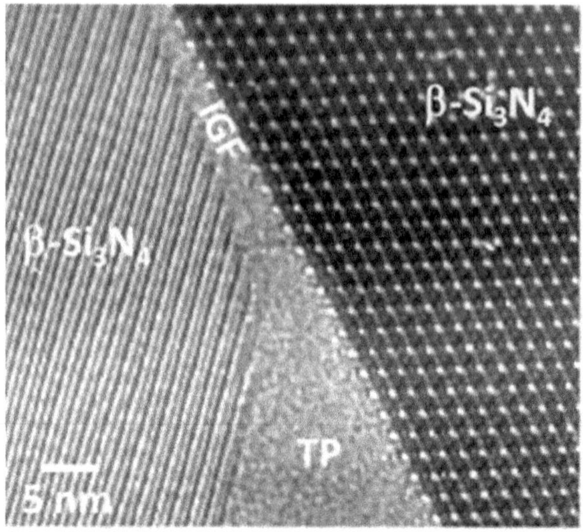

Fig. 5 High-resolution transmission electron micrograph (HRTEM) showing intergranular glass film (IGF) between β-Si_3N_4 grains and a triple point (TP) containing Y-Si-Al-O-N glass

IGF thickness varies in the range of 0.5–1.5 nm, depending on chemical composition [22]. Rare earth (RE) oxides may be used in place of yttria [23, 24] and some silicon nitrides are sintered with RE oxides plus MgO [24, 25] in place of alumina. The grain boundary chemistry (RE, Al or Mg content, and O:N ratio) and volume fraction of these glass phases control the microstructural development, which determines the overall properties of the ceramic. This is referred to as "microstructural engineering." The choice of a rare earth (RE) oxide used as a sintering additive crucially affects the rate of the $\alpha \rightarrow \beta$ phase transformation and the aspect ratios of β-Si_3N_4 grains [26]. Lanthanum (La) with its large ionic radius results in a lower transformation rate and higher aspect ratio β-Si_3N_4 grains, while lutetium (Lu) with its much smaller ionic radius leads to a higher rate of $\alpha \rightarrow \beta$ transformation and lower aspect ratio β grains.

The impact of various rare-earth and related doping elements (RE = Lu, Sc, Yb, Y, Sm, La) on the grain growth anisotropy in silicon nitride ceramics has been studied using model experiments [23], in which Si_3N_4 particles can grow freely in a RE-Si-Mg-O-N glass matrix. With increasing ionic radius of the RE, grain anisotropy increases due to non-linear growth kinetics.

The segregation of RE ions at the β-Si_3N_4/IGF interface was observed at the atomic scale by high-angle annular dark-field (HAADF) scanning transmission electron microscopy (STEM) [27–29]. Three RE ions (La, Gd, and Lu) showed distinctively different segregation positions at the interface, which was consistent with the results of first-principles calculations predicting that RE ions with larger ionic radii (La) segregate more preferentially to grain surfaces [28]. Ziegler et al. [27] observed no specific segregation position for La whereas other RE ions such as Sm, Er, Yb, and Lu showed clear segregation. Using various rare earth oxides as sintering additives has allowed the development of different types of Si_3N_4 microstructures by modifying the chemistry of the sintering liquid and the subsequent grain boundary phase.

4.2 *Sintered Reaction Bonded Silicon Nitride (SRBSN)*

Dense silicon nitride is also produced by mixing sintering additives, such as MgO + Y_2O_3, with silicon powder and nitriding as with RBSN [30], followed by sintering in the range 1750–1900 °C. The resulting material, Sintered Reaction Bonded Silicon Nitride (SRBSN), consists of β-Si_3N_4 and an amorphous grain-boundary phase. RBSN has a higher starting density (~2.2–2.7 g.cm^{-3}) than compacted Si_3N_4 powder compacts (~1.6–2.0 g.cm^{-3}) and therefore full densification is accompanied by a linear shrinkage of only ~5–10%. Thus, less post-sintering machining is required, which is a major advantage of this method along with lower-cost raw materials [30–32].

5 Mechanical Properties

Table 2 lists several physical, mechanical, and thermal properties of sintered silicon nitrides compared with RBSN and SRBSN. For sintered silicon nitride, flexural strengths vary from 600 to 1300 MPa. The lower strengths for RBSN are a feature of the level of porosity within the ceramic. For SRBSN, strengths (450–800 MPa) are more comparable with those of sintered silicon nitrides. The fracture toughness of dense silicon nitride also varies significantly from ~3.5 MPa·m$^{1/2}$, similar to other engineering ceramics, up to >13 MPa·m$^{1/2}$, which is only exceeded by transformation-toughened zirconia. Higher toughness values mean better resistance to machining damage and improved fatigue behavior. Because of high flexural strength, low thermal expansion coefficient, and reasonably high Young's modulus, silicon nitride has excellent thermal shock resistance with critical temperature changes (ΔT_c) to cause fracture (Table 2) in the range R – 400–1050 °C. Significant variations in mechanical and thermal properties are due to major differences in microstructural characteristics, even within similar classes of silicon nitrides. These differences have a critical influence on properties. The amounts and types of additives (Y_2O_3 or RE_2O_3 + Al_2O_3 or MgO) determine the quantity and chemistry of the sintering liquid, which leads to microstructural characteristics (e.g., β-grain size and aspect ratio) and the volume and distribution of intergranular glass on cooling. These features determine properties [21, 24, 33–36] such as flexural strength and fracture toughness as shown in Table 3 for specific additive combinations.

For silicon nitrides sintered with different amounts and ratios of Y_2O_3:Al_2O_3 to produce ceramics with the same level of porosity and the same grain size, fracture toughness increases with increasing Y:Al ratio as shown in Fig. 6 [37] confirming that fracture behavior is crucially dependent on grain boundary glass composition. As this changes, the size and aspect ratios of ß-grains vary with sintering time or temperature. The aspect ratio of the β grains and the fracture toughness both change in the same way with composition, and thus microstructures with more elongated rod-like grains have higher resistance to crack propagation. An advancing crack

Table 2 Mechanical and thermal properties of silicon nitrides

Property	Units	Dense silicon nitride (SSN)	Reaction-bonded silicon nitride (RBSN)	Sintered reaction-bonded silicon nitride (SRBSN)
Young's modulus *E*	GPa	300–330	125–230	280–300
Microhardness *Hv* (Vickers)	GPa	16.5–20	10–13.5	–
Flexural strength σ_f	MPa	600–1300	150–350	450–850
Fracture toughness K_{IC}	$MPa{\cdot}m^{1/2}$	3.5–13.5	1.5–2.8	3.5–11.0
Weibull modulus		14–30	10–15	15–40
Thermal expansion coefficient α (20–1500 °C)	$10^6{\cdot}K^{-1}$	3.0–3.4	2.8–3.0	3.0–3.3
Thermal conductivity κ (RT)	$W{\cdot}m^{-1}{\cdot}K^{-1}$	13–140	4–30	40–182
Thermal stress resistance parameters:				
$R = \Delta T_c = \sigma_f(1 - \nu)/\alpha E$	°C	400–1050	170–500	350–650
$R' = R\kappa$	$10^3{\cdot}W{\cdot}m^{-1}$	6–140	0.6–13	14–115

must take a more complex high-energy route through the ceramic and continued crack propagation is restricted by grain bridging and pull-out in the crack wake. Bridging grains are formed when a crack is deflected around Si_3N_4 grains, leaving grains intact [38]. This reduces tensile stresses at the crack tip resulting in increasing resistance to crack propagation, a phenomenon referred to as *R*-curve behavior [39].

Ultimately, improved fracture resistance occurs by activation of toughening mechanisms such as crack deflection and crack bridging via interfacial debonding. They are affected by the presence of residual stresses in the Si_3N_4, which interact with the propagating crack tip. These are caused by thermal expansion mismatch between the grain boundary glass and the β-Si_3N_4 grains. Residual tensile stress in the weaker glass would be expected to draw the crack along the grain boundary, enhancing debonding and leading to intergranular fracture as shown in Fig. 7. Improvements in steady-state fracture toughness values and *R*-curve behavior have been reported in silicon nitride seeded with preformed β-crystals [33, 34]. This is referred to as in situ *toughened* Si_3N_4 or *self-reinforced silicon nitride*. This involves the controlled incorporation of large (> 1 μm) diameter elongated ß-seeds, which leads to a bimodal microstructure of large elongated β-Si_3N_4 grains, grown from the seeds, dispersed within a fine-grained matrix. This material has strengths >1 GPa and toughness >10 $MPa{\cdot}m^{1/2}$, while Si_3N_4 with equiaxed (0.5 μm) β-grain size has strength of ~600 MPa and toughness of ~3.5 $MPa{\cdot}m^{1/2}$ [33].

With high Y:Al ratio sintering additives, more extensive debonding at grain boundary interfaces was observed [34] than in silicon nitrides sintered with additives with low Y:Al ratios, resulting in increased intergranular fracture. The Y:Al ratios in

Table 3 Flexural strength and fracture toughness of silicon nitrides sintered with mixed $RE_2O_3.Al_2O_3$ and $RE_2O_3.MgO$ additives (RE = Y, La, Gd, Lu)

Sintering additives	Sintering process	Flexural Strength σ_f (MPa) (4-point)	Fracture Toughness K_{IC} (MPa·m$^{1/2}$)	References
$4Y_2O_3.2.8Al_2O_3$ (wt.%) + β seeds	GPS 1800 °C	1096 ± 48	7.5 ± 0.2	Sun et al. [34]
$5Y_2O_3.2Al_2O_3$ (wt.%) + β seeds	GPS 1800 °C	1011 ± 78	8.7 ± 0.4	Sun et al. [34]
$5Y_2O_3.2Al_2O_3$ (wt.%) + β seeds	1850 °C/6 h 0.9 MPa N_2	~1050–1120	~11–11.5	Hirao [36]
$6.25Y_2O_3.1Al_2O_3$ (wt.%) + β seeds	GPS 1800 °C	1144 ± 126	10.6 ± 0.2	Sun et al. [34]
$8La_2O_3.2Al_2O_3$ (wt.%) + β seeds	HP 1800 °C	710 ± 159	~ 7.5	Becher et al. [24]
$8Gd_2O_3.2Al_2O_3$ (wt.%) + β seeds	HP 1800 °C	858 ± 143	–	Becher et al. [24]
$8Lu_2O_3.2Al_2O_3$ (wt.%) + β seeds	HP 1800 °C	710 ± 204	~ 9.5	Becher et al. [24]
$8La_2O_3.2MgO$ (wt.%) + β seeds	HP 1800 °C	1145 ± 20	12.5–13.5	Becher et al. [35]
$8Gd_2O_3.2MgO$ (wt.%) + β seeds	HP 1800 °C	1130 ± 40	11–12	Becher et al. [35]
$8Lu_2O_3.2MgO$ (wt.%) + β seeds	HP 1800 °C	1105 ± 10	12–13	Becher et al. [35]
$8Y_2O_3.2MgO$ (wt.%) no seeds	HP 1825 °C	729	6.8 ± 1.2	Mikijelj et al. [40]
$8La_2O_3.2MgO$ (wt.%) no seeds	HP 1825 °C	974	9.1 ± 0.8	Mikijelj et al. [40]
$8Gd_2O_3.2MgO$ (wt.%) no seeds	HP 1825 °C	954	9.2 ± 0.4	Mikijelj et al. [40]
$8Lu_2O_3.2MgO$ (wt.%) no seeds	HP 1825 °C	648	4.9 ± 0.9	Mikijelj et al. [40]

the additives influence the atomic bonding structure across the β-Si_3N_4/grain boundary glass interface. Higher Al contents result in stronger bonding at this interface whereas reduction of the Al concentration results in a more abrupt chemical gradient and structural interface.

With RE oxide additives, the selection of additives is essential to tailor microstructure and mechanical properties. Silicon nitrides with equivalent grain sizes and morphologies have higher fracture toughness values with increasing size of the RE^{3+} ion, reflecting a more predominate intergranular crack path. Becher et al. [24] compared RE_2O_3 + MgO additives (RE = La, Gd, Lu) in silicon nitrides containing β seeds and found that Lu_2O_3 generates larger diameter reinforcing β-grains grown from seeds, while La_2O_3 results in reinforcing grains with higher aspect ratio and results for Gd_2O_3 are intermediate. Nevertheless, a substantial fraction of large reinforcing grains was observed for each RE. As a result, the mechanical properties of the resultant ceramics (Table 3) have similar flexural strengths >1 GPa, fracture

Fig. 6 Fracture toughness (K_{IC}) as a function of Y_2O_3: Al_2O_3 additive molar ratio used to densify different silicon nitrides with similar relative densities and microstructures. (Based on data from Ref. [34])

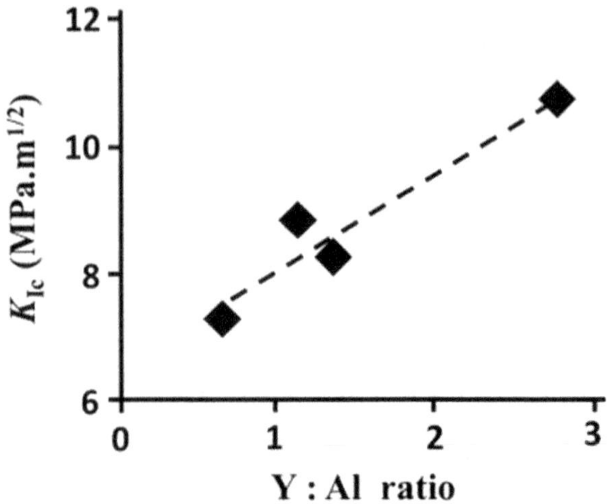

Fig. 7 Scanning electron micrograph (SEM) showing crack propagating along grain boundary glass (light phase) between β-Si_3N_4 grains (dark crystals)—intergranular fracture

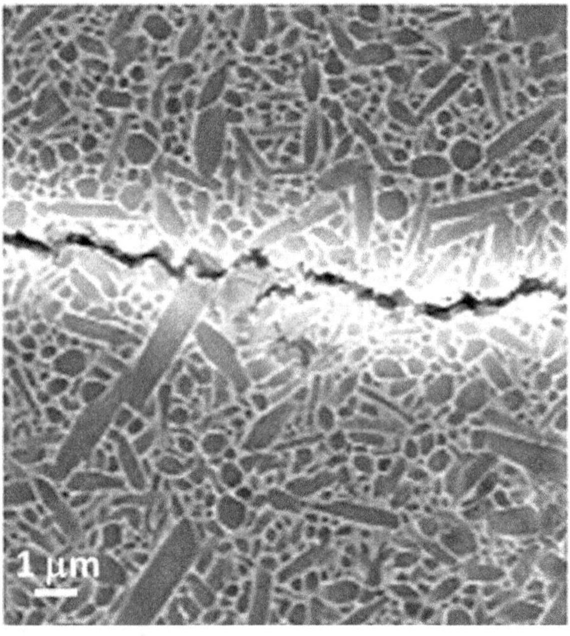

toughness >10 MPa·m$^{1/2}$, and substantially retained strength at 1200 °C (~800 MPa) [24, 35]. In contrast, a more recent study of silicon nitrides with the same additive combinations (RE_2O_3 / MgO; RE = Y, La, Gd, Lu) but no β seeds [40] resulted in lower strengths (648–974 MPa) and lower fracture toughness (4.9–9.2 MPa·m$^{1/2}$) (Table 3).

The strengths of ceramics degrade over time due to subcritical crack growth (SCCG) under stress, assisted by environmental factors, particularly moisture and

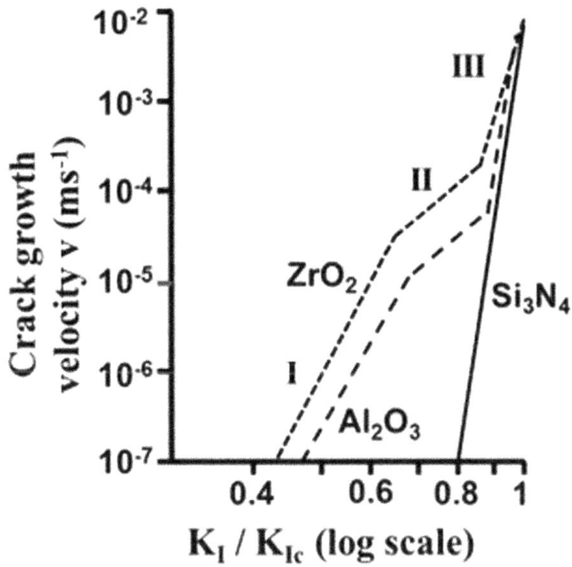

Fig. 8 A plot of crack growth velocity v. normalized K_I/K_{IC} ratio for zirconia (ZrO_2), alumina (Al_2O_3), and silicon nitride (Si_3N_4) ceramics. (After Chevalier and Fantozzi [42])

humid conditions. A constant stress will induce a particular stress-intensity factor K_I on a flaw, depending on its size, and a crack then grows at a velocity v, which depends on the $\nu - K_I$ relationship:

$$\nu = AK_I^{\ n} \tag{5}$$

where A and n are constants for a particular material. The value n is known as the subcritical crack growth susceptibility coefficient, while A and n are constants for a particular material; n can vary in the range of 10–100 [41]. Accurate determination of these constants allows lifetime predictions for ceramic components, which is important for long-term use. A comparison of the sensitivity to SCCG can be made using a normalized ν-K_I/K_{IC} diagram shown in Fig. 8. The higher the slope of the plot, i.e., the higher the K_I/K_{IC} ratio, the lower the sensitivity to SCCG by stress-assisted corrosion.

Zirconia (ZrO_2) and alumina (Al_2O_3) ceramics are susceptible to three regions of crack propagation evaluated in laboratory air whereas silicon nitride does not exhibit the same type of behavior [42]. In region I, where $\nu = \alpha_1 K_I^{\ n}$ (α_1 and n are constants), a stress-induced corrosion mechanism operates where the rate of crack growth is reaction-rate controlled. It depends on the rate of chemical attack of Zr-O or Al-O bonds by H_2O at the crack tip to form -OH species. In region II, ν is more constant so it is not as dependent on K_I. The rate of crack growth depends on the diffusion of corrosive species to the crack tip. In region III, fast fracture is observed at higher K_I values, and n is much higher than in region I. This is intrinsic to the material and insensitive to the environment. Finally, there is a critical stress intensity factor, K_{IC}

where fracture is instantaneous, and this value is what is commonly known as fracture toughness. Si_3N_4 exhibits only region III behavior at higher values of K_I and, thus, it is much less sensitive to SCCG and attack by environmental H_2O than ZrO_2 and Al_2O_3.

Sintering additive combinations have a strong influence on high-temperature strength, creep, and oxidation resistance as well as thermal properties. At temperatures exceeding 1200 °C, strengths decrease due to the softening of the intergranular glass phase. Grain boundary chemistry, effective viscosity, and the volume fraction of the intergranular glass control the creep rate and the formation and growth of cavities in the amorphous phase [43]. Observations of intergranular films in silicon nitride [22] show that their width shrinks with decreasing RE ion radius. The viscous flow of these films contributes to the initial stage of tensile creep deformation [43]. Smaller RE ions (Lu, Er) prefer to segregate at the triple points [28, 37]. Larger RE ions prefer N and remain concentrated in the intergranular films. Creep behavior is thus dependent on both intergranular film and triple-point glass viscosities.

6 Thermal Properties

Silicon nitride, particularly in its reaction-bonded form, may be considered a thermal insulator with values of thermal conductivity as low as 4 $W \cdot m^{-1} \cdot K^{-1}$. Even dense silicon nitride initially had values as low as 13–20 $W \cdot m^{-1} \cdot K^{-1}$, which is quite low compared with silicon carbide and aluminum nitride. Both have thermal conductivity values over 200 $W \cdot m^{-1} \cdot K^{-1}$ [44]. High thermal conductivity ceramics are increasingly being used as substrates for electronic power devices in alternative energy technologies, electric vehicles, and other industrial applications where thermal dissipation is a critical issue. High thermal stresses induced by elevated levels of electrical power pose serious challenges to devices with brittle ceramic substrates, which provide both electrical insulation and heat dissipation [45]. Aluminum nitride is particularly susceptible to failure and cracking and silicon carbide is a semiconductor, which would make its application as an electrical circuit substrate unsuitable. Silicon nitride was reported [46] to have an intrinsic thermal conductivity of 200–320 $W \cdot m^{-1} \cdot K^{-1}$, similar to AlN. Given its much higher strength, fracture toughness, and thermal shock resistance, research is concentrated on improving the thermal conductivity of Si_3N_4.

As shown in Table 4, there are significant variations in thermal conductivity for silicon nitrides. These variations are due to major differences in microstructural characteristics, which, in turn, depend on the sintering process and the types of additives that have evolved since the 1980s. Tsukuma et al. [47] hot-pressed Si_3N_4 at 1900 °C with 4 wt.% MgO, Y_2O_3, or Al_2O_3 and without additives. They reported that thermal conductivity was highest without additive at 28 $W \cdot m^{-1} \cdot K^{-1}$ and reduced in the order MgO > Y_2O_3 > Al_2O_3 (13 $W \cdot m^{-1} \cdot K^{-1}$). Ziegler and Hasselman [48] achieved a value of 20 $W \cdot m^{-1} \cdot K^{-1}$ by reaction-bonding and 40 $W \cdot m^{-1} \cdot K^{-1}$ by hot-pressing at 1700 °C with 3 wt.% MgO. Watari et al. [49]

Table 4 Thermal conductivity of silicon nitrides sintered with different additives

Sintering additives	Sintering process	RT thermal conductivity ($W \cdot m^{-1} \cdot K^{-1}$)	Mechanical properties, σf (MPa) /K_{IC} ($MPa \cdot m^{1/2}$)	References
No additive 4 wt.% MgO 4 wt.% Y_2O_3 4 wt.% Al_2O_3	HP 1900 °C	28 26 20 13	–	Tsukuma et al. [45]
3 wt.% MgO No additive + wax pore former	HP 1700 °C RBSN 1400 °C/ 100 h 90v%N_2/ 10v%H_2	40 10 20	– σf = 205 σf = 124	Ziegler and Hasselman [46]
3.5 wt.% Al_2O_3 3.5 wt.% Y_2O_3 Al_2O_3 + Y_2O_3	HIP 1750 °C	16 72 27	–	Watari et al. [47]
5 wt.% Y_2O_3 + 5 vol.% ß seeds + tape-casting	HP + GPS 1850 °C/66 h	121 (//) 60 (⊥)	–	Hirao et al. [48]
3.5 wt.% Y_2O_3	HP 1800 °C HIP 2400 °C	78 (// HP) 93 (⊥ HP) 80 (// HP) 102 (⊥ HP)	–	Watari et al. [50]
2 mol.% Yb_2O_3 + 5 mol.% $MgSiN_2$ 2 mol.% Yb_2O_3 + 5 mol.% MgO	GPS 1900 °C/ 2 h 48 h GPS 1900 °C/ 2 h 48 h	103 142 87 118	–	Hayashi et al. [51]
7 mol.% $MgSiN_2$ + 3 mol.% Y_2O_3	GPS 1900 °C Post-Annealing 1400 °C 1600 °C	81–97 105 117	σf = 501–689 K_{IC} = 8.1–9.7 σf = 698 σf = 517	Zhang et al. [52]
2 mol.% Y_2O_3 + 5 mol.% $MgSiN_2$ + Al impurity	SRBSN post-sintered at 1900 °C 6 h	wt.% Al 0 – 92 0.01 – 84 0.1 – 74 0.2 – 66 0.4 – 58	K_{IC} ~ 6–7.3 σf = 840 σf = 605 σf = 495 σf = 820 σf = 740	Kusano et al. [52]
2 mol.% Y_2O_3 + 5 mol.% MgO	SRBSN post-sintered at 1900 °C 3–48 h SSN sintered 1900 °C 3–48 h	3h – 100 12h – 117 48h – 142 3h – 87 12h – 96 48h – 111	σf = 840 σf = 605 σf = 495 σf = 820 σf = 740 σf = 615	Zhou et al. [54]
2 mol.% RE_2O_3 + 5 mol.% MgO	SRBSN post-sintered at 1900 °C 2 h	RE La – 80 Y – 81 Er – 90	K_{IC} ~ 8.3–10.6 σf = 733 σf = 758 σf = 953	Duan et al. [55]

(//) Parallel to ß grain orientation; (⊥) Perpendicular to ß grain orientation produced Si_3N_4 by hot isostatic pressing at 1750 °C and also recorded a low value of 16 $W \cdot m^{-1} \cdot K^{-1}$ using Al_2O_3 as the additive whereas with Y_2O_3 alone, a thermal conductivity of 72 $W \cdot m^{-1} \cdot K^{-1}$ was achieved. Following this remarkable increase in conductivity, further research using Y_2O_3 was carried out by Hirao et al. [50] using 5 wt.% Y_2O_3 + 5 vol.% β seeds and tape-casting, hot-pressing, and annealing at 1850 °C for up to 66 h under 0.9 MPa N_2. This produced a highly anisotropic microstructure with large elongated β grains, grown from the seed particles, oriented parallel to the casting direction they achieved a value of 121 $W \cdot m^{-1} \cdot K^{-1}$ whereas in the perpendicular direction, it was 60 $W \cdot m^{-1} \cdot K^{-1}$.

The thermal conductivity of secondary grain boundary phases in Si_3N_4 is quite low (< 5 $W \cdot m^{-1} \cdot K^{-1}$), thus causing reductions in the overall thermal conductivity of the sintered ceramic. The effect of the intergranular films is greater than that of the triple-point glasses. Kitayama et al. [51] showed that grain boundary phases have much less influence on thermal conductivity when the β grain sizes are on the order of several micrometers. Therefore, thermal conductivity can be increased by promotion of grain growth although this usually leads to lower mechanical strength.

Watari et al. [52] found that, by hot-pressing at 1800 °C with 3.5 wt.% Y_2O_3, β grains can be oriented such that thermal conductivity perpendicular to the hot-pressing direction is higher (78 // HP and 93 ⊥ HP). HIPing at 2400 °C achieves slightly higher values (80 // HP and 102 ⊥ HP). Hayashi et al. [53] produced the highest thermal conductivity silicon nitrides so far using 2 mol.% Yb_2O_3 + 5 mol.% $MgSiN_2$ additives and gas pressure sintering at 1900 °C for up to 48 h, achieving a value of 142 $W \cdot m^{-1} \cdot K^{-1}$. Zhang et al. [54] gas pressure sintered Si_3N_4 at 1900 °C using mixed $Y_2O_3/MgSiN_2$ additives and achieved 81–97 $W \cdot m^{-1} \cdot K^{-1}$, which could be increased by annealing at 1600 °C.

Kusano et al. [55] used mixed $Y_2O_3/MgSiN_2$ additives to RBSN, post-sintered at 1900 °C for 6 h, and added small amounts of Al impurity using AlN. All SRBSNs exhibited flexural strengths in the range of 800–850 MPa. With no Al, the thermal conductivity was 92 $W \cdot m^{-1} \cdot K^{-1}$, which decreased with further Al addition to 58 $W \cdot m^{-1} \cdot K^{-1}$ with 0.4 wt.% Al. Eighty percent (80%) of the Al was confirmed to be in solid solution within β-Si_3N_4 grains, which means that an equal amount of oxygen was also in solid solution to form β-SiAlON, which acts as a phonon scatterer. Lattice defects such as impurity atoms, vacancies, dislocations, stacking faults, etc. can induce phonon scattering [56], thereby reducing thermal conductivity; therefore, it is necessary to lower lattice oxygen content.

Other researchers [57, 58] also sintered reaction-bonded silicon nitride (SRBSN) using high-purity Si as the starting material. Compared with conventionally sintered silicon nitride (SSN), the amount of oxygen dissolved in the β-Si_3N_4 lattice is lower and the SRBSN ceramics attain high thermal conductivities and high mechanical strength (Table 4). As the β/α ratio increases during nitridation and β-grain growth is enhanced during post-sintering, thermal conductivity is increased, with values of 142 $W.m^{-1} \cdot K^{-1}$ achieved [57].

7 Tribological Properties

Silicon nitride components are being used in many wear-resistant applications because of their superior mechanical and thermal properties as outlined above. They exhibit low friction coefficients and have good corrosion resistance over a wide temperature range along with mass densities and thermal expansion coefficients that are lower than metals. Applications include automotive bearings, cam rollers, high-pressure pump parts, seals, tool dies, turbocharger rotors, and ball and roller bearings. Aircraft engine manufacturers are actively considering silicon nitride bearings to meet their requirements for advanced engines including better tribological performance with excellent surface durability, superior corrosion resistance, and extended lifetimes. All-ceramic Si_3N_4 ball or roller bearings can operate against Si_3N_4 rolling elements and rings at temperatures up to 1000 °C and extremely high speeds. Hybrid bearings, which utilize silicon nitride balls and steel raceways, have been found to have longer fatigue lives, have superior thermal behavior [59], perform well under corrosive conditions, and last up to five times longer without oil lubrication [60]. Thus, Si_3N_4 ceramic bearings have met major requirements of greater stiffness, higher efficiency, speed, reliability, accuracy, lower friction, good corrosion resistance, and non-conductivity [61].

Friction and wear data reported for several types of silicon nitrides have shown variable results because of their different microstructures. These studies focused on relationships between sliding conditions (sliding speed, load, temperature, and atmosphere) and the performance of the Si_3N_4 [62]. Few studies have correlated wear resistance to the microstructure of silicon nitride. Those that have are often contradictory, although this may be because the silicon nitrides were produced using different sintering additives and assessed under different test conditions. Zutshi et al. [63] investigated Si_3N_4 hot-pressed at 1725 °C with 2 wt.% Y_2O_3 and 5 wt.% Al_2O_3 additives. Strength and fracture toughness increased with hot-pressing time as did the Al content of the elongated grains to form a low-z β-SiAlON solid solution, causing a slight decrease in microhardness. The wear rate increased significantly with hot-pressing time due to enhanced third-body wear in the later stages of the wear process. Grain pullout was greater when both the aspect ratio and the β-Si_3N_4 grain size increased, but the exact mechanism that caused this was unclear. Wang and Mao [64] used Si_3N_4 hot-pressed at 1700 °C with 2.5 wt.% Y_2O_3, 7.5 wt.% La_2O_3, and 5 wt.% Al_2O_3 additives and found that wear resistance improved as a result of crack deflection and the pullout of β-Si_3N_4 grains, again a consequence of high strength and fracture toughness.

Carrasquero et al. [65] produced "modified" HP Si_3N_4 at 1850 °C with 3 wt.% La_2O_3 and 3 wt.% Y_2O_3 additives and investigated the wear of Si_3N_4 discs in sliding wear tests against AISI 52100 steel and WC/6%Co balls. For the conditions evaluated, the silicon nitride exhibited very high wear resistance, particularly against steel, and could be recommended for several wear protection applications. It was concluded that the "self-reinforced" microstructure, with interlocking β-grains, hinders the propagation of cracks and contributes to the "in situ" toughening of the material.

The studies cited above were conducted under dry conditions and the wear mechanisms involved mechanical abrasion through a small-scale surface fracture process. However, different results are observed in situations involving humid atmospheres and aqueous liquids. Fischer and Tomizawa [66] reported that silicon nitride sliding in water resulted in an exceptionally low friction coefficient and the worn surface had a mirror finish. The low friction coefficient was thought to be related to oxidation or hydration of the worn surface.

Saito et al. [67] conducted friction and wear experiments on silicon nitride in three types of aqueous solutions: pure water, 1% hydrogen peroxide, and 3% NaCl. During the sliding of silicon nitride in water, a tribochemical reaction resulted in a SiO_2 layer on the worn surface and NH_3 was evolved; eventually, the friction coefficient and wear rate were lowered. By increasing the contact pressure and decreasing the sliding speed, the low friction coefficient suddenly increased to 0.8. A low value was recovered slowly as the lubricating layer was reconstructed during subsequent sliding. The 1% H_2O_2 solution accelerated the reaction to form SiO_2 on the sliding surface and, subsequently, low-friction sliding was attained after shorter sliding distances than in pure water. In 3% NaCl solution, low-friction sliding was severely disturbed. Quantitative analysis indicated that Na^+ ions were introduced into the surface structure and the composition of the resulting wear debris was equivalent to Na_2O-$nSiO_2$ ($n = 4$–5).

In other studies, a wide range of values for coefficients of friction have been reported for Si_3N_4 articulating against itself, from 0.002 [68] to 0.8 [67, 69]. Some studies suggest that the wear rate for Si_3N_4 is higher than other ceramics while the opposite has also been reported. This may be due to different specimen preparation and testing methods. For silicon nitride, the coefficient of friction and wear rate depend markedly on the surrounding environment and velocity indicating that surface tribochemical changes are important for overall performance. The wear mechanism for silicon nitride is quite complex with two principal wear modes: mechanical and tribochemical and, thus, differences in friction and wear are dependent upon which mechanism is dominant [69].

The mechanical mode operates under high loads, low speeds, and stop-start conditions, which typically results in the observed high friction and high wear. Conversely, tribochemical wear occurs through the dissolution of the Si_3N_4 by the following reactions:

$$Si_3N_4 + 6H_2O \rightarrow 3SiO_2 + 4NH_3 \quad (6)$$

$$SiO_2 + H_2O \rightarrow Si(OH)_4 \quad (7)$$

Reaction (6) involves the breaking of Si-N bonds with the formation of ammonia. A hydrated silica lubricating film then forms (reaction 7) between the articulating surfaces. Tribochemical wear is dominant at lower loads, higher speeds, and under continuous motion. The observed low friction and wear in the tribochemical mode is attributed to a low-friction colloidal boundary film between the two Si_3N_4 components that have exceptionally smooth articulating surfaces [70]. Tests were conducted nominally at 25 °C but no indication was given of heating of the ceramics during friction and wear. The rate of tribochemical film formation was higher than the removal rate and the retention of this lubricating hydrated silica film between the sliding contacts protects the surfaces and this improves the friction behavior.

Olofsson et al. [71] investigated Si_3N_4 sliding against Si_3N_4 in both bovine serum and phosphate-buffered saline (PBS) to assess its potential in hip-joint replacements. Low wear rates were observed with tribochemical films formed on the Si_3N_4 surfaces in both solutions, which contributed to a decrease in friction. Human synovial fluid is also an excellent lubricant [72] making this environment very favorable and, thus, Si_3N_4 against Si_3N_4 is a material combination with good potential for hip joints, possibly applied as a coating on CoCr [71].

As fully dense silicon nitride bearings can function under very demanding conditions, Bal et al. [73] concluded that, because of their superior mechanical and tribological properties, Si_3N_4 ceramics easily exceed the demands placed on orthopedic bearings, whether articulating against polyethylene, metals, or Si_3N_4 itself. Bal et al. [74] tested prototype total hip bearings, fabricated using "in situ" toughened Si_3N_4 with 6 wt.% Y_2O_3 and 4 wt.% Al_2O_3, which had elongated β-grains with an average diameter of 1.5 μm, flexural strength of 923 ± 70 MPa, Weibull modulus of 19, and fracture toughness ~10 MPa·m$^{1/2}$. These values of strength and toughness were two to three times higher than medical-grade Al_2O_3, the most common ceramic used in orthopedic bearings. Results from a hip simulator [74] showed that both CoCr and Si_3N_4 femoral heads exhibited low wear rates against Si_3N_4 acetabular liners comparable with Al_2O_3–Al_2O_3 wear rates, which are among the lowest of any orthopedic bearing. The much higher flexural strength and fracture toughness combined with the high Weibull modulus are indicative of a far higher mechanical reliability of silicon nitride compared with that of alumina, making it a suitable material for hip and knee bearings.

8 Chemical Stability In Vivo

A key aspect of silicon nitride for in vivo applications is its long-term stability. It must be corrosion and oxidation-resistant and chemically stable over many years of service life. From this perspective, the first question that arises is: How stable is silicon nitride in air at ambient temperatures? It is well known that Si_3N_4 powders emit ammonia (NH_3) when the fine α-Si_3N_4 particles react with atmospheric moisture (see Eq. (6) above). It is considered that the powder surface consists of -NH_x species, a mixture of Si–NH_2 and Si_2NH formed by hydrolysis of Si_3N_4 [75]. Further hydrolysis of these surface species gives rise to the formation of ammonia:

$$\text{Si–NH}_2 + \text{H}_2\text{O} \rightarrow \text{Si–OH} + \text{NH}_3 \tag{8}$$

The surface-layer compositions of three different powders were identified as intermediate states between silica (SiO_2) and silicon oxynitride (Si_2N_2O). However, while bulk ceramics are quite different from fine powders, corrosion and hydrothermal degradation may have a significant effect on their in vivo performance and longevity.

Silicon nitride has better resistance to subcritical crack growth than oxide ceramics such as Al_2O_3 and ZrO_2, which, because of their ionic bonding, are more sensitive to attack by moisture at crack tips leading to the formation of Al-OH and Zr-OH species. In Si_3N_4, Si-N bonds at the crack tips are more covalent and more resistant to attack by moisture [76]. Bal et al. [74] concluded that dense Si_3N_4 was chemically stable when subjected to accelerated aging at 120 °C in 0.1 MPa steam for 100 h. After this treatment, there were no observed phase changes and flexural strength remained within 1 standard deviation (~36 MPa) of the mean strength (936 MPa) suggesting no deleterious moisture attack even at this higher temperature.

Hermann [77] reviewed the literature on corrosion of Si_3N_4 in aqueous solutions. Si_3N_4 in highly acidic solutions, below pH of 2, is very stable at 80–90 °C. For pH between 2 and 14, the dissolution rate is very low. For sintered silicon nitrides, chemical stability is related to the grain boundaries. They are the weakest parts of Si_3N_4 ceramics in corrosive media below 150–200 °C. Reducing the amount of sintering additives improves corrosion stability. In acids, even after the grain-boundary glass phase has been leached out, strength levels of 400–550 MPa are retained, which is a result of the strong network formed by the Si_3N_4 grains and protective SiO_2 layers around the grains. The overall conclusion is that at ambient temperatures up to 37 °C, silicon nitride is very stable and resistant to oxidation and corrosion in aqueous liquids. Tailoring the grain boundary composition is the best approach to improving overall corrosion resistance [77].

Silicon nitride is quite stable in dry air at ambient temperatures and up to ~750–900 °C, which is related to the glass transition temperature of the grain boundary amorphous phases [78]. Above this temperature, a SiO_2-rich oxide scale starts to form [79]. The oxidation rate is controlled by the diffusion of additive and impurity cations from the bulk Si_3N_4 grain boundaries to the reaction interface between the ceramic and the oxide scale. As oxidation continues, enrichment of the oxide scale by additive cations occurs with a corresponding depletion in the surface layers of the bulk ceramic.

For implantation in the body, materials must be biocompatible. Given the excellent combination of mechanical and tribological properties of silicon nitride, other properties critical in biomedical environments have been examined such as cytotoxicity, wetting behavior, chemical stability in simulated body fluid (SBF), and other physiological solutions [71, 80, 81]. Cell growth, viability, and morphology assessment were performed, applying the L929-mice fibroblast cell culture model in a direct contact assay. Silicon nitride was found to be non-toxic and also encouraged cell adhesion, normal proliferation, and differentiation, comparable with Ti [80]. Polished Si_3N_4 surfaces appeared to promote cell growth. Further investigation compared industrial-grade Si_3N_4, using the L929-cell culture model with Al_2O_3 and

Ti as controls. Again, Si_3N_4 ceramics showed no cytotoxicity, and it was concluded that Si_3N_4 is stable and suitable for in vivo applications.

Thermal, chemical, and mechanical treatments, to yield ultra-smooth or highly fibrous surface structures, have been applied to commercial silicon nitrides sintered with Y_2O_3/Al_2O_3 additives [82]. These are designed to lead to enhanced osseointegration and bacterial resistance without compromising bulk mechanical properties. Chemical modification of Si_3N_4 surfaces results in the stable formation of a balance of positively and negatively charged off-stoichiometric sites [83] such as N vacancies and $N - N$ bonds (+ve) and silanol groups (SiO^-), leading to enhanced in vitro cell adhesion and apatite formation. Pezzotti [84] reported that the unusual surface chemistry of bulk Si_3N_4 ceramic in an aqueous environment has the potential to reduce implant-associated bacterial infections. This is like the antibacterial effect exerted by reactive oxygen species (ROS) from TiO_2. In the case of Si_3N_4, reactive nitrogen species (RNS), such as nitrous oxide (N_2O), nitric oxide (NO), and peroxynitrite ($O{=}NOO^-$), severely affect bacterial metabolism, leading to their lysis. Modification of the stoichiometry of silicon nitride surfaces enhances cell metabolism, affects bacteria metabolism, and opens the possibility of promoting faster healing in potential bone repair devices.

9 Concluding Remarks

Silicon nitride has been developed over many decades of intensive research. This has resulted in a ceramic with unique properties that arise through the main processing route for the material. This involves liquid-phase sintering of Si_3N_4 powder compacts with oxide additives and the development of complex microstructures in which high aspect ratio β-Si_3N_4 grains and intergranular glass phase increase the resistance to crack propagation leading to the highest fracture toughness, combined with high strength, of any ceramic available. Many different additive compositions have been investigated but the favored combination is still Y_2O_3/Al_2O_3. The other processing route for Si_3N_4 is reaction-bonding (nitridation) of Si, which results in a lower strength but porous ceramic.

Silicon nitride has been used extensively in wear-resistant applications including cutting tools, bearings, pump components, turbocharger rotors, and other areas where thermal shock resistance is crucial. Given its combination of properties and its success in these specialist applications, silicon nitride has been investigated as a bioceramic for use in bearings for hip and knee replacements. Si_3N_4 has very favorable tribological properties when it articulates against polyethylene, the dominant material for bearing surfaces in prosthetic joints. It is biocompatible, non-toxic, and corrosion-resistant, all of which are crucial in medical implants. The porous form of Si_3N_4 may be suitable for devices where bone ingrowth is necessary for durable biological fixation.

Further developments in silicon nitride ceramics for biomedical applications may be expected through a better understanding of grain boundary chemistry and its effects on microstructure, surface chemistry, and biological interactions.

References

1. F.L. Riley (2000), "Silicon Nitride and Related Materials," *J. Am. Ceram. Soc.* 83: 245–265.
2. G. Petzow and M. Herrmann (2002), "Silicon Nitride Ceramics," in *High Performance Non-Oxide Ceramics II*, ed. M. Jansen, Berlin and NY: Springer, pp. 50–171.
3. H. Sainte-Claire Deville and F. Wöhler (1859), "On the Direct Compound Silicon Nitride," *Liebigs Ann. Chem. Pharm.* 110: 248–50.
4. J.F. Collins and R.W. Gerby (1955), "New Refractory Uses for Silicon Nitride Reported," *J. Met.* 7: 612–615.
5. N.L. Parr, G.F. Martin and E.R.W. May (1960), "Preparation, Microstructure, and Mechanical Properties of Silicon Nitride," in *Special Ceramics 1*, ed. P. Popper, London: Heywood, pp. 102–135.
6. G.G. Deeley, J.M. Herbert and N.C. Moore (1961), "Dense Silicon Nitride," *Powder Metall.* 8: 145–151.
7. R.F. Coe, R.J. Lumby, and M.F. Pawson (1972), "Some Properties and Applications of Hot-Pressed Silicon Nitride," in *Special Ceramics 5*, ed. P. Popper, Stoke-on-Trent, UK: British Ceramic Research Association, pp. 361–376.
8. U.S. Department Of Energy/National Aeronautics and Space Administration (DOE/NASA) *Ceramic Turbine Engine Demonstration Program* (1978).
9. S. Hampshire (2007), "Silicon Nitride Ceramics – Review of Structure, Processing and Properties," *J. Achiev. Mater. Manuf. Eng.* 24: 43–50.
10. A. Çelik and S. Turan (2021), "Si-Al-O-N Ceramics, Structure and Properties," in *Encyclopedia of Materials: Technical Ceramics and Glasses*, Vol.2, ed. M.J. Pomeroy, Oxford: Elsevier, pp. 128–143.
11. D. Hardie and K.H. Jack (1957), "Crystal structures of silicon nitride," *Nature* 180: 332–333.
12. A. Zerr, G. Miehe, G. Serghiou, M. Schwarz, E. Kroke, R. Riedel, H. Fuess, P. Kroll, R. Boehler (1999), "Synthesis of cubic silicon nitride," *Nature* 400: 340–342.
13. S. Hampshire, H.K. Park, D.P. Thompson, K.H. Jack, (1978), "α'-sialon ceramics," *Nature* 274: 880–882.
14. S. Wild, P. Grieveson and K.H. Jack (1972), "The crystal structures of alpha and beta silicon and germanium nitrides," in *Special Ceramics 5*, ed. P. Popper, Stoke-on-Trent, UK: British Ceramic Research Association, pp. 385–395.
15. D.P. Thompson (1977), "Unit cell dimensions of α-silicon nitrides," in *Nitrogen Ceramics*, ed. F.L. Riley, Leyden: Noordhoff, pp. 129–133.
16. Y. Oyama and O. Kamigaito (1971), "Solid solubility of some oxides in Si_3N_4," *Jap. J. Appl. Phys.* 10: 1637.
17. K.H. Jack and W.I. Wilson (1972), "Ceramics based on the Si-Al-O-N and related systems," *Nature Phys. Sci.* 238 [80]: 28–29.
18. A.J. Moulson (1979), "Review – Reaction-bonded silicon nitride: its formation and properties," *J. Mater. Sci.* 14: 1017–1051.
19. S. Hampshire and K.H. Jack (1981), "The Kinetics of Densification and Phase Transformation in Nitrogen Ceramics," *Proc. Brit. Ceram. Soc.*, 31: 37–49.
20. S. Hampshire (2009), "Silicon Nitride Ceramics," in *Advances in Ceramic Materials*, ed. B Ralph, P. Xiao, Switzerland: Trans Tech Publications, *Mater. Sci. Forum*, 606: 27–41.

21. H.J. Kleebe, G. Pezzotti and G. Ziegler (1999), "Microstructure and Fracture Toughness of Si_3N_4 Ceramics: Combined Roles of Grain Morphology and Secondary Phase Chemistry," *J. Am. Ceram. Soc.* 82: 1857–1867.
22. H.-J. Kleebe, M.K. Cinibulk, R.M. Cannon and M. Rühle (1993), "Statistical analysis of the intergranular film thickness in silicon nitride ceramics," *J. Am. Ceram. Soc.*, 76: 1969–1977.
23. R.L. Satet, M.J. Hoffmann and R.M. Cannon (2006), "Experimental evidence of the impact of rare-earth elements on particle growth and mechanical behaviour of silicon nitride," *Mater. Sci. Eng. A*, 422: 66–76.
24. P.F. Becher, N. Shibata, G.S. Painter, F. Averill, K. van Benthem, H-T. Lin and S.B. Waters (2010), "Observations on the Influence of Secondary Me Oxide Additives (Me = Si, Al, Mg) on the Microstructural Evolution and Mechanical Behavior of Silicon Nitride Ceramics Containing RE_2O_3 (RE = La, Gd, Lu)," *J. Am. Ceram. Soc.* 93 [2]: 570–580.
25. P.F. Becher, G.S. Painter, N. Shibata, R.L. Satet, M.J. Hoffmann and S.J. Pennycook (2006), "Influence of additives on anisotropic grain growth in silicon nitride ceramics," *Mater. Sci. Eng. A*, 422: 85–91.
26. M. Kitayama, K. Hirao and S. Kanzaki (2006), "Effect of rare earth oxide additives on the phase transformation rates of Si_3N_4," *J. Am. Ceram. Soc.* 89: 2612–2618.
27. A. Ziegler, J.C. Idrobo, M.K. Cinibulk, C. Kisielowski, N.D. Browning, R.O. Ritchie (2004), "Interface structure and atomic bonding characteristics in silicon nitride ceramics," *Science* 306: 1768–1770.
28. N. Shibata, S.J. Pennycook, T.R. Gosnell, G.S. Painter, W.A. Shelton, P.F. Becher (2004), "Observation of rare-earth segregation in silicon nitride ceramics at sub-nanometre dimensions," *Nature* 428: 730–733.
29. G.B. Winkelman, C. Dwyer, C. Marsh, T. S. Hudson, D. Nguyen-Manh, M. Doblinger, D.J.H. Cockayne (2006), "The crystal/glass interface in doped Si_3N_4," *Mater Sci Eng, A.* 422: 77–84.
30. J.A. Mangels, G.J. Tennenhouse (1980), "Densification of reaction-bonded silicon nitride," *Amer. Ceram. Soc. Bull.* 59: 1219–1222.
31. X. Zhu, Y. Zhou, K. Hirao (2004), "Post-densification behavior of reaction-bonded silicon nitride (RBSN): Effect of various characteristics of RBSN," *J. Mater. Sci.* 39: 5785–5797.
32. B.R. Golla, J.W. Ko, H.-D. Kim (2017), "Processing and characterization of sintered reaction bonded Si_3N_4 ceramics," *Int. J. Refr. Met. & Hard Mater.* 68: 75–83.
33. P.F. Becher, E.Y. Sun, K.P. Plucknett, K.B. Alexander, C.-H. Hsueh, H.-T. Lin, S.B. Waters, C.G. Westmoreland, E.-S. Kang, K. Hirao and M.E. Brito (1998), "Microstructural Design of Silicon Nitride with Improved Fracture Toughness: I, Effects of Grain Size and Shape," *J. Am. Ceram. Soc.* 81:2821–2830.
34. E.Y. Sun, P.F. Becher, K.P. Plucknett, C.-H. Hsueh, K.B. Alexander, S.B. Waters, K. Hirao and M. E. Brito (1998), "Microstructural Design of Silicon Nitride with Improved Fracture Toughness: II, Effects of Yttria and Alumina Additives," *J. Am. Ceram. Soc.* 81: 2831–2840.
35. P.F. Becher, G.S. Painter, N. Shibata, S.B. Waters, H.-T. Lin (2008), "Effects of Rare-Earth (RE) Intergranular Adsorption on the Phase Transformation, Microstructure Evolution, and Mechanical Properties in Silicon Nitride with RE_2O_3 + MgO Additives: RE = La, Gd, and Lu," *J. Am. Ceram. Soc.* 91: 2328–2336.
36. K. Hirao (2006), "Microstructure control of silicon nitride ceramics by seeding and their enhanced mechanical and thermal properties," *J. Ceram. Soc. Jap.* 114: 665–671.
37. S. Hampshire and M.J. Pomeroy (2013), "Silicon Nitride – Grain Boundary Oxynitride Glass Interfaces: Deductions From Glass Bulk Properties," *Int. J. Appl. Ceram. Tech.* 10 [5]: 747–755.
38. P. Šajgalík, J. Dusza, M.J. Hoffmann (1995), "Relationship Between Microstructure, Toughening Mechanisms and Fracture Toughness of Reinforced β- Si_3N_4 Ceramics," Journal of the American Ceramic Society, 78 [10] (1995) 2619–24.

39. Fünfschilling, S., Fett, T., Hoffmann, M. J., Oberacker, R., Jelitto, H., Schneider, G. A., Härtelt M. and Riesch-Oppermann, H. (2009). Bridging stresses from R-curves of silicon nitrides. *Journal of Materials Science* 44, 3900–3904.
40. B. Mikijelj, Z. Nawaz, J.J. Kruzic, J. Idrobo, J.J. Swab, H. Ozcoban, H. Jelitto, G.A. Schneider, T. Fett and Y. Liu (2015), "Intergranular Nanostructure Effects on Strength and Toughness of Si_3N_4," *J. Am. Ceram. Soc.* 98: 1650–1657.
41. M. Saâdaoui (2021), "Subcritical Crack Growth: Basic Relations and Experimental Evaluation," in *Encyclopedia of Materials: Technical Ceramics and Glasses*, Vol.1, ed. M.J. Pomeroy, Oxford: Elsevier, pp. 797–810.
42. J. Chevalier, G. Fantozzi (2005), "Slow crack propagation in ceramics at the nano- and microscale: Effect of microstructure," in *Fracture Mechanics of Ceramics*, Vol. 14, eds. R.C. Bradt, D. Munz, M. Sakai, K.W. White, New York: Springer, pp. 173–190.
43. W.E. Luecke, S.M. Wiederhorn, B.J. Hockey, R.F. Krause, G.G. Long (1995), "Cavitation contributes substantially to tensile creep in silicon nitride," *J. Am. Ceram. Soc.* 78: 2085–2096.
44. M. Kitayama (2021), "High Thermal Conductivity Ceramics," in *Encyclopedia of Materials: Technical Ceramics and Glasses*, Vol.2, ed. M.J. Pomeroy, Oxford: Elsevier, pp. 165–181.
45. M. Yamagiwa (2010), "Packaging Technologies of Power Module for Hybrid Electric Vehicles and Electric Vehicles (Recent Developments and Future Issues)," *Bull. Ceram. Soc. Jap.* 45: 432—37.
46. J.S. Haggerty and A. Lightfoot (1995), "Opportunities for Enhancing the Thermal Conductivity of SiC and Si_3N_4 Ceramics Through Improved Processing," *Ceram. Eng. Sci. Proc.* 16 [4]: 475–487.
47. K. Tsukuma, M. Shimada, M. Koizumi (1981), "Thermal conductivity and microhardness of Si_3N_4 with and without additives," *Am. Ceram. Soc. Bull.* 60: 910–912.
48. G. Ziegler, D.P.H. Hasselman (1981), "Effect of phase composition and microstructure on the thermal diffusivity of silicon nitride," *J. Mater. Sci.* 16:495–503.
49. K. Watari, Y. Seki, K. Ishizaki (1989), "Temperature Dependence of Thermal Coefficients for HIPped Silicon Nitride," *J. Ceram. Soc. Jap.* 97: 174–181.
50. K. Hirao, K. Watari, M.E. Brito, M. Toriyama, S. Kanzaki (1996), "High Thermal Conductivity in Silicon Nitride with Anisotropic Microstructure," *J. Am. Ceram. Soc.* 79: 2485–2488.
51. M. Kitayama, K. Hirao, M. Toriyama, S. Kanzaki (1999), "Thermal Conductivity of β-Si_3N_4: I, Effects of Various Microstructural Factors," *J. Am. Ceram. Soc.* 82: 3105–3112.
52. K. Watari, K. Hirao, M. Toriyama, K. Ishizaki (1999), "Effect of Grain Size on the Thermal Conductivity of Si_3N_4," *J. Am. Ceram. Soc.* 82: 777–779.
53. H. Hayashi, K. Hirao, M. Toriyama, S. Kanzaki, K. Itatani (2001), "$MgSiN_2$ Addition as a Means of Increasing the Thermal Conductivity of β-Silicon Nitride," *J. Am. Ceram. Soc.* 84: 3060–3062.
54. J. Zhang, W. Cui, F. Li, S. Du, Z. Tian, S. Sun, K. Chen, G. Liu (2020), "Effects of $MgSiN_2$ addition and post-annealing on mechanical and thermal properties of Si_3N_4 ceramics," *Ceram. Int.* 46: 15719–15722.
55. D. Kusano, H. Hyuga, Y. Zhou, K. Hirao (2014), "Effect of Aluminum Content on Mechanical Properties and Thermal Conductivities of Sintered Reaction-Bonded Silicon Nitride," *Int. J. Appl. Ceram. Tech.* 11: 534–542.
56. M. Kitayama, K. Hirao, A. Tsuge, K. Watari, M. Toriyama, S. Kanzaki (2000), "Thermal Conductivity of β-Si_3N_4: II, Effect of Lattice Oxygen," *J. Am. Ceram. Soc.* 83: 1985–1992.
57. Y. Zhou, H. Hyuga, D. Kusano, Y.-I. Yoshizawa, T. Ohji, K. Hirao (2015), "Development of high-thermal-conductivity silicon nitride ceramics," *J. Asian Ceram. Socs.* 3: 221–229.
58. Y. Duan, N. Liu, J. Zhang, H. Zhang, X. Li (2020), "Cost effective preparation of Si_3N_4 ceramics with improved thermal conductivity and mechanical properties," *J. Euro. Ceram. Soc.* 40: 298–304.
59. W. Wang, M. Hadfield, A.A. Wereszczak (2009), "Surface strength of silicon nitride in relation to rolling contact performance," *Ceram. Int.* 35: 3339–3346.

60. J.R. Miner, J. Dell, A. Galbato, M.A. Ragen (1996), "F-117-PW-100 hybrid bearing ceramic technology insertion," *J. Eng. Gas Turb. Power* 118, 434–442.
61. L. Wang, R.W. Snidle, L. Gu (2000), "Rolling contact silicon nitride bearing technology: a review of recent research," *Wear* 246: 159–173.
62. W. Liu, J.L. Duda, E.E. Klaus (1996), "Wear property of silicon nitride in steel-on-Si_3N_4 and Si_3N_4-on-Si_3N_4 systems," *Wear* 199: 217–221.
63. A. Zutshi, R.A. Haber, D.E. Niesz, J.W. Adams, J.B. Wachtman, K. Ferber, S.M. Hsu (1994), "Processing, Microstructure, and Wear Behavior of Silicon Nitride Hot-Pressed with Alumina and Yttria," *J. Am. Ceram. Soc.* 77: 883–890.
64. D. Wang, Z. Mao (1995), "Studies on Abrasive Wear of Monolithic Silicon Nitride and a Silicon Carbide Whisker-Reinforced Silicon Nitride Composite," *J. Am. Ceram. Soc.* 78:2705–2708.
65. E. Carrasquero, A. Bellosi, M.H. Staia (2005), "Characterization and wear behavior of modified silicon nitride," *Int. J. Refr. Met. Hard Mater.* 23: 391–397.
66. T.E. Fischer, H. Tomizawa (1985), "Interaction of tribochemistry and microfracture in the friction and wear of silicon nitride," *Wear* 105: 29–45.
67. T. Saito, Y. Imada, F. Honda (1997), "An analytical observation of the tribochemical reaction of silicon nitride sliding with low friction in aqueous solutions," *Wear* 205: 153–159.
68. D.S. Park, S. Danyluk, M.J. McNallan (1992), "Influence of tribochemical reaction products on friction and wear of silicon nitride at elevated temperatures in reactive environments," *J. Am. Ceram. Soc.* 75: 3033–3039.
69. J. Xu, K. Kato (2000), "Formation of tribochemical layer of ceramics sliding in water and its role for low friction," *Wear* 245: 61–75.
70. D. Amutha Rani, Y. Yoshizawa, M.I. Jones, H. Hyuga, K. Hirao, Y. Yamauchi (2005), "Comparison of tribological behavior between α-Sialon/Si_3N_4 and Si_3N_4/Si_3N_4 sliding pairs in water lubrication," *J. Am. Ceram. Soc.* 88: 1655–1658.
71. J. Olofsson, T.M. Grehk, T. Berlind, C. Persson, S. Jacobson, H. Engqvist (2012), "Evaluation of silicon nitride as a wear resistant and resorbable alternative for total hip joint replacement," Biomatter 2:2: 94–102.
72. D. Mazzucco, M. Spector (2004), "The role of joint fluid in the tribology of total joint arthroplasty," (The John Charnley Award Paper), Clin. Orthop. Relat. Res. 429: 17–32.
73. B.S. Bal, A. Khandkar, R. Lakshminarayanan, I. Clarke, A.A. Hoffman, M.N. Rahaman (2008), "Testing of silicon nitride ceramic bearings for total hip arthroplasty," *J. Biomed. Mater. Res. B* 87: 447–454.
74. B.S. Bal, A. Khandkar, R. Lakshminarayanan, I. Clarke, A.A. Hoffman, M.N. Rahaman (2009), "Fabrication and testing of silicon nitride bearings in total hip arthroplasty," (Winner of the 2007 "HAP" PAUL Award) *J. Arthroplasty* 24: 110–116.
75. L. Bergström, R.J. Pugh (1989), "Interfacial Characterization of Silicon Nitride Powders," *J. Am. Ceram. Soc.* 72: 103–109.
76. B.J. McEntire, B.S. Bal, M.N. Rahaman, J. Chevalier, G. Pezzotti (2015), "Ceramics and ceramic coatings in orthopaedics," *J. Euro. Ceram. Soc.* 35: 4327–4369.
77. M. Herrmann (2013), "Corrosion of Silicon Nitride Materials in Aqueous Solutions," *J. Am. Ceram. Soc.* 96: 3009–3022.
78. M. Sebai, J. Sjöberg, P. Goursat, E. Nestor, R. Flynn, R. Ramesh, S. Hampshire (1995), "Oxidation Behaviour of Yttrium and Neodymium Oxynitride Glasses," *J. Euro. Ceram. Soc.* 15: 1015–1024.
79. M. Pomeroy, S. Hampshire (1989), "Oxidation Processes in Silicon-Nitride-based Ceramics," *Mater. Sci. Eng. A* 109: 389–394.
80. A. Neumann, T. Reske, M. Held, K. Jahnke, C. Ragoss, H.R. Maier (2004), "Comparative investigation of the biocompatibility of various silicon nitride ceramic qualities *in vitro*," *J. Mater. Sci. Mater. Med.* 15: 1135–1140.
81. B.S. Bal, M.N. Rahaman (2012), "Orthopedic applications of silicon nitride ceramics. *Acta Biomater.* 8: 2889–2898.

82. R.M. Bock, J.B. McEntire, B.S. Bal, M.N. Rahaman, M. Boffelli, G. Pezzotti (2015), "Surface modulation of silicon nitride ceramics for orthopaedic applications," *Acta Biomater.* 26: 318–330.
83. G. Pezzotti, J.B. McEntire, R.M. Bock, M. Boffelli, W.L. Zhu, E. Vitale, L. Puppulin, T. Adachi, T. Yamamoto, N. Kanamura, B.S. Bal (2016), "Silicon Nitride: A Synthetic Mineral for Vertebrate Biology," *Sci. Rep.* 6: 31717.
84. G. Pezzotti (2018), "A spontaneous solid-state NO donor to fight antibiotic resistant bacteria," *Mater. Today Chem.* 9: 80–90.

Processing of Silicon Nitride Bioceramics

Bryan J. McEntire

1 Introduction

Silicon nitride (Si_3N_4) is unique among advanced ceramics. Its chemistry, bonding, and microstructure provide it with a nearly ideal combination of properties. Composed of two of the most abundant terrestrial elements—silicon and nitrogen—it is not only biocompatible but also bioactive. Multiple independent investigations have demonstrated its ability to promote osseous integration [1–10] while resisting pathogens (i.e., bacteria [5, 10–15], viruses [16–20], and fungi [21, 22]). Its covalent bonding structure, although difficult to sinter, provides it with improved properties including high strength, toughness, and hardness [23]. When densified and fully converted to its stable ß-phase, these mechanical qualities are due to its interlocking rod-like grain structure. Because of its unique microstructure, no other advanced ceramic has been the subject of more composition and process development than Si_3N_4. After its discovery in 1857 [24], it remained a curiosity for over a hundred years because it could not be easily densified. However, during the 1970s through the 1990s, multiple government-funded programs were directed at developing Si_3N_4 for various heat engine applications (e.g., aero and industrial gas turbines [25–30], and piston engines [31–34]). These programs laid the foundation for several industrial processes that were subsequently refined and adopted. It was found that the densification of Si_3N_4 required the interdependent use of raw materials and sintering additives along with high temperatures, pressures, and special atmospheres. This chapter will focus on these processes and their respective raw materials, manufacturing methods, physical and mechanical properties, inspection practices, and test standards.

B. J. McEntire (✉)
SINTX Technologies, Inc., Salt Lake City, UT, USA
e-mail: bmcentire@sintx.com

B. S. Bal et al. (eds.), *Silicon Nitride Bioceramics*,
https://doi.org/10.1007/978-3-031-67047-3_3

2 Silicon Nitride Powders

The need for high-quality Si_3N_4 raw materials is critical for the achievement of acceptable chemical, physical, mechanical, and biological properties. Current requirements include high purity (>99.95%), non-agglomerated powders, with small mean crystallite sizes (<200 nm), narrow particle size distributions (0.1 to <10 μm), high intrinsic surface area (9–13 m^2/g), and high α-phase contents (>95%) [35]. Despite the terrestrial abundance of silicon and nitrogen, the compound Si_3N_4 is extremely rare. It has only been found in the dust grains of primitive meteorites ejected from supernovas after the formation of the solar system [36], but it is also likely present deep in the earth's crust, created when the world was young due to an ammonia-rich atmosphere and high temperatures. Although it lacks native mineralogy, Si_3N_4 powder can be chemically synthesized, and tonnage quantities are produced annually by several manufacturers. Processes that have been used to prepare Si_3N_4 powders include direct nitridation of silicon, carbothermal reduction of silica, diimide decomposition, vapor phase synthesis, and pyrolysis of organic compounds [37–40]. Of these methods, only direct nitridation of silicon and diimide decomposition have been fully commercialized. Carbothermal reduction of silica was an early favored process, and it was extensively developed, but it has yet to become commercially viable.

2.1 *Direct Nitridation of Silicon*

Nitridation of silicon (Si) can be used to either prepare Si_3N_4 in situ as a semi-final component (i.e., reaction-bonded silicon nitride, RBSN) or as a fine powder that can be utilized in other densification processes [41]. Nitridation is a gas-solid exothermic reaction between N_2 and Si, performed in continuous rotary and flowing atmosphere kilns. The reaction initiates at about 1100 °C. It is kinetically enhanced with increasing temperature, but the process is generally regulated to keep the reaction below the melting point of silicon (~1412 °C); See Eq. 1.

$$3Si(s) + 2N_2(g) \rightarrow Si_3N_4(s); \quad \Delta H = -750\ \mathrm{kJ/mol} \tag{1}$$

A flowchart of the process is provided in Fig. 1. High-purity silicon powders (>99.99%) with a fine particle size distribution (≤8 μm) and pure N_2 gas (>99.9+%) are the key raw materials.

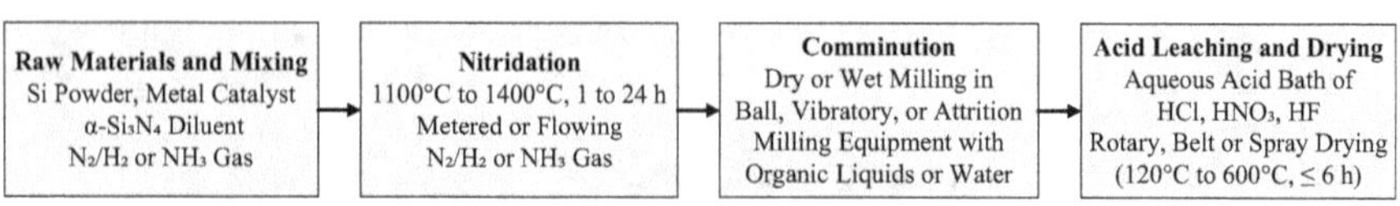

Fig. 1 Process for producing Si_3N_4 powder by the direct nitridation of silicon

Early studies examined the effects of silicon impurities on reaction kinetics. It was found that iron (Fe) contaminated powders promoted nitridation while aluminum (Al) inhibited the reaction [42, 43]. Both elements played roles but by different mechanisms. Fe catalyzed the reaction by removing the protective oxide film (i.e., silica or SiO_2) on the Si particles via the formation of a $FeSi_x$ liquid phase. This released SiO gas, thereby exposing more Si to N_2. The reaction produced Si_3N_4 powders with reasonably high α-phase contents (i.e., 75–95%). Conversely, Al combined with N_2 to form AlN. Then, the AlN reacted with native SiO_2 to form Al_2O_3 and ß-Si_3N_4 both of which suppressed further Si exposure to N_2. Later, it was also found that high-purity Si powders could be rapidly nitrided provided the SiO_2 layer was completely removed [44, 45]. Elimination of the oxide coating was performed in situ by exposing the Si powder to 100% H_2 gas at 1400 °C for 1 h. At this temperature, hydrogen reacts with silica to release SiO gas. Then, a mixture of 95% N_2 and 5% H_2 was introduced. With the silica layer removed, nitridation to ≥98% was completed within 30 min and the resulting phase composition was about 83% α-Si_3N_4 [45]. The utilization of N_2/H_2 mixtures has since become a common industrial practice. H_2 gas also serves to modulate the exothermic reaction. Ammonia (NH_3) has also been reportedly used as a modulating gas [37]. While Fe is still utilized as a catalyst (particularly for lower-quality powders), it can create large microstructural defects in densified components [46]. Consequently, researchers have continued to examine the catalytic effects of various other elements in promoting nitridation and conversion of Si to the preferred α-phase of Si_3N_4. In particular, they found that Ca, Y_2O_3, and other rare-earth oxides are effective in catalyzing nitridation and promoting α-Si_3N_4 [47–49]. To further control the exothermic reaction, more recent studies have demonstrated that the addition of a diluent effectively regulates the process [50–52]. Jin et al. utilized both α- and ß-Si_3N_4 powders (2 μm) as diluents. The Si_3N_4 powder coated the larger Si particles (4.5 μm) and acted as a heat sink and oxygen getter [52]. They found the optimum concentration to preferably be ~6 wt.% α-Si_3N_4 during a nitridation cycle of 3 h at 1400 °C. This resulted in 100% nitridation of Si and 95% conversion to α-Si_3N_4. The presence of the Si_3N_4 diluent minimized localized temperature spikes during the reaction, which reduced melting and prevented agglomeration, evaporation, and condensation of Si. The resulting nitrided powder was finer with fewer large and hard agglomerates.

After nitridation, the Si_3N_4 raw stock has to be crushed and finely ground. Early work that utilized steel-lined mills and tungsten-carbide milling media led to significant Fe contamination [53, 54]. Ironically, when polymer-lined mills and silicon nitride media replaced steel and tungsten-carbide, it was serendipitously discovered that iron catalyzed the nitridation reaction. This finding led to the manufacturing practice of adding iron, which is still in use today. A series of milling studies on raw and nitrided silicon powder was published as a collection of NASA Memorandums during the 1970s–80s [55–60]. These progressive experiments compared ball, vibratory, and attrition milling processes for raw and nitrided silicon in dry nitrogen and organic fluids (heptane and ethanol). They also progressively improved the equipment from steel drums and media to Si_3N_4-liners, rotating arms, and media. In the end, they found attrition milling in ethanol to be the most attractive method. It produced a reasonable particle size distribution (0.17–5.3 μm) in the shortest time (6–12 h) with minimal impurity contamination (0.23–0.8 wt.% oxygen pick-up, and no change in Fe) [59, 60]. In a more recent paper, Rabe and Linke [61] also utilized attrition milling in ethanol. Starting with a commercially available nitrided powder, they were able to reduce the average particle size from 1.4 μm to 0.3 μm and the upper limit from about 50 μm to below 2 μm in 6 h of milling. The surface area of the powder remarkably increased from 9 to 55 m^2/g after 24 h of operation. Because their mill had Si_3N_4 rotating arms, a polyamide liner, and Si_3N_4 milling media, oxygen was the only significant contaminant. It rose from 2.1 to 5.9 wt.% after 24 h. Although the use of ethanol as a milling fluid is still practiced, it is commonly known that commercially nitrided silicon can also be comminuted in deionized water [62]. The concomitant increase in oxygen by milling in water was found to be no greater than with ethanol.

Depending on the purity of the silicon raw material and the quality of the manufacturer's milling equipment, an additional step of acid leaching was employed to remove unwanted impurities, principally iron and oxygen. Although hydrochloric acid (HCl) was typically used, other acids were tested including nitric (HNO_3), hydrofluoric (HF), and combinations. Yeh, et al. were among the first to propose chemical cleaning of silicon powder using HCl [57]. They reported dramatic reductions in Fe content (from ~4.2 to ~1.5 wt.%) after milling but before nitridation. Hermansson et al. followed with a study using HNO_3 and combinations of HNO_3 and HF or HCl and HF [63]. The HCl plus HF combination proved to be the most effective in reducing metallic impurities, but residual fluorine in densified samples negatively affected high-temperature properties. Itoh utilized a two-step treatment to reduce iron to ~0.02 wt.% using HCl, and then eliminated unreacted silicon with a 1: 1 mixture of HNO_3 and HF [41]. More recent studies used an HCl treatment followed by thermal oxidation to improve the rheology of aqueous Si_3N_4 suspensions [62, 64, 65]. The HCl bath (5 wt.%) was effective in reducing Ca and Mg impurities, while calcination (600 °C, 6 h) removed volatile organics and ammonia and increased silanol surface groups [65]. However, they also found that rheological changes were highly dependent on the commercial powder source. After milling and acid treatments (if necessary), drying of nitrided powders was performed using several industrial processes including rotary or belt dryers and spray drying.

2.2 *Carbothermal Reduction of Silica*

Synthesis of Si_3N_4 by carbothermal reduction of silica was first described in a 1896 German patent by Melner [66]; but it wasn't until the 1970s that this low-cost process was used by Inoue et al. to produce a sinterable α-Si_3N_4 powder [67–69]. At about the same time, Lee and Cutler et al. proposed using carbothermal reduction for making Si_3N_4 and silicon carbide (SiC) powders [70] and later they also prepared a SiALON powder by reacting kaolin clay and coal in an N_2

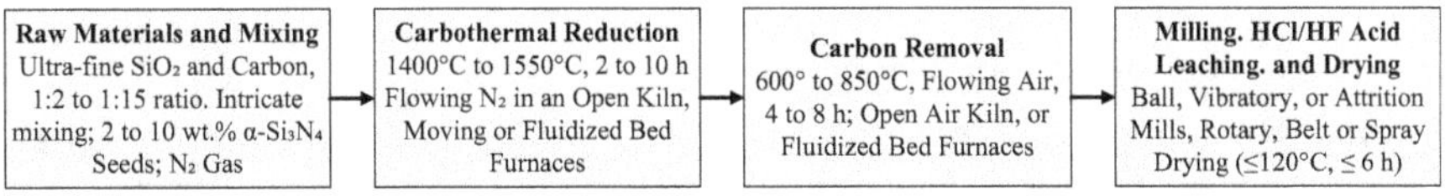

Fig. 2 Process for producing α-Si_3N_4 powder by carbothermal reduction of silica (SiO_2)

environment [71]. Numerous process refinements were subsequently completed by several researchers [72–81]. Kaolin and coal were replaced by higher-purity submicron silica and carbon. Improved wet and dry mixing methods were developed. Sol-gel processing was investigated. Reactant compositions, temperatures, and times were optimized. Fluidized or moving beds and rotary furnaces replaced static kilns. The endothermic nature of the reaction provided improved process control and produced weakly agglomerated powders. Seeding with submicron α-Si_3N_4 particles was utilized to accelerate the reaction. Excess carbon was thermally removed by flowing air. Various wet and dry comminution methods were implemented. Acid leaching to remove impurities and silica was examined. Powders were dried using conventional methods, rotary, or fluidized-bed dryers, and spray drying. The final powder product had impressive properties: (1) An α-Si_3N_4 content between 95% and 100%, with little or no ß-Si_3N_4 or SiC; (2) An average particle size between 0.3 μm and 1.2 μm; (3) Powder surface areas of 8.7–13 m^2/g; and (4) Purity >97% with residual oxygen and carbon contents of <2.5% and < 0.5%, respectively. After incorporating sintering aids (Al_2O_3, Y_2O_3), fired densities of ≥97.5% were achieved with a flexural strength of 795 MPa, Weibull modulus of 18, and fracture toughness of 7.3 ± 0.2 MPa·$m^{1/2}$ [80]. Carbothermal reduction was declared to be superior to both direct nitridation of silicon and diimide decomposition, and it was considered to be the most viable process for the production of fine Si_3N_4 powder [67–69, 79, 80, 82]. Carbothermal reduction of silica is given in Eq. 2, although it is acknowledged that this reaction consists of multiple steps, wherein both silica and carbon are first converted to their gaseous states (i.e., $SiO_{(g)}$ and $CO_{(g)}$ before forming Si_3N_4 [72, 73, 82, 83]). A flowchart of the process is provided in Fig. 2.

$$\begin{aligned} &3SiO_2(s) + 2N_2(g) + 6C(s) \rightarrow Si_3N_4(s) + 6CO(g); \\ &\Delta H = 1268\ \text{kJ/mol}\ [82] \end{aligned} \tag{2}$$

In this relatively simple method, high purity (>99.95%), fine particle size (~10–150 nm), and high surface area (up to 200 m^2/g) SiO_2 (e.g., fumed silica) is intricately mixed with activated carbon or lamp black. The carbon sources are also high in purity (>99.95%), possessing a small narrow particle size range (~10–20 nm), and large specific surface areas (up to 600 m^2/g), but particle size and intimate mixing appear to be more important in promoting the reaction than surface area. Although a broad range of carbon contents have been tested, ratios of 4:1 C:SiO_2 are preferred, and amounts greater than about 5:1 lead to undesirable SiC formation [75, 76]. The addition of between 2 and 10 wt.% α-Si_3N_4 seeds was found to accelerate the reaction [77, 79, 82, 84].

Mixing was conducted by various wet or dry mechanical methods or by sol-gel techniques [76]. The powder mixtures were preferably placed in graphite crucibles or within fluidized bed reactors [79]. To prevent the formation of SiC and Si_2N_2O, a constant copious flow of high-purity N_2 (99.95%) and intricate gas mixing was required to remove CO and O_2 reaction products [75, 76]. Preferred reaction temperatures were between 1450 °C and 1500 °C, for 2–10 h [72, 79]. Although higher temperatures accelerated the reaction, they also increased the formation of unwanted ß-Si_3N_4 and SiC. Impurities, particularly iron, catalyzed the reaction similar to direct nitridation of silicon, but they also increased the unwanted presence of SiC [85]. Upon completion, conversion to α-Si_3N_4 was typically ≥95%, but residual unreacted carbon and silica were often left in the fired cake. Carbon was subsequently removed by exposing the cake to flowing air at between 600 °C and 850 °C for 1–2 h [74, 76, 79]. Comminution by dry or wet methods (e.g., ball, attrition, or vibratory mills) was conducted to reduce the average particle size to <1 μm. Depending upon the quality of the raw materials, acid leaching was used to remove impurities and unreacted silica. The powder was then conventionally dried using rotary or belt dryers, or by spray drying.

In the 1980s–90s, Toshiba, Alusuisse, and Dow Chemical were key companies involved in the commercial development of Si_3N_4 powders by carbothermal reduction [68, 79, 80]. These organizations invested heavily in pilot and production equipment and made remarkable progress [86]. However, Dow Chemical's effort, which was principally supported by the US Department of Energy, ended with the withdrawal of government funding [80]; and the Alusuisse operation was shelved when Alcan Incorporated acquired Alusuisse-Lonza in 2000. While Toshiba still has limited powder production capability, its output is solely for internal use. No new manufacturing facilities have been brought online to produce α-Si_3N_4 by carbothermal reduction in the past two decades. However, academic institutions continue to investigate various aspects of the process, including its mechanisms and kinetics [87], the use of alternative raw materials and methods [87–95], and for the potential production of porous ß-Si_3N_4 ceramics [91, 96–98]. So, although carbothermal reduction was touted to be the most efficient method for producing fine α-Si_3N_4 powders, this prediction has yet to be realized. The most significant recent interest in this process appears to be in the development of high-thermal-conductivity Si_3N_4 materials, substrates, nanofibers, and nanobelts for aviation and transportation, including high-power IGBT modules for EVs and HEVs [99–101]. This process has never been used to produce powders that have been incorporated into medical devices, but that's not to indicate that it's incapable. Indeed, the physical and chemical properties of powders produced by carbothermal reduction of silica rival the best raw materials produced by direct nitridation or diimide decomposition.

2.3 Diimide Decomposition

Production of α-Si_3N_4 powder by pyrolysis of silicon diimide ($Si(NH)_2$) is actually among the oldest known methods [24, 102]; but commercial interest in this process only materialized in the 1970s–90s concurrent with increased interest in Si_3N_4 for gas-turbine engines [103–117]. Two reaction steps are necessary in the process:

$$SiCl_4 + 6NH_3 \rightarrow Si(NH)_2 + 4NH_4Cl; \quad \Delta H_{298} = -676.0 \text{ kJ/mol} \ [108] \tag{3}$$

$$3Si(NH)_2 \rightarrow Si_3N_4 + 2NH_3; \quad \Delta H_{298} = 230.8 \text{ kJ/mol} \ [118] \tag{4}$$

The first step forms the diimide and the second step converts it to α-Si_3N_4. Development of the process involved mixing combinations of $SiCl_4$ and NH_3, either as two liquids or as separate gases and liquids at various temperatures [37]. The process initially proved to be difficult to control. Due to its highly exothermic nature, the reaction occurred immediately. This resulted in plugging supply tubes or nozzles within the reaction vessel. Temperature control, removal of ammonium chloride (NH_4Cl) fumes, and collection of the diimide powder were also challenging [109]. These difficulties were uniquely overcome by combining liquified NH_3 with an immiscible aliphatic solvent within the reaction vessel. Of the various diimide decomposition techniques, only this immiscible liquid method was scaled to full production. Today, tonnage quantities are routinely produced; and the quality of this powder is now considered to be superior to raw materials produced by direct nitridation or carbothermal reduction. This process, which was developed and refined by Ube Industries (Ube, Japan), is shown in Fig. 3 [112, 114].

In their method, silicon tetrachloride ($SiCl_4$) and ammonia (NH_3) are the raw materials. Both are reasonably inexpensive and used in large quantities for various industrial products. An aliphatic solvent (e.g., n-heptane, C_7H_{16}) is also used, but it plays no reactive role; it is simply a carrier. It is charged into the reaction vessel along with liquid NH_3. The two liquids are immiscible and essentially insoluble. The denser aliphatic solvent resides at the bottom of the vessel. The amount of NH_3 present in the reaction vessel is stoichiometrically set to be at least ten times that of the $SiCl_4$. To initiate the reaction, gaseous $SiCl_4$ is injected only into the aliphatic liquid where it is vigorously dispersed by mechanical agitation. It reacts with a very small amount of dissolved NH_3 present in the organic solvent, but most of the reaction occurs at the immiscible interface between the solvent and liquid NH_3. Silicon diimide is immediately formed at the interface with the release of heat and

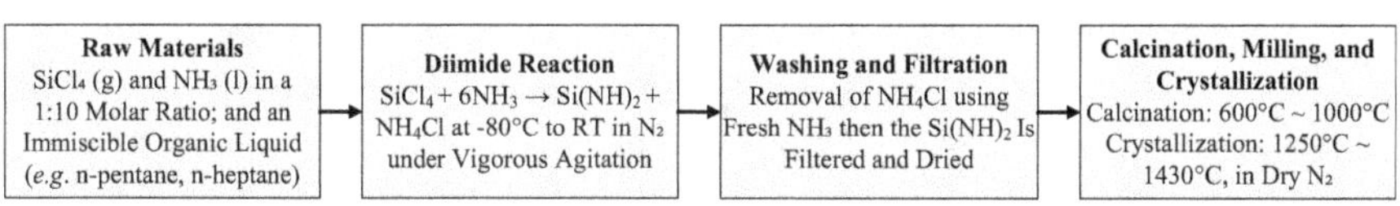

Fig. 3 Process for producing α-Si_3N_4 powder using silicon diimide decomposition

NH_4Cl, but both the heat of the reaction and NH_4Cl are absorbed into the cold ammonia. An external refrigerant maintains the vessel's temperature during the reaction. This unique production method solved the problems of plugged nozzles, excessive fumes, and temperature control. Upon completion of the reaction, an exit valve is opened, and the temperature of the vessel is slowly raised. The resulting increase in pressure forces the solid diimide into a calcination tube or tank where it is filtered, repeatedly rinsed with fresh liquid NH_3 and dried with N_2. In the next step of the process, the diimide is collected and placed in a furnace where it is calcined at between 600 °C and 1000 °C under high-purity dry N_2. This step changes $Si(NH)_2$ to amorphous Si_3N_4 (cf., Eq. 4). Afterward, a milling step may be included in the operation to reduce the needle-like morphology of the amorphous Si_3N_4. This milling operation reportedly improves the rate of its crystallization to α-Si_3N_4 in the next heat treatment step and allows for the formation of equiaxed particles. The last step at higher temperatures (1250–1430 °C, 2–30 h) is needed to form the alpha phase. Although temperatures up to 1600 °C have been used, they result in the undesirable formation of ß-Si_3N_4.

Since the turn of the century, most of the biomedical research on Si_3N_4 and all implantable biomedical products have utilized Si_3N_4 powder produced using the diimide decomposition process. It was preferred due to the consistency and quality of the resulting data and devices. In summary, listed in Table 1 are the current physical and chemical properties of selected powders produced by the three most

Table 1 Chemical and physical properties of commercially available Si_3N_4 powders

Powder production process	Direct nitridation of silicon				Diimide decomposition	Carbothermal reduction of silica
Manufacturer	Höganäs[a]	Vesta[b]	Denka[c]	AlzChem[d]	Ube[e]	Toshiba[f]
Trade name	M11 HP	P95H	SN-9FWS	Silzot HQ[Plus]	SN-E10	A-200
% N	≥38.5	NR	NR	>38.7	>38.0	37.5
% α-Si_3N_4	≥90.0	>91.0	≥91.0	≥95.0	>95%	98.0
Impurities wt.%						
C	<0.30	<0.50	<0.10	<0.20	<0.10	0.90
O	<2.0	<1.50	<0.80	<0.60	<2.00	2.00
Al	<0.005	<0.08	<0.09	<0.07	0.0001	0.20
Ca	<0.001	<0.02	<0.10	<0.03	<0.0001	0.01
Fe	<0.001	<0.06	<0.018	<0.03	<0.001	0.007
Si	NR	<0.2	NR	<0.02		NR
Cl	NR	<0.002	NR	<0.02	<0.01	NR
Physical properties						
Surface area (m^2/g)	9.0–12.0	10.0–12.0	11	NR	9.0–13.0	NR
Particle size (d_{50} μm)	0.7–0.9	0.9	0.7	1.9	0.6	0.9

Sources: [a]hoganas.com/en; [b]Vesta Si Europe AB, Ljungaverk, Sweden; [c]denka.co.jp/eng/; [d]alzchem.com/en/; [e]ube.com/contents/en/; [f]Lange et al. Ref. [38]

popular manufacturing methods. Although the diimide decomposition method is considered superior, the quality of powders produced by direct nitridation and carbothermal reduction is not lacking. In this author's opinion, all of these raw materials are considered to be acceptable and useful for the production of quality medical devices.

3 Silicon Nitride Processing

Table 2 provides an overview of the various raw materials, forming, and densification methods used to produce Si_3N_4 components along with typical properties (compiled from references provided in this chapter). Component fabrication follows a general route of powder preparation, forming, green machining, densification, post-firing machining, and final inspection. Refer to Fig. 4. While the process steps before and after densification are common to most ceramic materials, Si_3N_4 differs in its various sintering methods. These generally include reaction-bonding (RBSN), sinter reaction-bonding (SRBSN), hot-pressing (HPSN), sintering (SSN), gas-pressure sintering (GPSSN), and hot-isostatic pressing (HIPSN); but there are several variations in each of these categories.

3.1 Raw Materials

Ultra pure silicon and nitrogen are the essential raw materials for the RBSN and SRBSN processes. Both methods rely on the in situ formation of Si_3N_4 during firing. However, to achieve higher densities, the SRBSN process includes sintering aids—rare-earth oxides (Re_2O_3), MgO, SiO_2, Al_2O_3, or CaO. The other densification techniques (i.e., HPSN, SSN, GPSSN, and HIPSN) start with high-purity α-Si_3N_4 powder combined with various sintering aids. Numerous densification compounds have been tested; but those most commonly used include MgO, CaO, Y_2O_3, and Al_2O_3 [35]. These additives or their precursors are also of high purity ($\geq$99.95%) and fine particle size (typically $\leq$1.0 μm). They comprise $\leq$10 wt.% of the total, but detailed batch percentages often remain trade secrets with each manufacturer. Their amount is specifically kept low to minimize detrimental effects on mechanical properties. They facilitate densification by the formation of a liquid phase which allows for particle rearrangement, dissolution of α-Si_3N_4, and precipitation of ß-Si_3N_4 [119]. Some of these additives are incorporated into the ß-Si_3N_4 lattice (e.g., Al_2O_3) while others (e.g., MgO and Y_2O_3) combine with alumina, silica, and nitrogen to form an amorphous or crystalline intergranular glass [119]. Higher concentrations of Al_2O_3 are used to form a complete solid solution with ß-Si_3N_4 yielding SiAlONs [120], but at lower fractions, the pristine structure of ß-Si_3N_4 is hardly altered [121]. Several review articles have emphasized the importance of purity, particle size, and clean processing for all essential raw materials [35, 38, 40, 122–124].

Table 2 Comparison of materials, processes, and properties of silicon nitride ceramics

Materials, operational conditions, and properties	RBSN	SRBSN	HPSN	SSN	GPSSN	HIPSN
Raw materials (powders and gases)	Si and N_2	Si, Re_2O_3, MgO, SiO_2, Al_2O_3, CaO, CeO_2, ZrO_2, N_2	α-Si_3N_4, MgO, CaO, Li_2O Re_2O_3, Al_2O_3, AlN, N_2		α-Si_3N_4, MgO, Re_2O_3, Al_2O_3, AlN, N_2	α-Si_3N_4, MgO, Re_2O_3, Al_2O_3, AlN, N_2
Powder processing	Attrition, vibratory, or ball milling typically in alcohols or deionized H_2O, freeze drying, spray drying					
Component forming	Pressing, slip casting, extrusion, gel casting, injection molding, 3D printing		Not generally applicable	Pressing, slip casting, extrusion, gel casting, injection molding, 3D printing		
Green machining	Traditional CNC turning and milling			Traditional CNC turning and milling		
Binder removal	~500–750 °C in Air ~1–2 h			~500–750 °C in Air ~1–2 h		
Pre-sintering	1000–1200 °C, N_2, ~1–2 h			1000–1600 °C, N_2, ~1–2 h		
Reaction bonding, hot pressing, sintering, and/or HIPing	1250–1450 °C, N_2/H_2, Ar, 1 to >300 h	1750–1925 °C, ≥ 60 min, 0.1–10 MPa N_2	1300–1750 °C. ~20 to ~30 MPa, ≤ 60 min, N_2	1650–1780 ° C, 1–3 h, Static or Flowing N_2	(1) 1525 ° C ~ 2050 °C at 0.5~10 MPa N_2; (2) 1800 ° C ~ 2050 °C at 2–100 MPa N_2	(1) 1600 °C ~ 1800 °C, 1–3 h, ambient N_2 (2) 1650 °C ~ 1750 °C, 1–3 h, 100 ~ 300 MPa N_2 or Ar
Machining, grinding	Traditional CNC grinding, minimal material removal		Extensive CNC grinding required	Net shape for some features, but CNC grinding may be required		
Density (g/cc)	1.60 ~ 2.54	3.21 ~ 3.30	3.07 ~ 3.37	3.18 ~ 3.30	3.21 ~ 3.30	3.18 ~ 3.30
% Porosity	20 ~ 80	≤ 2	< 0.1	< 0.1	< 0.1	< 0.1
% Linear shrinkage	≤ 0.2	10 ~ 12	Not applicable	12 ~ 20	12 ~ 20	12 ~ 20
Microstructure	Equiaxed and anisotropic	Equiaxed and Anisotropic	Equiaxed and anisotropic	Anisotropic	Anisotropic	Anisotropic
Phase Composition	Mixture of α- and ß-Si_3N_4 and Si-intermetallic	Principally ß-Si_3N_4	Principally ß-Si_3N_4	Principally ß-Si_3N_4	Principally ß-Si_3N_4	Principally ß-Si_3N_4

Flexural Strength, σ_R (MPa)	≤ 350	650~830	383~1210	275~1200	600~1200	800~1200
Weibull Modulus, m	≤ 20	15~25	8~20			
Compressive Strength (MPa)	650	3500~4000	3500~4500	3500~4000	3000	4000
Fracture Toughness, K_{IC} (MPa·m$^{1/2}$)	1.8~4.8	5.5~6.5	2.8~12	3.0~10	6.5~8.3	6.6~10.6
Hardness, H_v (GPa)	8~10	15	10.3~18.7	15~16	14~16.5	15~16
Elastic modulus, E (GPa)	170~220	~300	217~354	195~315	300	300
Thermal conductivity (W/m·°K)	10~15	25~150	20~100	15~20	21~80	15~30
Resistivity (ohm-cm)	10^{10}	10^{14}	10^{14}	10^{14}	10^{10}~10^{12}	10^{12}
Maximum service temperature (°C)	1300	1370	1200	1200	1400	1200

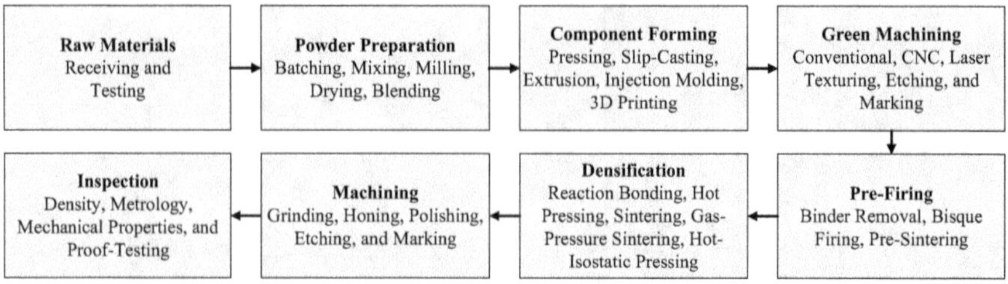

Fig. 4 Processing of silicon nitride bioceramics

3.2 Powder Preparation

Powder processing involves weighing, batching, mixing, milling, drying, blending, and compounding of the Si_3N_4 raw materials. The selection of these operations is dependent upon the subsequent component-forming method. For instance, dry-pressing generally requires a flowable granulated powder. Preparation of this powder typically involves aqueous mixing and milling of the raw materials with dispersants, anti-foaming agents, lubricants, and binders followed by spray drying. Slip casting is a similar aqueous preparation method, but it obviates the need for spray drying. Conversely, extrusion, injection molding, or 3D printing usually only requires mixing of the raw materials with water and/or appropriate polymers using single or twin screw compounders. All these processes are industrialized and while they continue to be improved, they are not unique to a specific material. Si_3N_4 is no exception. They are commonly used for most oxide and nitride ceramics. An extensive discussion of these processes is beyond the scope of this chapter. The interested reader is referred to relevant textbooks which provide both principles and examples [125–128].

3.3 Forming

The heart of any ceramic manufacturing operation is component forming. This is where powders are crafted into useful functional devices. Some parts can be formed with "net-shape" features, while others require green-state machining. High-tolerance parts can only be produced to "near-net-shape" in the green state and require hard grinding after firing. Numerous forming methods have been utilized to produce net-shape or near-net-shape Si_3N_4 components [129]. Detailed reviews of each method are also beyond the scope of this chapter; they are extensively described in the literature. However, selected references are included here for dry- or isostatic-pressing and green machining [130–133], slip-casting [134–138], extrusion [139–142], injection molding [143–146], and tape-casting [147–150]. Additive manufacturing has also been utilized for the fabrication of Si_3N_4 components. It encompasses a broad number of innovative methods including powder bed fusion, selective laser sintering, stereolithography, direct ink-jet printing, robocasting, fused deposition modeling, and laminated object manufacturing. The advantages and limitations of these fabrication processes for Si_3N_4 were discussed in a recent systematic review [151].

To date, essentially all implantable Si_3N_4 medical devices have been formed using the traditional method of pressing and green machining. However, substantial work is ongoing with more advanced methods, particularly additive manufacturing. Undoubtedly, future implants will benefit from the design flexibility and cost reduction that these newer technologies provide.

3.4 *Pre-firing and Densification*

Once components are formed, they are subjected to a series of firing operations that may include binder removal, de-oxidation or oxidation, and various densification methods. As indicated in Table 2 and Fig. 4, each of these processes varies based on the selected densification technique which broadly includes RBSN, SRBSN, HPSN, SSN, GPSSN, and HIPSN.

3.4.1 Reaction Bonding (RBSN) and Sintered Reaction Bonding Silicon Nitrides (SRBSN)

RBSN and SRBSN are essentially identical in their component fabrication operations except for densification. Both uniquely use silicon powder as the primary raw material. While RBSN relies solely on an exothermic reaction between silicon and nitrogen to form Si_3N_4 and achieve partial densification (i.e., $3Si + 2\,N_2 \rightarrow Si_3N_4$; $\Delta G° = -723$ kJ/mol), SRBSN uses sintering additives and an additional high-temperature firing step to increase density.

RBSN was the first process used to fabricate Si_3N_4 components. Under the direction of the British Admiralty during the late 1950s, Popper and Ruddlesden formed and indexed the crystal structures of both α- and ß-Si_3N_4 by heating Si metal powder in the presence of high-purity nitrogen gas (N_2) at temperatures of 1350–1450 °C [152, 153]. Although this seemed to be a simple process, almost 30 years elapsed before its mechanisms and kinetics were reasonably understood [154–162]. To manufacture RBSN, ultra-fine to coarse (Ø0.05 to Ø50μm), high purity (99.9 + %), Si powders are consolidated and prefired (~500–750 °C, air, ~1–2 h), then pre-sintered (1000–1200 °C, flowing Ar or N_2, ~1–2 h), and nitrided (1250–1450 °C, flowing N_2 with Ar or H_2, 1 to >72 h) to produce porous components (~50% to ~80% dense) that are mixtures of α- and ß-Si_3N_4, typically with residual silicon intermetallic phases [24, 35]. Flexural strengths of ≤350 MPa, Weibull moduli of up to 20, and fracture toughness of ≤4.8 MPa·m$^{1/2}$ have been reported [122]. The RBSN process has several advantages. The combination of silicon and nitrogen raw materials is less costly than using pre-synthesized Si_3N_4 powder. In situ formation of Si_3N_4 during nitridation occurs by reaction and growth of the α- and ß-phases within the existing pore structure of the green body [154, 155]. This minimizes dimensional changes and distortion. Typical shrinkage values can be less than about 0.2%. The need for green machining or expensive final grinding is often eliminated. An apparent disadvantage of RBSN is its residual porosity, but this can be beneficial for certain applications including filtration and medical devices.

Researchers desiring to attain higher densities than were achievable by RBSN developed the SRBSN method in the late 1970s and early 1980s [163, 164]. Investigators found that addition of sintering additives (e.g., Al_2O_3, SiO_2, CaO, MgO, CeO_2, ZrO_2, and/or rare-earth oxides) to Si powder, followed by component forming, then nitridation (~1400 °C. 150–300 h, in ambient N_2 with H_2 or Ar), and a subsequent high-temperature firing (1700–1950 °C, 1–2 h, 0.21–2.1 MPa N_2) were able to improve densities to ≥86% of theoretical values [165]. Later, they also discovered that increasing the N_2 gas pressure to >10 MPa during the sintering step resulted in densities ≥98% and flexural strengths >750 MPa [33]. Advocates of the SRBSN method suggest that it preserves the low-cost and near-net-shape advantages of RBSN while achieving physical and mechanical properties comparable to sintered Si_3N_4 [166, 167]. Due to low thermal expansion, RBSN and SRBSN ceramics have exceptional resistance to thermal shock. They are not easily wetted by molten metals. This makes them excellent materials for casting forms and crucibles. Although the RBSN and SRBSN processes have declined in favor of methods that produce stronger fully dense components, they still have a loyal following for many industrial parts such as crucibles, thermocouple tubes, welding jigs, filters, and high-temperature fixtures. Both RBSN and SRBSN are considered to be of particular interest for arthrodesis devices and bone scaffolds due to their engineered porosity, acceptable mechanical properties, and net-shape capabilities.

While most research on reaction bonded Si_3N_4 was directed at the development of components for advanced turbine engines, early medical studies also examined RBSN and SRBSN. For instance, Howlett, et al. investigated the effect of porous (55–70%) RBSN Si_3N_4 samples on rabbit marrow stromal cells in vitro and in vivo [168]. Samples were fabricated by mixing silicon powder with a pore-forming polymer (beads of polymethylmethacrylate, PMMA), and then shaping the mixture into discs and plugs. This was followed by firing in an electric furnace under flowing 80% N_2/20% H_2 with gradual temperature increases between 100 °C and 1200 °C for about 15 h. Nitridation occurred at 1450 °C for 20 h. The crystal structures of the samples were a mixture of α- and ß-Si_3N_4. Their in vitro experiments suggested that Si_3N_4 was biocompatible, but progenitor cells only attached and differentiated at the surfaces of the samples. However, RBSN samples implanted into femoral marrow cavities of mature rabbits promoted osteogenesis; and at 3 months, mature woven bone had encapsulated and permeated the porous structure. Three of the animals retained their implants for their natural lives of 2.5, 3.25, and 5 years. Post-mortem evaluation showed a 5–10 μm layer of fibrous tissue with regions of fibrocartilage and bone growth directly into the implants. In-growth was confirmed for pore sizes

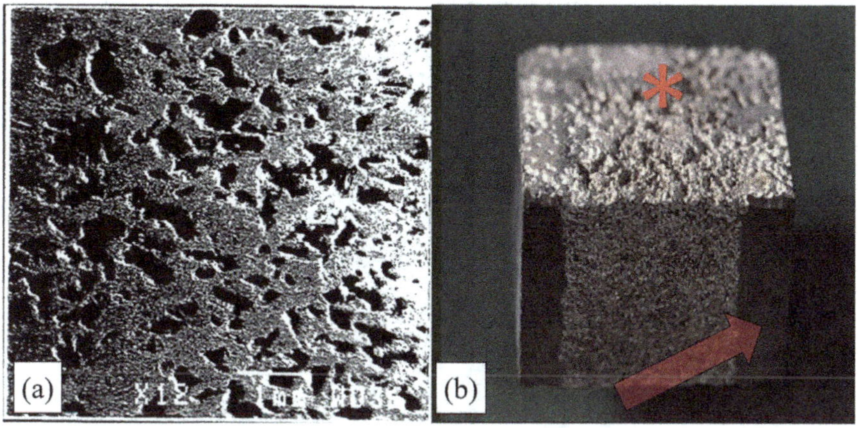

Fig. 5 The composite structure of silicon nitride implant: (**a**) Microstructure of porous core of the implant at 12× magnification; and (**b**) Macrostructure showing the load-bearing outer shell (arrow) and porous core (asterisk). Implant size was ~30 mm × 15 mm × 15 mm. (Used with permission [9])

≥Ø170 μm and it was prevalent for pores > Ø255 μm. While a concurrent study questioned the bioefficacy of Si_3N_4 in comparison to Al_2O_3 [169], subsequent more exhaustive evaluations confirmed the biocompatibilities of both RBSN and SRBSN [170–174].

At about the same time, the first clinical trial using implantable RBSN for spinal arthrodesis was initiated in Australia [175]. This was a unique one-of-a-kind study that involved 30 patients diagnosed with lumbar spondylolisthesis and discogenic pain unresponsive to conservative therapies. The RBSN implants were produced by Sialon Ceramics Pty. Ltd. (Australia). Figure 5 provides structural details of one of these implants. It consisted of an outer higher-density RBSN rectangular shell which contained a porous RBSN core. To increase the porosity of the core, Si powder was mixed with polyethylene glycol and naphthalene. Composite powders were compacted using double-stage pressing. Then, after the removal of organics, the components were nitrided at ~1400 °C for ~50 h in flowing 95 vol.% N_2/5 vol.% H_2 gas. The phase composition of the implants was primarily ß-Si_3N_4, with a pore size range of about 200–1000 μm, and an apparent porosity of ~25%. The reported compressive strength was about 200 MPa. Mobbs et al. published a 30+ year review of outcomes for surviving patients and found that these implants were still performing well. They had high fusion rates, minimal subsidence, no adverse reactions, and no other notable complications [9]. A radiographic example of one of the surviving patients is provided in Fig. 6.

Since then, various forming methods have been utilized to produce both RBSN and SRBSN structures that may be useful as functional bone scaffolds. Rambo et al. developed a low-cost replication process using the biomorphic properties of natural sea sponges [62]. Their process involved the infiltration of a Si slurry into the sponge by dip-coating, followed by drying, air firing at 600 °C for 30 min to remove carbon.

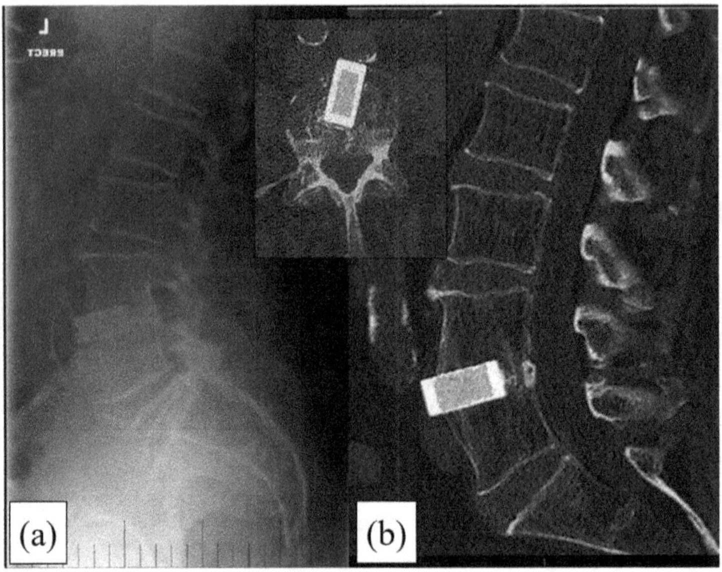

Fig. 6 Standalone L4/5 anterior lumbar interbody fusion 30 years postop. (**a**) Standing radiograph with solid fusion and acceptable sagittal balance, with minor anterior slip of the implant from its original position (1986). (**b**) Midsagittal computed tomography with nonlucency and excellent osseointegration. (Used with permission [9])

Nitridation was conducted at 1450 °C for 5 h in flowing N_2. They produced RBSN structures with an α/ß ratio of 67% and no residual Si or C phases.

Net-shape test components had a porosity of 88 vol.% with an average pore size of between 1 and 3 mm, and compressive strengths of ~30 MPa. Later, Alem, et al. improved the permeability and strength of porous RBSN structures by including PMMA beads in Si slurries [176, 177]. After gel casting and pre-firing, they achieved conversion to RBSN (predominately α-Si_3N_4) for nitridation times of ~4 h at 1400 °C in an N_2/H_2 gas mixture. Their components had from about 40 to 87 vol.% interconnected porosity and inversely corresponding strengths of between ~2 and ~ 40 MPa. A similar study was conducted by Park et al. [178], but they used an SRBSN process. Y_2O_3 and Al_2O_3 were added as densification aids. First, they mixed different size fractions of PMMA into RBSN ethanol-based slurries containing the sintering additives. Then, after drying, pressing pellets, and pre-firing, the various compositions were nitrided at 1350 °C for 3 h. This was followed by sintering at 1700 °C for 2–10 h in 0.1 MPa of N_2. Comparable porosity values were observed for all test specimens (~60%), with pore sizes ranging from about 0.8 to 1.3 μm. Remarkably, they found that the addition of a smaller-sized pore former coupled with the longest sintering time resulted in the greatest permeability.

In another series of studies conducted between 2013 and 2015, Šajgalík and coworkers utilized the polyurethane sponge replica method to increase SRBSN permeability [179–184]. After dip coating sponges in an aqueous suspension of Si powder, the excess slurry was removed by roll-compression, followed by drying and

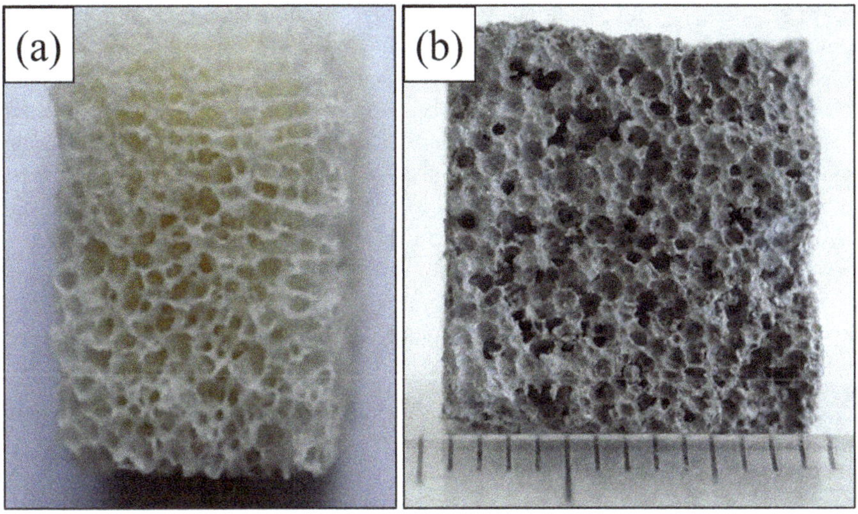

Fig. 7 Macrostructures of (**a**) trabecular bone and (**b**) a porous SRBSN bone scaffold. (Used with permission [184])

pre-firing at 600 °C for 1 h. Nitridation was conducted at 1400 °C for 3 h, and post-nitridation densification was performed at 1750 °C for 2 h in N_2. The investigators postulated that this process would produce SRBSN scaffolds that mimic natural trabecular bone. Test components had densities of 0.8–1.1 g/cm^3 and porosities of between 65% and 75%, which are within the range of trabecular bone [185]. Comparative macrostructure examples are provided in Fig. 7.

A similar set of studies investigated the effects of oxide sintering additives, powder beds, and PMMA particles (<40 μm) on SRBSN porosity [186–188]. CaO, MgO, or Al_2O_3 were mixed with Si powder along with PMMA and nitrided at temperatures between 1200 °C and 1400 °C, then sintered at 1700 °C for 2–4 h. Samples had porosity values of between 78 and 85 vol.%, a bimodal distribution of pores between about 0.5 and 10 μm and crush strengths of ≤10 MPa. They uniquely found that the use of MgO and PMMA enabled the formation of nearly spherical interconnected porosity due to the vapor phase dissolution of α-Si_3N_4 within the pores, and its reprecipitation outside of the pores as ß-Si_3N_4.

Lastly, fully dense high-strength SRBSN was produced and qualified per FDA guidance for use as spinal arthrodesis implants as part of a joint commercialization agreement between Kyocera and Amedica Corporation during the period 2013–2015.[1] While specifics of the agreement remain confidential, the SRBSN

[1] Amedica Signs Agreement with Kyocera for Commercial Manufacture of Silicon Nitride Medical Devices, https://www.prnewswire.com/news-releases/amedica-signs-agreement-with-kyocera-for-commercial-manufacture-of-silicon-nitride-medical-devices-233494661.html, Nov. 26, 2013.

was a standard Kyocera material; but its composition and manufacturing method remain proprietary [189].

RBSN and SRBSN continue to be exceptional processes for producing functional medical devices even though they have been superseded by more advanced densification methods. This is unfortunate because, as discussed earlier, RBSN was the first abiotic material used for lumbar spinal arthrodesis [9]; and 30+ years later, these implants are still serving surviving patients. At the time of their implantation, the favorable osteoconductive and antimicrobial properties of Si_3N_4 were unknown. Given these properties and Si_3N_4's superior strength and toughness, perhaps it's time to revisit RBSN and SRBSN methods for medical devices. There are clear advantages in low cost, net-shape forming, engineered porosity, biocompatibility, bioactivity, and adequate mechanical properties.

3.4.2 Hot-Pressed Si_3N_4 (HPSN)

Hot-pressing was another early method used to produce Si_3N_4 components. Hot-pressing simultaneously applies pressure and heat to compact a powder mass into a dense shape. The process is mostly limited to solid regular geometries such as rods, cylinders, and polygons. Deeley et al. were the first to publish a phenomenological study on HPSN in 1961 [190]. After numerous attempts to hot-press Si_3N_4 without sintering aids, they conducted a series of experiments using a range of oxide and nitride additives. They discovered that when Mg-based compounds (Mg_3N_2 and MgO, at 3.3–5.0 wt.%) were added to α-Si_3N_4 powders and hot-pressed at between 1600 °C and 1850 °C for 4–18 min at pressures of 23 MPa, components with less than 0.1% open porosity and flexural strengths of up to ~690 MPa were obtained. This was a remarkable achievement because up to that point in time, RBSN was the only useful manufacturing method. Numerous hot-pressing studies ensued using various additives [191–195] but one specific hot-pressed composition developed by Norton Company became a standard by which all subsequent Si_3N_4 materials were evaluated [196, 197]. Known commercially as NC-132, it was initially targeted for use in gas turbine engines [198]. However, hot-pressing was found to be a cost-prohibitive process, particularly due to extensive post-densification machining. Several studies explored both its manufacturing feasibility and biocompatibility. For instance, Amaral et al. prepared a Si_3N_4 bioglass composite (70 wt.% Si_3N_4 and 30 wt.% 45S5 bioglass) by hot-pressing pre-mixed powders in graphite dies at temperatures between 1300 °C and 1500 °C for 40 min at 30 MPa [199]. They achieved ~97% density and reasonable mechanical properties: H_v = 10.3 GPa; K_{IC} = 4.4 MPa·m$^{1/2}$; E = 197 GPa; σ_R = 383 MPa; and m = 8.3. Mazzocchi et al. prepared three hot-pressed compositions of Si_3N_4: (a) 90 wt.% Si_3N_4 with 10 wt.% bioglass; (b) 91.3 wt.% Si_3N_4 with 8.7 wt.% MgO; and (c) 60.45 wt.% Si_3N_4 with 1.3 wt.% Al_2O_3, 3.25 wt.% Y_2O_3, and 35.0 wt.% TiN. The addition of TiN to the third composition was performed to make it an electroconductive composite. Hot-pressing was conducted under vacuum using induction heating at a pressure of 30 MPa and temperatures of 1750 °C for 20 min, 1650 °C for 60 min, and 1800 °C for 10 min for compositions (a), (b), and (c), respectively [200]. All samples achieved densities of >99% with excellent mechanical properties: H_v = 14.7–18.7 GPa; E = 276~354 GPa; and σ_R = 835–969 MPa. In vitro cytotoxicity using L929 mouse fibroblast cells was evaluated for raw materials, powder mixtures, and dense components. None of the compositions showed any toxicity, suggesting that Si_3N_4 and TiN, along with the sintering aids were biocompatible compounds. Later, Cappi et al. produced HPSN specimens and examined their cytotoxicity toward mouse fibroblast cells (L929) and osteogenic differentiation using human mesenchymal stem cells (hMSC) [201]. The HPSN samples were made from 90 wt.% Si_3N_4 and 10 wt.% yttrium aluminum garnet ($Y_3Al_5O_{12}$). Hot-pressing was conducted inside graphite dies for 60 min at 1700 °C. Density values were >99%. Reported mechanical properties were: H_v = 16.4 GPa; K_{IC} = 6.2 MPa·m$^{1/2}$; σ_R = 1210 MPa; and m = 10.7. When compared to alumina and zirconia ceramics as controls, they could not detect differences in cyto-biocompatibility for the L929 cell line; and noted that hMSC differentiation to osteoblasts was equivalent for all the ceramics. Hot-pressing continues to be utilized as an important densification method for laboratory samples and pre-production evaluation of various silicon nitride compositions and composites.

3.4.3 Sintered Silicon Nitride (SSN)

Following the developments of SRBSN and HPSN, investigators initiated studies on ambient pressure sintering of Si_3N_4 (SSN) using Mg-based compounds [202]. However, they found that the decomposition of Si_3N_4 competed with densification. Terwilliger and Lange reported that the best density values (i.e., 72–86%) for a 5 wt.% Mg-doped SSN were achieved either at lower temperatures (1515–1545 °C) for longer times (20 mins) or at moderately high temperatures (1580–1630 °C) for short periods (5 mins) [203]. This effect is shown in Fig. 8. Decomposition is a common competing mechanism to densification for all Si_3N_4 compositions. It is particularly prevalent for alkali-earth sintering additives, but concurrent research by Messier and Gazza showed that it was also operative for Si_3N_4 mixtures with added Al_2O_3 [204]. To subdue decomposition while promoting densification, components were typically packed in setter powders consisting of mixtures of Si_3N_4 and BN. Gas-pressure sintering (GPSSN) and hot-isostatic pressing (HIPSN) were also found to be effective suppression techniques [205].

To improve ambient pressure densification, work at the US Army's laboratories in Watertown, MA, USA, utilized rare-earth additives, particularly yttrium oxide (Y_2O_3) [206]. Hot-pressing confirmed that Y_2O_3 was an effective sintering agent. After this discovery, researchers focused on combining Y_2O_3 and MgO, or Y_2O_3, Al_2O_3, or AlN sintering aids for pressureless sintering of Si_3N_4 [207, 208], leveraging off of prior work on SiAlON ceramics by Jack [209]. Various ratios of these additives were evaluated, and it was found that densification with minimal decomposition could be achieved at lower temperatures [210–212]. Other rare-earth oxides

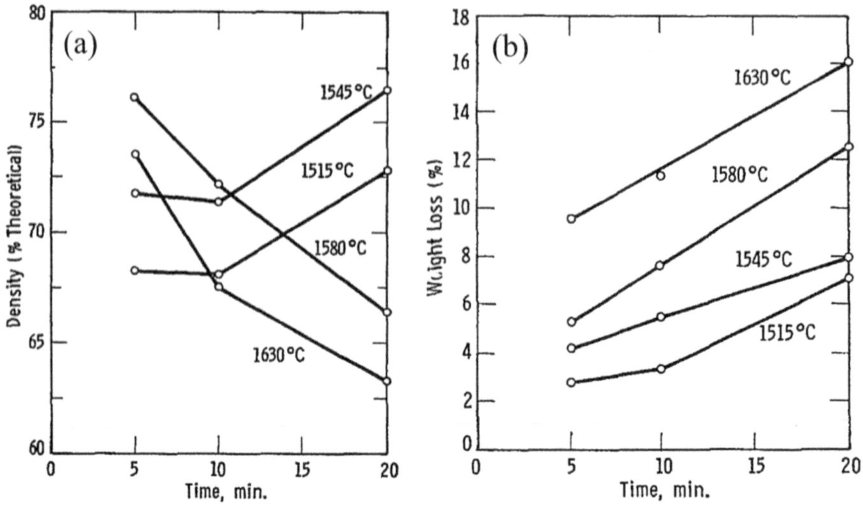

Fig. 8 Densities (**a**) and weight loss (**b**) obtained for isothermal experiments at three time periods and four temperatures. (Used with permission [203])

were also explored with Al_2O_3 or various alkali earth additives [213], but Y_2O_3 and Al_2O_3 or Y_2O_3 and MgO eventually became preferred commercial compositions.

Typical sintering conditions involved firing at between 1650 °C and 1780 °C for periods of 1–3 h in flowing or static high-purity N_2. Components were often packed in setter powders of the same composition. Improved densification (e.g., >99%) and mechanical properties were achieved with these additives, but residual flaws limited flexural strengths to less than about 800 MPa [214]. More recent research using improved raw materials, mixing, and comminution methods markedly enhanced pressureless sintered Si_3N_4. Flexural strengths of >900 MPa are now more common with isolated reports of values over 1 GPa and fracture toughness between 5.5 and 7.5 MPa·m$^{1/2}$ [215–217].

Despite these advancements, pressureless sintered Si_3N_4 has yet to make significant inroads as a unique biomaterial. This is partly because the competing technologies of SRBSN, GPSSN, and HIPSN achieved substantial improvements in their physical, mechanical, and biocompatibility properties in advance of SSN; but it was also because these early alternative processes markedly increased the reliability of Si_3N_4 [123, 218, 219]. Today, appropriate processing and non-destructive testing (NDT) remove critical defects. These advancements should enable SSN to be utilized for high-reliability medical devices; and given its compositional similarity, SSN is likely to be as biocompatible as any other Si_3N_4 with several studies validating this assumption [3, 150, 173, 220–224].

3.4.4 Gas-Pressure Sintered Silicon Nitride (GPSSN)

Using Le Chatelier's principle, the GPSSN process was developed primarily to suppress the decomposition of silicon nitride during densification. Providing an overpressure of N_2 during sintering shifts the decomposition reaction in the reverse direction (e.g., Si_3N_4 (s) ← 3Si (l) + 2 N_2 (g) with ↑ N_2 pressure). In 1977, Priest et al. were the first researchers in the USA to report on the effectiveness of N_2 gas pressure sintering [225], although similar work was ongoing in Japan by Mitimo [226]. Their initial studies, respectively, utilized CeO_2 (20 wt.%) or MgO (5 wt.%) additives at temperatures of 1450–1900 °C and N_2 pressures of about 2 MPa for 2–7 h. Weight loss during sintering was reduced from ~6 wt.% to between 1 and 2 wt.%. Densities of 81% to 99 + % with flexural strengths between ~140 MPa and ~430 MPa were achieved. Ten years of GPSSN research ensued, where experiments examined a multiplicity of factors associated with pressurized densification. Notably, in 1979, Galasso and Veltri used Y_2O_3 and Al_2O_3 additives with ~2 MPa N_2 at between 1730 °C and 1950 °C to achieve >90% density and flexural strengths up to ~480 MPa [227, 228]. Gazza et al. then followed in 1981 with a study that utilized a two-step pressure densification process using Y_2O_3 and Al_2O_3 [229]. The first step involved sintering samples to closed porosity (~92–94%) at 1770 °C to 1780 °C for 90 min under 0.1 MPa N_2, followed by a second step that simply raised gas pressure to ~2 MPa. Near-theoretical densities were achieved with the best flexural strength of ~780 MPa. By 1994, GPSSN compositions consisting of lower concentrations of sintering additives (~4 wt.% Y_2O_3, 2 wt.% Al_2O_3) coupled with the use of high-purity diimide-derived Si_3N_4 powders were routinely sintered using a two-step GPS process to achieve densities >99% with flexural strengths and fracture toughness values of ~1 GPa and 5.3 ~ 6.6 MPa·$m^{1/2}$, respectively [230]. The two-step firing process is depicted in Fig. 9. The first step involves gradually increasing the temperature and pressure to between 1700 °C and 2050 °C and 0.5–10 MPa N_2, respectively. These conditions are maintained for a sufficient period to achieve closed porosity within the components (e.g., 88–95% density). Then, in the second step, temperature and pressure are increased to between 1800 °C and 1950 °C with 10–100 MPa N_2, respectively, and these conditions are held for 15–60 min. This second step applies external pressure to the self-sealed partially dense components to collapse internal porosity through liquid phase viscous flow, grain boundary sliding, and diffusion. With some variations, GPSSN processes continue to be used today for both research [231, 232] and manufacturing [233, 234].

Like SSN, there are few reports on biomedical testing or the use of GPSSN. This is not to suggest that it is inappropriate or not applicable to medical devices. Indeed, the similarity of GPSSN to other Si_3N_4 materials suggests that GPSSN should produce effective and safe biomaterials. Limited biocompatibility studies have been conducted. Perhaps the most prominent is the work by Neumann et al. [172, 173]. They examined the biocompatibility of five commercially available silicon nitrides, one of which was a GPSSN composition. Two of the remaining four were SSN and SRBSN materials, while the final two were not disclosed.

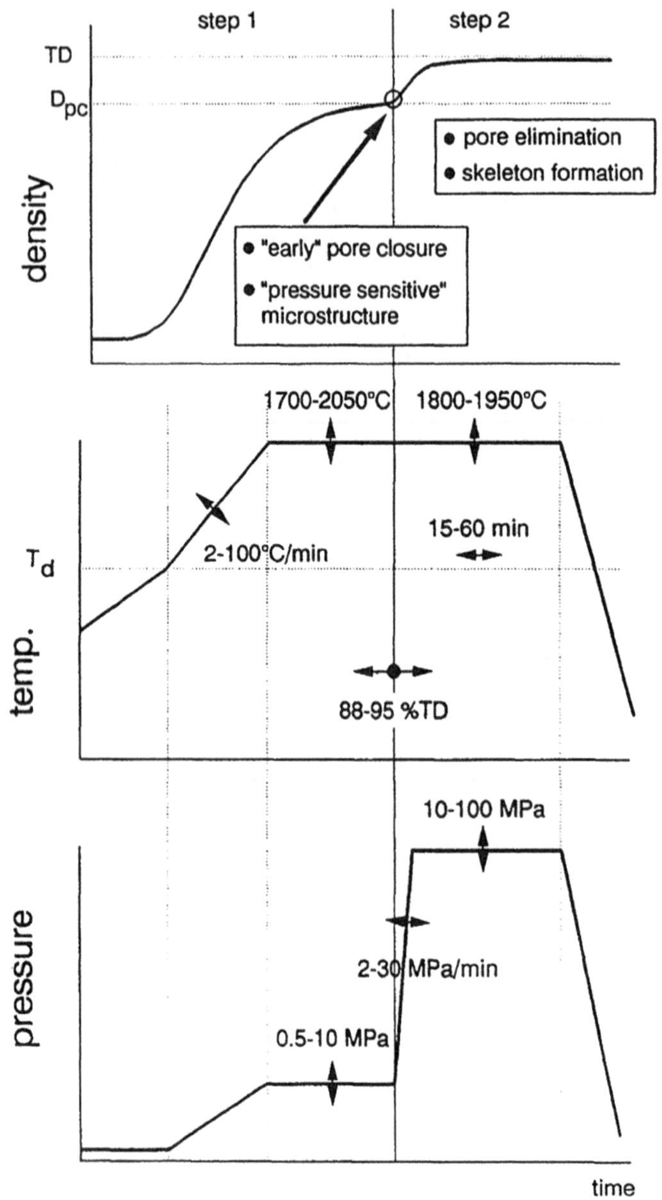

Fig. 9 Two-step schedule for gas-pressure sintering of Si_3N_4 with Y_2O_3 and Al_2O_3 additives. (Used with permission [230])

Biocompatibility testing was conducted following ISO 10993-12. They also performed cell viability tests using L-929 mice fibroblasts. In the end, they showed that none of the Si_3N_4 materials exhibited any cytotoxic effects; and there were no differences between the Si_3N_4 materials in fibroblast cell viability or growth. They therefore concluded that all Si_3N_4 test materials should be considered for biomedical applications. Similar independent studies have subsequently confirmed that GPSSN processes produce materials that can be biocompatible, anti-infective, and osteoconductive [235, 236].

3.4.5 Hot-Isostatically Pressed Silicon Nitride (HIPSN)

To densify complex-shaped components, glass-encapsulated hot-isostatic pressing (one variation of HIPSN) was developed during the 1970s–80s [237–239]. This process is essentially multi-directional hot-pressing. Green preformed parts are placed in evacuated glass ampules or within a glass powder bed inside a heated pressure vessel. The hot zone of the vessel is increased in temperature to melt the glass and coat the Si_3N_4 pieces. Then, nitrogen pressure (~100–400 MPa) and temperature (1500–1800 °C) are simultaneously applied to densify the encapsulated preforms. The melted glass acts as the pressing medium. After cooling, the solidified glass encapsulant is removed from the parts by sandblasting or acid etching. Using this method, Si_3N_4 compositions which were previously only densified by hot-pressing (e.g., NC-132) could now be sintered into complex near-net-shape parts [240]. Glass-encapsulated HIPSN also achieved new early standards in high temperature properties [238] (e.g., flexural strength ~628 MPa at 1337 °C). Yet, NC-132 and many other Si_3N_4 compositions were never used for hot-stage turbine engine components due to their poor oxidation and creep resistance. However, NC-132 and its successor composition (NBD-200) were subsequently commercialized for industrial roller and ball bearings, and are still available today under the trademark Cerbec® silicon nitride [198].[2]

Glass-encapsulated HIPSN has not been commercialized for medical devices either. Work by Westman et al. for Si_3N_4 compositions doped with Y_2O_3 and Al_2O_3 demonstrated that the borosilicate glass used as the encapsulant penetrates the surface of green Si_3N_4 parts during the initial stages of sintering (i.e., ~1200 °C, 130 MPa N_2, 30 min), and upon application of full temperature and pressure (i.e., 1750 °C, 250 MPa, 1 h), a surface reaction layer of ~200 μm forms [241, 242]. This defective layer has to be removed by post-densification machining in order to achieve adequate mechanical properties and dimensional tolerances. Thus, even though glass-encapsulated HIPSN was capable of producing near-net-shape parts, it was not a cost-effective manufacturing process for complex high-strength or tight-tolerance components.

Consequently, a new method of hot-isostatic pressing was developed during the late 1970s through the 1990s [243–247]. It had its origin in GPSSN, and it also

[2]Cerbec® "Silicon Nitride, Genuine Cerbec Ceramic Balls, Marketing Brochure Accessed Dec. 6, 2022, at, https://www.coorstek.com/en/products/cerbec-ceramic-bearing-elements/, CoorsTek, 14143 Denver West Pkwy, Golden, CO 80401 USA.

utilized two steps in densification. However, it differed from GPSSN because it involved special HIP equipment and higher pressures. Known as "sinter-HIP" (another variation of HIPSN), the first step sinters Si_3N_4 preforms to closed porosity (95–98%) within a conventional or continuous furnace under static or flowing N_2 at temperatures of 1600–1800 °C for 1–3 h. Most of the components' shrinkage occurs in this step. Upon cooling, parts are removed from their crucibles. They are then densely packed in the same or separate crucibles and transferred to the hot-isostatic press. In this second step they are subjected to high gas pressures (100–310 MPa Ar or preferably N_2), temperatures (1650–1750 °C), and times (1–3 h). This final step increases the density of the parts to >99% and completes the α- to ß-phase conversion. After this second step, it was discovered that continuous application of N_2 pressure was essential to prevent bloating of the ceramic upon cooling [248]. During sintering, the intergranular glass phase (IGP) becomes supersaturated with nitrogen due to its enhanced solubility at high temperatures and pressures. Adiabatic cooling allows nitrogen evolution within the IGP resulting in pores and gas channels. These microstructural defects negatively affect mechanical properties. Maintaining pressure during cooldown (i.e., non-adiabatic cooling) suppresses gas evolution and prevents these defects from forming. For compositions incorporating Y_2O_3 and Al_2O_3 as sintering aids, the sinter-HIP process typically resulted in flexural strength of 800~1200 MPa and fracture toughness values of 6~10 $MPa{\cdot}m^{1/2}$ [132, 246].

Si_3N_4 components produced by this process have been used in medical devices since 2008. Shown in Figs. 10 and 11 are production "run charts" of the physical and mechanical properties of a Y_2O_3-Al_2O_3-doped HIPSN exclusively used for medical devices. Note that overall average density is >99%, average flexural strength of ~1 GPa, and average fracture toughness of 10.6 $MPa{\cdot}m^{1/2}$. A SEM microstructural image of the Si_3N_4 produced by this process is provided in Fig. 12. A significant portion of the cited studies within this compilation utilized Si_3N_4 parts produced by this process.

3.5 Machining

Due to its hardness and fracture toughness, dense Si_3N_4 presents a formidable challenge to machinists. Achievement of acceptable design features and dimensional tolerances is typically only accomplished by fine diamond grinding, honing, lapping, and polishing. Over the past 50 years, these unique material removal operations have been developed for Si_3N_4 ceramics. Certain medical devices, particularly total joint arthroplasty components, have greatly benefited from the development of these material removal methods.

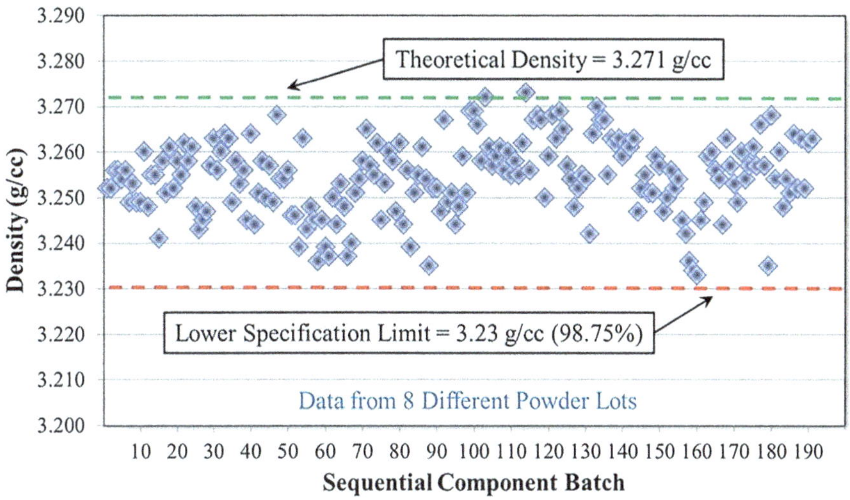

Fig. 10 Mass density results from 194 sequential component batches comprising eight different powder lots of a biomedical Si_3N_4. (Used with permission under Creative Commons License 4.0 [132])

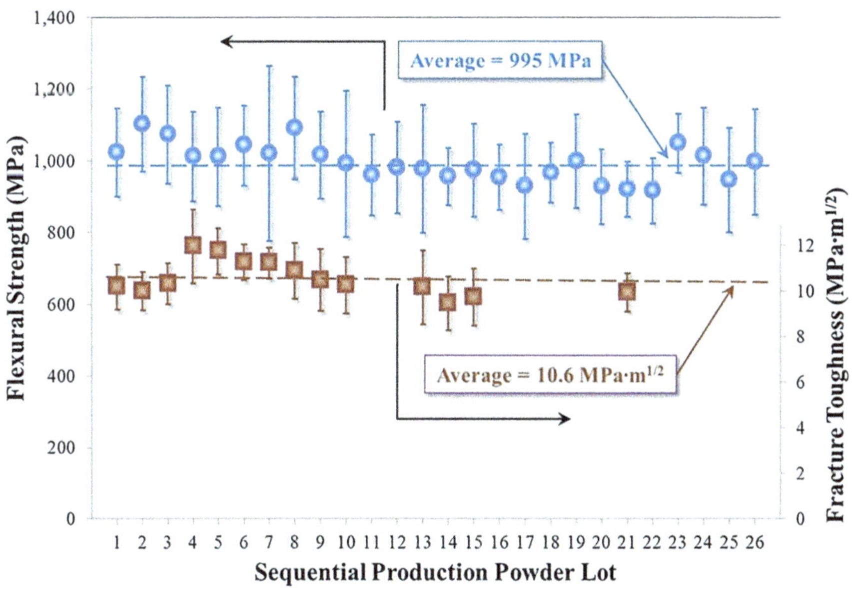

Fig. 11 Flexural strength and fracture toughness results from sequentially produced production Si_3N_4 powder lots. (Used with permission under Creative Commons License 4.0 [132])

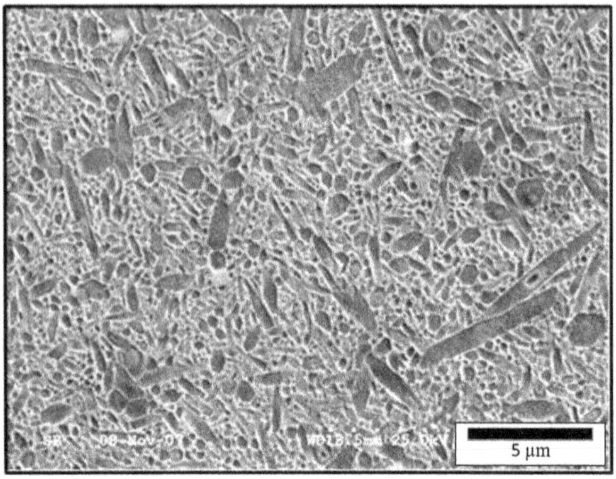

Fig. 12 Representative microstructure of dense Si_3N_4 bioceramic. (Used with permission under Creative Commons License 4.0 [132])

3.5.1 Diamond Grinding

The selection of appropriate diamond tools, grits, feeds, speeds, and lubricants for grinding Si_3N_4 is critical to prevent surface damage which compromises mechanical properties. An early indication of this phenomenon was included in a 1976 report to the US Army on the grinding and finishing of Si_3N_4 bearings. Twenty-seven different finishing procedures were investigated followed by physical and mechanical property assessments [249].

It was found that surface and subsurface damage induced by coarser grinding wheels (120 or 220 versus 320 grit) reduced flexural strengths by up to 45% despite equivalent surface finishes. The researchers documented the use of fine grit diamonds, slow infeed rates, high wheel speeds, and flood coolant conditions as essential parameters for the removal of excess material along with surface and subsurface flaws. At about the same time, Rabinowicz conducted a similar study using abrasive wear to measure surface damage in Si_3N_4. He emphasized the importance of small infeed rates, suggesting a two-step process of rough and finish grinding at no greater than 25 and 12 μm/pass, respectively [250]. Multiple parametric studies followed over the next 50 years [251–256]. Mechanistic studies were concurrently performed [257–259] and follow-on papers confirmed their validity using resin [260], brazed [261], and electroplated [262, 263] diamond wheels. The governing grinding parameters include material hardness, fracture toughness, porosity, workpiece velocity, diamond wheel velocity, grit size and concentration, depth of cut, and grinding direction. General guidance on selection and the influence of these parameters on grinding force and flexural strength is summarized in Table 3.

While traditional diamond grinding continues to be a preferred method for material removal, other techniques have also been developed including ultrasonic, rotary ultrasonic, laser-assisted, water-jet, and electrochemical discharge machining [264–267]. These machining practices may be useful for special component features or to increase overall grinding efficiency.

Table 3 Influence of material and machining parameters on grinding force (f) and post-grinding material strength (σ)

Material or machining parameter	Change in parameter	Effect of parameter change on: Grind. Force (F)	Strength (σ)
Hardness (H)	↑H	↑	–
Fracture Toughness (K_{IC})	↑K_{IC}	↑	↑
Porosity (P)	↓P	↑	↑
Workpiece Velocity (V_W)	↑V_W	↑	↓
Diamond Wheel Velocity (V_D)	↑V_D	↓	↑
Diamond Grit Size (D_G)	↑D_G	↑	↓
Diamond Concentration (D_C)	↑D_C	↑	–
Depth of Cut (d)	↑d	↑	↓
Longitudinal Grinding (F_L)	↓F_L	↓	↑
Transverse Grinding (F_T)	↑F_T	↑	↓

Adapted from Ref. [257]

3.5.2 Lapping, Honing, and Polishing

Lapping, honing, and polishing are free-abrasive machining methods that are used to improve the surface finish of components into the submicron range [126]. Studies performed in 1976 on monolithic Si_3N_4 bearing elements under US Department of Defense contracts showed that an average surface finish of less than 0.008 μm (8 nm or 80 Å) was achievable for hot-pressed NC-132 Si_3N_4 balls using free diamond abrasives in an oil medium [249, 268]. Furthermore, in rolling contact fatigue tests, the Si_3N_4 balls had ≥9.5x the life of comparably finished M50 tool steel balls. These pioneering efforts opened the door for the use of Si_3N_4 in high-speed and heavy-load industrial and aerospace bearings, and these applications remain significant commercial markets for Si_3N_4 to this day. Shortly afterward, it was demonstrated that mechanochemical polishing (CMP) of Si_3N_4 was also effective in producing highly polished surfaces. This process used softer powdered abrasives (e.g., iron oxides, Fe_2O_3, Fe_3O_4, SiO_2, CeO_2, Cr_2O_3) dispersed in an aqueous medium [269]. Mechanistically, the application of pressure and rotational friction coupled with the presence of the abrasive accelerates the naturally occurring chemical reaction (i.e., $Si_3N_4 + 12H_2O \rightarrow 3Si(OH)_4 + 4NH_3$) resulting in the creation of damage and scratch-free surfaces [270–272]. As discussed elsewhere in this volume, this reaction is also the reason Si_3N_4 has become an effective biomaterial. Because of Si_3N_4's growing popularity as a bearing component and its application in semiconductor devices, numerous CMP studies were conducted in the ensuing decades [273–282]. These studies not only provided mechanistic insight into fine polishing but also contributed significant knowledge about Si_3N_4's surface chemistry and reactivity. Fundamentally, nitride hydrolysis releases ammonia with the formation of surface silanols. Sialons are softer and weaker compounds that are readily abraded by the soft abrasive slurry—leaving a fresh ultra-smooth surface for further hydrolysis and removal. Reported optimization of the CMP process resulted in average surface finishes of ≤0.004 μm (4 nm or 40 Å) [274, 277]. However, because CMP material removal rates are low, actual production uses multistep processes. Mechanical abrasion by coarse and intermediate abrasives (e.g., diamond or boron carbide) is followed by CMP [279].

3.5.3 Machining of Si_3N_4 Biomedical Components

Because of Si_3N_4's inherent hardness, fracture toughness, and strength, net-shape forming of biomedical components is always preferred. Spinal spacers are a prime example (see Sect. 3.4.5). However, this isn't always possible. Some medical devices, such as those used in total joint arthroplasty, require machining and polishing. Articulation surfaces need precise features and an ultrafine finish. Several tribological studies and review articles have recognized the potential of Si_3N_4 in these applications [283–286]. Specific studies on the development of Si_3N_4 for hip arthroplasty heads and liners have been conducted [287–290], and components in these in vitro studies were machined and polished using the methods described above. Spherical and surface roughness values of these test parts were typically less than 0.003 mm and 0.01 μm, respectively. As discussed in a later chapter, the wear performance of machined and polished Si_3N_4 femoral heads articulating against polyethylene (PE) was essentially equivalent to other commonly used metal- or ceramic-on-polyethylene couples [289, 290].

3.6 *Inspection and Non-destructive Evaluation (NDE)*

To ensure product consistency, it is essential to establish in-process and final inspection criteria and then conduct routine or periodic testing, Additionally, non-destructive inspection techniques (e.g., proof-testing) can be used to further assure the quality and reliability of individual components. Significant efforts have been directed at developing these important steps for Si_3N_4.

3.6.1 Inspection

A listing of both routine and periodic inspection tests and standards utilized for Si_3N_4 is provided in Table 4. Routine tests conducted on each powder lot or component batch generally include mass density, grain size, flexural strength, fracture toughness, Weibull modulus, and phase composition. These tests are typically performed by each manufacturer. Routine metrology is also conducted by the producer to ensure that component dimensions and surface finish conform to specific customer requirements. The remaining tests listed in Table 4 are conducted periodically or as needed. Third-party services are often utilized for some of these tests.

Table 4 Physical, mechanical, biocompatible, thermal, and electrical properties, and test methods for hot-isostatically pressed Si_3N_4 having a nominal composition of 90.0 wt.% Si_3N_4, 6.0 wt.% Y_2O_3, and 4.0 wt.% Al_2O_3

Property	Units	Test method	Typical[a]	Specification[a]
Mass density	g/cc	ASTM C373-18	3.27	≥3.23
Grain size	μm	ASTM E112-13	0.5 × 5.0	≤25
Flexural strength	MPa	ASTM C1161-18	1000	≥900
Weibull modulus	m	ASTM C1239-13	10	≥6
Elastic modulus	GPa	ASTM C623-21	296	≥290
Poisson's ratio	–	ASTM C623-21	0.27	–
Compressive strength	MPa	ASTM C1424-15	4000	–
Hardness	GPa	ASTM C1327-15	15.0	≥14.3
Fracture toughness	$MPa{\cdot}m^{1/2}$	ASTM C1421-18	10.5	≥9.0
Phase composition	%	JCPDS 82-0697	100% ß-Si_3N_4	≥95% ß-Si_3N_4
Biocompatibility	–	ISO 10993	Pass	Pass
Thermal conductivity	W/(m·°K)	ASTM C408-88	15–30	–
Thermal expansion	1×10^{-6}/°C	ASTM C1300-95	2.2	–
Specific heat	J/(kg·°K)	ASTM C1470-20	170	–
Volume resistivity	ohm-cm	ASTM D1829	10^{12}	–

[a]Typical values and specifications from Ref. [132]

Due to the broad range of compositions and processing, general ASTM, ISO, or country-specific standards have yet to be established for Si_3N_4. However, component-specific standards have either been published or are under consideration. For instance, ASTM F2904 was established in 2008 as the first Si_3N_4 standard for quality assessment testing of Si_3N_4 bearing balls [291]. It defines minimum dimensional and surface finish requirements for three grades of bearing balls and stipulates compositional limits and microstructural anomalies, along with physical and mechanical properties. It also describes or references appropriate testing methods. Since its publication, it has served as a model for the development of other Si_3N_4 standards, one of which is a new draft standard for medical devices (ISO/AWI 18368: "Implants for surgery – Nitride ceramic materials — Monolithic materials made of beta silicon-nitride").

3.6.2 Proof Testing and Non-destructive Evaluation

Proof testing was recognized by early researchers as an important adjunct method of ensuring Si_3N_4 mechanical reliability. Most of the work was directed at truncating the strength distribution of components to remove low-stress failures. This methodology is depicted in Fig. 13 from early work by Wiederhorn and Tighe [292].

The proof test is simple. The procedure subjects 100% of the components to a predetermined load which is above the maximum predicted service stress of the component. It assures that the stress intensity (K_I) during service does not exceed the critical stress intensity (K_{IC}) for components that survive the proof test [293]. Parts

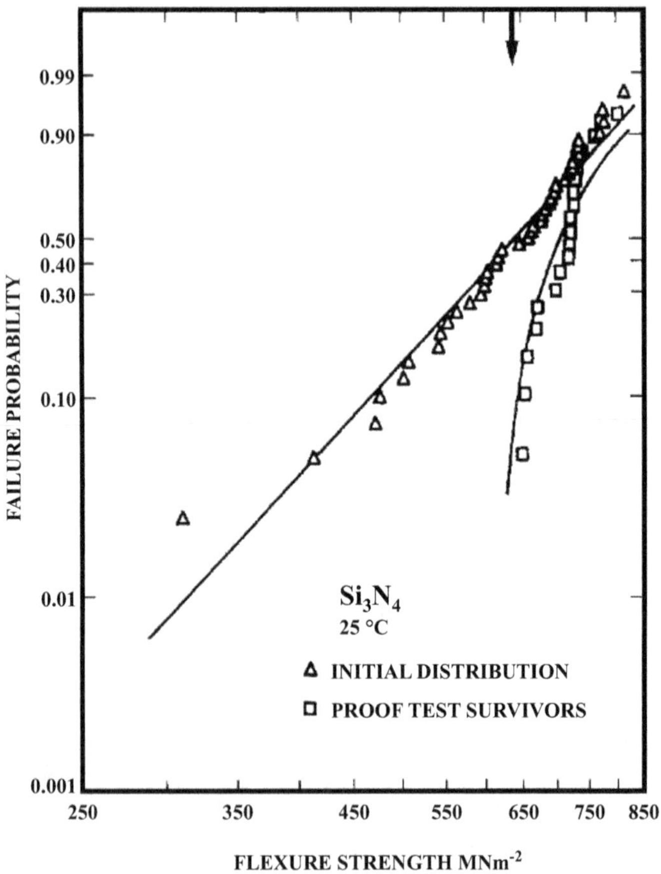

Fig. 13 Proof test conducted at room temperature on hot-pressed Si_3N_4 (NC-132) test bars. The proof test of 650 MPa is indicated by the arrow. (Redrawn and used with permission from Ref. [292])

with large flaws fail and are culled from the batch before their delivery to the customer. Proof-testing was conceptually developed and demonstrated for high-stress engine components during the 1970s and 80s. [292–294]. Although that market never fully materialized, compression proof-testing of ceramics is common for some medical devices. It is particularly pertinent and practiced for femoral heads made from Al_2O_3 and ZrO_2 [295, 296], but the technology is also directly applicable to Si_3N_4 femoral heads. Compression-proof testing coupled with acoustic emission NDE of Si_3N_4 spinal spacers is also routinely practiced [132, 297].

Even though significant advances have been made in other non-destructive evaluation methods, none are accepted as valid in assuring the functionality of ceramic medical devices. Microfocus x-ray radiography (MFXR), ultrasonics, computed tomography, and other techniques have demonstrated detection capabilities for gross defects in simple shapes, but their detectability limits are still insufficient.

4 Summary

As attested by this chapter, the processing and testing of Si_3N_4 bioceramics are multifaceted technologies. Multiple methods of producing high-quality Si_3N_4 raw materials have been developed and commercialized. A broad range of Si_3N_4 compositions with different densification aids have been demonstrated. Raw materials processing along with traditional and advanced forming methods are readily available. Firing conditions have been optimized to produce unique structures that possess a range of properties—from highly porous devices to fully dense high-strength components. Machining, inspection, and proof-testing have been adapted and successfully implemented. Today, more than ever before, Si_3N_4 is destined to become a significant biomaterial with the requisite physical, mechanical, and biological properties to extend both the health span and lifespan of future generations.

References

1. Neumann, A.; Kramps, M.; Ragoß, C.; Maier, H.R.; Jahnke, K. Histological and Microradiographic Appearances of Silicon Nitride and Aluminum Oxide in a Rabbit Femur Implantation Model. *Materwiss. Werksttech.* **2004**, *35*, 569–573, https://doi.org/10.1002/mawe.200400778.
2. Guedes e Silva, C.C.; König, B.; Carbonari, M.J.; Yoshimoto, M.; Allegrini, S.; Bressiani, J.C. Bone Growth Around Silicon Nitride Implants—An Evaluation by Scanning Electron Microscopy. *Mater. Charact.* **2008**, *59*, 1339–1341, https://doi.org/10.1016/j.matchar.2007.11.007.
3. Guedes e Silva, C.C.; Konig, B.; Carbonari, M.J.; Yoshimoto, M.; Allegrini, S.; Bressiani, J.C. Tissue Response Around Silicon Nitride Implants in Rabbits. *J. Biomed. Mater. Res.* **2008**, *84A*, 337–343, https://doi.org/10.1002/jbm.a.
4. Anderson, M.C.; Olsen, R. Bone Ingrowth into Porous Silicon Nitride. *J. Biomed. Mater. Res.* **2010**, *92A*, 1598–1605, https://doi.org/10.1002/jbm.a.32498.
5. Webster, T.J.; Patel, A.A.; Rahaman, M.N.; Bal, B.S. Anti-Infective and Osteointegration Properties of Silicon Nitride, Poly (Ether Ether Ketone), and Titanium Implants. *Acta Biomater.* **2012**, *8*, 4447–4454.
6. Pezzotti, G.; Oba, N.; Zhu, W.; Marin, E.; Rondinella, A.; Boschetto, F.; McEntire, B.J.; Yamamoto, K.; Bal, B.S. Human Osteoblasts Grow Transitional Si/N Apatite in Quickly Osteointegrated Si_3N_4 Cervical Insert. *Acta Biomater.* **2017**, *64*, 411–420.
7. Pezzotti, G.; Bock, R.M.; Adachi, T.; Rondinella, A.; Boschetto, F.; Zhu, W.; Marin, E.; McEntire, B.; Bal, B.S.; Mazda, O. Silicon Nitride Surface Chemistry: A Potent Regulator of Mesenchymal Progenitor Cell Activity in Bone Formation. *Appl. Mater. Today* **2017**, *9*, 82–95, https://doi.org/10.1016/j.apmt.2017.05.005.
8. Pezzotti, G.; Marin, E.; Adachi, T.; Rondinella, A.; Boschetto, F.; Zhu, W.-L.; Sugano, N.; Bock, R.M.; McEntire, B.J.; Bal, B.S. Bioactive Silicon Nitride: A New Therapeutic Material for Osteoarthropathy. *Sci. Rep.* **2017**, *7*, 44848.

9. Mobbs, R.J.; Rao, P.J.; Phan, K.; Hardcastle, P.; Choy, W.J.; McCartney, E.R.; Druitt, R.K.; Mouatt, C.A.L.; Sorrell, C.C. Anterior Lumbar Interbody Fusion Using Reaction Bonded Silicon Nitride Implants: Long Term Case Series of the First Synthetic ALIF Spacer Implanted in Humans. *World Neurosurg.* **2018**, *120*, 256–264, https://doi.org/10.1016/j.wneu.2018.08.237.
10. Pezzotti, G. Silicon Nitride: A Bioceramic with a Gift. *ACS Appl. Mater. Interfaces* **2019**, *11*, 26619–26636, https://doi.org/10.1021/acsami.9b07997.
11. Gorth, D.J.; Puckett, S.; Ercan, B.; Webster, T.J.; Rahaman, M.; Bal, B.S. Decreased Bacteria Activity on Si_3N_4 Surfaces Compared with PEEK or Titanium. *Int. J. Nanomedicine* **2012**, *7*, 4829–4840, https://doi.org/10.2147/IJN.S35190.
12. Bock, R.M.; Jones, E.N.; Ray, D.A.; Bal, B.S.; Pezzotti, G.; McEntire, B.J. Bacteriostatic Behavior of Surface-Modulated Silicon Nitride in Comparison to Polyetheretherketone and Titanium. *J. Biomed. Mater. Res. Part A* **2017**, *105*, 1521–1534, https://doi.org/10.1002/jbm.a.35987.
13. Pezzotti, G.; Bock, R.M.; McEntire, B.J.; Jones, E.; Boffelli, M.; Zhu, W.; Baggio, G.; Boschetto, F.; Puppulin, L.; Adachi, T.; et al. Silicon Nitride Bioceramics Induce Chemically Driven Lysis in Porphyromonas Gingivalis. *Langmuir* **2016**, *32*, 3024–3035, https://doi.org/10.1021/acs.langmuir.6b00393.
14. Pezzotti, G. A Spontaneous Solid-State NO Donor to Fight Antibiotic Resistant Bacteria. *Mater. Today Chem.* **2018**, *9*, 80–90, https://doi.org/10.1016/j.mtchem.2018.05.004.
15. McEntire, B.J.; Bal, B.S.; Pezzotti, G. Antimicrobial Nitric Oxide Releasing Compounds and Scaffolds. In *Antimicrobial Combination Devices*; Urish, K.L., Mihalko, W.M., Eds.; ASTM International: West Conshohocken, PA, USA, 2020; pp. 105–137.
16. Pezzotti, G. Rapid Inactivation of SARS-CoV-2 by Silicon Nitride, Copper, and Aluminum Nitride. *BioRxiv Prepr.* **2020**, 1–16, https://doi.org/10.1101/2020.06.19.159970.
17. Pezzotti, G.; Ohigitani, E.; Shin-Ya, M.; Adachi, T.; Marin, E.; Boshetto, F.; Zhu, W.; Mazda, O. Instantaneous "Catch-and-Kill" Inactivation of SARS-CoV-2 by Nitride Ceramics. *Clin. Transl. Med.* **2020**, 1–4, https://doi.org/10.1002/ctm2.212.
18. Lehman, C.W.; Flur, R.; Kehn-hall, K.; McEntire, B.J.; Bal, B.S.; Bock, R.M. Silicon Nitride Inactivates SARS-CoV-2 *in vitro*. *BioRxiv* **2020**, 1–11, https://doi.org/10.1101/2020.08.29.271015.
19. Pezzotti, G.; Boschetto, F.; Ogitani, E.; Fujita, Y.; Zhu, W.; Marin, E.; McEntire, B.J.; Bal, B.S.; Mazda, O. Silicon Nitride: A Potent Solid-State Bioceramic Inactivator of ssRNA Viruses. *Sci. Rep.* **2021**, *11*, 2977, https://doi.org/10.1038/s41598-021-82608-3.
20. Pezzotti, G.; Boschetto, F.; Ohgitani, E.; Fujita, Y.; Shin-Ya, M.; Adachi, T.; Yamamoto, T.; Kanamura, N.; Marin, E.; Zhu, W.; et al. Mechanisms of Instantaneous Inactivation of SARS-CoV-2 by Silicon Nitride Bioceramic. *Mater. Today Bio* **2021**, 100144, https://doi.org/10.1016/j.mtbio.2021.100144.
21. Pezzotti, G.; Fujita, Y.; Boschetto, F.; Zhu, W.; Marin, E.; Vandelle, E.; McEntire, B.J.; Bal, B.S.; Giarola, M.; Makimura, K.; et al. Activity and Mechanism of Action of the Bioceramic Silicon Nitride as an Environmentally Friendly Alternative for the Control of the Grapevine Downy Mildew Pathogen Plasmopara viticola. *Front. Microbiol.* **2020**, *11*, 1–17, https://doi.org/10.3389/fmicb.2020.610211.
22. Pezzotti, G.; Asai, T.; Adachi, T.; Ogitani, E.; Yamamoto, T.; Kanamura, N.; Boschetto, F.; Zhu, W.; Zanocco, M.; Marin, E.; et al. Antifungal Activity of Polymethyl Methacrylate/Si_3N_4 Composites against Candida Albicans. *Acta Biomater.* **2021**, *126*, 259–276, https://doi.org/10.1016/j.actbio.2021.03.023.
23. McEntire, B.J.; Bal, B.S.; Rahaman, M.N.; Chevalier, J.; Pezzotti, G. Ceramics and Ceramic Coatings in Orthopaedics. *J. Eur. Ceram. Soc.* **2015**, *35*, 4327–4369, https://doi.org/10.1016/j.jeurceramsoc.2015.07.034.
24. Sangster, R.C. *Formation of Silicon Nitride from the 19th to the 21st Century: A Comprehensive Summary and Guide to the World Literature*; Trans Tech Publications: Geneva, Switzerland, 2015; Vol. 84.

25. Tsuji, I.; Aoki, S.; Miyazaki, S.; Hara, Y.; Furuse, Y.; Hamada, S. Cascade Testing of Ceramic Vanes for Industrial Gas Turbines. In Proceedings of the International Gas Turbine and Aeroengine Congress and Exposition; ASME Press: New York City, NY USA, 1991; Vol. 78989, pp. 91-GT-311.
26. McEntire, B.J.; Taglialavore, A.P.; Heichel, D.N.; Bright, E.; Yeckley, R.L.; Holoczak, J.E.; Quackenbush, C.L. Silicon Nitride Component Development for Advanced Gas Turbine Engines. In Proceedings of the Proceedings of the Annual Automotive Technology Development Contractors' Coordination Meeting; SAE Press: Dearborn, MI USA, 1990; pp. 341–356.
27. Hengst, R.R.; Heichel, D.N.; Holowczak, J.E.; Taglialavore, A.P.; McEntire, B.J. A Comparison of Forming Technologies for Ceramic Gas-Turbine Engine Components. In Proceedings of the 35th International Gas Turbine and Aeroengine Congress and Exposition, 35th, Brussels, Belgium, June 11–14, 1990; ASME Press: Brussels, Belgium, 1990; p. PAPER 90-GT-184.
28. McEntire, B.J.; Hengst, R.R.; Collins, W.T.; Taglialavore, A.P.; Yeckley, R.L.; Bright, E.; Bingham, M.G. Ceramic Component Processing Development for Advanced Gas-Turbine Engines. In Proceedings of the Proceedings of the Annual Meeting of the Automotive Technology Development Contractors' Coordination Meeting; ASME Press: Dearborn, MI USA, 1991; pp. 91-GT-120.
29. Taglialavore, A.P.; Hengst, R.R.; Bright, E.; Lynch, L.D.; McEntire, B.J. Advanced Ceramics for Gas Turbine Applications. In Proceedings of the Proceedings of the Annual Automotive Technology Development Contractors' Coordination Meeting; SAE Press: Dearborn, MI USA, 1992; pp. 71–82.
30. McEntire, B.J.; Hengst, R.R.; Collins, W.T.; Taglialavore, A.P.; Yeckley, R.L. Ceramic Component Processing Development for Advanced Gas Turbine Engines. *J. Eng. Gas Turbines Power* **1993**, *115*, 1–8.
31. McEntire, B.J.; Wills, R.W.; Southam, R.E. *The Development and Testing of Ceramic Components in Piston Engines, Final Report, Prepared for the U.S. Department of Energy under Contract with Oak Ridge National Laboratories, Contract No. DE-AC05-840R21400*; Oak Ridge, TN 37831-6285, 1994;
32. Bright, E.; Eckalbar, J.F.; McEntire, B.J.; Pujari, V.K.; Tricard, M. Advanced Ceramic Manufacturing of SiALON Exhaust Valves. *Adv. Diesel Engine Compon. Des.* **1996**, 19–30, Paper # 960051, https://doi.org/10.4271/960051.
33. Mikijelj, B.; Mangels, J. SRBSN Material Development for Automotive Applications. In Proceedings of the Seventh International Symposium on Ceramic Materials and Components for Engines; 2000; pp. 1–6.
34. Mikijelj, B.; Mangels, J.; Belfield, E.; MacQueen, A. Silicon Nitride Applications in Modern Diesel Engines. *SAE Int.* **2004**, 2004-01–1448, https://doi.org/10.4271/2004-01-1448.
35. Riley, F.L. Silicon Nitride and Related Materials. *J. Am. Ceram. Soc.* **2000**, *83*, 245–265.
36. Nittler, L.R.; Hoppe, P.; Alexander, C.M.O.D.; Amari, S.; Eberhardt, P.; Gao, X.; Lewis, R.S.; Strebel, R.; Walker, R.M.; Zinner, E. Silicon Nitride from Supernovae. *Astrophys. J.* **1995**, *453*, L25, https://doi.org/10.1086/309743.
37. Schwier, G.; Nietfeld, G.; Franz, G. Production and Characterization of Silicon Nitride Powders. *Mater. Sci. Forum* **1989**, *47*, 1–20, https://doi.org/10.4028/www.scientific.net/msf.47.1.
38. Lange, H.; Wötting, G.; Winter, G. Silicon Nitride—From Powder Synthesis to Ceramic Materials. *Angew. Chemie Int. Ed. English* **1991**, *30*, 1579–1597, https://doi.org/10.1002/anie.199115791.
39. Rostislav A Andrievskii Silicon Nitride: Synthesis and Properties. *Russ. Chem. Rev.* **1995**, *64*, 291, https://doi.org/10.1070/RC1995v064n04ABEH000151.
40. Petzow, G.; Herrmann, M. Silicon Nitride Ceramics. In *High Performance Non-Oxide Ceramics II SE – 2*; Jansen, M., Ed.; Structure and Bonding; Springer Berlin Heidelberg, 2002; Vol. 102, pp. 47–167 ISBN 978-3-540-43132-9.

41. Itoh, T. Preparation of Pure α-Silicon Nitride from Silicon Powder. *J. Mater. Sci. Lett.* **1991**, *10*, 19–20, https://doi.org/10.1007/BF00724419.
42. Mitomo, M. Effect of Fe and Al Additions on Nitridation of Silicon. *J. Mater. Sci.* **1977**, *12*, 273–276, https://doi.org/10.1007/BF00566267.
43. Boyer, S.M.; Moulson, A.J. A Mechanism for the Nitridation of Fe-Contaminated Silicon. *J. Mater. Sci.* **1978**, *13*, 1637–1646, https://doi.org/10.1007/BF00548727.
44. Atkinson, A.; Moulson, A.J.; Roberst, E.W. Nitridation of High-Purity Silicon. *J. Am. Ceram. Soc.* **1976**, *59*, 285–289.
45. Rahaman, M.N.; Moulson, A.J. The Removal of Surface Silica and its Effect on the Nitridation of High-Purity Silicon. *J. Mater. Sci.* **1984**, *19*, 189–194, https://doi.org/10.1007/BF00553008.
46. Pasto, A.E. Causes and Effects of Fe-Bearing Inclusions in Sintered Si_3N_4. *J. Am. Ceram. Soc.* **1984**, *67*, C-178-C-180, https://doi.org/10.1111/j.1151-2916.1984.tb19612.x.
47. Pavarajarn, V.; Kimura, S. Catalytic Effects of Metals on Direct Nitridation of Silicon. *J. Am. Ceram. Soc.* **2001**, *84*, 1669–1674, https://doi.org/10.1111/j.1151-2916.2001.tb00897.x.
48. Pavarajarn, V.; Vongthavorn, T.; Praserthdam, P. Enhancement of Direct Nitridation of Silicon by Common Metals in Silicon Nitride Processing. *Ceram. Int.* **2007**, *33*, 675–680, https://doi.org/10.1016/j.ceramint.2005.12.004.
49. Hyuga, H.; Zhou, Y.; Kusano, D.; Hirao, K.; Kita, H. Nitridation Behaviors of Silicon Powder Doped with Various Rare Rarth Oxides. *J. Ceram. Soc. Japan* **2011**, *119*, 251–253, https://doi.org/10.2109/jcersj2.119.251.
50. Yang, F.; Wang, L.; Yin, S.; Li, Y.; Liu, C.; Tong, L. Effect of the Addition of α-Si_3N_4 on the Direct Nitridation of Silicon Powder at Atmospheric Pressure. *Adv. Mater. Res.* **2012**, *486*, 247–252, https://doi.org/10.4028/www.scientific.net/AMR.486.247.
51. Yin, S.; Wang, L.; Tong, L.; Yang, F.; Li, Y. Kinetic Study on the Direct Nitridation of Silicon Powders Diluted with α-Si_3N_4 at Normal Pressure. *Int. J. Miner. Metall. Mater.* **2013**, *20*, 493–498, https://doi.org/10.1007/s12613-013-0756-x.
52. Jin, X.; Xing, P.; Zhuang, Y.; Kong, J.; Jiang, S.; Wei, D. Effect of Si_3N_4 Diluent on Direct Nitridation of Silicon Powder. *Ceram. Int.* **2019**, *45*, 10943–10950, https://doi.org/10.1016/j.ceramint.2019.02.175.
53. Lange, F.F. Dense Silicon Nitride Ceramics: Fabrication and Interrelations with Properties. In *Processing of Crystalline Ceramics*; Palmour, H., Davis, R.F., Hare, T.M., Eds.; Springer US: Boston, MA, 1978; pp. 597–614 ISBN 978-1-4684-3378-4.
54. Riley, F.L. Silicon Nitridation. In *Progress in Nitrogen Ceramics*; Riley, F.L., Ed.; Springer Netherlands: Dordrecht, 1983; pp. 121–133 ISBN 978-94-009-6851-6.
55. Herbell, T.P.; Glasgow, T.K.; Yeh, H.C. Effect of Attrition Milling on the Reaction Sintering of Silicon Nitride. *NASA Tech. Memo. TM-78965* **1978**, 1–20.
56. Herbell, T.P.; Glasgow, T.K. Bend Strengths of Reaction Bonded Silicon Nitride Prepared from Dry Attrition Milled Silicon Powder. *NASA Tech. Memo. TM-79230* **1979**, 1–16.
57. Yeh, H.C.; Glasgow, T.K.; Herbell, T.P. Effect of Milling and Leaching on the Structure of Sintered Silicon. *NASA Tech. Memo. TM-81602* **1980**, 1–17.
58. Herbell, T.P.; Glasgow, T.K.; Shaw, N.J. Reaction Bonded Silicon Nitride Prepared from Wet Attrition-Milled Silicon. *NaSA Tech. Memo. TM-81428* **1980**, 1–22.
59. Herbell, T.P.; Glasgow, T.K.; Orth, N.W. Demonstration of a Silicon Nitride Attrition Mill for Production of Fine Pure Si and Si_3N_4 Powders. *Am. Ceram. Soc. Bull.* **1984**, *63*, 104–107.
60. Herbell, T.P.; Freedman, M.R.; Kiser, J.D.; Beach, C.; Sa, I.J.I. Particle Size Reduction of Si_3N_4 Powder With Si_3N_4 Milling Hardware. *NASA Tech. Memo. TM-86864* **1986**, 1–20.
61. Rabe, T.; Linke, D. Attrition Milling of Silicon Nitride Powder Under Conditions for Minimal Impurity Pickup. *Ceram. Int.* **1992**, *18*, 161–166, https://doi.org/10.1016/0272-8842(92)90090-Z.
62. Dai, J.-Q.; Huang, Y.; Xie, Z.-P.; Xu, X.-L.; Yang, J.-L. Effect of Acid Cleaning and Calcination on Rheological Properties of Concentrated Aqueous Suspensions of Silicon

Nitride Powder. *J. Am. Ceram. Soc.* **2002**, *85*, 293–298, https://doi.org/10.1111/j.1151-2916.2002.tb00087.x.
63. Hermansson, L.A.G.; Burstrom, M.; Burströ, M.; Johansson, T.; Hatcher, M.E. High-Temperature Behavior of Chemically Treated Silicon Nitride Powders. *J. Am. Ceram. Soc.* **1988**, *71*, C183–C184, https://doi.org/10.1111/j.1151-2916.1988.tb05868.x.
64. Liu, X.J.; Huang, L.P.; Xu, X.; Gu, H.C.; Fu, X.R. Effect of Acid Leaching on Rheological Behavior of Silicon Nitride Aqueous Suspension. *J. Mater. Sci. Lett.* **2000**, *19*, 177–178, https://doi.org/10.1023/A:1006732719196.
65. Dai, J.-Q.; Huang, Y.; Ma, J.-T. Surface Characteristics and Aqueous Dispersibility of Commercial Silicon Nitride Powders: Effects of Acid Leaching, Surface Hydrolysis, and Thermal Oxidation. *Ceram. Int.* **2008**, *34*, 1835–1842, https://doi.org/10.1016/j.ceramint.2007.06.011.
66. Melner, H. Verfahren zur Darstellung von Nitriden. *Ger. Pat. 88 999* **1896**.
67. Komeya, K.; Inoue, H. Synthesis of the α Form of Silicon Nitride from Silica. *J. Mater. Sci.* **1975**, *10*, 1243–1246, https://doi.org/10.1007/BF00541410.
68. Inoue, H.; Komeya, K.; Tsuge, A. Synthesis of Silicon Nitride Powder from Silica Reduction. *J. Am. Ceram. Soc.* **1982**, *65*, C-205-C-205, https://doi.org/10.1111/j.1151-2916.1982.tb09951.x.
69. Mori, M.; Inoue, H.; Ochiai, T. Preparation of Silicon Nitride Powder from Silica. In *Progress in Nitrogen Ceramics*; Riley, F.L., Ed.; Springer Netherlands: Dordrecht, 1983; pp. 149–156 ISBN 978-94-009-6851-6.
70. Lee, J.G.; Miller, P.D.; Cutler, I.B. Carbothermal Reduction of Silica. In *Reactivity of Solids*; Wood, J., Lindqvist, O., Helgesson, C., Vannerberg, N.-G., Eds.; Springer US: Boston, MA, 1977; pp. 707–711 ISBN 978-1-4684-2340-2.
71. Lee, J.; Cutler, I.B. Sinterable Sialon Powder by Reaction of Clay with Carbon and Nitrogen. *Bull. Am. Ceram. Soc.* **1979**, *58*, 869–871.
72. Zhang, S.-C.; Cannon, W.R. Preparation of Silicon Nitride from Silica. *J. Am. Ceram. Soc.* **1984**, *67*, 691–695, https://doi.org/10.1111/j.1151-2916.1984.tb19684.x.
73. Van Dijen, F.K.; Metselaar, R.; Siskens, C.A.M. Reaction-Rate-Limiting Steps in Carbothermal Reduction Processes. *J. Am. Ceram. Soc.* **1985**, *68*, 16–19, https://doi.org/10.1111/j.1151-2916.1985.tb15244.x.
74. Durham, S.J.P.; Shanker, K.; Drew, R.A.L. Process Parameters for the Production of Silicon Nitride by Carbothermal Reduction. In *Proceedings of Metallurgical Society of Canadian Institute of Mining and Metallurgy*; Plumpton, A.J.B.T.-P. and P. of F.P., Ed.; Pergamon: Amsterdam, 1988; pp. 549–557 ISBN 978-0-08-036448-3.
75. Cho, Y.W.; Charles, J.A. Synthesis of Nitrogen Ceramic Powders by Carbothermal Reduction and Nitridation Part 1 Silicon Nitride. *Mater. Sci. Technol.* **1991**, *7*, 289–298, https://doi.org/10.1179/mst.1991.7.4.289.
76. Durham, S.J.P.; Shanker, K.; Drew, R.A.L. Carbothermal Synthesis of Silicon Nitride: Effect of Reaction Conditions. *J. Am. Ceram. Soc.* **1991**, *74*, 31–37.
77. Ličko, T.; Figusch, V.; Púchyová, J. Synthesis of Silicon Nitride by Carbothermal Reduction and Nitriding of Silica: Control of Kinetics and Morphology. *J. Eur. Ceram. Soc.* **1992**, *9*, 219–230, https://doi.org/10.1016/0955-2219(92)90008-2.
78. Ekelund, M.; Forslund, B. Carbothermal Preparation of Silicon Nitride: Influence of Starting Material and Synthesis Parameters. *J. Am. Ceram. Soc.* **1992**, *75*, 532–539, https://doi.org/10.1111/j.1151-2916.1992.tb07838.x.
79. Hofmann, H.; Vogt, U.; Kerber, A.; van Dijen, F. Silicon Nitride Powder from Carbothermal Reaction. *MRS Online Proc. Libr.* **1992**, *287*, 105–120, https://doi.org/10.1557/PROC-287-105.
80. Cochran, G.A.; Conner, C.L.; Eismann, G.A.; Weimer, A.W.; Carroll, D.F.; Dummead, S.D.; Hwang, C.J. The Synthesis of a High Quality, Low Cost Silicon Nitride Powder by the Carbothermal Reduction of Silica. *Key Eng. Mater.* **1994**, *89–91*, 3–8, https://doi.org/10.4028/www.scientific.net/KEM.89-91.3.

81. Koc, R.; Kaza, S. Synthesis of α-Si_3N_4 from Carbon Coated Silica by Carbothermal Reduction and Nitridation. *J. Eur. Ceram. Soc.* **1998**, *18*, 1471–1477, https://doi.org/10.1016/S0955-2219(98)00005-3.
82. Weimer, A.W.; Eisman, G.A.; Susnitzky, D.W.; Beaman, D.R.; McCoy, J.W. Mechanism and Kinetics of the Carbothermal Nitridation Synthesis of α-Silicon Nitride. *J. Am. Ceram. Soc.* **1997**, *80*, 2853–2863, https://doi.org/10.1111/j.1151-2916.1997.tb03203.x.
83. Liou, T.-H.; Chang, F.-W. Kinetics of Carbothermal Reduction and Nitridation of Silicon Dioxide/Carbon Mixture. *Ind. Eng. Chem. Res.* **1995**, *34*, 118–127, https://doi.org/10.1021/ie00040a009.
84. Kang, H.; Komeya, K.; Meguro, T.; Natto, M.; Hayakawa, O. Effect of Silicon Nitride Seeds Addition on the Particle Size and the Crystal Form of Resulting Powder in Carbothermal Reduction-Nitridation of Silica. *J. he Ceram. Soc. Japan* **1996**, *104*, 471–475.
85. Siddiqi, S.A.; Hendry, A. The influence of Iron on the Preparation of Silicon Nitride from Silica. *J. Mater. Sci.* **1985**, *20*, 3230–3238, https://doi.org/10.1007/BF00545189.
86. Weimer, A.W. From a Laboratory Curiosity to a Commercial Powder Processing Plant – A Personal Perspective. *Powder Technol.* **2023**, *418*, 118279, https://doi.org/10.1016/j.powtec.2023.118279.
87. Ortega, A.; Alcalá, M.D.; Real, C. Carbothermal Synthesis of Silicon Nitride (Si_3N_4): Kinetics and Diffusion Mechanism. *J. Mater. Process. Technol.* **2008**, *195*, 224–231, https://doi.org/10.1016/j.jmatprotec.2007.05.004.
88. Real, C.; Alcalá, M.D.; Criado, J.M. Synthesis of Silicon Nitride from Carbothermal Reduction of Rice Husks by the Constant-Rate-Thermal-Analysis (CRTA) Method. *J. Am. Ceram. Soc.* **2004**, *87*, 75–78, https://doi.org/10.1111/j.1551-2916.2004.00075.x.
89. Demir, A.; Tatli, Z.; Caliskan, F.; Kurt, A.O. Carbothermal Reduction and Nitridation of Quartz Mineral for the Production of Alpha Silicon Nitride Powders. *Mater. Sci. Forum* **2007**, *554*, 163–168, https://doi.org/10.4028/www.scientific.net/MSF.554.163.
90. Alcalá, M.D.; Criado, J.M.; Real, C. Preparation of Si_3N_4 from Carbothermal Reduction of SiO Employing the CRTA Method. *Mater. Sci. Forum* **2001**, *383*, 25–30, https://doi.org/10.4028/www.scientific.net/MSF.383.25.
91. Luyjew, K.; Tonanon, N.; Pavarajarn, V. Mesoporous Silicon Nitride Synthesis Via the Carbothermal Reduction and Nitridation of Carbonized Silica/RF Gel Composite. *J. Am. Ceram. Soc.* **2008**, *91*, 1365–1368, https://doi.org/10.1111/j.1551-2916.2008.02285.x.
92. Chung, S.-L.; Chang, C.-W. Carbotheral Reduction and Nitridation Synthesis of Silicon Nitride by Using Solution Combustion Synthesized Precursors. *J. Mater. Sci.* **2009**, *44*, 3784–3792, https://doi.org/10.1007/s10853-009-3509-5.
93. Jiang, J. A New Synthesis Method of α-Silicon Nitride Powder-Reductive Combustion Synthesis from Silicon and Silicon Dioxide. *J. Am. Ceram. Soc.* **2009**, *92*, 3095–3097, https://doi.org/10.1111/j.1551-2916.2009.03314.x.
94. Li, B.; Feng, Y.; Li, G.; Chen, H.; Chen, J.; Hou, X. Preparation of High-Purity α-Si_3N_4 Nano-Powder by Precursor-Carbothermal Reduction and Nitridation. *Ceram. Int.* **2019**, *45*, 6335–6339, https://doi.org/10.1016/j.ceramint.2018.12.118.
95. Wang, Y.; Wu, H.; Jia, B.; Zhang, D.; Zhang, Y.; Zhang, Z.; Liu, C.; Tian, J.; Qin, M.; Qu, X. Synthesis of Monodisperse and High-Purity α-Si_3N_4 Powder by Carbothermal Reduction and Nitridation. *Adv. Powder Technol.* **2021**, *32*, 3101–3106, https://doi.org/10.1016/j.apt.2021.06.023.
96. Shan, S.-Y.; Yang, J.-F.; Gao, J.-Q.; Zhang, W.-H.; Jin, Z.-H.; Janssen, R.; Ohji, T. Porous Silicon Nitride Ceramics Prepared by Reduction–Nitridation of Silica. *J. Am. Ceram. Soc.* **2005**, *88*, 2594–2596, https://doi.org/10.1111/j.1551-2916.2005.00444.x.
97. Shan, S.-Y.; Yang, J.-F.; Lu, Y.; Gao, J.-Q.; Niihara, K. Effects of Carbon Source on Microstructural Evolution and Sintering Behavior of Porous Silicon Nitride Ceramics. *Scr. Mater.* **2007**, *56*, 193–196, https://doi.org/10.1016/j.scriptamat.2006.10.012.

98. Shan, S.; Jia, Q.; Jiang, L.; Wang, Y.; Yang, J. Microstructure Control and Mechanical Properties of Porous Silicon Nitride Ceramics. *Ceram. Int.* **2009**, *35*, 3371–3374, https://doi.org/10.1016/j.ceramint.2009.06.009.
99. Wang, K.; Wang, H.; Cheng, Y.B. Large-Scale Synthesis of α-Si_3N_4 Nanofibers and Nanobelts from Mesoporous Silica-Carbon Nanocomposites. *J. Ceram. Sci. Technol.* **2017**, *8*, 259–264, https://doi.org/10.4416/JCST2016-00118.
100. Hu, F.; Xie, Z.-P.; Zhang, J.; Hu, Z.-L.; An, D. Promising High-Thermal-Conductivity Substrate Material for High-Power Electronic Device: Silicon Nitride Ceramics. *Rare Met.* **2020**, *39*, 463–478, https://doi.org/10.1007/s12598-020-01376-7.
101. Wang, B.; Huang, X.; Zhou, X.-N.; Zhi, Q.; Hao, L.-C.; Li, Z.-X.; Zhao, S.; Hou, B.-Q.; Yang, J.-F.; Ishizaki, K. Synthesis and Carbothermal Nitridation Mechanism of Ultra-Long Single Crystal α-Si_3N_4 Nanobelts. *Nanotechnology* **2020**, *31*, 194001, https://doi.org/10.1088/1361-6528/ab6fd7.
102. Satoh, S. Silicon Amide and Silicon Nitride Production. *Inst. phys. chem. Res* **1938**, *34*, 144–149.
103. Mazdiyasni, K.S.; Cooke, C.M. Synthesis, Characterization, and Consolidation of Si_3N_4 Obtained from Ammonolysis of SiCl4. *J. Am. Ceram. Soc.* **1973**, *56*, 628–633, https://doi.org/10.1111/j.1151-2916.1973.tb12440.x.
104. Mazdiyasni, K.S.; Cooke, C.M. Synthesis of High Purity, Alpha Phase Silicon Nitride Powder. *US Pat. 3,959,446* **1976**, 1–5.
105. Kasai, K.; Tsukidate, T.; Arakawa, T. Process for Preparation of High-Type Silicon Nitride Powder. *US Pat. 4,716,028* **1987**, 1–4.
106. Beckwith, E.C.; Williams, R.M. Method of Making Ultrapure Silicon Nitride Precursor. *US Pat. 4,686,095* **1987**, 1–7.
107. Crosbie, G.M.; Predmesky, R.L. Method of Making High Purity Silicon Nitride Precursor. *US Pat. 4,732,746* **1988**, 1–11.
108. Crosbie, G.M.; Predmesky, R.L.; Nicholson, J.M.; Stiles, E.D. Prepilot Scale Synthesis of Silicon Nitride under Pressure. *Am. Ceram. Soc. Bull.* **1989**, *68*, 1010–1014.
109. Tetsuo Yamada Preparation and Evaluation of Sinterable Silicon Nitride Powder by Imide Decomposition Method. *Am. Ceram. Soc. Bull.* **1993**, *72*, 99–106.
110. Prochazka, S.; Geskovich, C.D. Preparation of Silicon Nitride Powder. *US Pat. 4,122,155* **1978**, 1–4.
111. Kasai, K.; Kubota, Y.; Tsukidate, T. Synthesis of α-Si_3N_4 Powders. *J. Ceram. Assoc. Japan* **1980**, *88*, 62–63.
112. Iwai, T.; Kawahito, T.; Yamada, T. Process for Producing Metallic Nitride Powder. *US Pat. 4,196,178* **1980**, 1–6.
113. Kasai, K.; Tsukuma, K.; Tsukidate, T. Method of Manufacturing High-Purity Silicon Nitride Powder. *US Pat. 4,387,079* **1983**, 1–6.
114. Iwai, T.; Kawahito, T.; Yamada, T. Process for Producing Silicon Nitride Powder. *US Pat. 4,405,589* **1983**, 1–7.
115. Yamada, T.; Kawahito, T.; Iwai, T. Crystallization of Amorphous Si_3N_4 Prepared by the Thermal Decomposition of Si(NH)2. *J. Mater. Sci. Lett.* **1983**, *2*, 275–278, https://doi.org/10.1007/BF00723254.
116. Kasai, K.; Nagata, S.; Arakawa, T.; Tsukidate, T. Synthesis of Si_3N_4 Powder by Thermal Decomposition of Si(NH)2 and Sintering Properties. In *Proceedings of the Conference on Raw Materials for Advanced and Engineered Ceramics: Ceramic Engineering and Science Proceedings*; Ceramic Engineering and Science Proceedings; 1985; pp. 1278–1288 ISBN 9780470320297.
117. Kawahito, T.; Himukashi, Y.; Yamada, T. Grinding Amorphous Silicon Nitride Powder Prior to Heating to Produce Crystalline Silicon Nitride Powder. *US Pat. 4,517,168* **1985**, 1–7.
118. NIST-JANAF Thermochemical Tables. https://janaf.nist.gov/ **2023**.

119. Hampshire, S.; Pomeroy, M.J. Grain Boundary Glasses in Silicon Nitride: A Review of Chemistry, Properties and Crystallisation. *J. Eur. Ceram. Soc.* **2012**, *32*, 1925–1932, https://doi.org/10.1016/j.jeurceramsoc.2011.12.023.
120. Hampshire, S. Silicon Nitride Ceramics – Review of Structure, Processing and Properties. *J. Achiev. Mater. Manuf. Eng.* **2007**, *24*, 43–50.
121. Zhu, W.; McEntire, B.J.; Enomoto, Y.; Boffelli, M.; Pezzotti, G. Point-Defect Populations as Induced by Cation/Anion Substitution in β-Si_3N_4 Lattice – A Cathodoluminescence Study. *J. Phys. Chem. C* **2015**, *119*, 3279–3287, https://doi.org/10.1021/jp509995c.
122. Ziegler, G.; Heinrich, J.; Wötting, G. Relationships Between Processing, Microstructure and Properties of Dense and Reaction-Bonded Silicon Nitride. *J. Mater. Sci.* **1987**, *22*, 3041–3086, https://doi.org/10.1007/BF01161167.
123. Greil, P. Processing of Silicon Nitride Ceramics. *Mater. Sci. Eng. A* **1989**, *109*, 27–35, https://doi.org/10.1016/0921-5093(89)90561-3.
124. Pasto, A.E. Synthesis/Processing of Silicon Nitride Ceramics. In *Comprehensive Hard Materials*; Sarin, V.K.B.T.-C.H.M., Ed.; Elsevier: Oxford, 2014; pp. 73–88 ISBN 978-0-08-096528-4.
125. Onoda Jr., G.Y.; Hench, L.L. *Ceramic Processing Before Firing*; John Wiley & Sons: New York City, NY USA, 1978; ISBN 0-471-65410-8.
126. Reed, J.S. Principles of Ceramic Processing. **1995**, 1–658.
127. International, A. Engineered Materials Handbook, Ceramics and Glasses. *Eng. Mater. Handb.* 1991.
128. Rahaman, M.N. *Ceramic Processing*; 2nd Editio.; CRC Press, 2017; ISBN 9781498716413.
129. Bocanegra-Bernal, M.H.; Matovic, B. Dense and Near-Net-Shape Fabrication of Si_3N_4 Ceramics. *Mater. Sci. Eng. A* **2009**, *500*, 130–149, https://doi.org/10.1016/j.msea.2008.09.015.
130. Naito, N.; DeJonghe, L.C.; Rahaman, M.N. Pore Size Distribution During Compaction and Early Stage Sintering of Silicon Nitride. *J. Mater. Sci.* **1990**, *25*, 1686–1689.
131. McEntire, B.J. Dry Pressing. In *Engineered Materials Handbook*; Schneider, S.J.J., Ed.; ASM International: Metals Park, OH, USA, 1991; Vol. 4, pp. 141–146.
132. McEntire, B.J.; Lakshminarayanan, R.; Thirugnanasambandam, P.; Seitz-Sampson, J.; Bock, R.; O'Brien, D. Processing and Characterization of Silicon Nitride Bioceramics. *Bioceram. Dev. Appl.* **2016**, *6*, 1000093, https://doi.org/10.4172/2090-5025.100093.
133. Grigoriev, S.N.; Soe, T.N.; Malakhinsky, A.; Makhadilov, I.; Romanov, V.; Kuznetsova, E.; Smirnov, A.; Podrabinnik, P.; Khmyrov, R.; Solís Pinargote, N.W.; et al. Granulation of Silicon Nitride Powders by Spray Drying: A Review. *Materials (Basel).* **2022**, *15*, 1–23, https://doi.org/10.3390/ma15144999.
134. Albano, M.P.; Garrido, L.B. Processing of Concentrated Aqueous Silicon Nitride Slips by Slip Casting. *J. Am. Ceram. Soc.* **1998**, *81*, 837–844, https://doi.org/10.1111/j.1151-2916.1998.tb02417.x.
135. Senturk, U.; Timucin, M. Slip-Casting Properties of Si_3N_4 with Y_2O_3 and Al_2O_3 as Sintering Additives. *J. Mater. Sci.* **1998**, *33*, 1881–1885, https://doi.org/10.1023/A:1004361506723.
136. Moreno, R.; Salomoni, A.; Mello Castanho, S. Colloidal Filtration of Silicon Nitride Aqueous Slips. Part I: Optimization of the Slip Parameters. *J. Eur. Ceram. Soc.* **1998**, *18*, 405–416, https://doi.org/10.1016/S0955-2219(97)00133-7.
137. Moreno, R.; Salomoni, A.; Stamenkovic, I.; Castanho, S.M. Colloidal Filtration of Silicon Nitride Aqueous Slips, Part II: Slip Casting and Pressure Casting Performance. *J. Eur. Ceram. Soc.* **1999**, *19*, 49–59, https://doi.org/10.1016/S0955-2219(98)00137-X.
138. Penas, O.; Zenati, R.; Dubois, J.; Fantozzi, G. Processing, Microstructure, Mechanical Properties of Si_3N_4 Obtained by Slip Casting and Pressureless Sintering. *Ceram. Int.* **2001**, *27*, 591–596, https://doi.org/10.1016/S0272-8842(01)00006-2.
139. Park, D.S.; Lee, M.W.; Kim, H.D.; Park, Y.J.; Jung, Y.-G. Fabrication and Properties of Porous RBSN. *Key Eng. Mater.* **2005**, *287*, 277–281, https://doi.org/10.4028/www.scientific.net/KEM.287.277.

140. Jiang, G.-P.; Yang, J.-F.; Gao, J.-Q.; Niihara, K. Porous Silicon Nitride Ceramics Prepared by Extrusion Using Starch as Binder. *J. Am. Ceram. Soc.* **2008**, *91*, 3510–3516, https://doi.org/10.1111/j.1551-2916.2008.02701.x.
141. Jiang, G.-P.; Yang, J.-F.; Gao, J.-G. Extrusion of porous silicon nitride using different binders. *J. Ceram. Process. Res.* **2010**, *11*, 126–128.
142. Yu, F.; Xie, H.; Bai, Y.; Xu, F.; Han, P.; Tian, L.; Ni, S. Synthesis of Porous Silicon Nitride with Unidirectionally Aligned Channels Using Extrusion Method. *Ceramics–Silikáty* **2018**, *62*, 248–252.
143. Quackenbush, C.L.; French, K.; Neil, J.T. Fabrication of Sinterable Silicon Nitride by Injection Molding. In *A Collection of Papers Presented at the 1981 New England Section Topical Meeting on Nonoxide Ceramics: Ceramic Engineering and Science Proceedings*; Ceramic Engineering and Science Proceedings; 1982; pp. 20–34 ISBN 9780470318140.
144. Bandyopadhyay, G.; French, K.W. Fabrication of Near-Net-Shape Silicon Nitride Parts for Engine Application. *J. Eng. Gas Turbines Power* **1986**, *108*, 536–539, https://doi.org/10.1115/1.3239943.
145. Atre, S. V.; Weaver, T.J.; German, R.M. Injection Molding of Metals and Ceramics. *SAE Tech. Pap.* **1998**, https://doi.org/10.4271/982417.
146. Enneti, R.K.; Shivashankar, T.S.; Park, S.-J.; German, R.M.; Atre, S. V Master Debinding Curves for Solvent Extraction of Binders in Powder Injection Molding. *Powder Technol.* **2012**, *228*, 14–17, https://doi.org/10.1016/j.powtec.2012.04.027.
147. Gutiérrez, C.A.; Moreno, R. Tape Casting of Non-Aqueous Silicon Nitride Slips. *J. Eur. Ceram. Soc.* **2000**, *20*, 1527–1537, https://doi.org/10.1016/S0955-2219(99)00216-2.
148. Bitterlich, B.; Heinrich, J.G. Aqueous Tape Casting of Silicon Nitride. *J. Eur. Ceram. Soc.* **2002**, *22*, 2427–2434, https://doi.org/10.1016/S0955-2219(02)00029-8.
149. Duan, Y.; Zhang, J.; Li, X.; Shi, Y.; Xie, J.; Jiang, D. Optimization of the Tape Casting Process for the Development of High Performance Silicon Nitride Substrate. *Int. J. Appl. Ceram. Technol.* **2017**, *14*, 712–718, https://doi.org/10.1111/ijac.12679.
150. Cecen, B.; Topates, G.; Kara, A.; Akbulut, S.O.; Havitcioglu, H.; Kozaci, L.D. Biocompatibility of Silicon Nitride Produced via Partial Sintering and Tape Casting. *Ceram. Int.* **2021**, *47*, 3938–3945, https://doi.org/10.1016/j.ceramint.2020.09.257.
151. Dong, X.; Wu, J.; Yu, H.; Zhou, Q.; Wang, W.; Zhang, X.; Zhang, L.; Li, L.; He, R. Additive Manufacturing of Silicon Nitride Ceramics: A Review of Advances and Perspectives. *Int. J. Appl. Ceram. Technol.* **2022**, *19*, 2929–2949, https://doi.org/10.1111/ijac.14162.
152. Popper, P.; Ruddlesden, S.N. Structure of the Nitrides of Silicon and Germanium. *Nature* 1957, *179*, 1129.
153. Ruddlesden, S.N.; Popper, P. On the Crystal Structure of the Nitrides of Silicon and Germanium. *Acta Crystallogr.* **1958**, *11*, 465–468.
154. Jennings, H.M.; Richman, M.H. Structure, Formation Mechanisms and Kinetics of Reaction-Bonded Silicon Nitride. *J. Mater. Sci.* **1976**, *11*, 2087–2098, https://doi.org/10.1007/bf02403359.
155. Moulson, A.J. Reaction-Bonded Silicon Nitride: Its Formation and Properties. *J. Mater. Sci.* **1979**, *14*, 1017–1051, https://doi.org/10.1007/BF00561287.
156. Jennings, H.M. On Reactions between Silicon and Nitrogen. *J. Mater. Sci.* **1983**, *18*, 951–967, https://doi.org/10.1007/BF00551961.
157. Pigeon, R.G.; Varma, A.; Miller, A.E. Some Factors Influencing the Formation of Reaction-Bonded Silicon Nitride. *J. Mater. Sci.* **1993**, *28*, 1919–1936, https://doi.org/10.1007/BF00595765.
158. Pigeon, R.G.; Varma, A. Quantitative Kinetic Analysis of Silicon Nitridation. *J. Mater. Sci.* **1993**, *28*, 2999–3013, https://doi.org/10.1007/BF00354705.
159. Lei, B.; Babushkin, O.; Warren, R. Nitridation Study of Reaction-Bonded Silicon Nitride in situ by High Temperature X-Ray Diffraction. *J. Eur. Ceram. Soc.* **1997**, *17*, 1113–1118, https://doi.org/10.1016/S0955-2219(97)00016-2.

160. Li, W.B.; Lei, B.Q.; Lindbäck, T. A Kinetic Model for Reaction Bonding Process of Silicon Powder Compact. *J. Eur. Ceram. Soc.* **1997**, *17*, 1119–1131, https://doi.org/10.1016/S0955-2219(96)00212-9.
161. Chang, F.-W.; Liou, T.-H.; Tsai, F.-M. The Nitridation Kinetics of Silicon Powder Compacts. *Thermochim. Acta* **2000**, *354*, 71–80, https://doi.org/10.1016/S0040-6031(00)00432-9.
162. Lee, B.-T.; Kim, H.-D. Effect of Sintering Additives on the Nitridation Behavior of Reaction-Bonded Silicon Nitride. *Mater. Sci. Eng. A* **2004**, *364*, 126–131, https://doi.org/10.1016/j.msea.2003.07.005.
163. Mangels, J.A.; Tennenhouse, G.J. Densification of Reaction Bonded Silicon Nitride. *Bull. Am. Ceram. Soc.* **1980**, *59*, 1216.
164. Mangels, J.A.; Tennenhouse, G.J. Method of Densifying A Reaction Bonded Silicon Nitride Article. *US Pat. 4,285,895* **1981**, 1–5.
165. Mangels, J.A.; Baer, J.R. Method of Forming a Densified Silicon Nitride Article. *US Pat. 4,377,542* **1983**, 1–4.
166. Tiegs, T.N.; Kiggans, J.O.; Ploetz, K.L. Cost-Effective Sintered Reaction-Bonded Silicon Nitride for Structural Ceramics. *Ceram. Eng. Sci. Proc.* **1993**, *14*, 378–388, https://doi.org/10.1002/9780470314050.ch37.
167. Nunn, S.D.; Kiggans Jr., J.O.; Simpson II, R.E.; Maria, J.P. Gel Casting of Silicon Compositions for SRBSN. In Proceedings of the CONF-950401-32; Oak Ridge National Lab. (ORNL): Oak Ridge, TN, USA, 1995; pp. 1–7.
168. Howlett, C.R.; McCartney, E.; Ching, W. The Effect of Silicon Nitride Ceramic on Rabbit Skeletal Cells and Tissue. *Clin. Orthop. Relat. Res.* **1989**, *244*, 293–304.
169. Orth, J.; Ludwig, M.; Piening, W.; Wilke, A.; Griss, P. Biocompatibility of Silicon Carbide and Silicon Nitride Ceramics. Results of an Animal Experiment. In *Bioceramics and the Human Body*; Ravaglioli, A., Krajewski, A., Eds.; Springer: Dordrecht, Netherlands, 1992; pp. 372–377 ISBN 978-1-85166-748-2.
170. Kue, R.; Sohrabi, A.; Nagle, D.; Frondoza, C.; Hungerford, D. Enhanced Proliferation and Osteocalcin Production by Human Osteoblast-Like MG63 Cells on Silicon Nitride Ceramic Discs. *Biomaterials* **1999**, *20*, 1195–1201.
171. Sohrabi, A.; Holland, C.; Kue, R.; Nagle, D.; Hungerford, D.S.; Frondoza, C.G. Proinflammatory Cytokine Expression of IL-1β and TNF-α by Human Osteoblast-Like MG-63 Cells upon Exposure to Silicon Nitride *in vitro*. *J. Biomed. Mater. Res.* **2000**, *50*, 43–49, https://doi.org/10.1002/(SICI)1097-4636(200004)50:1<43::AID-JBM7>3.0.CO;2-A.
172. Neumann, A.; Jahnke, K.; Maier, H.R.; Ragoss, C. Biocompatibilty of Silicon Nitride Ceramic *in vitro*. A Comparative Fluorescence-Microscopic and Scanning Electron-Microscopic Study. *Laryngo-Rhino-Otologie* **2004**, *83*, 845–851.
173. Neumann, A.; Reske, T.; Held, M.; Jahnke, K.; Ragoss, C.; Maier, H.R. Comparative Investigation of the Biocompatibility of Various Silicon Nitride Ceramic Qualities *in vitro*. *J. Mater. Sci. Mater. Med.* **2004**, *15*, 1135–1140, https://doi.org/10.1023/B:JMSM.0000046396.14073.92.
174. Yamamoto, A.; Honma, R.; Sumita, M.; Hanawa, T. Cytotoxicity Evaluation of Ceramic Particles of Different Sizes and Shapes. *J. Biomed. Mater. Res. Part A* **2004**, *68A*, 244–256, https://doi.org/10.1002/jbm.a.20020.
175. Sorrell, C.C.; Hardcastle, P.H.; Druitt, R.K.; Howlett, C.R.; McCartney, E.R. Results of 15-Year Clinical Study of Reaction Bonded Silicon Nitride Intervertebral Spacers. *Proc. 7th World Biomater. Conf.* **2004**, 1872.
176. Alem, A.; Pugh, M.D.; Drew, R.A.L. Open-Cell Reaction Bonded Silicon Nitride Foams: Fabrication and Characterization. *J. Eur. Ceram. Soc.* **2014**, *34*, 599–609, https://doi.org/10.1016/j.jeurceramsoc.2013.09.011.
177. Alem, A.; Pugh, M.D.; Drew, R.A.L. Reaction Bonded Silicon Nitride Foams: The Influence of Iron Disilicide on Microstructure and Mechanical Strength. *Ceram. Int.* **2015**, *41*, 4966–4974, https://doi.org/10.1016/j.ceramint.2014.12.060.

178. Park, Y.-J.; Kim, H.-D.; Halloran, J.W. Permeability Enhancement in Porous-Sintered Reaction-Bonded Silicon Nitrides. *Int. J. Appl. Ceram. Technol.* **2011**, *8*, 809–814, https://doi.org/10.1111/j.1744-7402.2010.02511.x.
179. Bodišová, K.; Kašiarová, M.; Domanická, M.; Hnatko, M.; Lenčéš, Z.; Nováková, Z.V.; Vojtaššák, J.; Gromošová, S.; Šajgalík, P. Porous Silicon Nitride Ceramics Designed for Bone Substitute Applications. *Ceram. Int.* **2013**, *39*, 8355–8362, https://doi.org/10.1016/j.ceramint.2013.04.015.
180. Kašiarová, M.; Vilčeková, Z.; Bodišová, K.; Domanická, M.; Hnatko, M.; Dusza, J.; Šajgalík, P.; Gromošová, S. Mechanical Properties of Porous Si_3N_4 Ceramics. *Key Eng. Mater.* **2013**, *586*, 166–169, https://doi.org/10.4028/www.scientific.net/KEM.586.166.
181. Vilcekova, Z.; Kasiarova, M.; Domanicka, M.; Hnatko, M.; Sajgalik, P. Influence of the Preparation of Si_3N_4 Based Foams on the Compressive Strength and Processing Flaws. *Powder Metall. Prog.* **2013**, *13*, 139–146.
182. Bodišová, K.; Kašiarová, M.; Vilčeková, Z.; Domanická, M.; Lenčéš, Z.; Hnatko, M.; Gromošová, S.; Šajgalík, P. Mechanical Properties of Macroporous Silicon Nitride-Based Ceramics Designed for Bone Substitutes. *Ceram. – Silikaty* **2014**, *58*, 99–105.
183. Vilčeková, Z.P.; Kašiarová, M.; Domanická, M.P.; Hnatko, M.; Šajgalík, P. Local Mechanical Properties of Highly Porous Si_3N_4 for Trabecular Bone Replacement. *Key Eng. Mater.* **2015**, *662*, 142–146, https://doi.org/10.4028/www.scientific.net/KEM.662.142.
184. Precnerová, M.; Bodišová, K.; Frajkorová, F.; Galusková, D.; Varchulová Nováková, Z.; Vojtaššák, J.; Lenčéš, Z.; Šajgalík, P. *In vitro* Bioactivity of Silicon Nitride–Hydroxyapatite Composites. *Ceram. Int.* **2015**, *41*, 8100–8108, https://doi.org/10.1016/j.ceramint.2015.03.011.
185. Ciarelli, M.J.; Goldstein, S.A.; Kuhn, J.L.; Cody, D.D.; Brown, M.B. Evaluation of Orthogonal Mechanical Properties and Density of Human Trabecular Bone from the Major Metaphyseal Regions with Materials Testing and Computed Tomography. *J. Orthop. Res.* **1991**, *9*, 674–682, https://doi.org/10.1002/jor.1100090507.
186. Nikonam M.R.; Pugh, M.D.; Drew, R.A.L. Microstructural Evolution Mechanism of Porous Reaction Bonded Silicon Nitride Ceramics Heat-Treated in Two Powder Beds. *Ceram. Int.* **2019**, *45*, 21986–21997, https://doi.org/10.1016/j.ceramint.2019.07.213.
187. Nikonam M.R.; Pugh, M.D.; Drew, R.A.L. Pore Structure, Porosity and Compressive Strength of Highly Porous Reaction-Bonded Silicon Nitride Ceramics with Various Grain Morphologies. *J. Mater. Sci.* **2020**, *55*, 509–523, https://doi.org/10.1007/s10853-019-04078-3.
188. Nikonam-Mofrad, R.; Pugh, M.D.; Drew, R.A.L. A Comparative Study on Nitridation Mechanism and Microstructural Development of Porous Reaction Bonded Silicon Nitride in the Presence of CaO, MgO and Al_2O_3. *J. Asian Ceram. Soc.* **2020**, *8*, 873–890, https://doi.org/10.1080/21870764.2020.1793471.
189. Horiuchi, A.; Yokoyama, K.; Yoshida, M. Method for Production of Silicon Nitride Sintered Body. *US Pat. 4,716,133* **1987**.
190. Deeley, G.G.; Herbert, J.M.; Moore, N.C. Dense Silicon Nitride. *Powder Metall.* **1961**, *4*, 145–151.
191. Lumby, R.J.; Coe, R.F. The Influence of Some Process Variables on the Mechanical Properties of Hot-Pressed Silicon Nitride. In Proceedings of the Proc. Brit. Ceram. Soc; 1970; Vol. 15, pp. 91–101.
192. Coe, R.F.; Lumby, R.J.; Pawson, M.F. Some Properties and Applications of Hot-Pressed Silicon Nitride. *Spec. Ceram.* **1972**, *5*, 361–376.
193. Torti, M.L.; Alliegro, R.A.; Richerson, D.W.; Washburn, M.E.; Weaver, G.Q. High-Temperature Properties of Silicon Carbide and Silicon Nitride. *High-Temperature Mater. Gas Turbines* **1974**, 184–186.
194. Gazza, G.E. Effect of Yttria Additions on Hot-Pressed Si_3N_4. *Am. Ceram. Soc. Bull.* **1975**, *54*, 778–781.
195. Ahn, C.C.; Thomas, G. Microstructure and Grain-Boundary Composition of Hot-Pressed Silicon Nitride with Yttria and Alumina. *J. Am. Ceram. Soc.* **1983**, *66*, 14–17.

196. Washburn, M.E. Production of Silicon Oxynitride. *US Pat. 3,356,513* **1967**, 1–6, https://doi.org/10.1145/178951.178972.
197. Richerson, D.W.; Washburn, M.E. Hot Pressed Silicon Nitride. *US Pat. 3,836,374* **1974**, 1–9.
198. Torti, M.L. Processing Hot Pressed Silicon Nitride for Improved Reliability: HS-110 to NC-132. In *Ceramics for High-Performance Applications III: Reliability*; Lenoe, E.M., Katz, R.N., Burke, J.J., Eds.; Springer US: Boston, MA, 1983; pp. 261–273 ISBN 978-1-4684-3965-6.
199. Amaral, M.; Lopes, M.A.; Silva, R.F.; Santos, J.D. Densification Route and Mechanical Properties of Si_3N_4-Bioglass Biocomposites. *Biomaterials* **2002**, *23*, 857–62.
200. Mazzocchi, M.; Bellosi, A. On the Possibility of Silicon Nitride as a Ceramic for Structural Orthopaedic Implants. Part I: Processing, Microstructure, Mechanical Properties, Cytotoxicity. *J. Mater. Sci. Mater. Med.* **2008**, *19*, 2881–2887, https://doi.org/10.1007/s10856-008-3417-2.
201. Cappi, B.; Neuss, S.; Salber, J.; Telle, R.; Knüchel, R.; Fischer, H. Cytocompatibility of High Strength Non-Oxide Ceramics. *J. Biomed. Mater. Res.* **2010**, *93A*, 67–76, https://doi.org/10.1002/jbm.a.32527.
202. G R Terwilliger; F F Lange Hot-Pressing Behavior of Si_3N_4. *J. Am. Ceram. Soc.* **1974**, *57*, 25–29.
203. Terwilliger, G.R.; Lange, F.F. Pressureless Sintering of Si_3N_4. *J. Mater. Sci.* **1975**, *10*, 1169–1174, https://doi.org/10.1007/BF00541399.
204. Messier, D.R.; Gazza, G.E. Thermal Decomposition of Al_2O_3-Si_3N_4 Mixtures. *J. Am. Ceram. Soc.* **1975**, *58*, 538–540.
205. Lange, F.F. Silicon Nitride Polyphase Systems: Fabrication, Microstructure, and Properties. *Int. Met. Rev.* **1980**, *25*, 1–20, https://doi.org/10.1179/imtr.1980.25.1.1.
206. Gazza, G.E. Hot Pressed, High Strength Silicon Nitride. *US Pat. 3,830,652* **1974**, 1–3.
207. Dutta, S. Pressureless Sintered ß'-Si_3N_4 Solid Solution: Fabrication, Microstructure, and Strength. *NASA Tech. Memo. 78950; Fall Meet. Basic Sci. Div. Am. Ceram. Soc.* **1977**, 1–19.
208. Milberg, M.E.; Miller, W.M. High Temperature Ceramic Material Suitable for Gas-Turbine Applications and a Process for Producing Same. *US Pat. 4,071,371* **1978**, 1–7.
209. Jack, K.H. Sialons and Related Nitrogen Ceramics. *J. Mater. Sci.* **1976**, *11*, 1135–1158, https://doi.org/10.1007/BF02396649.
210. Loehman, R.E.; Rowcliffe, D.J. Sintering of Si_3N_4-Y_2O_3-Al_2O_3. *J. Am. Ceram. Soc.* **1980**, *63*, 144–148.
211. Suttor, D.; Fischman, G.S. Densification and Sintering Kinetics in Sintered Silicon Nitride. *J. Am. Ceram. Soc.* **1992**, *75*, 1063–1067, https://doi.org/10.1111/j.1151-2916.1992.tb05538.x.
212. Yang, J.-F.; Ohji, T.; Niihara, K. Influence of Yttria–Alumina Content on Sintering Behavior and Microstructure of Silicon Nitride Ceramics. *J. Am. Ceram. Soc.* **2000**, *83*, 2094–2096.
213. Hirosaki, N.; Okada, A.; Matoba, K. Sintering of Si_3N_4 with the Addition of Rare-Earth Oxides. *J. Am. Ceram. Soc.* **1988**, *71*, 144–147.
214. Themelin, L.; Desmaison-Brut, M.; Billy, M.; Crampon, J. Post-Densification Behaviour of Pressureless Sintered Silicon Nitride Materials: Relationship Between Properties and Microstructure. *Ceram. Int.* **1992**, *18*, 119–130, https://doi.org/10.1016/0272-8842(92)90041-B.
215. Yang, J.; Oliveira, F.J.; Silva, R.F.; Ferreira, J.M.F. Pressureless Sinterability of Slip Cast Silicon Nitride Bodies Prepared from Coprecipitation-Coated Powders. *J. Eur. Ceram. Soc.* **1999**, *19*, 433–439, https://doi.org/10.1016/s0955-2219(98)00222-2.
216. Ling, G.; Yang, H. Pressureless Sintering of Silicon Nitride with Magnesia and Yttria. *Mater. Chem. Phys.* **2005**, *90*, 31–34, https://doi.org/10.1016/j.matchemphys.2004.08.041.
217. Liu, X.J.; Huang, Z.Y.; Ge, Q.M.; Sun, X.W.; Huang, L.P. Microstructure and Mechanical Properties of Silicon Nitride Ceramics Prepared by Pressureless Sintering with MgO-Al_2O_3 -SiO2 as Sintering Additive. *J. Eur. Ceram. Soc.* **2005**, *25*, 3353–3359, https://doi.org/10.1016/j.jeurceramsoc.2004.08.025.
218. Tajima, Y. Development of High Performance Silicon Nitride Ceramics and their Applications. In Proceedings of the Silicon Nitride-Scientific and Technological Advances: MRS

Symposium Proceedings; Chen, I.W., Ed.; Cambridge Univ Press: Pttsburgh, PA, 1992; Vol. 287, p. 189.
219. Matsuo, Y. Development of Gas Pressure Sintered Silicon Nitride Ceramics. In *Hor Isostatic Pressing – Theory and Applications*; Springer: Dordrecht, 1992; pp. 49–54.
220. Pettersson, M.; Pakdaman, Z.; Engqvist, H.; Liu, Y.; Shen, Z.; Östhols, E. Spark Plasma Sintered β-phase Silicon Nitride with Sr and Ca as a Sintering Aid for Load Bearing Medical Applications. *J. Eur. Ceram. Soc.* **2012**, *32*, 2705–2709, https://doi.org/10.1016/j.jeurceramsoc.2011.12.027.
221. Wananuruksawong, R.; Wasanapiarnpong, T.; Dhanesuan, N.; Didron, P.P. Microhardness and Biocompatibility of Silicon Nitride Ceramic Developed for Dental Applications. *Mater. Sci. Appl.* **2014**, *5*, 1034, https://doi.org/10.4236/msa.2014.514106.
222. Das, M.; Bhimani, K.; Balla, V.K. *In vitro* Tribological and Biocompatibility Evaluation of Sintered Silicon Nitride. *Mater. Lett.* **2018**, *212*, 130–133, https://doi.org/10.1016/J.MATLET.2017.10.061.
223. Fu, L.; Engqvist, H.; Xia, W. Spark Plasma Sintering of Biodegradable Si_3N_4 Bioceramic with Sr, Mg and Si as Sintering Additives for Spinal Fusion. *J. Eur. Ceram. Soc.* **2018**, *58*, 2110–2119, https://doi.org/10.1016/j.jeurceramsoc.2017.10.003.
224. Li, M.; Zhang, L.; Zhang, C.; Xu, E.; Liu, X.; Zhao, F.; Sun, X.; Chen, H.; Gao, J. Effect of Y_2O_3 on the Physical Properties and Biocompatibility of β–SiAlON Ceramics. *Ceram. Int.* **2020**, *46*, 23427–23432, https://doi.org/10.1016/j.ceramint.2020.06.112.
225. Priest, H.F.; Priest, G.L.; Gazza, G.E. Sintering of Silicon Nitride Under High Nitrogen Pressure. *J. Am. Ceram. Soc.* **1977**, *60*, 81.
226. Mitomo, M. Pressure Sintering of Si_3N_4. *J. Mater. Sci.* **1976**, *11*, 1103–1107, https://doi.org/10.1007/BF00553119.
227. Galasso, F.; Veltri, R. Sintering of Si_3N_4-$15Y_2O_3$ under High N_2 Pressure. *Am. Ceram. Soc. Bull.* **1979**, *58*, 793–794.
228. Galasso, F.S.; Veltri, R.D.; Galsso, F.S.; Veltri, R.D. Sintering of Si_3N_4-Y_2O_3 Under Nitrogen Pressure. *J. Am. Ceram. Soc.* **1981**, *64*, C-15-C-16, https://doi.org/10.1111/j.1151-2916.1981.tb09543.x.
229. Gazza, G.E.; Katz, R.N.; Priest, H.F. Densification of Si_3N_4-Y_2O_3/Al_2O_3 by a Dual N_2 Pressure Process. *J. Am. Ceram. Soc.* **1981**, *64*, C-161-C-162, https://doi.org/10.1111/j.1151-2916.1981.tb15882.x.
230. Oberacker, R.; Grathwohl, G. High Temperature Resistant Silicon-Nitride Ceramics by Optimized Two-Step Gas Pressure Sintering. In *Tailoring of Mechanical Properties of Si_3N_4 Ceramics*; Hoffmann, M.J., Petzow, G., Eds.; Springer Netherlands: Dordrecht, 1994; pp. 43–55 ISBN 978-94-011-0992-5.
231. Liu, S.; Ye, F.; Hu, S.; Yang, H.; Liu, Q.; Zhang, B. A New Way of Fabricating Si_3N_4 Ceramics by Aqueous Tape Casting and Gas Pressure Sintering. *J. Alloys Compd.* **2015**, *647*, 686–692, https://doi.org/10.1016/j.jallcom.2015.05.134.
232. Hu, J.; Zhang, B.; Li, C.; Wang, L.; Wang, S.; Shi, Z.; Yang, J. Fabrication of Si_3N_4 Ceramics with High Thermal Conductivity and Flexural Strength via Novel Two-Step Gas-Pressure Sintering. *J. Eur. Ceram. Soc.* **2022**, *42*, 4846–4854, https://doi.org/10.1016/j.jeurceramsoc.2022.04.049.
233. Arakawa, T.; Matsumoto, Y. Silicon Nitride Sintered Body and Process for Preparation Thereof. *US Pat. 5,017,530* **1991**, 1–4.
234. Mizuno, K.; Yabuta, K.; Watanabe, M. Sintered Body of Silicon Nitride and Its Manufacture. *US Pat. 5,056,513* **1991**, 1–5.
235. Kumar, A.; Mallik, A.K.; Acikbas, N.C.; Yaygıngol, M.; Kara, F.; Mandal, H.; Basu, D.; Biswas, K.; Basu, B.; Yaygingol, M.; et al. Cytocompatibility Property Evaluation of Gas Pressure Sintered SiAlON–SiC Composites with L929 Fibroblast Cells and Saos-2 Osteoblast-Like Cells. *Mater. Sci. Eng. C* **2012**, *32*, 464–469, https://doi.org/10.1016/j.msec.2011.11.019.

236. Zhou, H.; Yang, S.; Wei, D.; Liang, C.; Yang, Q.; Yang, H.; Wang, D.; Li, M.; Yang, L. Development of Hydrofluoric Acid-Cleaned Silicon Nitride Implants for Periprosthetic Infection Eradication and Bone Regeneration Enhancement. *Mater. Sci. Eng. C* **2021**, *127*, 112241, https://doi.org/10.1016/j.msec.2021.112241.
237. Larker, H. Hot Isostatic Pressing of Shaped Silicon Nitride Parts. In *High-Pressure Science and Technology: Volume 1: Physical Properties and Material Synthesis/Volume 2: Applications and Mechanical Properties*; Timmerhaus, K.D., Barber, M.S., Eds.; Springer US: Boston, MA, 1979; pp. 1683–1687 ISBN 978-1-4684-7470-1.
238. Richerson, D.W.; Wimmer, J. Properties of Isostatically Hot-Pressed Silicon Nitride. *J. Am. Ceram. Soc.* **1983**, *66*, c173–c176, https://doi.org/10.1111/j.1151-2916.1983.tb10633.x.
239. Adlerborn, J.; Burström, M.; Hermansson, L.; Larker, H.T. Development of High Temperature High Strength Silicon Nitride by Glass Encapsulated Hot Isostatic Pressing. *Mater. Des.* **1987**, *8*, 229–232, https://doi.org/10.1016/0261-3069(87)90138-5.
240. Larker, H.T.; Lundberg, R. Near Net Shape Production of Monolithic and Composite High Temperature Ceramics by Hot Isostatic Pressing (HIP). *J. Eur. Ceram. Soc.* **1999**, *19*, 2367–2373, https://doi.org/10.1016/S0955-2219(99)00109-0.
241. Westman, A. Hot Isostatic Pressing of Ceramics and Influence of Glass Encapsulation, Luleå University of Technology, Luleå Sweden, 1998.
242. Westman, A.-K.; Wei, L.-Y. TEM Study of the Interface between HIPed Silicon Nitride and Encapsulation Borosilicate Glass. *J. Mater. Sci.* **2000**, *35*, 2847–2854, https://doi.org/10.1023/A:1004759505321.
243. Yeh, H.C.; Sikora, P.F. Consolidation of Si_3N_4 by Hot Isostatic Pressing. *Bull. Am. Ceram. Soc.* **1979**, *58*, 444–447.
244. Greskovich, C. Preparation of High-Density Si_3N_4 by a Gas-Pressure Sintering Process. *J. Am. Ceram. Soc.* **1981**, *64*, 725–730.
245. Ziegler, G.; Wötting, G. Post-Treatment of Pre-Sintered Silicon Nitride by Hot Isostatic Pressing. *Int. J. High Technol. Ceram.* **1985**, *1*, 31–58, https://doi.org/10.1016/0267-3762(85)90023-2.
246. Horiuchi, Y.; Ichinoseki, K.; Nakamura, Y. Effects of Hip Conditions Upon the Changes of Mechanical Properties with Sintered Silicon Nitride. In *Hot Isostatic Pressing—Theory and Applications*; Koizumi, M., Ed.; Springer Netherlands: Dordrecht, 1992; pp. 61–66 ISBN 978-94-011-2900-8.
247. Oda, T.; Tanaka, K.; Iwaida, T.; Yamamoto, S.; Kohsaka, S.; Sato, M.; Uchimura, H.; Tajima, K. Sintered Product of Silicon Nitride. *US Pat. 5,804,523* **1998**.
248. Kim, S.S.; Baik, S. Hot Isostatic Pressing of Sintered Silicon Nitride. *J. Am. Ceram. Soc.* **1991**, *74*, 1735–1738.
249. Baumgartner, H.R.; Cowley, P.E. *Finishing Techniques for Silicon Nitride Bearings*; Norton Company (Worcester, MA) and Federal Mogul Corp (Ann Arbor, MI): Watertown, MA USA, 1976;
250. Rabinowicz, E. Grinding Damage of Silicon Nitride Determined by Abrasive Wear Tests. *Wear* **1976**, *39*, 101–107, https://doi.org/10.1016/0043-1648(76)90225-8.
251. Miyasato, H.; Okamoto, H.; Usui, S.; Miyamoto, A.; Ueno, Y. The Effect of Grinding on Strength of Hot-Pressed Silicon Nitride. *ISIJ Int.* **1989**, *29*, 726–733.
252. Tomlinson, W.J.; Newton, R.C. Effect of Grinding, Lapping and Various Surface Treatments on the Strength of Silicon Nitride. *Ceram. Int.* **1990**, *16*, 253–257, https://doi.org/10.1016/0272-8842(90)90036-F.
253. Zhang, B.I.; Yang, F.; Wang, J.; Zhu, Z.; Monahan, R. The Effect of Machine Stiffness on Strength of Ground Silicon Nitride. In Proceedings of the NAMRC XXVI; SME: Atlanta, Georgia, 1998; pp. MR98–248.
254. Strakna, T.; Jahanmir, S.; Allor, R.; Kumar, K. Effect of Grinding on Strength of Silicon Nitride. In Proceedings of the NAMRC XXIII; SME Press: Houghton, MI, 1995; pp. MR95–138.

255. Zhang, B.; Zheng, X.L.; Tokura, H.; Yoshikawa, M. Grinding Induced Damage in Ceramics. *J. Mater. Process. Technol.* **2003**, *132*, 353–364, https://doi.org/10.1016/S0924-0136(02)00952-4.
256. Liu, W.; Deng, Z.; Shang, Y.; Wan, L. Effects of Grinding Parameters on Surface Quality in Silicon Nitride Grinding. *Ceram. Int.* **2017**, *43*, 1571–1577, https://doi.org/10.1016/j.ceramint.2016.10.135.
257. Hakulinen, M. Residual Strength of Ground Hot Isostatically Pressed Silicon Nitride. *J. Mater. Sci.* **1985**, *20*, 1049–1060, https://doi.org/10.1007/BF00585749.
258. Malkin, S.; Ritter, J.E. Grinding Mechanisms and Strength Degradation for Ceramics. *J. Eng. Ind.* **1989**, *111*, 1–8.
259. Mayer, J.E.; Fang, G.-P. Effect of Grinding Parameters on Strength and Finish of Ceramics. In Proceedings of the 1st International Machining and Grinding Conference; SME: Dearborn, Michigan, 1995; pp. MR95–170.
260. Huang, H.; Yin, L.; Zhou, L. High speed Grinding of Silicon Nitride with Resin Bond Diamond Wheels. *J. Mater. Process. Technol.* **2003**, *141*, 329–336, https://doi.org/10.1016/S0924-0136(03)00284-X.
261. Chen, J.; Shen, J.; Huang, H.; Xu, X. Grinding Characteristics in High Speed Grinding of Engineering Ceramics with Brazed Diamond Wheels. *J. Mater. Process. Technol.* **2010**, *210*, 899–906, https://doi.org/10.1016/j.jmatprotec.2010.02.002.
262. Hwang, T.W.; Evans, C.J.; Whitenton, E.P.; Malkin, S. High Speed Grinding of Silicon Nitride with Electroplated Diamond Wheels, Part 1: wear and Wheel Life. *J. Manuf. Sci. Eng.* **2000**, *122*, 32–41.
263. Hwang, T.W.; Evans, C.J.; Malkin, S. High Speed Grinding of Silicon Nitride with Electroplated Diamond Wheels, Part 2: Wheel Topography and Grinding Mechanisms. *J. Manuf. Sci. Eng.* **2000**, *122*, 42–50.
264. Lei, S.; Shin, Y.C.; Incropera, F.P. Experimental Investigation of Thermo-Mechanical Characteristics in Laser-Assisted Machining of Silicon Nitride Ceramics. *J. Manuf. Sci. Eng.* **2001**, *123*, 639–646.
265. Sarkar, B.R.; Doloi, B.; Bhattacharyya, B. Parametric Analysis on Electrochemical Discharge Machining of Silicon Nitride Ceramics. *Int. J. Adv. Manuf. Technol.* **2006**, *28*, 873–881, https://doi.org/10.1007/s00170-004-2448-1.
266. Ghosh, D.; Doloi, B.; Das, P.K. Parametric Analysis and Optimisation on Abrasive Water Jet Cutting of Silicon Nitride Ceramics. *Int. J. Precis. Technol.* **2015**, *5*, 294–311, https://doi.org/10.1504/IJPTECH.2015.073833.
267. Bharathi, V.; Anilchandra, A.R.; Sangam, S.S.; Shreyas, S.; Shankar, S.B. A Review on the Challenges in Machining of Ceramics. *Mater. Today Proc.* **2021**, *46*, 1451–1458, https://doi.org/10.1016/j.matpr.2021.03.019.
268. Dalal, H.M.; Rosenlieb, J.W.; Sibley, L.B. Effect of Lapping Parameters on Generation of Damage on Silicon NItride Ball Surfaces. *U.S. Navy Contract No. N00019-76-0147, SKF Rep. AL76T026* **1976**, 1–44.
269. Vora, H.; Orent, T.W.; Stokes, R.I. Mechanochemical Polishing of Silicon Nitride. *J. Am. Ceram. Soc.* **1982**, *65*, c140–c141, https://doi.org/10.1111/j.1151-2916.1982.tb10517.x.
270. Sugita, T.; Ueda, K.; Kanemura, Y. Material Removal Mechanism of Silicon Nitride During Rubbing in Water. *Wear* **1984**, *97*, 1–8, https://doi.org/10.1016/0043-1648(84)90076-0.
271. Fischer, T.E.; Tomizawa, H. Interaction of Tribochemistry and Microfracture in the Friction and Wear of Silicon Nitride. *Wear* **1985**, *105*, 29–45.
272. Tomizawa, H.; Fischer, T, E. Friction and Wear of Silicon Nitride and Silicon Carbide in Water: Hydrodynamic Lubrication at Low Sliding Speed Obtained by Tribochemical Wear. *ASLE Trans.* **1987**, *30*, 41–46.
273. Bhagavatula, S.R.; Komanduri, R. On Chemomechanical Polishing of Si_3N_4 with Cr2O3. *Philos. Mag. A* **1996**, *74*, 1003–1017, https://doi.org/10.1080/01418619608242173.

274. Jiang, M.; Wood, N.O.; Komanduri, R. On Chemo-Mechanical Polishing (CMP) of Silicon Nitride (Si_3N_4) Workmaterial with Various Abrasives. *Wear* **1998**, *220*, 59–71, https://doi.org/10.1016/S0043-1648(98)00245-2.
275. Hu, Y.Z.; Gutmann, R.J.; Chow, T.P. Silicon Nitride Chemical Mechanical Polishing Mechanisms. *J. Electrochem. Soc.* **1998**, *145*, 3919, https://doi.org/10.1149/1.1838893.
276. Muratov, V.A.; Luangvaranunt, T.; Fischer, T.E. The Tribochemistry of Silicon Nitride: Effects of Friction, Temperature and Sliding Velocity. *Tribol. Int.* **1998**, *31*, 601–611, https://doi.org/10.1016/S0301-679X(98)00081-4.
277. Yuan, J.L.; Lü, B.H.; Lin, X.; Zhang, L.B.; Ji, S.M. Research on Abrasives in the Chemical–Mechanical Polishing Process for Silicon Nitride Balls. *J. Mater. Process. Technol.* **2002**, *129*, 171–175, https://doi.org/10.1016/S0924-0136(02)00604-0.
278. Maw, W.; Stevens, F.; Langford, S.C.; Dickinson, J.T. Single Asperity Tribochemical Wear of Silicon Nitride Studied by Atomic Force Microscopy. *J. Appl. Phys.* **2002**, *92*, 5103–5109, https://doi.org/10.1063/1.1510595.
279. Umehara, N.; Kirtane, T.; Gerlick, R.; Jain, V.K.; Komanduri, R. A New Apparatus for Finishing Large Size/Large Batch Silicon Nitride (Si_3N_4) Balls for Hybrid Bearing Applications by Magnetic Float Polishing (MFP). *Int. J. Mach. Tools Manuf.* **2006**, *46*, 151–169, https://doi.org/10.1016/j.ijmachtools.2005.04.015.
280. Oh, M.-H.; Nho, J.-S.; Cho, S.-B.; Lee, J.-S.; Singh, R.K. Polishing Behaviors of Ceria Abrasives on Silicon Dioxide and Silicon Nitride CMP. *Powder Technol.* **2011**, *206*, 239–245, https://doi.org/10.1016/j.powtec.2010.09.025.
281. Dante, R.C.; Kajdas, C.K. A Review and a Fundamental Theory of Silicon Nitride Tribochemistry. *Wear* **2012**, *288*, 27–38, https://doi.org/10.1016/j.wear.2012.03.001.
282. Xiao, X. lan; Li, G. xian; Mei, H. juan; Yan, Q. sheng; Lin, H. tay; Zhang, F. lin Polishing of Silicon Nitride Ceramic Balls by Clustered Magnetorheological Finish. *Micromachines* **2020**, *11*, 1–18, https://doi.org/10.3390/mi11030304.
283. Zhou, Y.S.; Ohashi, M.; Tomita, N.; Ikeuchi, K.; Takashima, K. Study on the Possibility of Silicon Nitride—Silicon Nitride as a Material for Hip Prostheses. *Mater. Sci. Eng. C* **1997**, *5*, 125–129, https://doi.org/10.1016/S0928-4931(97)00032-5.
284. Mazzocchi, M.; Gardini, D.; Traverso, P.L.; Faga, M.G.; Bellosi, A. On the Possibility of Silicon Nitride as a Ceramic for Structural Orthopaedic Implants. Part II: Chemical Stability and Wear Resistance in Body Environment. *J. Mater. Sci. Mater. Med.* **2008**, *19*, 2889–2901, https://doi.org/10.1007/s10856-008-3437-y.
285. Bal, B.S.; Rahaman, M.N. Orthopedic Applications of Silicon Nitride Ceramics. *Acta Biomater.* **2012**, *8*, 2889–2898, https://doi.org/10.1016/j.actbio.2012.04.031.
286. Heimann, R.B. Silicon Nitride, A Close to Ideal Ceramic Material for Medical Application. *Ceramics* **2021**, *4*, 208–223.
287. Bal, B.S.; Khandkar, A.; Lakshminarayanan, R.; Clarke, I.; Hofmann, A.A.; Rahaman, M.N. Testing of Silicon Nitride Ceramic Bearings for Total Hip Arthroplasty. *J. Biomed. Mater. Res. Part B Appl. Biomater.* **2008**, *87*, 447–454, https://doi.org/10.1002/jbm.b.31123.
288. Bal, B.S.; Khandkar, A.; Lakshminarayanan, R.; Clarke, I.; Hofmann, A.A.; Rahaman, M.N. Fabrication and Testing of Silicon Nitride Bearings in Total Hip Arthroplasty. *J. Arthroplasty* **2009**, *24*, 110–116, https://doi.org/10.1016/j.arth.2008.01.300.
289. McEntire, B.J.; Lakshminarayanan, R.; Ray, D.A.; Clarke, I.C.; Puppulin, L.; Pezzotti, G.; Bal, B.S.; Lakshminarayanan, A.; Bock, R. Silicon Nitride Bearings for Total Joint Arthroplasty. *Lubricants* **2016**, *4*, 1–24, https://doi.org/10.3390/lubricants4040035.
290. Yorifuji, M.; Affatato, S.; Tateiwa, T.; Takahashi, Y.; Shishido, T.; Marin, E.; Zanocco, M.; Zhu, W.; Pezzotti, G.; Yamamoto, K. Wear Simulation of Ceramic-on-Crosslinked Polyethylene Hip Prostheses: A New Non-Oxide Silicon Nitride versus the Gold Standard Composite Oxide Ceramic Femoral Heads. *Materials (Basel).* **2020**, *13*, 1–11, https://doi.org/10.3390/ma13132917.
291. Standard Specification for Silicon Nitride Bearing Balls. *ASTM F-2094 M-11* **2018**.

292. Wiederhorn, S.M.; Tighe, N.J. Proof-Testing of Hot-Pressed Silicon Nitride. *J. Mater. Sci.* **1978**, *13*, 1781–1793, https://doi.org/10.1007/BF00548742.
293. Evans, A.G.; Fuller, E.R. Proof Testing—The Effects of Slow Crack Growth. *Mater. Sci. Eng.* **1975**, *19*, 69–77, https://doi.org/10.1016/0025-5416(75)90009-9.
294. Evans, A.G.; Wiederhorn, S.M. Proof Testing of Ceramic Materials – An Analytical Basis for Failure Prediction. *Int. J. Fract.* **1984**, *26*, 355–368, https://doi.org/10.1007/BF00962969.
295. Weisse, B.; Affolter, C.; Koller, R.E.; Stutz, A. Proof Testing of Ceramic Femoral Heads for Hip Joint Implants. *Proc. Inst. Mech. Eng. Part H J. Eng. Med.* **2010**, *224*, 1051–1059, https://doi.org/10.1243/09544119JEIM743.
296. Richter, H. Application of Proof-Testing to Ceramic Hip Joint Heads. In Proceedings of the 3rd International BIOLOX Symposium on Ceramic Wear Couple; 1998; pp. 7–12.
297. Ray, D.A.; Lakshminarayanan, R.; McEntire, B.J.; Ley, O. The Novel Use of Acoustic Emission Monitoring during Proof-Testing of Ceramic Spinal Implants. In Proceedings of the Contributed Papers from Materials Science and Technology 2016 (MS&T16); American Ceramic Society Press: Salt Lake City, UT USA, 2016; pp. 687–704.

The Surface Chemistry of Silicon Nitride

Giuseppe Pezzotti

1 Introduction

Biomedical silicon nitride (Si_3N_4) is manufactured from fine high-purity submicron powders that are consolidated into near-net-shape green bodies. The green bodies are then densified by sintering [1]. The formation of a green body can be achieved through a colloidal process, which involves dispersing fine particles in an aqueous medium. The interfacial chemistry of the slip is critical in this initial powder consolidation step. If surface reactions are not controlled, agglomeration and microstructural inhomogeneities will occur within the green body [1, 2]. The initial inhomogeneities in fine-powder slurries can persist in the final sintered products, significantly degrading mechanical properties. Therefore, the stability and rheology of these slurries must be carefully controlled by tailoring the solid/liquid interfacial chemistry. Research conducted in the 1990s on the surface composition and chemical reactions of Si_3N_4 powders in aqueous environments led to the successful production of high-strength silicon nitride. This knowledge is now well-known, and methods for manufacturing high-strength silicon nitride ceramics are firmly established [3].

Three decades later, this body of specialized chemical knowledge took a fortuitous turn toward the biochemistry of silicon nitride ceramics. Si_3N_4 has now been shown to be a unique bioceramic due to its surface chemistry, which allows it to interact favorably with the human body [4]. The aqueous surface chemistry of Si_3N_4, which was originally studied to improve its mechanical properties, has become the key to understanding its biological properties. In this chapter, the surface chemistry of Si_3N_4 is reviewed to identify the chemical origins of its bio-functionality.

G. Pezzotti (✉)
Ceramic Physics Laboratory, Kyoto Institute of Technology, Sakyo-Ku,
Matsugasaki, Kyoto, Japan
e-mail: pezzotti@kit.ac.jp

B. S. Bal et al. (eds.), *Silicon Nitride Bioceramics*,
https://doi.org/10.1007/978-3-031-67047-3_4

2 Oxidation of Silicon Nitride

Silicon nitride is passively oxidized at room temperature. The oxide coating that forms spontaneously on its surface is thermodynamically favored because the nitride surface is unstable in the presence of oxygen. This occurs by the following reaction [5]:

$$\mathrm{Si_3N_4 + 3O_2 \rightarrow 3SiO_2 + 2N_2, \Delta G_{298\,^\circ K} = -1,272\ kJ/mol} \tag{1}$$

The oxidation of Si_3N_4 at room temperature is evident from the negative free energy of the reaction. The oxide layer that forms on its surface has been imaged by electron microscopy and found to be amorphous and variable in thickness, typically ranging from 1 to 4 nm [6, 7]. Its thickness increases with increasing exposure at 1000 °C, as shown in Fig. 1. X-ray photoelectron spectroscopy (XPS) studies by various authors have provided insight into the chemical structure of the oxidation layer [5, 8–12]. XPS core-level spectra and related binding energies at the Si*2p*, N*1s*, and O*1s* edges in Si_3N_4 are schematically shown in Fig. 2 [13]. The electronegativity of Si_3N_4 elements increases in the order Si < N < O due to the direct relationship between the binding energy of an element and the Pauling electronegativity of neighboring atoms. Accordingly, the binding energy of the Si*2p* peak in Si_3N_4 is lower than that in SiO_2 (i.e., 101.4 eV vs. 103.7 eV). The O*1s* in SiO_2 and Si_2N_2O were reported as 533.6 and 531.9 eV, respectively [14, 15]. Vacancies or hydrogen bonds can introduce energy shifts and additional peaks in the XPS spectrum. The presence of two or more types of bonds can also cause asymmetry in the XPS peak. This is indeed the case for all Si*2p*, N*1s*, and O*1s* XPS spectra of Si_3N_4. The Si*2p* binding energy in silicon oxynitride is between 101.7 and 102.2 eV, which is intermediate between that of Si_3N_4 and SiO_2. The N*1s* spectrum appears at ~398.2 eV, which is slightly higher than that of Si_3N_4 (397.3 eV). The O*1s* level of the superficial oxidation layer on Si_3N_4 appears at 532.8 eV, which is slightly higher than that of Si_2N_2O and close to that of amorphous silica. The asymmetry is clear in the XPS spectra of Fig. 2. It was measured on a commercially available Si_3N_4 powder synthesized from silicon diimide [13].

Ni et al. examined the effect of calcination in air (6 h at 200, 400, and 600 °C) on the XPS spectrum of a Si_3N_4 powder prepared by direct nitridation of silicon [16]. Figure 3a–h illustrate the evolution of the Si*2p* and N*1s* edges using different calcination conditions (cf. labels in the inset). Tables 1 and 2 provide quantitative analyses for Si*2p* and N*1s* edges. Consistent with the Si*2p* and N*1s* edges, the effect of calcination in air is twofold: a relative increase in bands including Si-O bonds and a decrease in bands containing the Si-N bonds. XPS studies of Si_3N_4 powders have estimated the thickness of the oxidation layer to be between 0.1 and 1 nm, assuming pure SiO_2, and between 0.3 and 3 nm, assuming an internal Si_2ON_2 oxynitride layer. XPS analyses of Si_3N_4 powders located an intermediate state between the SiO_2 and Si_2ON_2 layers, which was interpreted as the presence of amines at the surface based on shifts in the binding energy of the N*1s* peak [10, 12].

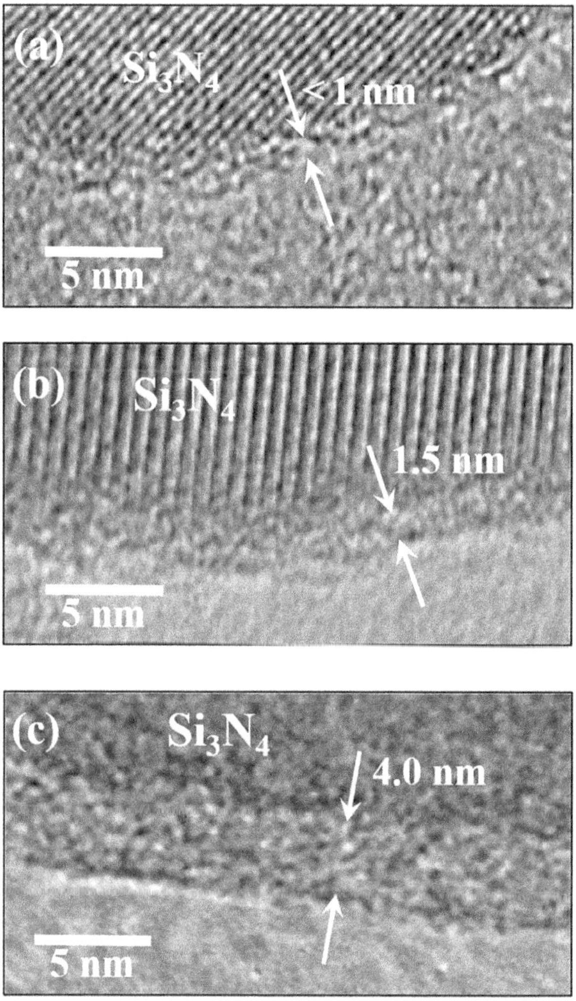

Fig. 1 High-resolution electron microscopy images of Si_3N_4 powder (UBE SN E-10) (**a**) as received and oxidized in air at 1000 °C for (**b**) 10 and (**c**) 20 h. the oxidation layer is hardly resolved in the as-received powder, and its thickness appears to be sub-nanometer in agreement with XPS data (cf. Table 1). After oxidation, the thickness increases exponentially to ~4 nm (cf. labels in inset). (The micrographs were rearranged and used with permission [7])

A controlled oxidation study by Raider et al. confirmed that the oxidized surface was a "graded" oxynitride layer, with steadily decreasing oxygen concentration with increased distance from the surface [5]. Using three successive procedures, Peukert and Greil analyzed ten commercial Si_3N_4 powders prepared by different methods for their chemical and structural characteristics [8]. The powders were first characterized by their specific surface areas (S_V) using the BET method to calculate their equivalent spherical particle sizes (D_S) [17].

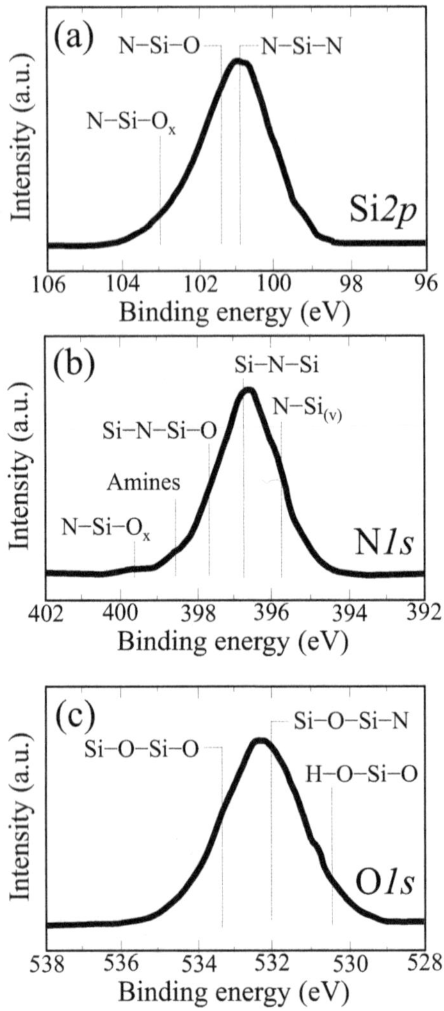

Fig. 2 The Si2*p*, N1*s*, and O1*s* edges in Si_3N_4 powder are synthesized by precipitation of silicon diimide in (**a**–**c**), respectively. Spectral locations for different bonds are given by labels in the inset. The abbreviation (*v*) means vacancy. (Redrawn and used with permission [13])

The equation relating D_S to S_V is:

$$D_S = \frac{6}{\rho_{SN} S_V} \tag{2}$$

where the density of Si_3N_4, ρ_{SN}, was taken as 3.18 g/cm^3. The measured S_V and the computed D_S values are listed in Table 3. Next, the total oxygen content, C_{TO}, of the powders was measured using an inert gas fusion technique [18]. The fictive thickness of the SiO_2 layer (t_S) was calculated under the assumption that the total oxygen content is at the SiO_2 surface, using the following equation:

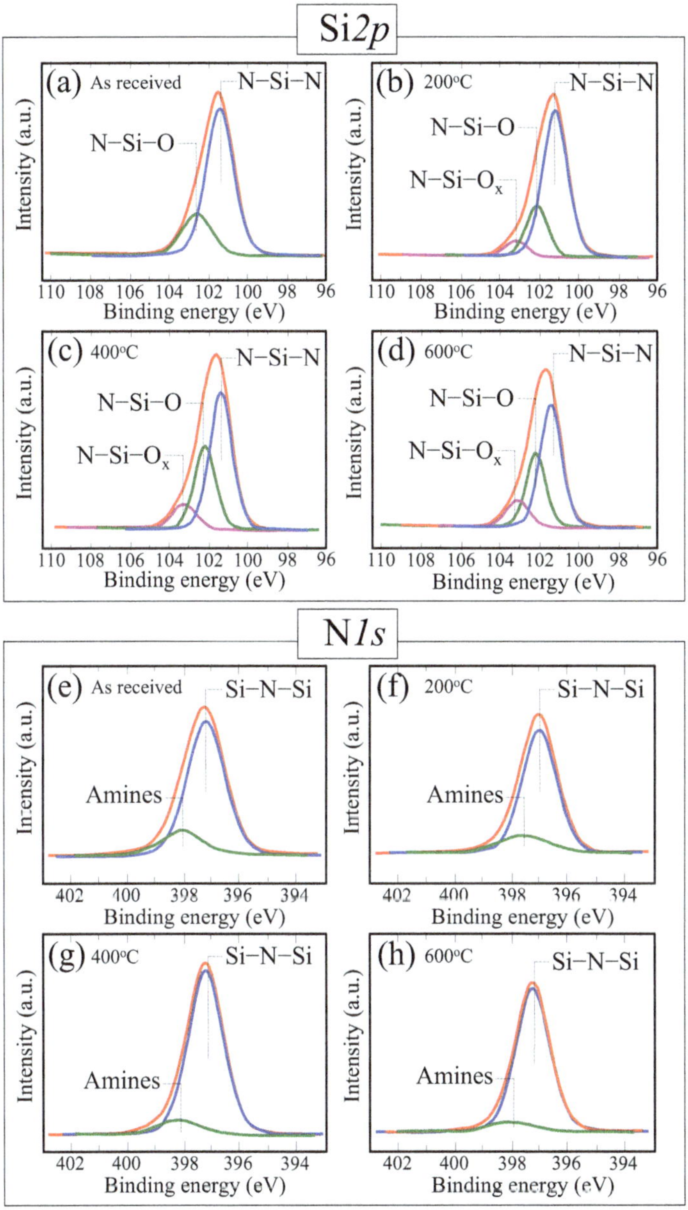

Fig. 3 The Si2*p* and N1*s* edges of Si_3N_4 powder prepared by direct nitridation of silicon before and after calcination. (Redrawn and used with permission [16])

Table 1 Results of XPS analysis at the Si2*p* edge of Si_3N_4 powder prepared by direct nitridation of silicon before and after calcination [16]

Si2*p*	Powder	Binding energy (eV)	Areal ratio (%)
SiN_4	As received	101.42	76.2
	200 °C	101.25	71.0
	400 °C	101.39	56.2
	600 °C	101.36	55.8
O–Si–N	As received	102.59	23.8
	200 °C	102.17	21.6
	400 °C	102.20	32.1
	600 °C	102.12	31.3
SiO_4	200 °C	103.20	7.4
	400 °C	103.28	11.7
	600 °C	103.07	12.9

Table 2 Results of XPS analysis at the N*1s* edge of Si_3N_4 powder prepared by direct nitridation of silicon before and after calcination [16]

N*1s*	Powder	Binding energy (eV)	Areal ratio (%)
NSi_3	As received	397.18	79.8
	200 °C	397.02	81.0
	400 °C	397.18	88.9
	600 °C	397.25	91.1
Amine group	As received	398.03	20.2
	200 °C	397.57	19.0
	400 °C	398.16	11.1
	600 °C	398.09	8.9

$$\begin{aligned} t_S &= r_t - r_i = \left[1 - (1 - 0.026C_{TO})^{1/3}\right] r_t \\ &= \left[1 - (1 - 0.026C_{TO})^{1/3}\right] \frac{D_S}{2} \end{aligned} \tag{3}$$

The D_S values calculated for the examined ten commercial powders are listed in Table 3.

Finally, the oxide thickness value (t_{XPS}) was calculated from XPS data using the following equation, based on the details and fractions of chemical bonds estimated from XPS analyses at the O*1s* and N*1s* edges.

$$t_{XPS} = m \ln\left(\frac{I_O I_N^0}{I_N I_O^0} + 1\right) \tag{4}$$

where m is the average escape depth of the probe (taken as 1.2 nm), I_O and I_N are the measured O*1s* and N*1s* signal intensities, respectively, while I_O^0 and I_N^0 are the theoretical XPS signal intensities found for pure SiO_2 and Si_3N_4, respectively. The

Table 3 Comparison of the thicknesses of oxidized layers in a series of ten commercially available Si_3N_4 powders according to Eqs. (3) and (4) (t_S and t_{XPS}, respectively [8]. Specific surface areas, S_V, was measured by the BET method [17], and total oxygen content, C_{TO}, was measured according to an inert gas fusion technique [18]

Powder	Fabrication method	S_V (m^2g^{-1})	C_{TO} (wt%)	t_S (nm)	t_{XPS} (nm)	F_{SO} (%)
Anzon (grade 1001) Warwick, GB	Nitridation of metallic silicon	7.3	1.4	1.6	0.4	25
Denka (9-FW) Tokyo, Japan	Nitridation of metallic silicon	12	0.9	0.6	0.1	17
KemaNord (P-95) Ljungaverk, Sweden	Nitridation of metallic silicon	8.8	1.3	1.2	0.7	58
H. C. Starck (H-1) Berlin, Germany	Nitridation of metallic silicon	10	1.4	1.2	0.2	17
H. C. Starck (LC-10) Berlin, Germany	Nitridation of metallic silicon	15	1.5	0.9	0.4	44
H. C. Starck (LC-12) Berlin, Germany	Nitridation of metallic silicon	21	1.3	0.5	0.14	28
Sumitomo Chem (TC-2) Osaka, Japan	Carbothermal reduction of silica in nitrogen atmosphere	15	2.4	1.4	0.7	51
Toshiba (Grade A) Tokyo, Japan	Carbothermal reduction of silica in nitrogen atmosphere	10	2.5	2.1	0.5	24
Toyo soda (TS-7) Tokyo, Japan	Thermal decomposition of silicon diimide	12	1.4	1.0	0.6	60
UBE (E-10) Tokyo, Japan	Thermal decomposition of silicon diimide	12	0.9	0.6	0.5	83

ratio of the XPS peak intensities of N*1s* to O*1s*, which is related to the relative atomic concentrations of nitrogen in Si_3N_4 and oxygen in SiO_2, was calculated as 1.03 [19]. The calculated t_{XPS} values for the ten powders (cf., Table 3) varied from 0.1 nm for the Denka 9-FW powder to 0.7 nm for the Sumitomo Chemical TC-2 powder, with an average value of 0.42 nm. Because a SiO_4 tetrahedron has a size of ~0.265 nm, the average oxide layer thickness corresponds to one to two silica monolayers. Only Denka 9-FW, H. C. Starck H-1, and H. C. Starck LC-12 powders had t_{XPS} values lower than the size of one SiO_4-tetrahedron (cf. Table 3). In these

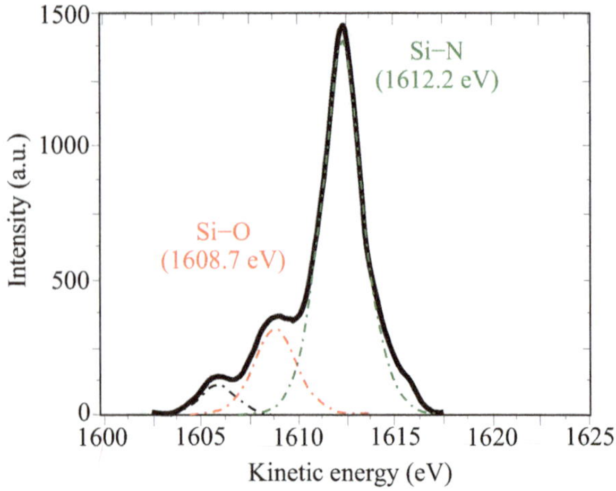

Fig. 4 XAES spectrum of Si_3N_4 powder prepared by nitridation of Si (kinetic energies for Si *KLL* peaks are shown in brackets). (Used with permission [20])

instances, rather than a defined oxide film, the oxidized thickness should be represented as a layer of adsorbed oxygen.

Comparison of t_S and t_{XPS} measurements demonstrates that the latter values are always smaller than the former. This disparity can be explained in part by assuming the presence of oxygen both on the surface and in the bulk. The ratio, $F_{SO} = t_{XPS}/t_S$, shown in the last column of Table 3, indicates that the powders formed by silicon diimide breakdown (i.e., Toyo Soda TS-7 and UBE E-10) contain extremely little oxygen content in the Si_3N_4 lattice. In contrast, the nitrided powders (i.e., Anzon 1001, Denka 9-FW, KemaNord P-95, H. C. Starck H-1, and LC-12) exhibit a significant proportion of oxygen in the bulk. The two powders prepared by carbothermal reduction of silica (i.e., Sumitomo Chemical TC-2 and Toshiba Grade A) have the most total oxygen, with a larger percentage in the bulk.

X-ray-induced Auger electron spectroscopy (XAES) is a different approach to characterize the oxide surface thickness in Si_3N_4 powders. XAES measures radiation-less internal rearrangement that occurs upon X-ray bombardment. In XAES, Auger transitions are named according to the electron levels involved in the transition. A *KLL* transition occurs when an initial *K*-level hole is created, an *L*-shell electron fills the *K*-shell hole, and another *L*-shell electron is ejected as the Auger electron. This method, which probes a depth of ~5 nm, can be used to determine the effective thickness (t_A) of the SiO_2 surface layer by computing it from the Si *KLL* intensity ratio of Si-O and Si-N deconvoluted peaks in the XAES spectrum. Figure 4 shows a representative XAES spectrum in which the two Si *KLL* peaks are separated after deconvolution [20].

Hackley et al. [20] measured the Si *KLL* peaks of five commercially available Si_3N_4 powders (i.e., KemaNord P95M, Norton NSNC-300, Ube SN E-10, H. C. Starck LC-10, and Denka P21B, and computed their t_A values using the following equation [21]:

Table 4 Surface composition by XPS and thickness of oxidized layer, t_A for a series of five commercially available Si_3N_4 powders; t_A is computed according to Eq. (5) [20]. The parameter R represents the ratio of the Si-N and Si-O Si *KLL* intensities (cf. Fig. 4)

Powder	Fabrication method	S_V (m^2g^{-1})	C_{TO} (wt%)	t_S (nm)	t_{XPS} (nm)	F_{SO} (%)
Anzon (grade 1001) Warwick, GB	Nitridation of metallic silicon	7.3	1.4	1.6	0.4	25
Denka (9-FW) Tokyo, Japan	Nitridation of metallic silicon	12	0.9	0.6	0.1	17
KemaNord (P-95) Ljungaverk, Sweden	Nitridation of metallic silicon	8.8	1.3	1.2	0.7	58
H. C. Starck (H-1) Berlin, Germany	Nitridation of metallic silicon	10	1.4	1.2	0.2	17
H. C. Starck (LC-10) Berlin, Germany	Nitridation of metallic silicon	15	1.5	0.9	0.4	44
H. C. Starck (LC-12) Berlin, Germany	Nitridation of metallic silicon	21	1.3	0.5	0.14	28
Sumitomo Chem (TC-2) Osaka, Japan	Carbothermal reduction of silica in nitrogen atmosphere	15	2.4	1.4	0.7	51
Toshiba (Grade A) Tokyo, Japan	Carbothermal reduction of silica in nitrogen atmosphere	10	2.5	2.1	0.5	24
Toyo Soda (TS-7) Tokyo, Japan	Thermal decomposition of silicon diimide	12	1.4	1.0	0.6	60
UBE (E-10) Tokyo, Japan	Thermal decomposition of silicon diimide	12	0.9	0.6	0.5	83

$$t_A = -2.04\ \ln[R/(R + 1.71)] \tag{5}$$

where R is the ratio of the Si_3N_4 and SiO_2 Si *KLL* intensities. The findings of these XAES analyses are shown in Table 4 along with the surface concentrations of O, N, Si, and adventitious C as measured by XPS. The thickness values, t_A, are consistent with the t_{XPS} values determined by XPS. They range from 0.1 to 1.7 nm (cf. Table 3).

Based on the above analyses, the passive oxidation of Si_3N_4 is assumed to be controlled by the slow inward diffusion of oxygen through a surface oxide film, which follows parabolic kinetics. The significantly slower oxidation rate of Si_3N_4

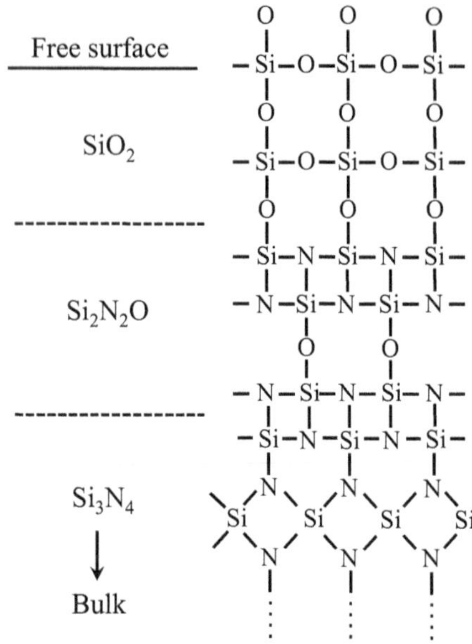

Fig. 5 Schematic representation of the compositional gradient for the oxidation layer on a Si_3N_4 particle. (Figure rearranged and used with permission [9])

compared to silicon suggests that the oxide layers have fundamental structural differences. Electron microscopy, XPS, and AES analyses suggest that these differences are because the oxide film on silicon is pure amorphous SiO_2, while the one on Si_3N_4 contains nitrogen. Therefore, the oxidized layer on Si_3N_4 has the structure of a duplex film consisting of an outer layer of amorphous SiO_2 and an interface layer of amorphous silicon oxynitride (Si_2N_2O) [9]. This structure is schematically shown in Fig. 5. The Si_2N_2O phase was initially believed to have an approximately constant SiO_2:Si_2N_2O ratio [22–24]; but it was later shown to be a graded composition according to AES and Rutherford backscattering spectroscopy studies [25]. The graded Si_2N_2O layer improves the oxidation resistance of Si_3N_4 when compared to Si. Oxygen permeation through the "sub-oxide layer" is the rate-limiting step. Oxygen diffusion and its substitution for nitrogen represent two interlinked steps. Both increase Si_3N_4's resistance to oxygen penetration [26–28]. However, the surface oxide layer on Si_3N_4 can be easily reduced or removed using a variety of methods, such as heating in a reducing atmosphere or chemically treating it with alkaline or hydrofluoric acid [6, 10]. Simple washing in an aqueous electrolyte solution is one mild therapy that can change the surface layer [10, 29], and high-energy mechanical abrasion during powder milling severely alters the surface [30].

Another important tool in the structural characterization of the surface layer of Si_3N_4 powders is Fourier transform infrared (FTIR) spectrometry. This technique has the unique ability to non-destructively identify several Si-N and H-related species, such as $SiNH_2$, Si_2NH, SiOH, and SiH [31, 32]. Figure 6a shows the FTIR of Si_3N_4 powder between 600 and 1300 cm^{-1}. Wavenumbers and assignments are shown in Table 5 [33–36]. FTIR can accurately differentiate Si-N and Si-O bonds embedded in various clusters within this spectral region. Hien et al. compared

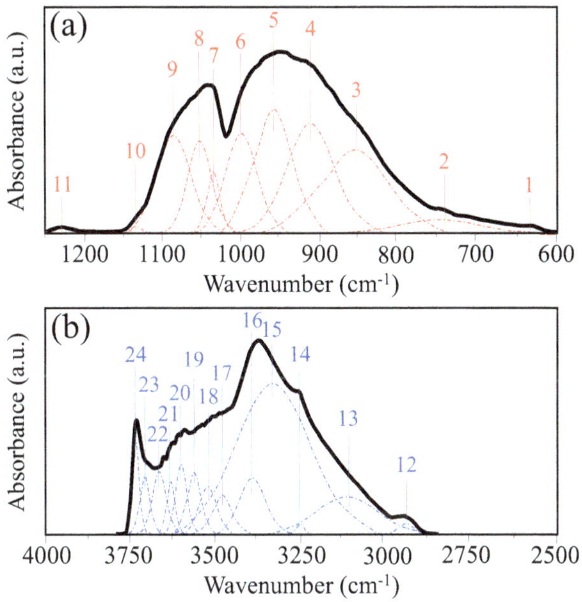

Fig. 6 (**a**) FTIR spectrum of Si_3N_4 powder between 600 and 1300 cm^{-1} and (**b**) DRIFT spectrum of Si_3N_4 powder between 2500 and 4000 cm^{-1}. The labels correspond to those in Tables 5 and 6, respectively. (Used with permission [37])

Table 5 Deconvoluted FTIR bands in the wavenumber interval 600~1300 cm^{-1} and their spectroscopic origins [33–36]

Label	Spectral location (cm^{-1})	Physical origin
1	630	Si–H wagging
2	750	Si–O–Si bending Si_3–Si–N asymm. stretching
3	850	Si–N stretching (TO)
4	920	Si–N_4 stretching (TO)
5	950	Si–N_x ($x<1$) stretching
6	1000	Si–N asymm. stretching in H–SiN_3
7	1042	Si–N_4 stretching (LO)
8	1056	Si–N stretching (LO)
9	1080	Si–O–Si stretching
10	1142	Si–O stretching (TO)
11	1248	Si–O stretching (LO)

Labels correspond to those shown in Fig. 6a

the diffuse reflectance infrared Fourier transform (DRIFT) spectra of Si_3N_4 powders produced by diimide precipitation and carbothermal reduction [37]. As shown in Fig. 6b, the DRIFT spectra were collected in the wavenumber range of 2500 to 4000 cm^{-1}. Within this range, NHx stretching modes can be found at

Table 6 List of deconvoluted FTIR bands in the interval 2500~4000 cm^{-1} and their spectroscopic origins [37, 38]

Label	Spectral location (cm^{-1})	Physical origin
12	2900	NH•••N symmetric stretching
13	3110	NH stretching
14	3252	NH•••N asymmetric stretching
15	3335	Si–NH symmetric stretching
16	3378	Si–NH asymmetric stretching
17	3482	Si–NH_2 symmetric stretching
18	3510	Si–NH_2 asymmetric stretching
19	3568	OH stretching
20	3603	OH stretching
21	3625	OH stretching
22	3657	OH stretching
23	3705	OH stretching
24	3748	OH stretching

The labels correspond to those shown in Fig. 6b

3280 to 3400 cm^{-1} and OH stretching modes can be found at 3500 to 3750 cm^{-1} (see labels in the inset). Regardless of the synthesis method, powders produced in an NH_3 medium showed bands that were one order of magnitude more intense than powders prepared with N_2 as the nitriding medium. Bands at 3480~3497 cm^{-1} and 3379~3390 cm^{-1} were assigned to Si-NH_2 and Si-NH, respectively. The highest intensity for all powders was the Si-NH stretching band at 3290~3330 cm^{-1} (cf. assignments in Table 6) [37, 38]. While multiple Si-OH stretching bands were observed, a band at 3048–3115 cm^{-1} was also thought to be the tail of the 3320 cm^{-1} Si-NH asymmetric band (cf. labels in Fig. 6). These FTIR results revealed a complex scenario of surface groups: silanols (Si-OH), imido (Si_2NH), silylamine ($SiNH_2$), and silylidyne (SiH), together with nitrogen bonded to three silicon atoms [31, 37, 38]. Figure 7 provides a schematic representation of the oxide surface layer on Si_3N_4.

This representation simplifies the surface population into silanols and silylamines. In its initial state, the Si_3N_4 surface is charge-balanced and contains nearly equal fractions of silylamine and silanol sites. Studies have shown that these two species, silanol and silylamine, are present in nearly equivalent proportions in the pH range of 6 to 8.5 [39]. However, oxidation shifts this balance toward a silanol-rich population (cf. Fig. 7). The structure of silanols on oxidized Si_3N_4 has been studied in detail by Sulpizi et al. [40]. They found that four distinct species are present on the surface: silicic acid with an acceptor character ($Si(OH)_4$), in-plane acceptor silanols (–SiOH), doubly bonded water molecules, and silicic acid with a donor character. Silanol groups on Si_3N_4 surfaces play a counteracting role in surface charging and acidity modulation. Neutral (in-plane) silanol groups accept hydrogen bonds, while silicic acid donates them. These different charge characteristics are crucial for biological interactions. Chemically modulating the Si_3N_4 surface to control these groups provides an important degree of freedom in optimizing Si_3N_4 for biological purposes.

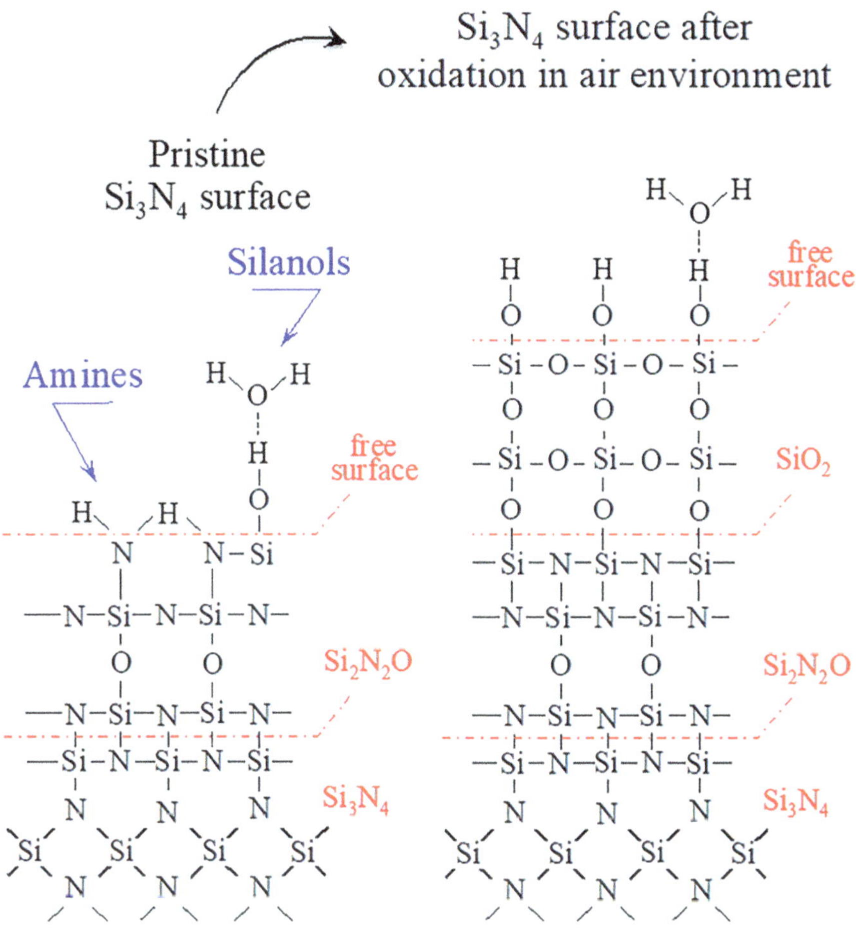

Fig. 7 Schematic representation of silanols and amine groups at the surface of Si_3N_4, and their modification upon oxidation in air. (Figure redrawn and used with permission [9])

3 Hydrolysis of Silicon Nitride

When silicon nitride comes into contact with water, it undergoes hydrolysis, a reaction that releases ammonia and forms a silicon oxynitride surface layer by the following reaction [41]:

$$Si_3N_4 + 6H_2O \rightarrow 3SiO_2 + 4NH_3 \ \Delta G = -146.0\,\text{kJ/mol}\ (298\,^\circ\text{K}) \tag{6}$$

Si_3N_4 powders are always partially oxidized when exposed to moist air. The pungent smell of ammonia is unmistakable due to the hydrolysis of Si-N covalent bonds. In an aqueous suspension, this reaction leads to local pH buffering at the

Fig. 8 (**a**) Graph of Eq. (9) showing the relative concentrations NH_3 and NH_4^+ at the Si_3N_4 surface; and (**b**) experimentally derived values of eluted nitrogen as a function of pH. (Reprinted with permission under Creative Commons License 4.0 [46])

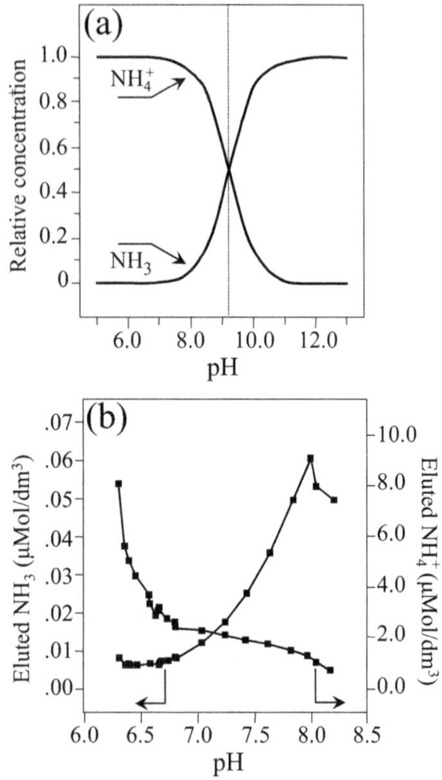

solid-liquid interface [4]. The release of nitrogen increases the electrophilicity of adjacent silicon sites and also makes them susceptible to nucleophilic attack. When water interacts with silicon sites, metastable complexes are first formed, but they decay rapidly, releasing Si ions in a ratio that depends on the environmental pH [42]. Ammonia (NH_3) or ammonium (NH_4^+) and a silicon oxynitride surface layer concurrently form; and continued hydrolysis produces orthosilicic acid ($Si(OH)_4$) according to the following reactions:

$$H^+ + NH_3 \rightarrow NH_4{}^+ \tag{7}$$

$$SiO_2 + 2H_2O \rightarrow Si(OH)_4 \tag{8}$$

At room temperature, the fraction of ammonia (NH_3) in suspension with the Si_3N_4 particles varies with pH in a sigmoidal manner [43, 44]:

$$[NH_3][H^+]/[NH_4{}^+] = 5.7 \times 10^{-10} \tag{9}$$

At physiologic pH (7.35~7.45), this equation shows that the amount of free NH_3 in solution is very low (i.e., 1 to 2%). Figure 8a is a plot of Eq. (9). Note that NH_3

increases to about 7% at a pH of ~8.3. It continues an exponential rise to ~50% at a pH of about 9.3 and becomes 100% at a pH > 11.

Using colorimetric assays, Fig. 8b shows the experimentally derived concentrations of the two species as a function of pH in a dilute aqueous slurry of Si_3N_4 particles [4]. Silanol (Si-OH) and silylamine (Si-NH_2) groups are always found on oxidized Si_3N_4 surfaces [37, 45]. A model of this structure was presented earlier in Fig. 7. The relative fractions of silanols and silylamines affect the wettability and isoelectric point of Si_3N_4. This, in turn, has a significant impact on its biological properties. Although different Si_3N_4 powders or dense samples may initially exhibit varying biological characteristics, a local increase in pH invariably dissolves the silanols and exposes the N-containing lattice (cf., Fig. 7) [9, 25, 40].

As can be seen in Fig. 8a, the pH at which the concentrations of NH_3 and NH_4^+ are equal (i.e., pH~9.3) is the isoelectric point of Si_3N_4, and this value is also within the expected range for surface amines (pK = 9~10) [20]. The surface charge density and polarity of Si_3N_4 powders are controlled by two hydrolyzed species, amorphous SiO_2 and amines [9, 20, 29, 42–45]. Using a zwitterionic disassociation model, Bergström and Bostedt developed disassociation constant for these two moieties [29], and Greil et al. linked the oxygen content of Si_3N_4 powders to its isoelectric point, pH_{iep} [9]. Given this information, it became apparent that the hydrolysis of Si_3N_4 is due to the formation of a gel layer (Δr_h) within the oxide ≡Si–O– and nitride ≡Si–N = groups according to the following equations:

$$[\equiv \text{Si–O–Si} \equiv] + H_2O \rightarrow 2[\text{Si–OH}] \quad (10)$$

$$\begin{bmatrix} \equiv \text{Si}\backslash & & \\ & \text{N—Si} \equiv \\ \equiv \text{Si}/ & & \end{bmatrix} + 2H_2O \rightarrow [\text{Si–NH}_2] + 2[\text{Si–OH}] \quad (11)$$

Surface charging of Si_3N_4 particles and their pH-dependent disassociation reactions can be described as follows [47–51]:

$$[\text{Si–OH}_2{}^+] \rightarrow [\text{Si–OH}] + H^+ \quad (12)$$

$$[\text{Si–OH}] \rightarrow [\text{Si–O–}] + H^+ \quad (13)$$

$$[\text{Si–NH}_3{}^+] \rightarrow [\text{Si–NH}_2] + H^+ \quad (14)$$

The isoelectric point for Si_3N_4 powders can vary widely (i.e., pH_{iep}~2 to >9). Its value depends on the amount of different hydrolyzable surface groups and how these groups dissociate at different pH levels [10, 52]. In Fig. 9, the surface oxygen content of various Si_3N_4 powders is plotted as a function of pH, normalized to their specific surface areas [9]. Regression analysis (R^2 = 0.97) provides this empirical relationship:

Fig. 9 Surface oxygen content for commercial Si_3N_4 powders normalized to surface area C_o/S_v versus pH_{iep} Eq. (10). (Redrawn and used with permission [9])

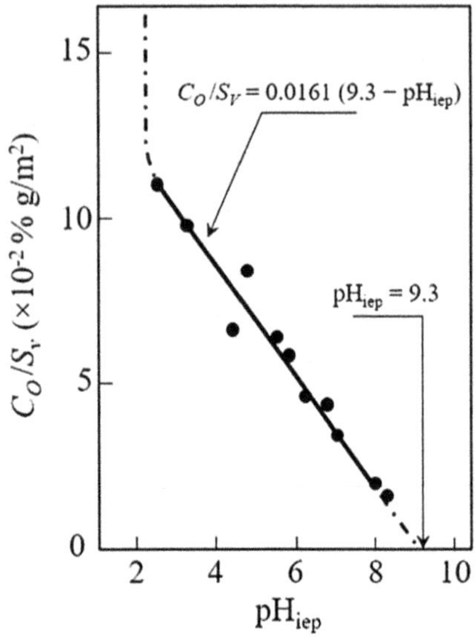

$$C_O/S_V = 0.0161 \times \left(9.3 - pH_{iep}\right) \tag{15}$$

Note that an oxygen-free surface has a pH_{iep} of ~9.3, whereas highly oxidized surfaces have values of between 1.5 and 3.5, which corresponds to the IEP of SiO_2 [49, 53, 54]. This relationship explains the dissociation of amines in comparison to silanols. Figure 10 provides a schematic of a Si_3N_4 particle interface in an aqueous environment. Constants for the hydrolytic dissociation reactions are given by the following expressions [55]:

$$K^+ = [\text{Si–XH}][H^+]_S/[\text{Si–XH}_2{}^+] \tag{16}$$

$$K^- = [\text{Si–X}^-][H^+]_S/[\text{Si–XH}] \tag{17}$$

where X = O or NH. The proton concentration on the particle surface, $[H^+]_S$, is a function of the number of bulk protons, $[H^+]$, and the Boltzmann equation:

$$[H^+]_S = [H^+]\ \exp[-q\Psi_O/kT] \tag{18}$$

where q is the surface charge, Ψ_O is the surface potential, k is Boltzmann's constant, and T is the absolute temperature. The surface potential, Ψ_O, is related to the pH by the Nernst equation [48–50, 56]:

$$2.303\left(pH_{iep}{}^{\text{Si–OH}} - pH\right) = [q\Psi_O/kT] + \ln\{[\text{Si–OH}^+]/[\text{Si–O}^-]\}/2 \tag{19}$$

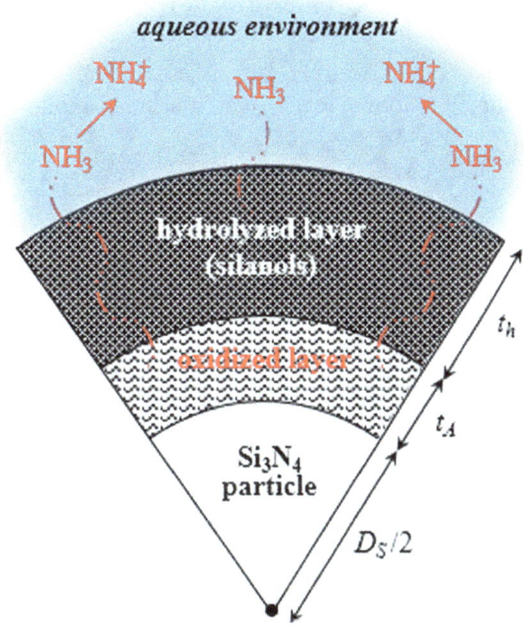

Fig. 10 Schematic of a Si_3N_4 particle in an aqueous environment. $D_S/2$, t_A, and t_h represent particle radius, the thickness of its oxidized layer, and the thickness of the hydrolyzed silanol layer. Redrawn and used with permission [9, 57]

These relationships describe how the surface parameters of amphoteric Si–OH sites, such as surface charge, q, and surface potential, Ψ_O, are linked to environmental pH [51]. When the pH is higher than the isoelectric point of Si–OH sites, the surface of the material becomes negatively charged due to the dissociation of Si–OH sites into negatively charged ions. The tendency of the Si–OH sites to dissociate can be calculated using Eqs. (16) and (17) [50] as:

$$K_S = K^+/K^- = [\text{Si–OH}_2{}^+][\text{H}^+]_S/[\text{Si–OH}]^2 \tag{20}$$

$$\Delta \text{p}K_S = \log K_S = \text{p}K_S{}^- - \text{p}K_S{}^+ \tag{21}$$

Theoretical and experimental studies have shown that the difference between the pKa values of silanol and amine sites on Si_3N_4 is in the range of 8 to 10 [49, 50]. This means that the pH values at which these two types of sites dissociate are very similar. This is an important characteristic of Si_3N_4, as it affects how the material interacts with cells and bacteria.

At the isoelectric point, the number of negatively and positively charged surface sites is equal. This is because the concentration of protons on the surface of the material ($[H^+]_S$) is equal to the concentration of protons in the solution ($[H^+]$). A study by Harame et al. derived a relationship that links the isoelectric point (pH_{iep}) to the surface fractions of [Si–OH] and [Si–NH_2] sites, which are referred to as f_S and f_A, respectively [51]:

$$pH_{iep} = pH_{iep}^{Si\text{–}OH} + \left\{ \sin h^{-1} \left[K_S^{1/2} f_A / 2 f_S \right] \right\} / 2.303 \quad (22)$$

Using Eqs. (19)~(21), the surface reaction constants and the pH_{iep} can then be combined to form a linear approximation, as follows:

$$pH_{iep} \approx f_S \left[(\Delta pK_S/2) + pK_S^{+} \right] + f_A \left[(\Delta pK_A/2) + pK_A^{+} \right] \quad (23)$$

with the fractional site concentrations obeying this relationship:

$$f_S + f_A = 1 \quad (24)$$

By taking specific values for the $pK_a^{+} = 10$, $\Delta pK_S = 8$, and $pK_S^{+} = -1.8$ [49, 50], and by using the value of $\Delta pK_A = -1.4$ which is obtained from $pH_{iep} \approx 9.3$ for an oxygen-free surface (cf., Fig. 9), a linear relationship linking the silanol concentration, f_s, to the isoelectric point, pH_{iep}, for Si_3N_4 powder is obtained:

$$f_S \approx 0.1408 \left(9.3\text{–}pH_{iep} \right) \quad (25)$$

For $f_S = 1$, it turns out that pH_{iep} of Si–OH = 2.2. This pH corresponds to full surface coverage by silanol groups [49, 55]. Fung et al. found that silica films absorb water and ions over time, which causes the Ψ_O—pH response to fluctuate [50]. This fluctuation is unique to silica films and is important for the long-term interactions of Si_3N_4 with biological systems.

Molecular simulations suggest that an unstable intermediate complex forms on Si_3N_4 when it is hydrolyzed [42]. This complex involves a penta-coordinated silicon atom, per the following reaction [58]:

$$Si(OH)_4 + (OH)^- \rightarrow \left[Si(OH)_5 \right]^- \rightarrow \left[(OH)_3 SiO \right]^- \cdot H_2O; \quad (26)$$

Figure 11a shows that the amount of dissolved silicon in solution remains constant once the O/Si ratio reaches ~0.2 [59]. At this point, the oxidized surface layer is at its minimum thickness.

Silicic acid dissolution only continues when new surface oxide forms by the ongoing hydrolysis of Si_3N_4 (cf., Eq. 6). The ability to periodically refresh its surface by restarting the elution of ammonia and silicic acid (Fig. 11b) is an important feature of Si_3N_4 for its long-term bioactivity. Additionally, residual Si_3N_4 particles or debris will not persist in the body. Two studies showed that about 65 to 75 mg/L of Si_3N_4 dissolves in simulated body fluid within a month. This suggests that Si_3N_4 wear debris will be resorbed without local saturation [60, 61]. The clearance of wear debris in a relatively short time is an attractive Si_3N_4 property for implant applications.

In summary, when Si_3N_4 interacts with water, it undergoes a series of chemical reactions that result in the formation of ammonia, silica, and other silicate species.

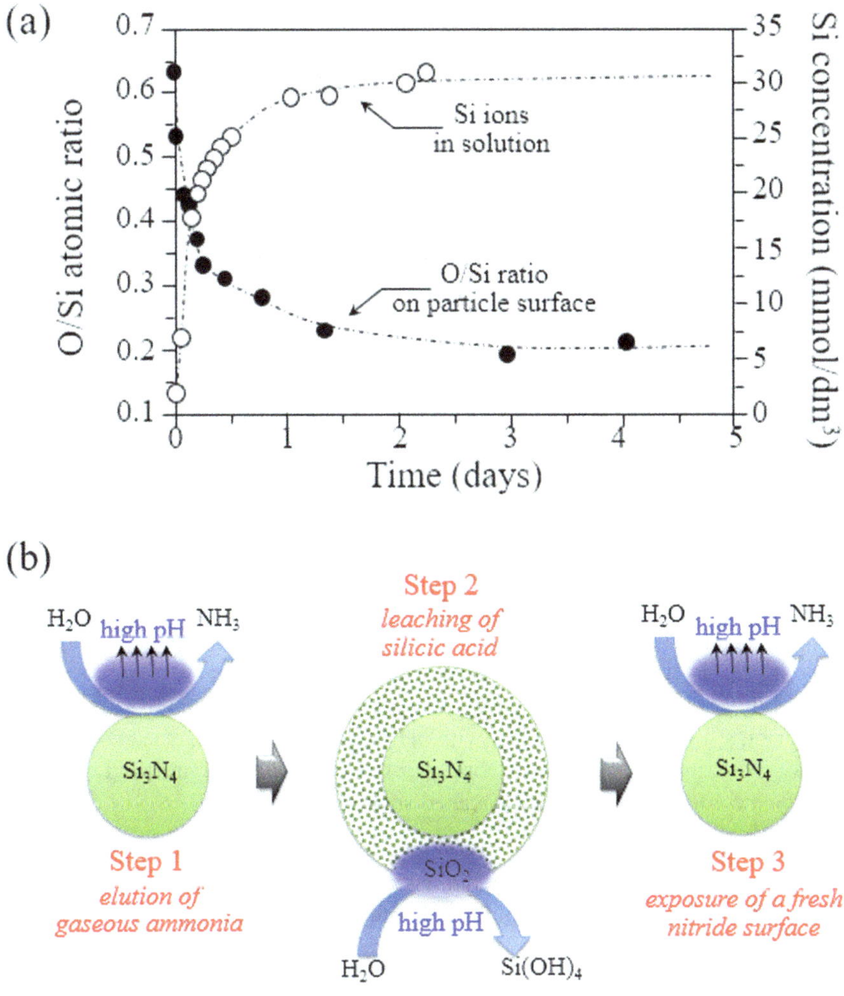

Fig. 11 (**a**) Elemental O/Si ratio on the Si_3N_4 particle and the Si concentration in solution versus time in an aqueous environment containing 5 vol.% Si_3N_4 in suspension at pH of 12 and a temperature of 25 °C. Replotted with permission from Ref. [59]; (**b**) Schematic diagram showing how Si_3N_4 undergoes a periodic cycle of leaching, oxidation, and dissolution. In Step 1, ammonia release and pH buffering occur. In Step 2, a layer of silicic acid is dissolved. In Step 3, a fresh Si_3N_4 surface is exposed, and the cycle starts over. (Redrawn and used with permission [59])

The leaching rate of the oxidized surface layer increases with pH. A thorough understanding of this underlying chemistry can help biomaterial engineers design implantable devices that modulate Si_3N_4's surface chemistry for different applications.

4 Summary

In conclusion, the interaction between Si_3N_4 and air or water environments is characterized by the following important phenomena:

- Oxidation: Exposure of Si_3N_4 powders to dry air results in the formation of a transitional oxide layer composed primarily of Si_2N_2O. In moist air, the transition includes a host of additional intermediates including silanol, imido, silylamine, and silylidyne groups. Nitrogen is released and immediately reacts to form ammonia. Continued oxidation forms silica at the surface boundary which is subsequently hydrolyzed to silicic acid.
- Hydrolysis: Si_3N_4 readily hydrolyzes when dispersed in water. The reactions are identical to those observed in moist air. Ammonia is released forming silica at the solid surface. The ammonia elution (and its subsequent protonation to ammonium) leads to robust pH buffering at the solid/liquid interface. The oxidized silica layer tends to dissolve, forming colloidal silica (i.e., silicic acid) and other silicate species.
- Interdependency of surface chemistry: There is a clear interdependency between the oxygen content at the surface, the relative amount of Si-OH and Si_2-NH surface groups, and the surface charge density. Understanding this interdependency provides biomaterial scientists with the ability to design and modulate Si_3N_4's surface chemistry to match biological needs.
- Solubility of the oxidized-surface layer: The leaching rate of the oxidized surface layer on silicon nitride increases with pH. Leaching studies have shown that not only oxidation and hydrolysis, but also dissolution can significantly alter the surface composition and the associated surface charge.

As described in detail within other chapters, silicon nitride's surface chemistry, although simple, is unique among biomaterials. It is its reaction with biological fluids that provides it with significant antipathogenic, osteogenic, and other therapeutic properties.

References

1. Bocanegra-Bernal, M.H.; Matovic, B. Dense and Near-Net-Shape Fabrication of Si_3N_4 Ceramics. *Mater. Sci. Eng. A* **2009**, *500*, 130–149, doi:https://doi.org/10.1016/j.msea.2008.09.015.
2. Lange, F.F. Powder Processing Science and Technology for Increased Reliability. *J. Am. Ceram. Soc.* **1989**, *72*, 3–15, doi: https://doi.org/10.1111/j.1151-2916.1989.tb05945.x.
3. McEntire, B.J.; Bal, B.S.; Rahaman, M.N.; Chevalier, J.; Pezzotti, G. Ceramics and Ceramic Coatings in Orthopaedics. *J. Eur. Ceram. Soc.* **2015**, *35*, 4327–4369, doi:https://doi.org/10.1016/j.jeurceramsoc.2015.07.034.
4. Pezzotti, G. Silicon Nitride: A Bioceramic with a Gift. *ACS Appl. Mater. Interfaces* **2019**, *11*, 26619–26636, doi:https://doi.org/10.1021/acsami.9b07997.

5. Raider, S.I.; Flitsch, R.; Aboaf, J.A.; Pliskin, W.A. Surface Oxidation of Silicon Nitride Films. *J. Electrochem. Soc.* **1976**, *123*, 560–565.
6. Rahaman, M.; Boiteux, Y.; DeJonghe, L. Surface Characterization of Silicon Nitride and Silicon Carbide Powders. *Bull. Am. Ceram. Soc.* **1986**, *65*, 1171–1176.
7. Qadir, A.; Fogarassy, Z.; Horváth, Z.E.; Balazsi, K.; Balazsi, C. Effect of the Oxidization of Si_3N_4 Powder on the Microstructural and Mechanical Properties of Hot Isostatic Pressed Silicon Nitride. *Ceram. Int.* **2018**, *44*, 14601–14609, doi: https://doi.org/10.1016/j.ceramint.2018.05.081.
8. Peuckert, M.; Greil, P. Oxygen Distribution in Silicon Nitride Powders. *J. Mater. Sci.* **1987**, *22*, 3717–3720, doi:https://doi.org/10.1007/BF01161483.
9. Greil, P.; Nitzsche, R.; Friedrich, H.; Hermel, W. Evaluation of Oxygen Content on Silicon Nitride Powder Surface from the Measurement of the Isoelectric Point. *J. Eur. Ceram. Soc.* **1991**, *7*, 353–359, doi:https://doi.org/10.1016/0955-2219(91)90058-8.
10. Bergstrom, L.; Pugh, R.J. Interfacial Characterization of Silicon Nitride Powders. *J. Am. Ceram. Soc.* **1989**, *72*, 103–109.
11. Brow, R.K.; Pantano, C.G. Compositionally Dependent Si 2p Binding Energy Shifts in Silicon Oxynitride Thin Films. *J. Am. Ceram. Soc.* **1986**, *69*, 314–316, doi: https://doi.org/10.1111/j.1151-2916.1986.tb04738.x.
12. Brow, R.K.; Panano, C.G. Thermochemical Nitridation of Microporous Silica Films in Ammonia. *J. Am. Ceram. Soc.* **1987**, *70*, 9–14, doi:https://doi.org/10.1111/j.1151-2916.1987.tb04845.x.
13. Hien, T.T.T.; Ishizaki, C. Surface Structure of Commercial Si_3N_4 Powders Analyzed by X-Ray Photoelectron Spectroscopy (XPS). *J. Ceram. Soc. Japan* **2005**, *113*, 647–653.
14. Wagner, C.D.; Passoja, D.E.; Hillery, H.F.; Kinisky, T.G.; Six, H.A.; Jansen, W.T.; Taylor, J.A. Auger and Photoelectron Line Energy Relationships in Aluminum–Oxygen and Silicon–Oxygen Compounds. *J. Vac. Sci. Technol.* **1982**, *21*, 933–944, doi:https://doi.org/10.1116/1.571870.
15. Klein, G.; Chun, H.-U. Determination of Optical Interband Transitions in Crystalline Quartz from X-ray Spectroscopical Data. *Phys. status solidi* **1972**, *49*, 167–172, doi: https://doi.org/10.1002/pssb.2220490115.
16. Ni, Z.; Yang, X.; Yang, X.; He, Q.; Xu, X.; Xie, H. Surface Modification of Ultrafine Silicon Nitride Powders by Calcination. *Int. J. Appl. Ceram. Technol.* **2019**, *16*, 1364–1372, doi: https://doi.org/10.1111/ijac.13232.
17. Malghan, S.G.; Hsu, S.M. Ceramic Powders Characterization: Results of an International Laboratory Study. *NIST Spec. Publ.* **1995**, 29–57.
18. Mangels, J.A. The Effect of Silicon Purity on the Strength of Reaction-Bonded Si3N4. *J. Mater. Sci.* **1980**, *15*, 2132–2135, doi:https://doi.org/10.1007/BF00550647.
19. Peuckert, M.; Bonzel, H.P. Characterization of Oxidized Platinum Surfaces by X-ray Photoelectron Spectroscopy. *Surf. Sci.* **1984**, *145*, 239–259, doi: https://doi.org/10.1016/0039-6028(84)90778-7.
20. Hackley, A. Effects of Soxhlet Extraction on the Surface Oxide Layer of Silicon Nitride Powders. *Mater. Phys.* **1993**, *36*, 112–118.
21. Wang, P. S.; Malghan, S.G.; Hsu, S.M.; Wittberg, T.N. The Oxidation of an Aluminum Nitride Powder Studied by Bremsstrahlung-Excited Auger Electron Spectroscopy and X-ray Photoelectron Spectroscopy. *J. Mater. Res.* **1995**, *10*, 302 305, doi:https://doi.org/10.1557/JMR.1995.0302.
22. Luthra, K.L. A Mixed Interface Reaction/Diffusion Control Model for Oxidation of Si3 N 4. *J. Electrochem. Soc.* **1991**, *138*, 3001, doi:https://doi.org/10.1149/1.2085355.
23. Ogbuji, L.U.J.T. Role of Si2N2O in the Passive Oxidation of Chemically-Vapor-Deposited Si_3N_4. *J. Am. Ceram. Soc.* **1992**, *75*, 2995–3000, doi: https://doi.org/10.1111/j.1151-2916.1992.tb04377.x.
24. Tressler, R.E. Theory and Experiment in Corrosion of Advanced Ceramics. In *Corrosion of Advanced Ceramics: Measurement and Modelling Proceedings of the NATO Advanced*

Research Workshop on Corrosion of Advanced Ceramics Tübingen, Germany August 30–September 3, 1993; Nickel, K.G., Ed.; Springer Netherlands: Dordrecht, 1994; pp. 3–23 ISBN 978-94-011-1182-9.
25. Ogbuji, L.U.J.T.; Bryan, S.R. The SiO_2-Si_3N_4 Interface, Part I: Nature of the Interphase. *J. Am. Ceram. Soc.* **1995**, *78*, 1272–1278, doi: https://doi.org/10.1111/j.1151-2916.1995.tb08481.x.
26. Ogbuji, L.U.J.T. The SiO_2-Si_3N_4 Interface, Part II: O_2 Permeation and Oxidation Reaction. *J. Am. Ceram. Soc.* **1995**, *78*, 1279–1284, doi: https://doi.org/10.1111/j.1151-2916.1995.tb08482.x.
27. Sheldon, B.W. Silicon Nitride Oxidation Based on Oxynitride Interlayers with Graded Stoichiometry. *J. Am. Ceram. Soc.* **1996**, *79*, 2993–2996, doi: https://doi.org/10.1111/j.1151-2916.1996.tb08742.x.
28. Riley, F.L. Silicon Nitride and Related Materials. *J. Am. Ceram. Soc.* **2000**, *83*, 245–265.
29. Bergstrow, L.; Bostedt, E. Surface Chemistry of Silicon Nitride Powders: Electrokinetic Behaviour and ESCA Studies. *Colloids and Surfaces* **1990**, *49*, 183–197.
30. Malghan, S.G.; Pei, P.T.; Minor, D.B. Characterization of Attrition Milled Silicon Nitride Powder. *Ceram. Trans.* **1991**, *22*, 297–304.
31. Busca, G.; Lorenzelli, V.; Porcile, G.; Baraton, M.I.; Quintard, P.; Marchand, R. FT-IR Study of the Surface Properties of Silicon Nitride. *Mater. Chem. Phys.* **1986**, *14*, 123–140.
32. Li, Y.-L.; Liang, Y.; Zheng, F.; Xiao, K.; Hu, Z.-Q.; Shun, T. Fourier Transformation Infrared Investigation of Surface Oxidation of Ultrafine Si_3N_4 Powders. *J. Mater. Sci. Lett.* **1995**, *14*, 713–715, doi:https://doi.org/10.1007/BF00253380.
33. Bugaev, K.O.; Zelenina, A.A.; Volodin, V.A. Vibrational Spectroscopy of Chemical Species in Silicon and Silicon-Rich Nitride Thin Films. *Int. J. Spectrosc.* **2012**, *2012*, 281851, doi:https://doi.org/10.1155/2012/281851.
34. Bermudez, V.M. Wet-Chemical Treatment of Si_3N_4 Surfaces Studied Using Infrared Attenuated Total Reflection Spectroscopy. *J. Electrochem. Soc.* **2005**, *152*, F31, doi:https://doi.org/10.1149/1.1851056.
35. Scardera, G.; Puzzer, T.; Conibeer, G.; Green, M.A. Fourier Transform Infrared Spectroscopy of Annealed Silicon-Rich Silicon Nitride Thin Films. *J. Appl. Phys.* **2008**, *104*, 104310, doi: https://doi.org/10.1063/1.3021158.
36. Kislyi, P.S.; Posun'ko, L.S.; Malogolovets, V.G. Phase Composition of Si_3N_4 Powders Synthesized by Various Methods. *Sov. Powder Metall. Met. Ceram.* **1988**, *27*, 407–410, doi:https://doi.org/10.1007/BF00797693.
37. Hien, T.T.T.; Ishizaki, C.; Ishizaki, K. NHx/OH Surface Groups in Commercial Si_3N_4 Powders Analyzed by Diffuse Reflectance Infrared Fourier Transform (DRIFT) Spectroscopy. *J. Ceram. Soc. Japan* **2003**, *111*, 28–32, doi:https://doi.org/10.2109/jcersj.111.28.
38. Yin, Z.; Smith, F.W. Optical Dielectric Function and Infrared Absorption of Hydrogenated Amorphous Silicon Nitride Films: Experimental Results and Effective-Medium-Approximation Analysis. *Phys. Rev. B* **1990**, *42*, 3666–3675, doi:https://doi.org/10.1103/PhysRevB.42.3666.
39. Senden, T.J.; Drummond, C.J. Surface Chemistry and Tip-Sample Interactions in Atomic Force Microscopy. *Colloids Surfaces A Physicochem. Eng. Asp.* **1995**, *94*, 29–51, doi:https://doi.org/10.1016/0927-7757(94)02954-Q.
40. Sulpizi, M.; Gaigeot, M.-P.; Sprik, M. The Silica–Water Interface: How The Silanols Determine The Surface Acidity And Modulate The Water Properties. *J. Chem. Theory Comput.* **2012**, *8*, 1037–1047, doi:https://doi.org/10.1021/ct2007154.
41. WebBook, N.C. NIST Standard Reference Database SRD Number 69 Available online: https://webbook.nist.gov/.
42. Laarz, E.; Zhmud, B.V.B.; Bergström, L. Dissolution and Deagglomeration of Silicon Nitride in Aqueous Medium. *J. Am. Ceram. Soc.* **2000**, *83*, 2394–2400.
43. Sawyer, C.N.; McCarty, P.L. *Chemistry for Environmental Engineering*; 3rd Ed.; McGraw-Hill Book Co: New York City, NY USA, 1978;
44. Parker, D.R.; Norvell, W.A.; Chaney, R.L. GEOCHEM-PC— A Chemical Speciation Program for IBM and Compatible Personal Computers. In *Chemical Equilibrium and Reaction Models*;

Loeppert, R.H., Schwab, A.D., Goldberg, S., Eds.; Soil Science Society of America and American Society of Agronomy: Madison, SI, USA, 1995; pp. 253–269.
45. Nakamatsu, T.; Saito, N.; Ishizaki, C.; Ishizaki, K. Silicon Nitride and Oxide Powder Surface Characterization by TPD. *J. Eur. Ceram. Soc.* **1998**, *18*, 1273–1279, doi: https://doi.org/10.1016/S0955-2219(98)00053-3.
46. Pezzotti, G.; Fujita, Y.; Boschetto, F.; Zhu, W.; Marin, E.; Vandelle, E.; McEntire, B.J.; Bal, S. B.; Giarola, M.; Makimura, K.; Polverari, A. Activity and Mechanism of Action of the Bioceramic Silicon Nitride as an Environmentally Friendly Alternative for the Control of the Grapevine Downy Mildew Pathogen*Plasmopara viticola. Front. Microbiol.* **2020,** *11, 610211.*
47. Healy, T.W.; White, L.R. Ionizable Surface Group Models of Aqueous Interfaces. *Adv. Colloid Interface Sci.* **1978**, *9*, 303–345, doi: https://doi.org/10.1016/0001-8686(78)85002-7.
48. Davis, J.A.; James, R.O.; Leckie, J.O. Surface Ionization and Complexation at the Oxide/Water Interface: I. Computation of Electrical Double Layer Properties in Simple Electrolytes. *J. Colloid Interface Sci.* **1978**, *63*, 480–499, doi: https://doi.org/10.1016/S0021-9797(78)80009-5.
49. Bousse, L.; Rooij, N.F. De; Bergveld, P. Operation of Chemically Sensitive Field-Effect Sensors as a Function of the Insulator-Electrolyte Interface. *IEEE Trans. Electron Devices* **1983**, *30*, 1263–1270, doi:https://doi.org/10.1109/T-ED.1983.21284.
50. Fung, C.D.; Cheung, P.W.; Ko, W.H. A Generalized Theory of an Electrolyte-Insulator-Semiconductor Field-Effect Transistor. *IEEE Trans. Electron Devices* **1986**, *33*, 8–18, doi: https://doi.org/10.1109/T-ED.1986.22429.
51. Harame, D.L.; Bousse, L.J.; Shott, J.D.; Meindl, J.D. Ion-Sensing Devices with Silicon Nitride and Borosilicate Glass Insulators. *IEEE Trans. Electron Devices* **1987**, *34*, 1700–1707, doi: https://doi.org/10.1109/T-ED.1987.23140.
52. Hartmann, W.; Van Dijen, F.; Metselaar, R.; Siskens, C. Slip Casting Parameters for Commercial Si_3N_4 Powders. *J. Phys. Colloq.* **1986**, *47*, C1-79-C1-83.
53. Ahmed, S.M. Electrical Double Layer at Metal/Oxide/Solution Interfaces. In *Oxide and Oxide Films*; J. W. Diggle, Ed.; Marcel Dekker: New York City, NY USA, 1973; pp. 319–517.
54. Parks, G.A. The Isoelectric Points of Solid Oxides, Solid Hydroxides, and Aqueous Hydroxo Complex Systems. *Chem. Rev.* **1965**, *65*, 177–198, doi:https://doi.org/10.1021/cr60234a002.
55. Lyklema, J. Interfacial Electrochemistry of Disperse Systems. In *Emergent Process Methods for High-Technology Ceramics*; Davis, R.F., Palmour, H., Porter, R.L., Eds.; Springer US: Boston, MA, 1984; pp. 1–24 ISBN 978-1-4684-8205-8.
56. Yates, D.E.; Levine, S.; Healy, T.W. Site-Binding Model of the Electrical Double Layer at the Oxide/Water Interface. *J. Chem. Soc. Faraday Trans. 1 Phys. Chem. Condens. Phases* **1974**, *70*, 1807–1818, doi:https://doi.org/10.1039/F19747001807.
57. Pezzotti, G. Silicon Nitride as a Biomaterial. *J. Ceram. Soc. Jpn.* **2023,** *131,* 398–428.
58. Iler, R.K. *The Surface Chemistry of Silica*; John Wiley & Sons: New York City, NY USA, 1979;
59. Laarz, E. Colloidal Processing of Non-Oxide Ceramic Powders in Aqueous Medium, Kemi, 2000.
60. Olofsson, J.; Grehk, T.; Berlind, T. Evaluation of Silicon Nitride as a Wear Resistant and Resorbable Alternative for Total Hip Joint Replacement. *Biomatter* **2012**, *2*, 94–102.
61. Pettersson, M.; Skjoldebrand, C.; Filho, L.; Engqvist, H.; Persson, C. Morphology and Dissolution Rate of Wear Debris from Silicon Nitride Coatings. *ACS Biomater. Sci. Eng.* **2016**, 1–28.

Osteoconductive Properties of Silicon Nitride

Giuseppe Pezzotti

1 Introduction

Silicon nitride has always been an intriguing material. Rarely occurring in nature, its synthesis was first achieved in 1857 by Henri Etienne Sainte-Claire Deville and Friedrich Wöhler [1], but its crystal chemistry was not fully understood until two decades later when Paul Schützenberger identified the stoichiometric compound Si_3N_4 [2]. It only became an industrial material in the 1950s when patents were granted for its use as a refractory in various applications, such as thermocouple tubes, furnace boats, insulators, and casting nozzles [3, 4], but 40 more years elapsed before dense silicon nitride was fully developed. It was found that small additions of selected oxide ceramics formed a stable high-temperature liquid phase [5]. This liquid facilitated the formation of a dense fine-grained acicular microstructure. With continued development, the unique combination of properties of Si_3N_4 became apparent. It had remarkable characteristics such as high strength and fracture toughness, low density, high thermal shock resistance, and a maximum operating temperature of ~1370 °C. Indeed, Si_3N_4 is now known as one of the toughest ceramic materials, capable of withstanding high mechanical stresses and temperatures without losing strength, toughness, or hardness [6–15]. Because of these advantages, interest in Si_3N_4 for various heat engine applications exponentially increased during the 1990s. Yet, despite substantial support from multiple governmental entities, only a limited number of Si_3N_4 engine components were eventually commercialized because of cost and reliability issues [16].

In the twenty-first century, silicon nitride has found a new application in the biomedical space. This is due in part to the increasing interdisciplinary nature of

G. Pezzotti (✉)
Ceramic Physics Laboratory, Kyoto Institute of Technology, Sakyo-ku, Matsugasaki, Kyoto, Japan
e-mail: pezzotti@kit.ac.jp

B. S. Bal et al. (eds.), *Silicon Nitride Bioceramics*,
https://doi.org/10.1007/978-3-031-67047-3_5

medical science, which has led to the development of new materials and techniques that are better suited for biomedical applications. For example, advances in cell and tissue engineering, genomics, nanotechnology, and computational biology have made it possible to create biomaterials that are more biocompatible and effective [17–26]. However, orthopedic implants, which were first introduced in the 1960s, have yet to fully embrace these advances. These devices have indeed benefited the health and well-being of countless patients who suffer from various conditions that affect their bones and joints, whether they are congenital, degenerative, or traumatic, but these implants have often been viewed as simple mechanical objects, rather than as complex and dynamic systems that interact with the biological environment [27]. Now, in the third decade of this new century, scientific perceptions are changing. Biomaterial innovations are opening new horizons for the prevention, diagnosis, and treatment of osseous diseases [28, 29], and Si_3N_4 is well-positioned to provide unique benefits. Although first recognized for its excellent mechanical and corrosion-resistant properties [30–32], Si_3N_4 is now known for its fortuitous osteoconductive characteristics, which are due to both its crystal structure and distinctive surface chemistry [33]. Similar to synthetic hydroxyapatite (HAp), it has a zwitterionic surface, which is beneficial to osteoblast attachment and proliferation. However, it also releases minute amounts of silicon and nitrogen, which are converted into orthosilicic acid (H_4SiO_4), ammonia (NH_3), and ammonium (NH_4^+). These moieties have been unequivocally recognized as the key factors that facilitate Si_3N_4's bone-promoting properties.

In 1989, Howlett et al. conducted a pioneering in vitro and in vivo studies on the osteoconductive potential of Si_3N_4 [34]. Using rabbit skeletal cells and live animals, they demonstrated that Si_3N_4 was not only biocompatible but also supportive of attachment and differentiation of osseous progenitor cells. Samples implanted into the femoral marrow cavities of mature rabbits promoted bone growth, and by 3 months, mature woven bone had encapsulated and permeated the porous structure of the implants. Around the same time, a first-of-its-kind clinical study of spinal arthrodesis using Si_3N_4 implants was conducted in Australia on 30 patients. A published review of this study, which initially occurred 15 years after implantation, showed good to excellent osteointegration of the implants [35]. A second review, published 30+ years after implantation, found that the implants were still performing well in the surviving patients [36]. Over the ensuing 40 years, more than 80 phenomenological studies investigated the biocompatibility and bone-forming effectiveness of silicon nitride. The vast majority of this work demonstrated that Si_3N_4 was not only biocompatible but also osteopromotive. Clinical studies that show the osteoconductive efficacy of Si_3N_4 are covered in a subsequent chapter. Although numerous in-depth experiments have demonstrated the osteopromotive or osteoconductive properties of Si_3N_4, there is a paucity of evidence to suggest that it is osteogenic. Osteogenesis is defined as the ability of a biomaterial to induce ectopic bone formation (i.e., bone growth in a non-native location) [37, 38]. Yet, while research in this area is still highly active, silicon nitride's osteopromotive properties have been attributed to two complementary mechanisms: (1) Si_3N_4 is a bio-friendly scaffold for bone attachment, and (2) Si_3N_4 releases ions that promote bone growth.

2 Si_3N_4 as a Biocompatible and Osteoconductive Scaffold for Bone Regeneration

Synthetic hydroxyapatite (HAp) is widely used as a biomimetic coating on orthopedic implants in clinical settings because it enhances bone fixation and growth. Its ability to promote bone integration through rapid cell proliferation has been the subject of intense research [39–44]. Electric polarization studies have shown that native hydroxyapatite is a piezoelectric material. Its crystal chemistry is composed of residual charges of opposite sign (i.e., electric dipoles). Proton mobility within its crystal structure, also known as nomadic polarization, occurs by the partial dehydration of the OH^- sublattice. This creates a swing of positively and negatively charged sites, which is thought to promote bone mineralization [45]. Additionally, the unique crystal structure of HAp can be greatly distorted to accommodate a wide range of cation/anion sizes. Up to 20% of these sites can be substituted with other elemental groups, such as Na^+, Mg^{2+}, K^+, Sr^{2+}, CO_3^{2-}, HPO_4^{2-}, Cl^-, and F^-. This remarkable stoichiometric flexibility enables the formation of a variety of charged structural sites at the material's surface. These sites facilitate bone remodeling by binding to biomolecule moieties in a zwitterionic fashion.

Remarkably, Si_3N_4 exhibits cell adhesion and improved osteoconductivity comparable to highly defective, non-stoichiometric native HAp. Si_3N_4 also displays zwitterionic-like behavior due to the presence of both positively and negatively charged off-stoichiometric sites, such as N vacancies and silanol groups [46]. It is analogous to the way defective Ca^{2+} and PO_4^{3-} sites act in HAp [47]. This unique chemistry was demonstrated in a series of experiments where the Si_3N_4 surface was modified to observe changes in osteoblast adhesion and proliferation [48, 49]. The tests involved a set of disk samples (90 wt.% Si_3N_4, 6 wt.% Y_2O_3, and 4 wt.% Al_2O_3), which were collectively densified (~1700 °C, ~3 h, N_2) and hot isostatically pressed (~1650 °C, ~200 MPa, N_2, ~2 h). Their resulting microstructure had two phases: anisotropic ß-Si_3N_4 and a partially crystallized grain boundary phase of silicon-yttrium-aluminum-oxynitride (SiYAlON). After firing, the samples were divided into two groups. The surfaces of one group were polished, while the other group retained their original, as-fired features. Three different surface treatments were then performed on both sets: wet-chemical etching in acid (5% HF solution for 45 s), thermal oxidation (TO, 7 h, 1070 °C, open-air kiln), and nitrogen annealing (NA, 1400 °C, 30 min, dry N_2, 115 kPa). A fourth set of samples from each set remained as-sintered (AS). The HF and TO treatments are antithetical, as they either eliminated or enhanced the amount of silica at their respective surfaces. The NA treatment increased the number of surface amines relative to the number of hydroxyls. Biomedical titanium (Ti-alloy) or alumina (Al_2O_3) disk samples with

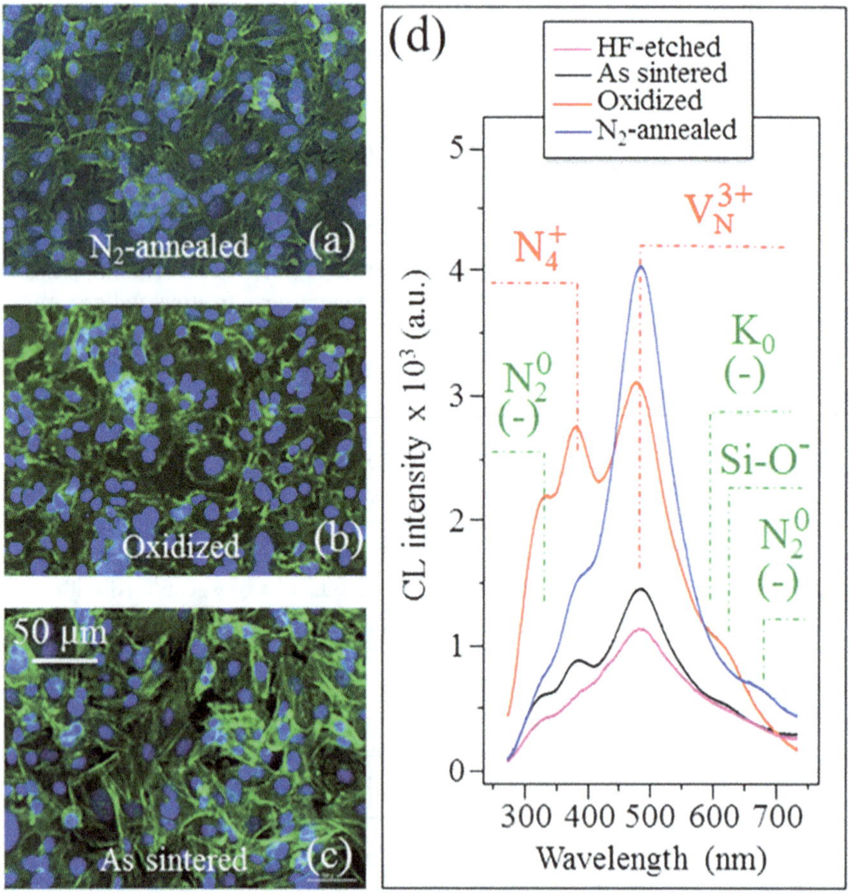

Fig. 1 Fluorescence microscopy visualization of the proliferation of SaOS-2 cells after 24 h on the surfaces of (**a**) NA (**b**) TO, and (**c**) AS Si_3N_4 samples. CL spectra for all samples are plotted in (**d**). The labels N_2^0, N_4^+, V_N^{3+}, K_0, Si-O$^-$, and N_2^0 refer to N dangling bonds, N-N bonds, N vacancies, Si dangling bonds, non-bridging oxygen hole centers in surface SiO_2, and N dangling bonds, respectively [50]. The minus in brackets labels the defects with a negative charge. (Used with permission under a Creative Commons Attribution 4.0 International License [48])

the same dimensions and similar surface roughness were produced for comparison testing. Cell proliferation (human osteosarcoma cells, SaOS-2, seeded at 5×10^5 cells/ml) was analyzed by fluorescence microscopy and then by cathodoluminescence (CL) and Raman spectroscopy. Figure 1a–c presents fluorescence micrographs of SaOS-2 proliferation after 24 h on polished samples [48]. While all three treatment conditions showed similar deposits, the NA samples had the strongest affinity for cell adhesion. This was confirmed by the CL spectra in (d), which indicated that the NA samples had the highest population of positively charged defects (i.e., nitrogen vacancies and N-N bonds).

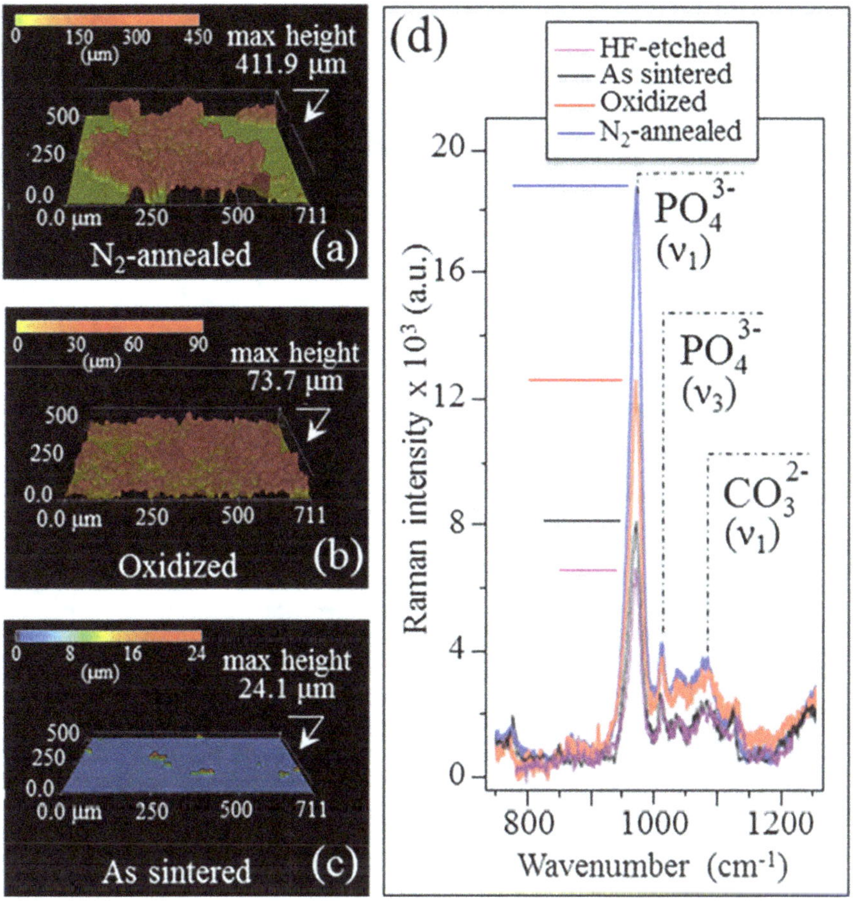

Fig. 2 Three-dimensional laser images of HAp grown on the surface of (**a**) NA, (**b**) TO, and (**c**) AS Si_3N_4 samples after 1 week of exposure to SaOS-2 cells. Raman spectra of all samples are shown in (**d**). The peaks at ~960 cm^{-1} and ~ 600 cm^{-1} are characteristic of HAp, while the peak at ~1080 cm^{-1} is characteristic of carbonate substitution. (Used with permission under a Creative Commons Attribution 4.0 International License [48])

Figure 2a–c shows three-dimensional laser micrographs of the HAp structure on the NA, TO, and AS Si_3N_4 samples after 1 week of exposure to SaOS-2 cells, respectively. The NA sample showed the most significant HAp growth, with deposits as high as 0.4 to 0.5 mm. Figure 2d summarizes the Raman characterization of their surfaces. These analyses confirmed the presence of HAp (as verified by the PO_4^{3-} stretching peak at ~960 cm^{-1}) with a higher average amount on the NA samples.

Cell proliferation and HAp formation for all Si_3N_4 samples in comparison with commercially available Al_2O_3 and Ti-alloy biomaterials are shown in Fig. 3a and b, respectively. The NA Si_3N_4 samples had the highest activity in both cell

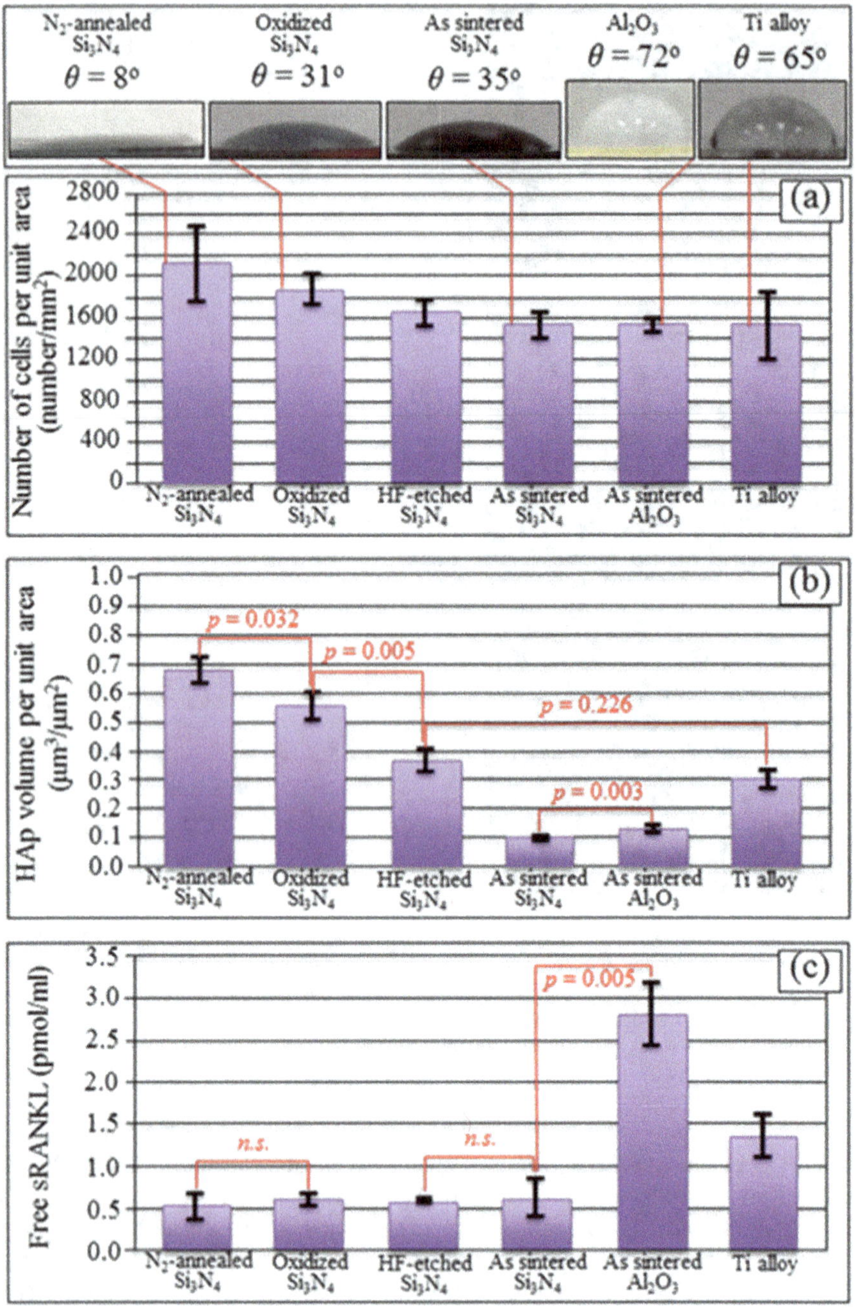

Fig. 3 Statistical evaluation of (**a**) cell proliferation and (**b**) HAp formation for all Si_3N_4 samples in comparison with commercially available Al_2O_3 and Ti-alloy biomaterials, and (**c**) an affinity for osteoblast cell attachment assessed by sRANKL experiments. Photographs of sessile water droplets on the different materials, along with the corresponding contact angles, are shown at the top. All samples were tested under the same conditions. (Used with permission under a Creative Commons Attribution 4.0 International License [48])

proliferation and HAp formation. This improvement was not only greater than that of the AS Si_3N_4 samples but also greater than that of the Al_2O_3 and Ti-alloy biomaterials. Water contact angles, shown at the top of Fig. 3, demonstrated improved hydrophilicity for the tested samples, which correlated with their biological enhancement [50]. The affinity of the Si_3N_4 samples for osteoblast cell attachment was confirmed by sRANKL assays (Fig. 3c). The observed deficiency in RANKL, a membrane-bound protein that is cleaved into soluble sRANKL by metalloproteinase-14, indicates that Si_3N_4 surfaces have a low propensity for osteoclast formation [51].

Although streaming potential measurements at homeostatic pH reveal that Si_3N_4 has an overall net negative surface charge [50], positive charges were found to be present. Consequently, it is posited that thermal treatment of Si_3N_4 produces a mixture of negatively and positively charged functional groups on its surface (e.g., silanols, silicon amines, and N vacancies). These groups can bind biomolecules through zwitterionic interactions. The zwitterionic-like properties of the Si_3N_4 samples are due to four factors: (1) a significant but minority population of positively charged amine sites; (2) substitutional Y and Al atoms on Si sites, which yield positively charged $Y\text{-}OH_2^+$ and $Al\text{-}OH_2^+$ groups; (3) an increased concentration of positively charged lattice defects; and (4) an intermixed layer of negatively charged $Si\text{-}O^-$ groups.

When a mostly negative surface becomes mixed with more positive charges, it becomes zwitterionic, hydrophilic, and osteopromotive. This is because lattice defects in the biomaterial's surface enhance interactions with both positively and negatively charged functional groups in biomolecules. Figure 4a–c shows results for Si_3N_4 samples with as-fired surfaces [49]. HAp deposition data are given in Fig. 4a, while sRANKL results are provided in Fig. 4b.

The results of these tests were similar to the outcomes previously presented for the Si_3N_4 samples with polished surfaces (cf., Fig. 3b–c). Figure 4a indicates that the nitrogen-annealed (NA) samples had the highest amount of HAp formation among the tested series. It was 100% greater than the Ti-alloy. While the as-sintered (AS) samples had a 40% lower amount of HAp on their surfaces, it was still significantly greater than the Ti-alloy, but the thermally oxidized (TO) and etched (HF) Si_3N_4 samples showed no significant differences in HAp formation compared to the Ti-alloy. However, the Ti-alloy samples had the highest amount of free sRANKL compared to all of the Si_3N_4 samples (Fig. 4b). In this series of samples, insulin-like growth factor (IGF-1) signaling was assessed as a marker of cell proliferation and differentiation efficiency. IGF-1 is essential for the growth and differentiation of bone. It acts through endocrine, paracrine, and autocrine mechanisms to modulate bone growth [52]. Differences in IGF-1 expression are shown in Fig. 4c. While the IGF-1 concentration was not significantly different between the Si_3N_4 samples, all IGF-1 values for the Si_3N_4 samples were statistically higher than that of the Ti-alloy. This is consistent with the results of the free sRANKL assays, which also showed that Si_3N_4 promoted cell proliferation and differentiation. Overall, these data suggest that, regardless of surface treatment conditions, Si_3N_4

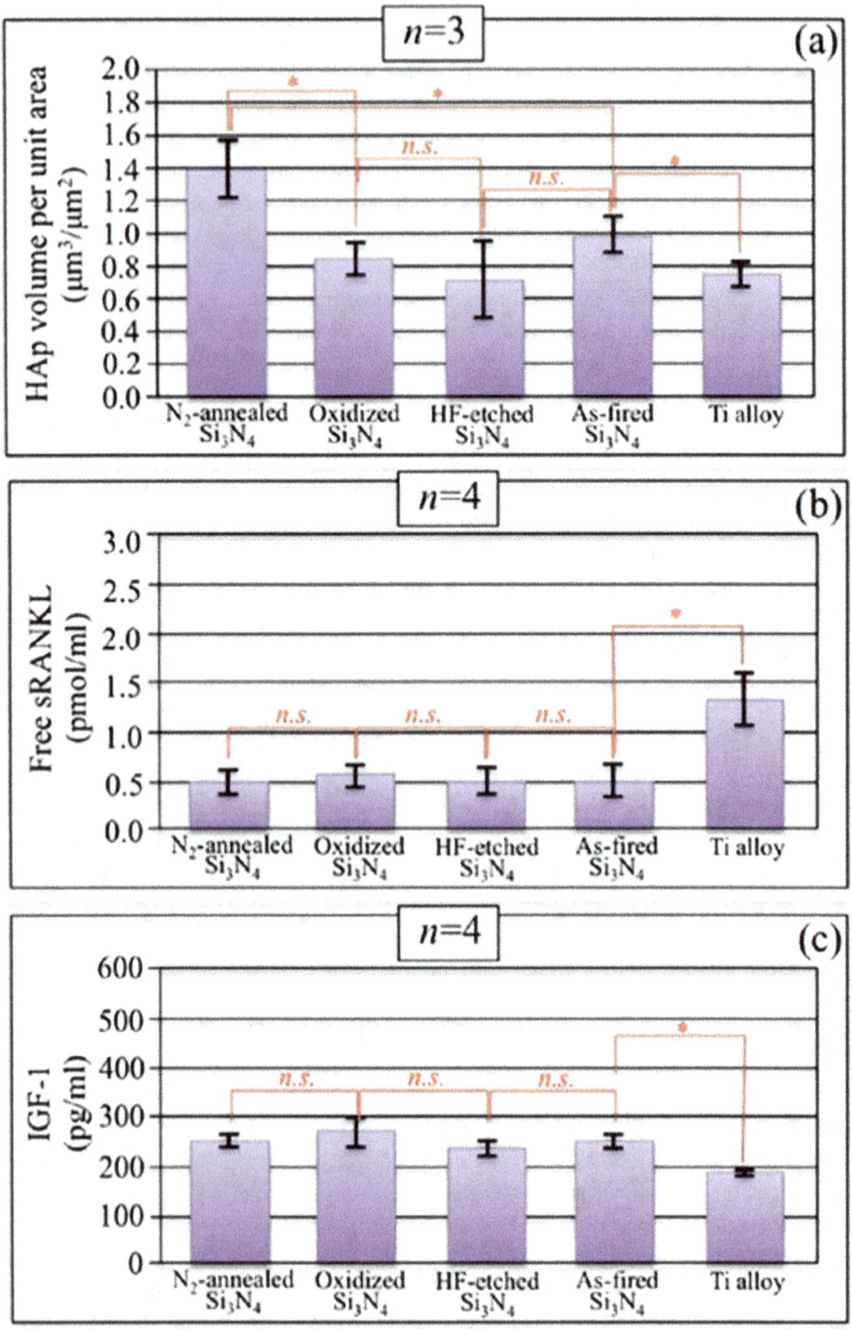

Fig. 4 Effects of surface-modulated Si_3N_4 samples on SaOS-2 cells. (**a**) Hydroxyapatite formation as measured by 3D laser scanning microscopy. (**b**) The propensity for osteoclast formation was measured by the concentration of free sRANKL. (**c**) The efficiency of cell proliferation and differentiation was measured by the concentration of IGF-1. The data show that surface-modulated Si_3N_4 samples promoted HAp formation, inhibited osteoclast formation, and enhanced cell

provided an environment in which SaOS-2 cells were upregulated to differentiate into osteoblasts.

FEG-SEM micrographs of the Si_3N_4 samples with as-fired surfaces, AS, HF, TO, and NA, are shown in Fig. 5a–d, respectively, while selected EDS analyses are given in Fig. 5e–f.

All of the samples had acicular grain structures with minimal morphological differences. However, one unique feature of the NA sample was the presence of a larger amount of heavier yttrium elements (spot B in Fig. 5d) than in the matrix (spot A in Fig. 5d. See EDS analyses for spots A and B in Fig. 5e and f, respectively). The larger amount of yttrium was accompanied by a significant increase in oxygen and a slight increase in aluminum to form a SiYAlON lattice. The presence of this off-stoichiometric ß-SiYAlON phase was subsequently confirmed by Raman spectroscopy (data not shown) [49]. All of the as-fired Si_3N_4 samples had a high degree of cell attachment, but the NA samples had more localized spots with pronounced apatite formation. These spots were found to be areas where the SiYAlON phase was present.

Highly resolved optical images of the osteosarcoma cells during in situ apatite formation are shown in Fig. 6a and b. The micrograph in (a) shows that SaOS-2 cells have a strong tendency to merge into large clusters on the Si_3N_4 surface. Hydroxyapatite nodules appeared as circular areas with dark gray centers, similar to the nodule in the micrograph Fig. 6b. The dark gray core within the cells represents a mature hydroxyapatite nodule and the initial stage for the formation of an osteoid on the mineralized surface. Raman spectroscopy confirmed the formation of the apatite phase at the center of the gray nodule shown in Fig. 6c, (data not shown) [50]. Figure 6d summarizes the three stages of hydroxyapatite formation on ß-SiYAlON: Stage I: Osteoblasts differentiate and aggregate on ß-SiYAlON; Stage II: Hydroxyapatite nodules form and grow; and Stage III: The apatite nodules continue to grow and eventually incorporate osteocytes to form osteoids, i.e., new bone formation occurs.

In situ monitoring of osteosarcoma cells on AS and NA Si_3N_4 substrates showed that both treatments promoted cell proliferation and apatite formation. However, the differences between the two treatment conditions were subtle. These electron microscopy and EDS analyses demonstrated that a SiYAlON phase formed on the Si_3N_4 samples; the presence of this phase coincided with increased cell proliferation and apatite formation. This phase was rich in yttrium and nitrogen vacancies. Si^{4+} cations were partially replaced by Y^{3+} cations, and N^{3-} anions were partially replaced by O^{2-} anions. The pristine hexagonal structure of ß-Si_3N_4 was retained, but the chemical formula changed to $Si_{3-3(m+n)/4}Al_mY_nO_xN_{4-2x/3}$, where the average values of m, n, and x were 0.051, 0.021, and 0.166, respectively [50]. This off-stoichiometric condition created a high concentration of N vacancies with a 3^+

Fig. 4 (continued) proliferation and differentiation compared to a biomedical Ti-alloy. Statistical validation of the results is shown in the inset of each figure. (Used with permission [49])

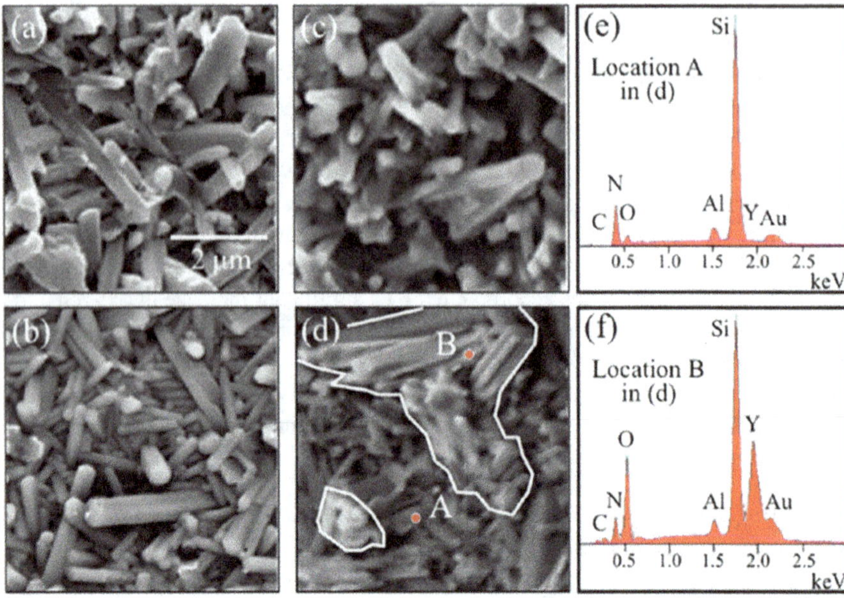

Fig. 5 Backscattered FEG-SEM micrographs of Si_3N_4 samples after different treatments: (**a**) AS, (**b**) HF, (**c**) TO, and (**d**) NA. Brighter areas are emphasized by white contours. (**e, f**) Elemental analyses by energy-dispersive spectroscopy (EDS) at spots A and B, respectively, as labeled in (d). Samples were sputter-coated with a thin layer of gold for EDS analysis. (Used with permission [49])

charge. This increase in V_N^{3+} was further enhanced by the formation of negatively charged defects Al_{Si}^- or Y_{Si}^-; substitutions of Al or Y on Si sites led to the abundant formation of N – N bonds, N_4^+. The formation of negatively charged sites (e.g., $AlSi^-$ or YSi^-) and nitrogen vacancies (V_N^{3+}) within the SiYAlON phase can enhance structural disorder and promote the formation of N – N bonds (with a positive charge) to re-establish electrical neutrality.

Si_3N_4 biologically functions similar to HAp. Figure 7a and b shows schematic diagrams of the cell-substrate interactions on Si_3N_4 and $Ca_{10}(PO_4)_6(OH)_2$, respectively. As shown in Fig. 7b, HAp's surface is composed of positively charged calcium ions (C-sites) and negatively charged phosphate groups (P-sites). At physiological pH, these C- and P-sites, along with neutral hydroxyl groups, are arranged in a fixed pattern. Biocellular amino groups are then attracted to P-sites, while carboxyl groups are repelled by C-sites [53–55]. This electrostatic dipolar surface charge condition leads to the formation of stronger coordination complexes [56–58]. For Si_3N_4, the cell-substrate interaction mechanism in Fig. 7a explains the difference in apatite formation on treated Si_3N_4 samples due to enhanced cell motility and proliferation. Chemically modulating as-fired Si_3N_4 surfaces by incorporating both positive and negative charges significantly improved the in vitro behavior of SaOS-2 cells. All of the Si_3N_4 samples supported cell differentiation, attachment, proliferation, and apatite formation. However, the NA samples, having a

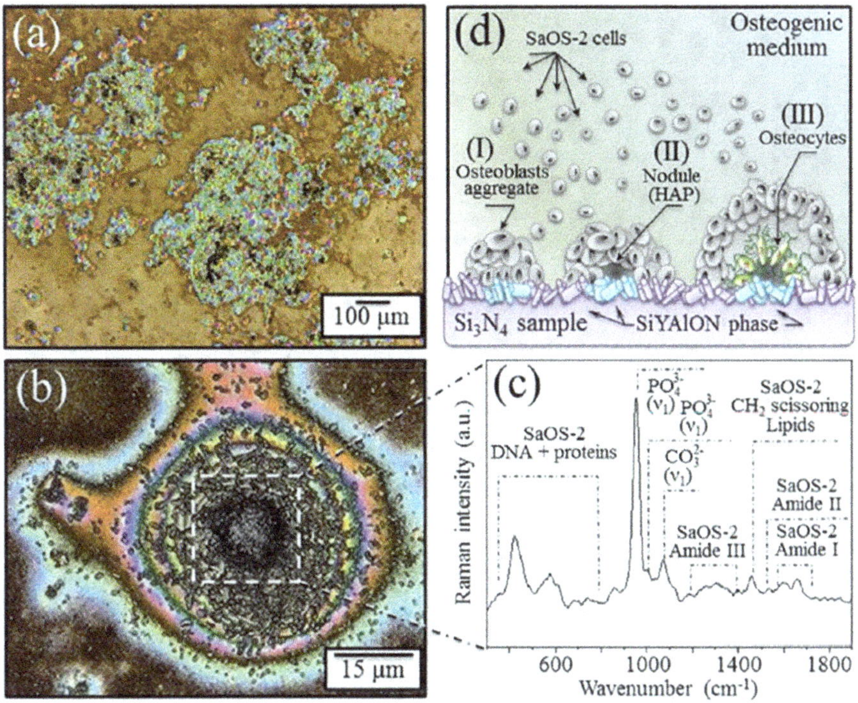

Fig. 6 High-resolution optical images of the process of hydroxyapatite formation on a Si_3N_4 substrate by osteosarcoma cells. In (**a**) and (**b**), in situ images of cells during confluence in the early stage of exposure to the Si_3N_4 substrate and subsequent HAp formation at a later stage. In (**c**), Raman spectrum of the dark gray feature confirms apatite phase formation. In (**d**), schematic diagram summarizing the process of hydroxyapatite formation, starting with osteoblast differentiation and aggregation (I), followed by hydroxyapatite formation and growth (II) and osteocyte entrapment (III). (Used with permission [49])

higher concentration of a SiYAlON phase on their surface with an abundance of positively charged N vacancies and N – N bonds, produced 40% and 100% increases in apatite formation when compared to as-sintered Si_3N_4 and a biomedical Ti-alloy, respectively. In situ Raman spectroscopy revealed that the plus-charge spots chemically induced onto the overall negative Si_3N_4 surface significantly increased SaOS-2 metabolic activity, which is believed to be due to an electrical charge mechanism that enhanced adhesion and adsorption of transmembrane proteins and migration of membrane phospholipids. The silanol groups (Si-OH) that are initially neutral in charge deprotonate and become negatively charged (SiO^-) at physiological pH. The presence of V_N^{3+} and N_4^+ sites with positive charges also creates a dipolar environment. These dipolar groups and associated electrostatic attractions function similarly to the C- and P-sites in HAp. These in vitro studies demonstrated that modifying the stoichiometry of Si_3N_4 surfaces through the

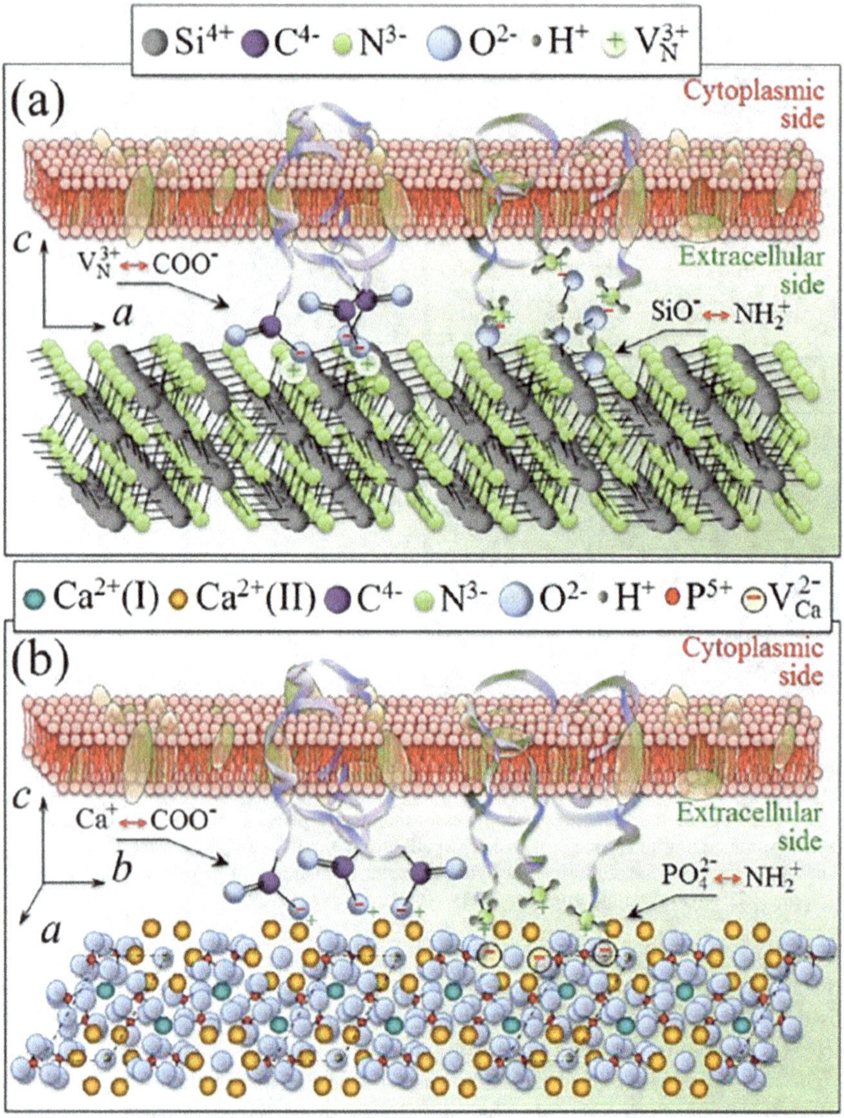

Fig. 7 Schematic diagrams of the interactions between cells and (**a**) Si_3N_4 or (**b**) $Ca_{10}(PO_4)_6(OH)_2$. (**a**) Cells interact with Si_3N_4 through positively charged V_N^{3+} and N_4^+ sites that form at the Si_3N_4 surface. These sites are similar to the C- and P-sites in hydroxyapatite. Y and Al ions are omitted from the diagram for simplicity. (**b**) Cells interact with $Ca_{10}(PO_4)_6(OH)_2$ through negatively charged C- and P-sites. These sites are attracted to positively charged amino acids on the cell surface. (Used with permission [49])

formation of plus-charged sites upregulates cell metabolism, and it opens the door for a new generation of Si_3N_4 bone repair devices that can accelerate osseous integration.

3 Si_3N_4 Scaffolds for Bone Regeneration: The Role of Ion Release

As described in the prior chapter, the hydrolysis of Si_3N_4 releases both nitrogen and silicon species into solution. These moieties interact with the biological milieu to upregulate osseous activity. Silicon is a trace mineral that is essential for many bodily functions, including bone health. It assists in the synthesis of glycosaminoglycans and proteoglycans, and it is directly incorporated into HAp by ionic substitution [59, 60]. In its most bioavailable form (i.e., orthosilicic acid ($Si(OH)_4$), it plays a role in osseous activity by enhancing the formation and cross-linking of collagen fibers, promoting the mineralization of bone, and increasing the activity of osteoblasts while suppressing osteoclasts [61–65].

Silicon-substituted HAp, porous silicon, and silicon/silica nanoparticles all enhance bioactivity by caveolae-mediated endocytosis of silicon [66–70]. Similarly, nitrogen is an essential element for protein synthesis, bone growth, and tissue repair [71]. Cells metabolically assimilate and transform water, oxygen, carbon dioxide, and inorganic nitrogen compounds—ammonia (NH_3), ammonium (NH_4^+), nitrate (NO_3^-), and nitrogen (N_2)—into complex biomolecules [72]. In addition, exogenous nitric oxide (NO), which can be enzymatically formed from these precursors, can have a variety of effects on bone cells, including promoting osteoblast growth and differentiation, increasing bone mineral density, inhibiting bone resorption, and protecting bone from damage [73–79]. One of the more intriguing aspects of using Si_3N_4 as a biomaterial is that these numerous chemical moieties—$Si(OH)_4$, NH_3, NH_4^+, NO_3^-, N_2, and NO—are all readily available for biological interactions [80]. The surface chemistry of Si_3N_4, which can be chemically modulated, facilitates a range of favorable metabolic interactions between both eukaryotic and prokaryotic cells at its inorganic surface [48, 80].

This can be seen by comparing Si_3N_4's surface chemistry to other well-known biomaterials. Chemical reactions that occur at bioceramic interfaces are critical to the fate of mammalian cells, and in this regard, Si_3N_4 has unique advantages over other bioceramics. For instance, when Al_2O_3 comes into contact with bodily fluids, hydrolysis generates reactive oxygen species (ROS), which can damage adjacent tissues. Si_3N_4 also hydrolyzes, but it forms reactive nitrogen species (RNS) instead. These RNS scavenge ROS from the local environment and neutralize hydrons (H^+ ions). This is a unique therapeutic property of Si_3N_4 that does not exist with oxide bioceramics. Si_3N_4's ability to eliminate ROS, neutralize hydrons, and enhance local pH potentially makes it an effective treatment for osteoarthritis and other conditions caused by oxidative stress [81].

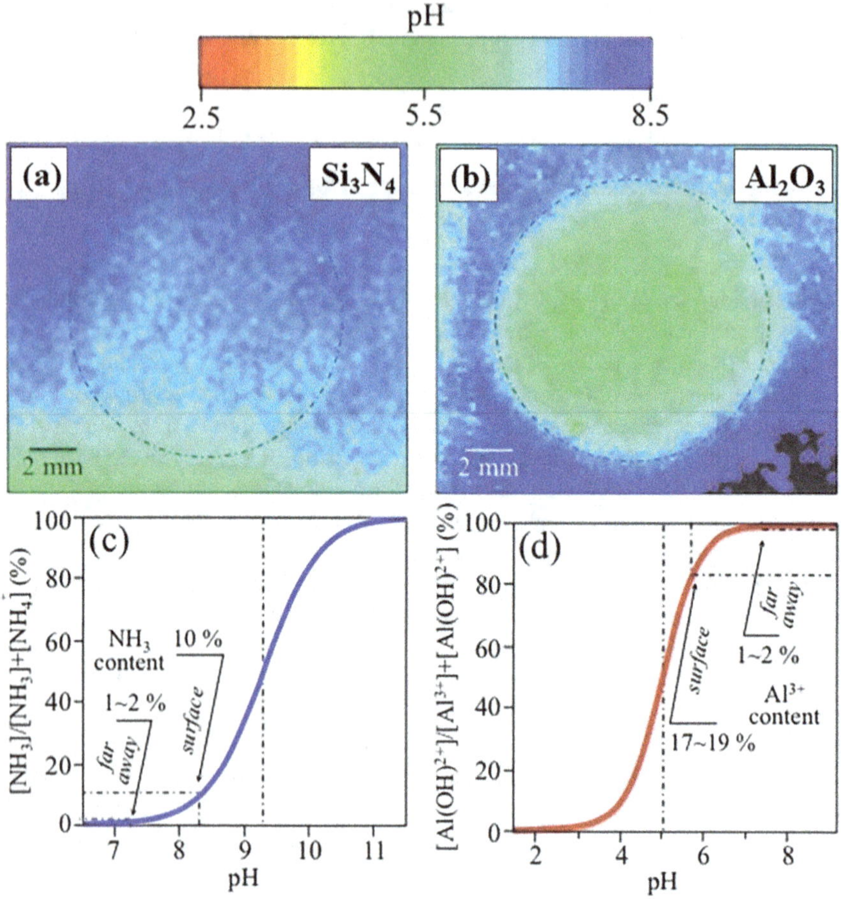

Fig. 8 pH imaging at room temperature of dense disk samples of (**a**) Si_3N_4 and (**b**) Al_2O_3 embedded into an electrolytic gel; the sigmoidal concentration profiles of eluted ionic species from (**c**) Si_3N_4 and (**d**) Al_2O_3 [82, 83]. (Used with permission [84])

To demonstrate these advantages, Fig. 8 shows images of Si_3N_4 and Al_2O_3 samples embedded in a pH-sensitive electrolytic gel at room temperature after 45 min of exposure.

The image of the Si_3N_4 sample (Fig. 8a) shows that the pH of the gel near the surface of the sample is buffered to ~8.3 ± 0.6, while the gel next to the Al_2O_3 sample (Fig. 8b) has an acidic pH of ~5.1 ± 0.5. This difference in pH is due to how these two biomaterials interact with hydrons. Si_3N_4 eliminates hydrons, while Al_2O_3 favors their production. Figure 8c and d shows pH-dependent trends for the ionic species eluted from Si_3N_4 and Al_2O_3, respectively. The fraction of ammonia (NH_3) from Si_3N_4 increases from 1 to 2% far away from the sample to ~10% near its surface. This corresponds to an increase in NH_3 concentration from ~0.15 to ~1 μMol/dm^3. The fraction of Al^{3+} ions at the Al_2O_3 surface is nearly ten-fold that of its

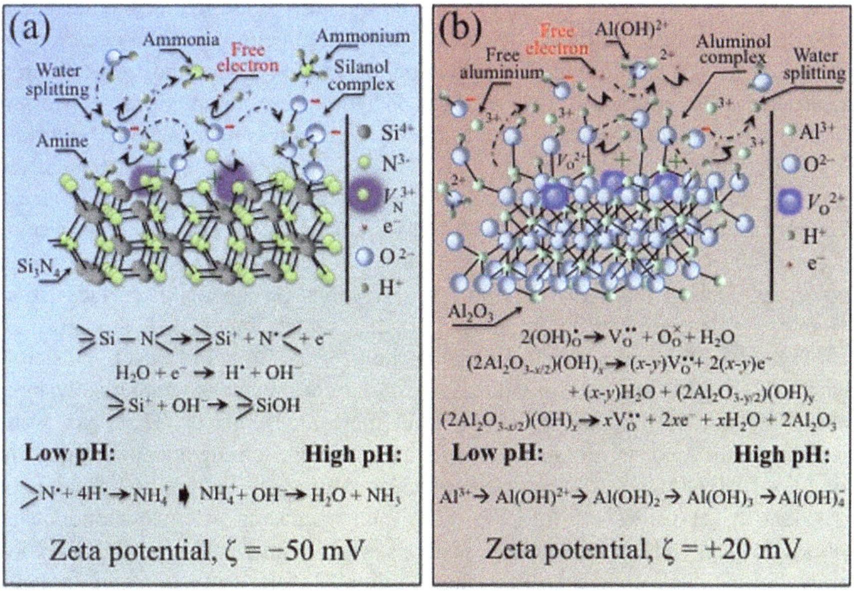

Fig. 9 Schematic diagrams of (**a**) Si_3N_4 and (**b**) Al_2O_3 surfaces in an aqueous environment, showing the off-stoichiometric reactions and ionic conditions of the respective surfaces. (Used with permission [84])

far-away region, increasing from 1–2% to 17–19%. This corresponds to an Al^{3+} concentration at its surface of ~0.2 mMol/dm^3.

The hydrolysis of Si_3N_4 and Al_2O_3 is shown schematically in Fig. 9a and b, respectively. In Fig. 9a, the covalent bonds of Si_3N_4 are broken by water, releasing nitrogen and creating vacancies, V_N^{3+}. This process also exposes silicon sites to the surface. The released nitrogen reacts spontaneously with hydrogen to form ammonium, NH_4^+, and ammonia, NH_3, which drives the observed pH buffering (cf., Fig. 8a).

The newly exposed Si ions then hydrate by linking to hydroxide ions (OH^-) to form silanols. The off-stoichiometric reactions shown in the lower part of Fig. 9a include the breaking of Si-N bonds, the splitting of water, and the generation of free radicals and electrons [85–87]. Note that the stability of surface silanols decreases with increasing pH [88]. The concentration of eluted silicon in the gel was typically about 1 mMol/dm^3, but it did not reach this level until several days had passed. In contrast, the amount of eluted Si ions after 12 h of exposure to water at physiological pH was ~0.5 mMol/dm^3. For fine Si_3N_4 powder in water, the concentration of eluted silicon can reach tens of mmol/dm^3. The oxygen atoms bound to silicon have low basicity and therefore act as weak acids. As discussed in the prior section, this gives Si_3N_4 zwitterionic character, which is favorable for cell attachment. Oxygen atoms react with water to form surface silanol groups (Si-OH). The acidity of these groups governs the surface charge, which is positive at low pH and negative at high pH (cf., bottom of Fig. 9a). Figure 9b shows a similar diagram of the free Al_2O_3 surface in water [82]. At physiologic pH, a series of off-stoichiometric reactions occur on the surface of Al_2O_3, including the formation of hydroxyl aluminols, dissociative adsorption of the aluminols, and removal of hydroxyl groups with the concurrent formation of oxygen vacancies. Both acidic and basic sites are present on the surface of Al_2O_3. The Al^{3+} cations act as Lewis acids, while the O^{2-} anions act as Lewis/Brønsted bases. Because Al_2O_3 contains only one type of cation, the hydroxyl groups at its surface are expected to be weakly acidic due to the high-ionic strength of the Al-O bond; these groups can only form weak hydrogen bonds with sorbed molecules. During the hydrolysis process, Al^{3+} ions on the Al_2O_3 surface attract electrons. This causes water molecules to disassociate when they are absorbed (cf. Fig. 9b) [89]. As a result, charge is transferred from the Al_2O_3 cluster to hydroxyl, oxygen, and proton radicals. This charge transfer is energetically more favorable than the adsorption of the H_2O molecule [90, 91]. Then, six water molecules surround an aluminum ion in an octahedral arrangement, forming the complex $[Al(H_2O)_6]^{3+}$. This species, which is commonly known as hexaaqua, is prevalent at pH values <5 (cf., Fig. 8d). With increasing pH, two mononuclear species (i.e., $Al(OH)^{2+}$ and $Al(OH)_2{}^{+}$) and a soluble-neutral species ($Al(OH)_3$) form due to the removal of a proton from the coordinated water, but some other polynuclear species can also form (e.g., $Al_2(OH)_2{}^{4+}$ or $Al_3(OH)_4{}^{5+}$) depending on the Al^{3+} concentration. $Al(OH)_3$ can precipitate at homeostatic pH, but at higher values, Al $(OH)_4{}^{-}$ remains soluble (cf. Fig. 9b). In Fig. 8d, the free Al^{3+} ion concentration away from the Al_2O_3 sample is ~20 μMol/dm^3. However, due to the acidic nature of the surface (i.e., pH ~5.1, cf., Fig. 8b), its concentration is an order of magnitude greater (i.e., ~0.2 mMol/dm^3). It should be noted that hydrolyzed aluminum has been implicated in various chronic health conditions, including cancer and Alzheimer's disease [92]. The most toxic form of aluminum, mononuclear Al^{3+}, typically occurs as a polymeric species. It forms when OH^- groups coordinate with two or more Al^{3+} ions.

The speciation plots in Fig. 8c and d show that the pH at the solid surface and far away from the surface can be quite different for both Si_3N_4 and Al_2O_3. The behavior of mesenchymal cells on these two bioceramics is governed by pH and their respective interfacial reactions. To accentuate this point, Fig. 10a and b presents fluorescence images of the proliferation of KUSA-A1 mesenchymal cells after 24 h exposure to Si_3N_4 and Al_2O_3, respectively. The fluorescence micrographs show the nuclei in blue and the F-actin in green. The insets show highly resolved images of individual cells at 12 h of exposure. The quantitative number of cells per unit area on each substrate and their statistical significance are shown in Fig. 10c. Note that KUSA-A1 cells proliferated on Si_3N_4 at a rate about twice that of Al_2O_3. This difference was supported by the high-resolution images. They showed that cells on Si_3N_4 exhibited actin activation and reorganization, with the appearance of filamentous F-actin (cf. inset in Fig. 10a and b). IGF-1 signaling (Fig. 10d) was used to independently confirm the efficiency of cell proliferation and differentiation. IGF-1 is a growth factor that stimulates the differentiation of mesenchymal cells into osteoblasts. Al_2O_3 showed a significantly lower propensity to induce osteoblast

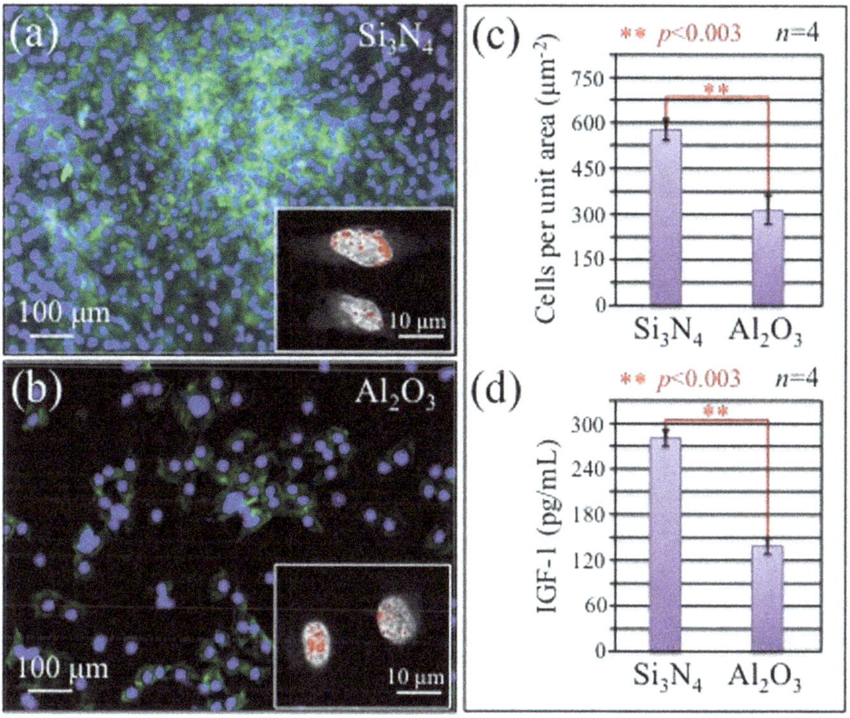

Fig. 10 Proliferation of KUSA-A1 mesenchymal cells on Si_3N_4 and Al_2O_3 substrates after 24 h. The nuclei are shown in blue, and the F-actin is shown in green. The insets show magnified views of individual cells at 12 h. The main figure also shows the number of cells per unit area on each substrate, as well as the statistical significance of these differences. The final panel shows the degree of cell differentiation and the propensity to synthesize bone tissue, as measured by the IGF-1 assay. (Used with permission [84])

differentiation and bone formation than Si_3N_4. This was confirmed by measuring the supernatant IGF-1 concentration, which was 93% lower for Al_2O_3 than for Si_3N_4. In contrast, the IGF-1 concentration for Si_3N_4 was 300% higher than the concentration for a positive control sample exposed to a silica glass substrate under the same conditions (data not shown).

Optical micrographs in Fig. 11a and b show the status of bone formation on Si_3N_4 and Al_2O_3 after 9 days of exposure. The insets in each figure are 3D laser micrographs of the respective samples. The optical micrographs indicate that the amount of bony tissue formed on Si_3N_4 is significantly greater than on Al_2O_3. This is further supported by the quantitative 3D assessment of deposited bony apatite in Fig. 11c. It shows that KUSA-A1 cells generated ~2.75 times the amount of apatite on Si_3N_4 compared to Al_2O_3 under the same culture conditions.

ALP assays were performed on the cells on both substrates after 10 days of exposure (Fig. 11d). It also demonstrated that the ALP content on Si_3N_4 was about twice that of Al_2O_3. This indicates that the KUSA-A1 cells were twice as fast at

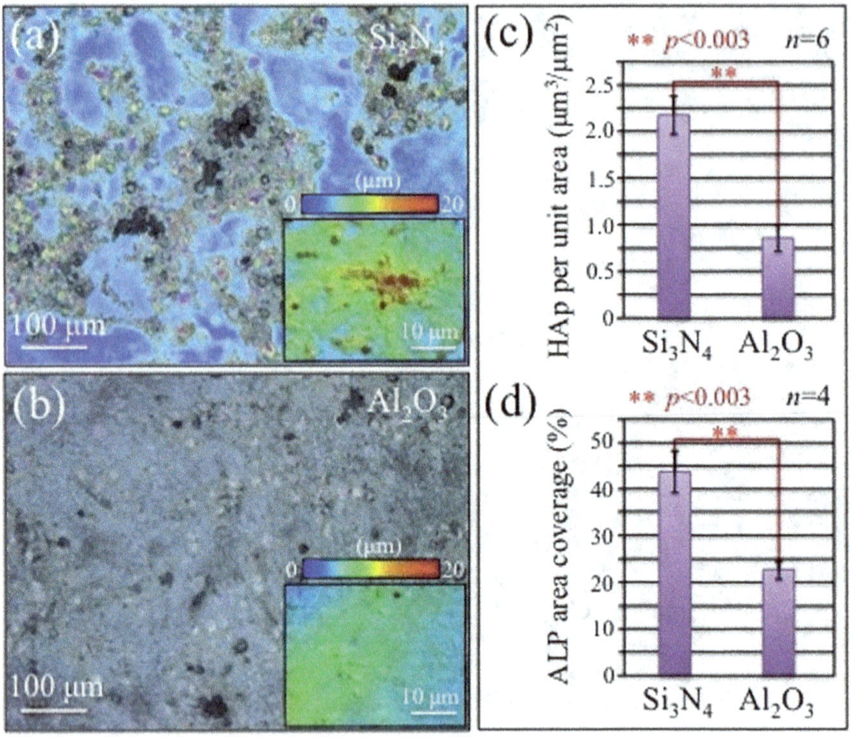

Fig. 11 Formation of osseous tissue on Si_3N_4 and Al_2O_3 substrates after 9 days. The optical micrographs in (**a**) and (**b**) show the 2D views of the tissue, while the laser micrographs in the insets show 3D views. The quantification of HAp volume per unit area and the results of ALP assays are shown in (**c**) and (**d**), respectively. Statistical differences between the two substrates were confirmed by one-way analysis of variance (ANOVA). (Used with permission [84])

osteoblastic differentiation on Si_3N_4. This result is consistent with the IGF-1 assays performed after 7 days (cf., Fig. 10d).

The positive effect of Si_3N_4 on bone tissue deposition was also seen when it was coated on an oxide surface. A bioceramic made from zirconia-toughened alumina (ZTA; trade name BIOLOX®*delta*) was coated with Si_3N_4 powder using laser cladding [93]. Optical and laser microscopy images of the coating after 10 days of exposure to uniformly distributed KUSA-A1 cells are shown in Fig. 12a and b. The cells migrated to the Si_3N_4 coating and produced a large amount of osseous tissue. Bony apatite deposits up to 113 μm in height were observed. Figure 12d shows that the Raman $(PO_4)^{3+}$ stretching band of the apatite on the Si_3N_4-coated region was nearly three times more intense than that on the uncoated region.

The effectiveness of Si_3N_4 laser cladding was also confirmed in an animal study using a Ti6Al4V biomedical alloy. Figure 13a and b shows X-ray images of uncoated and Si_3N_4-coated Ti6Al4V screws, respectively. The screws were inserted into the thighbones of 11-week-old male Sprague-Dawley rats. After 4 weeks

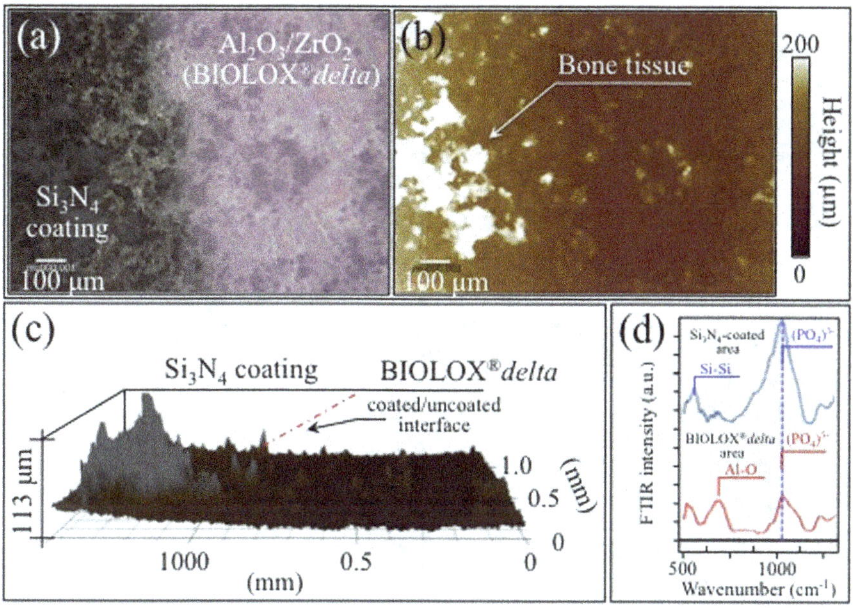

Fig. 12 (**a**) Optical and (**b**) laser images of the boundary between Si_3N_4-coated and uncoated ZTA surfaces after 10 days of exposure to KUSA-A1 cells. The 3D laser micrograph (**c**) and Raman spectroscopy data (**d**) show that the Si_3N_4 coating promoted the formation of bony apatite. (Used with permission [84])

in vivo, the animals were sacrificed, and the amount of bone-screw bonding was quantitatively analyzed using 3D micro-CT scanning. The images showed that the Si_3N_4-coated screws had significantly stronger bone-screw bonding than uncoated screws. The redder the image, the stronger the bond. The cross-sectional views of the screws and the statistical analysis of the data showed that there was a ~threefold improvement in interfacial bonding when the metal alloy was coated with Si_3N_4.

This in vivo experiment confirmed the results of previous in vitro studies, which showed that Si_3N_4 has remarkably superior osteoconductivity compared to commonly used orthopedic biomaterials.

To further verify Si_3N_4's osseous-promoting properties, a human spinal arthrodesis explant was investigated. This provided direct human evidence of the bone-promoting ability of this material. Presented in Fig. 14 are orthogonal views of a cervical implant that was retrieved after 11 months in vivo, during repeat cervical surgery for reasons unrelated to the implant. Note that there is a generous amount of bone residue adhering to the superior and inferior faces. On the right side of Fig. 14, a micro-CT image shows uniform bone growth throughout the graft hole (GH). The highest bone density was observed on the internal surface of the implant, rather than its central core. This suggests that osteoblasts preferentially formed bony apatite at the Si_3N_4 interface. This finding is not seen in comparable spinal implants made of polymeric materials, such as polyetheretherketone (PEEK) [94]. To gain insights

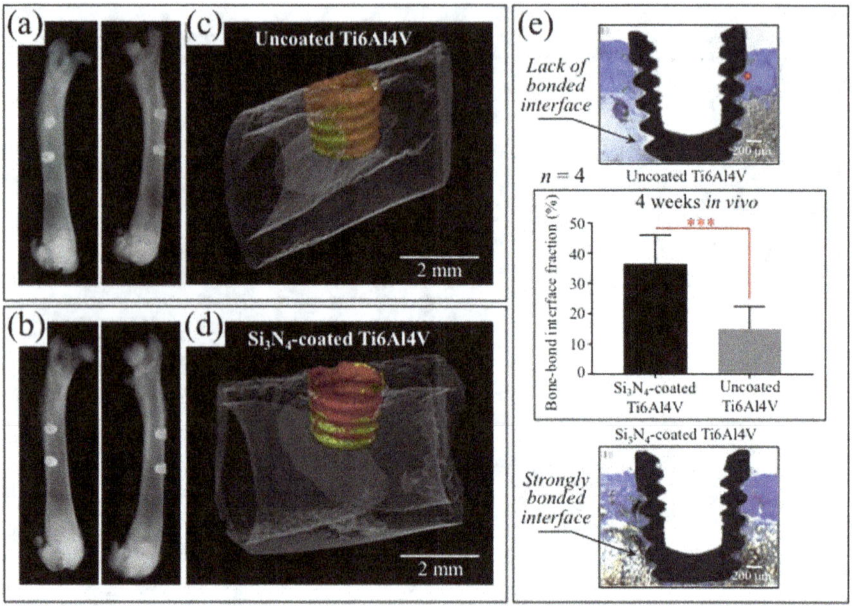

Fig. 13 X-ray images of (**a**) uncoated and (**b**) Si_3N_4-coated Ti6Al4V screws implanted in the thighbones of 11-week-old male rats. Micro-CT scans taken after 4 weeks showed the degree of bone-screw bonding in (**c**) and (**d**), respectively. The amount of bonded area was quantified in (**e**) from cross-sectional views of the screws (cf., labels in inset). (Used with permission [84])

into the unusual behavior of this spinal implant, the Si_3N_4/bone interface was examined using transmission electron microscopy (TEM) and high-resolution energy-dispersive spectroscopy (EDS).

The TEM images at low and high magnifications are shown in Fig. 14b and c, respectively, while the EDS analyses are presented in Fig. 14d. Figure 14b shows an initial cell-grown intermediate layer on the Si_3N_4 granular structure, upon which osteoblasts constructed bony tissue. The high-resolution image in Fig. 14c shows the presence of hydroxyapatite nanorods (red circles in the inset), which have a low degree of crystallinity (diffraction pattern in the inset). The most striking feature of the cell-grown apatite is its elemental composition. The amount of Si is double that of P (Fig. 14d). The EDS spectrum refers to a zone that is relatively far from the Si_3N_4 interface (red box in Fig. 14b). This observation suggests that the osteoblasts scavenged and endocytotically incorporated Si from the implant to use in bone synthesis [95].

To confirm this endocytosis process, polished Si_3N_4 samples, previously subjected to nitrogen annealing (NA) (cf., Figs. 1a and 2a) and seeded with SaOS-2 cells for up to 7 days, were examined using Raman, fluorescence, and cathodoluminescence spectroscopies. Figure 15 provides contrasting laser microscopy images and a surface profile of a sample after 3 days of exposure to SaOS-2 cells.

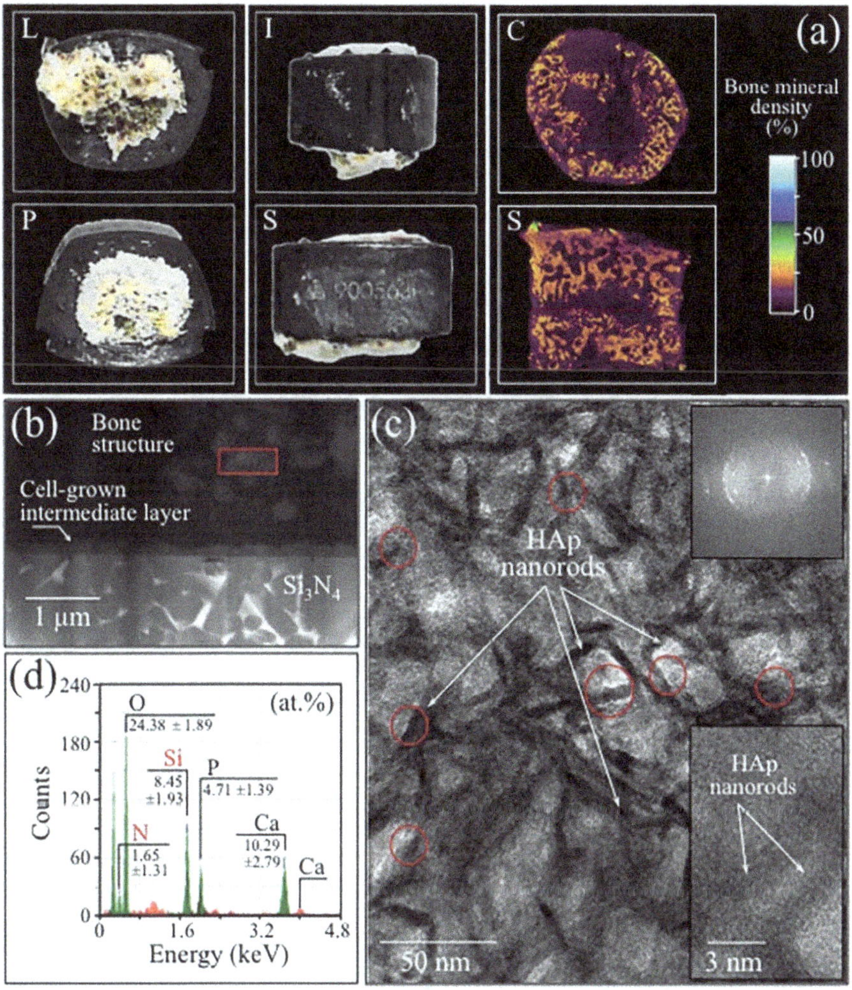

Fig. 14 (**a**) Different views of the Si_3N_4 cervical spacer after explantation. A anterior, P posterior, S superior, I inferior, and L lateral. Note the abundant amount of bone developed above and below the graft hole (GH); (**b**) low-magnification TEM image of the Si_3N_4/bone interface; (**c**) high-magnification TEM image of the Si_3N_4/bone interface, showing the presence of HAp nanorods; (**d**) EDS analyses of the Si_3N_4/bone interface, showing the elemental atomic fractions of HAp nanorods and surrounding collagen. EDS analyses were performed in the red square box in (b). The green and red spectra in (d) refer to HAp nanorods and surrounding collagen, respectively [94]. (Used with permission [84])

This micrograph revealed a large number of areas with pronounced HAp formation. These areas preferentially occurred on top of the SiYAlON phase. However, surface depressions were also seen in the vicinity of these HAp deposits (cf., Fig. 15c). These observations suggested that the SaOS-2 cells endocytotically scavenged molecules from the SiYAlON phase and used them in the genesis of

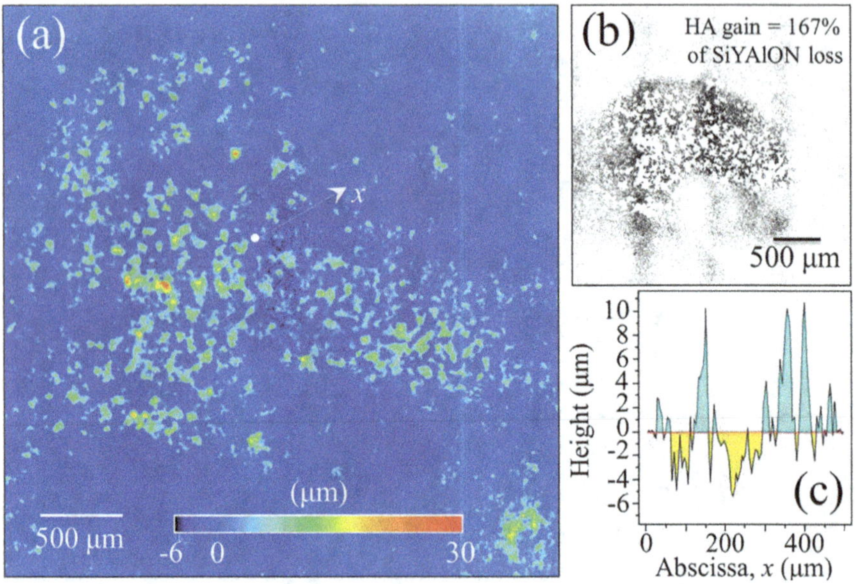

Fig. 15 Laser scanning microscopic images of N_2-annealed Si_3N_4 sample after osteoconductivity test. (**a**) The surface protrusions are sites of apatite formation. (**b**) The below surface zones in the same area as in (a) are highlighted with black contrast for improved visualization. (**c**) A typical profile line scan corresponding to the linear abscissa, x, drawn in (a). (Used with permission [95])

bony HAp. The average gain in HAp was ~167%, which corresponded to the loss of SiYAlON.

Photometric optical density analysis (Alizarin red stain) was used to monitor the formation of bony HAp. Results are shown in Fig. 16a. An almost ten-fold growth in HAp deposits was observed over the 7-day experimental period. This growth was then confirmed using in situ Raman spectroscopy of SaOS-2 cellular metabolism as given in Fig. 16b, (cf., HAp O-P-O symmetric stretching band at ~960 cm^{-1}). It also showed details of the structure of HAp, such as the presence of carbonate clusters ($(CO_3)^{2-}$at ~1100 cm^{-1}) during their initial growth stage. Compared to the nearly constant signal from symmetric ring breathing in phenylalanine (Phe at ~1005 cm^{-1}), the concentration of nucleic acids decreased significantly at the early stages of cell exposure (cf., DNA/RNA bands in the spectral region 780 ~ 880 cm^{-1}). Undifferentiated cells exhibit a high concentration of RNA, similar to early embryos [96, 97]. As embryogenesis progresses, non-translated RNA is reactivated through polyadenylation to synthesize specific proteins required for cell differentiation. Therefore, a smaller intensity of the RNA Raman peak reflects a higher degree of cell differentiation, as it results from RNA translation to produce the specific proteins needed for differentiation. The Raman spectrum in Fig. 16b shows a nearly constant emission from cytochrome c, suggesting that mitochondrial metabolism and apoptosis were not disrupted [98–101].

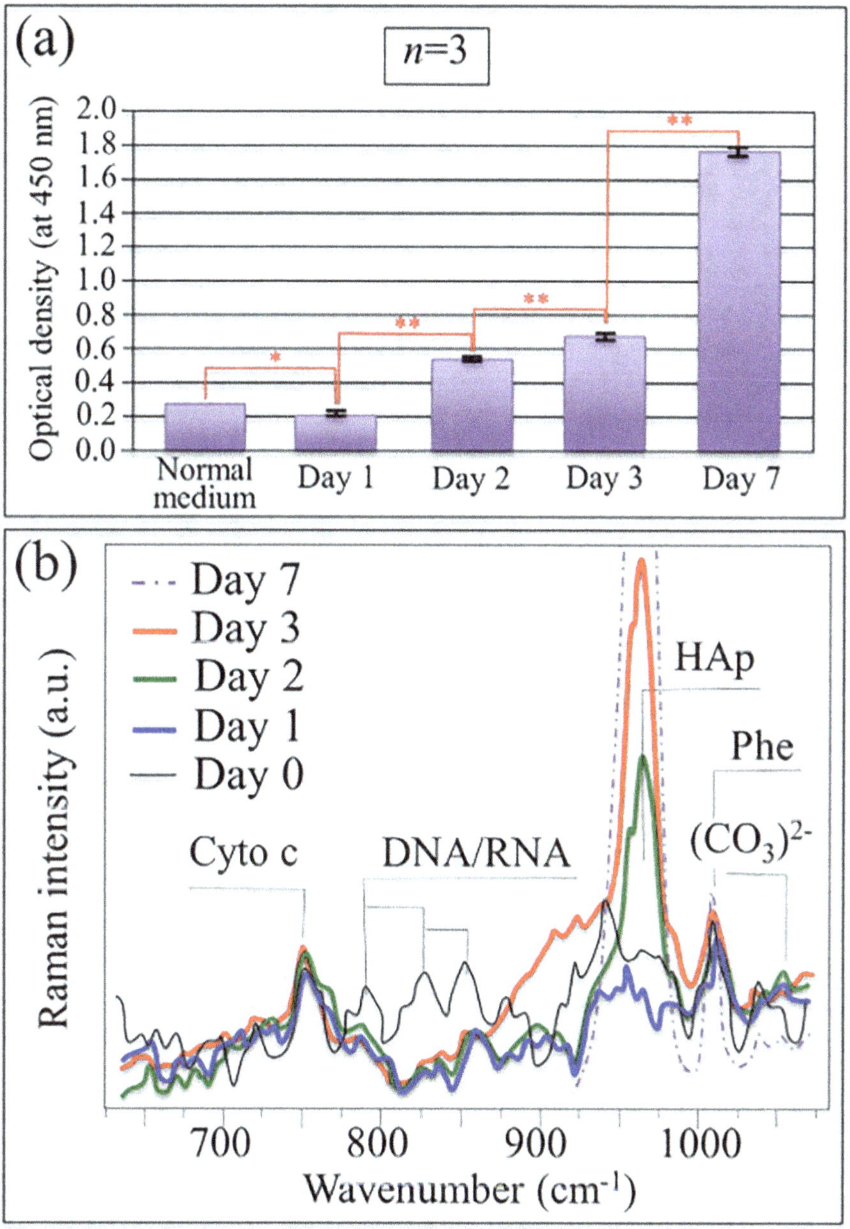

Fig. 16 Time-dependent optical density of Alizarin red stain (**a**) and in situ Raman spectroscopy (**b**) of living SaOS-2 cells exposed to the N_2-annealed Si_3N_4 substrate. (Used with permission [95])

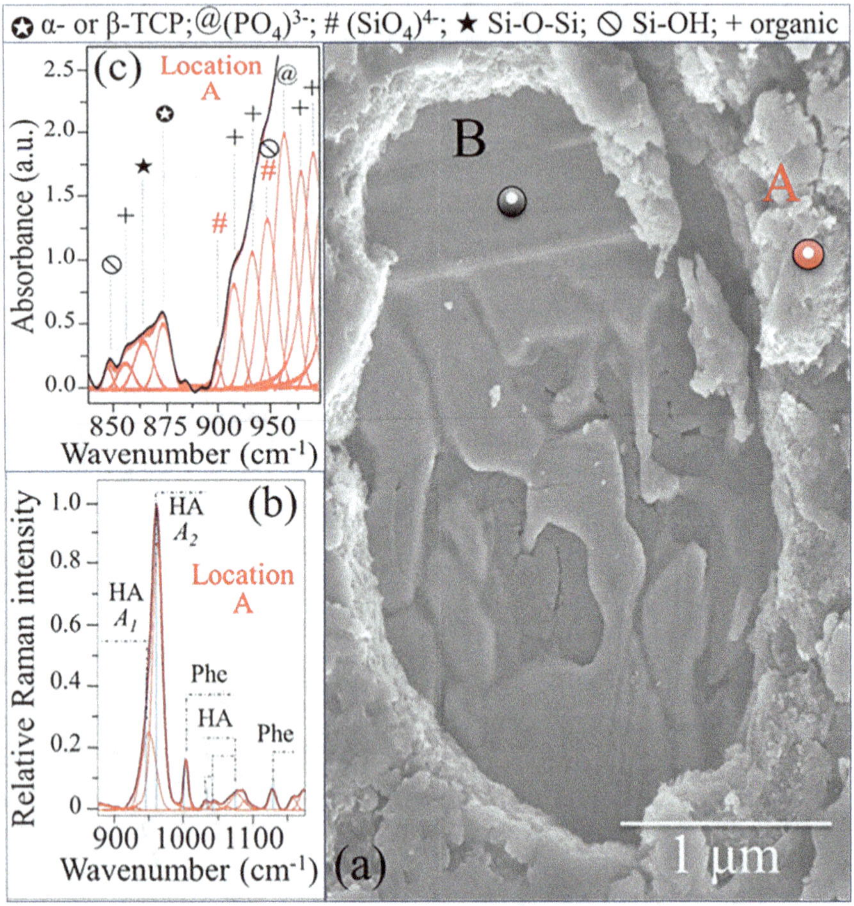

Fig. 17 Electron micrograph of an N_2-annealed Si_3N_4 sample after 7 days of exposure to SaOS-2 cells. (**a**) Area A shows a thick layer of HAp (cf., Raman spectrum in (**b**)), and area B shows where a layer of HAp was mechanically removed. In this area, note the "icing" SiYAlON phase on which the SaOS-2 cells were deposited and subsequently covered by a hydroxyapatite layer, (**c**) shows the FT-IR spectrum for $(SiO_4)^{4-}$ substitution in HAp. (Redrawn and used with permission [95])

After 7 days of exposure of the SaOS-2 cells to the Si_3N_4 surface, the samples were uniformly covered with deposited HAp. In one sample, a small portion of this HAp layer was mechanically removed to observe the buried interface (see Fig. 17a). Underneath were generous amounts of the SiYAlON phase. Raman spectroscopy was then used to examine different locations on the thick apatite (location A) and the buried layers (location B). The Raman band at location A consisted of two sub-bands, exhibiting significant asymmetry. This asymmetry is due to the presence of a HAp sub-band at approximately 946 cm^{-1} (labeled *A1* in Fig. 17b), in addition to the main component at approximately 960 cm^{-1} (labeled *A2*). Zou et al. discussed the physical meaning of the low-frequency sub-band *A1* [102]. They demonstrated that when silicon was incorporated into the apatite lattice structure to form $(SiO_4)^{4-}$ tetrahedra, the HAp Raman spectrum at around 960 cm^{-1} becomes asymmetric. It splits into a doublet with a low-frequency component. The bony apatite deposited by the SaOS-2 cells exhibited the same doublet (cf., Fig. 17b *A2*). Energy-dispersive X-ray (EDX) analysis at location A revealed a calcium-deficient structure (Ca/P atomic ratio of 1.55 ± 0.55), and this same analysis at thick hydroxyapatite locations also confirmed the presence of trace amounts of silicon (0.11 ± 0.06 at.%).

Nitric oxide production within the SaOS-2 cells was monitored using diaminofluorescein-2 diacetate and DAF-2. Figure 18 shows optical and fluorescence microscopy images of SaOS-2 cells on an N_2-annealed Si_3N_4 sample from 0 to 24 h. This sequence of images clearly shows an increase in NO concentration with exposure time. As discussed earlier, NO is a potent regulator of bone cell metabolism, differentiation, proliferation, and resorption. When combined with cytokines (e.g., interferon-gamma (IFN-γ)), NO suppresses both the formation and activity of osteoclasts [73]. Building on these previous findings, the results in Fig. 18 indicate that nitrogen scavenged from Si_3N_4 plays a role in osteoconduction and that NO activity is linked to the enhanced proliferation of SaOS-2 cells.

These studies found that the biochemical reactions at the surface of Si_3N_4 produced silicic acid (H_4SiO_4), ammonia (NH_3), and ammonium (NH_4^+), all of which played active roles in regulating cellular metabolism [84, 93]. It was posited that osteoblast cells actually scavenged Si from eluted silicic acid and endocytotically incorporated it into the defective HAp lattice. Furthermore, it is hypothesized that cellular uptake of NH_3 and/or NH_4^+ was enzymatically converted to nitric oxide (NO). NO serves as an important signaling molecule for osteogenesis [78].

Mechanistic details for these reactions were put forth in a subsequent in vitro study that examined KUSA-A1 cells deposited on identical Si_3N_4 samples. Data supporting these assumptions were obtained using fluorescence microscopy, XPS, and in situ Raman analyses [103]. A schematic model of the possible reactions and their intracellular pathways is presented in Fig. 19. When Si_3N_4 is placed in a biological environment, the covalent Si-N bonds on its surface dissociate. Nitrogen ions are released, creating vacancies (V_N^{3+}). The nitrogen immediately combines with hydrogen to form NH_4^+ and NH_3, which significantly increases the local pH at the solid-liquid interface [80, 104]. This exposes silicon sites to the free surface. They react with OH^- ions to form silanol complexes (Si–OH and $Si(OH)_4$). The dissociation of surface amine and silanol sites generates free electrons and oxygen−/nitrogen-free radicals [85–87]. Superoxide radicals form ($\equiv Si–O^\bullet$ and $\equiv Si–O_2^\bullet$), and unpaired electrons then react with adsorbed O_2 to form $O_2^{-\bullet}$ anions or other protonated radicals that are highly oxidative. These newly formed species react with the free electrons to produce $NO^\bullet$ radicals through various enzymatic pathways [95]. While extracellular NO can be independently derived from nitrates (NO_3^-) and/or nitrites (NO_2^-) under hypoxic conditions [105–107], NH_3 readily diffuses across cell membranes [108] and NH_4^+ only enters cells after deprotonation using carrier proteins in a highly regulated two-lane process [109]. Although the specific intracellular enzymes that convert NH_3 to NO have yet to be elucidated, the presence

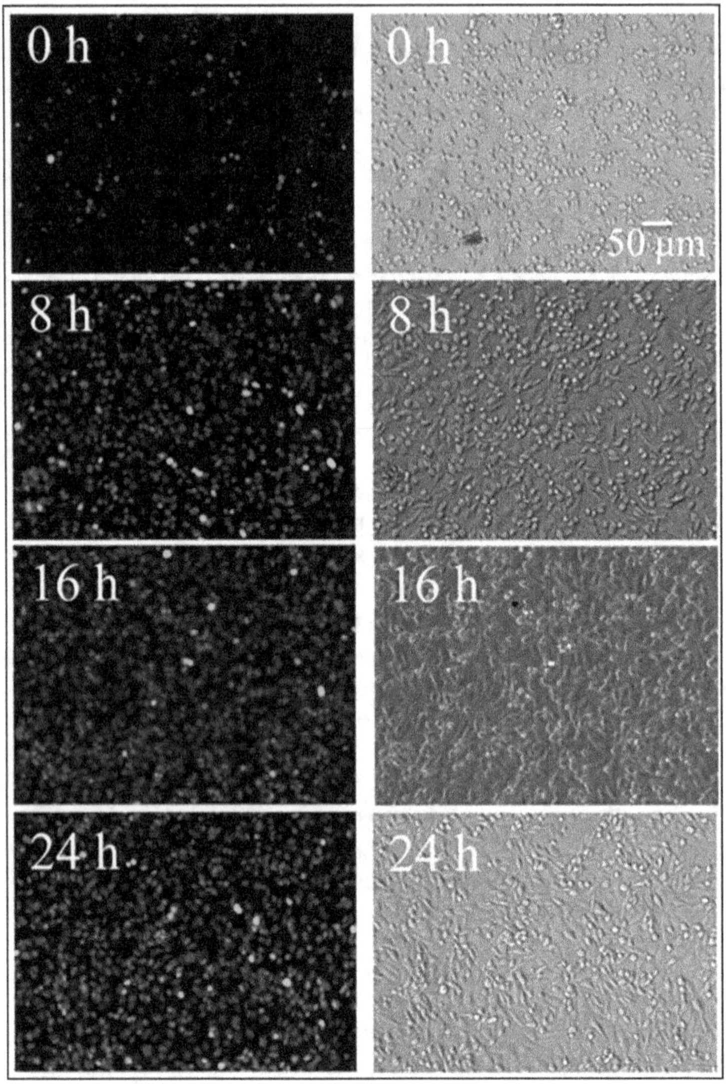

Fig. 18 Optical and fluorescence microscopy images of DAF-2 dye-loaded SaOS-2 cells on Si_3N_4 from 0 to 24 h, in 8-h intervals. (Used with permission [95])

of NO within osteoblast cells provides evidence of an activated endothelial nitric oxide synthase (eNOS) pathway. The monotonic increase in NO concentration with exposure time is positively correlated with increased osteoblast cell proliferation (cf., Fig. 18). However, another well-known enzymatic pathway for NH_3 to NO conversion occurs in nature. As an essential part of the nitrogen cycle, ammonia monooxygenase is recognized for this role in prokaryotes and plants [110]. This is the process partially depicted in the left half of Fig. 19.

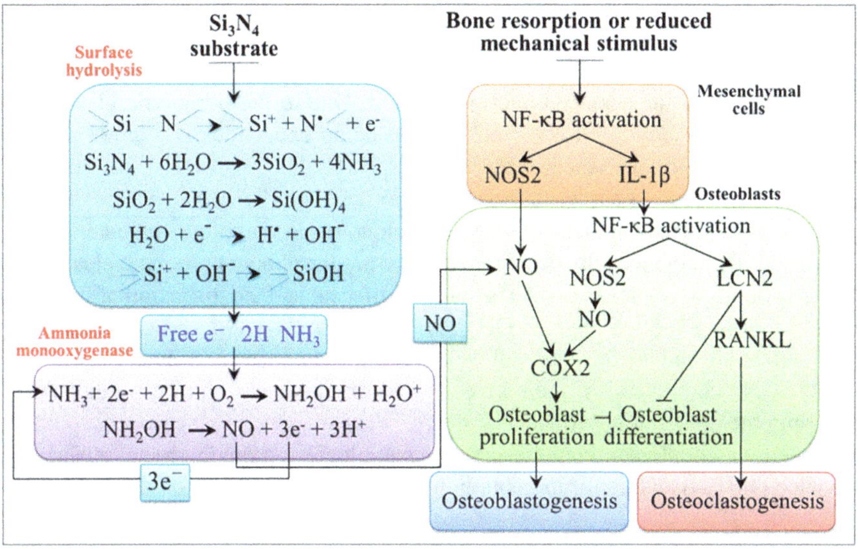

Fig. 19 Schematic diagram of the pathways involved with osteoblast-osteoclast crosstalk. IL-1β/LCN2/RANKL pathway leading to osteoclastogenesis and NOS2/NO/COX2 pathway leading to osteoblastogenesis. Increased exogenous NO is an effect of the reactive nitrogen species (RNS) formed at the cell/Si_3N_4 interface. (Used with permission under Creative Commons Attribution License [103])

While nitric oxide is well known for its fundamental role in cell signaling [111], exogenous NO release can either promote or disrupt osteoblastogenesis [108]. On the one hand, high concentrations of NO inhibit the growth and differentiation of osteoblasts. This may be partly due to the inhibitory effects of pro-inflammatory cytokines on bone formation. On the other hand, lower amounts of NO produced by constitutive nitric oxide synthase (eNOS) enzymes may play a role in regulating normal osteoblast growth [76, 112]. Remarkably, the amount of NO released from Si_3N_4 appears to be low and slow enough to be within the concentration range that promotes cell proliferation.

Mesenchymal cell differentiation into osteoblasts and osteoclasts is regulated by NF-κB, a set of multifunctional transcription factors that control the expression of genes involved in skeletal development [113]. The right half of Fig. 19 illustrates two distinct NF-κB activation pathways for osteoblastogenesis or osteoclastogenesis. According to a study by Veeriah et al. [114], the osteoclastogenesis pathway begins with the activation of the pro-inflammatory cytokine IL-1β by caspase-1. IL-1β then acts as a pro-osteoclastogenic factor [114, 115]. Lipocalin 2 (LCN2) is a pleiotropic and acute-phase protein that is upregulated by inflammation. While it does not affect osteoblast proliferation, it does significantly impair osteoclast differentiation. LCN2 stimulates osteoblasts to produce RANKL, which promotes osteoclastogenesis and bone resorption. In addition to the IL-1β/LCN2/RANKL pathway, NF-κB activation also induces

NOS2, which increases osteoblast proliferation through the formation of NO molecules. NO molecules then target cyclooxygenase-2 (COX2) in a downstream pathway (NOS2/NO/COX2). Maintaining the balance between these two pathways is essential for healthy bone metabolism. Due to the homolytic cleavage of its covalent Si-N bonds, Si_3N_4 is believed to be an exogenous NO donor and can be expected to enhance the NOS2/NO/COX2 pathway to promote osteoblastogenesis, as shown in the right half of Fig. 19.

Collectively, these studies support the unique osteopromotive characteristics of implanted Si_3N_4 devices. Its behavior results from the simple homolytic cleavage of the Si-N bond, which leads to interfacial pH buffering and the formation of favorable bone-promoting species (e.g., $Si(OH)_4$, NH_3, NH_4^+), the endocytotic incorporation of Si into HAp, and the formation of NO. While these moieties are inherently present in all Si_3N_4 compositions, their kinetics and effectiveness can be modulated by tailoring the material's surface chemistry. This engineering approach could be exploited to counteract bone resorption pathologies, promote bone repair, and advance other related applications in bone tissue engineering.

4 Summary and Conclusions

Far from being inert materials, bioceramics exhibit diverse interactions with living tissues that significantly impact cell metabolism and fate. This is because the immune system is designed to treat implants as potential hazards and react accordingly. The system prioritizes its defensive actions as dissolution—if possible; encapsulation—if necessary; and integration—if appropriate. Therefore, the interfacial phase structure and chemistry of a bioceramic determine whether a device is integrated or rejected [116]. This chapter reviewed the osteoconductive behavior and mechanisms of Si_3N_4 and contrasted them to other commonly used biomaterials. The long-held belief that bioceramics are bioinert was shown to be false at the microscopic scale. In contrast to oxide-based ceramics and Ti-alloys, Si_3N_4 actively promotes attachment, differentiation, and proliferation of osseous tissue. Similar to hydroxyapatite, Si_3N_4 has a bio-friendly zwitterionic surface which facilitates the deposition and motility of osteoblasts. Si_3N_4's unique surface chemistry is activated when immersed in the biological environment and a simple hydrolysis reaction releases silicon and nitrogen. These elements are converted to the osteopromotive moieties of silicic acid ($Si(OH)_4$), ammonium (NH_4^+), and ammonia (NH_3). On the one hand, Si is endocytotically scavenged by osteoblasts and incorporated into newly mineralized bone. On the other hand, NH_3 freely diffuses across cell membranes, while NH_4^+ is deprotonated and transported transcellularly by a dual path. NH_3 is subsequently enzymatically converted to exogenous NO. In this form, it serves as a signaling agent for the upregulation or dysregulation of osseous processes. Based on experimental evidence, it is released and converted at a concentration and kinetics that enhance osteoblasts and suppress osteoclasts. The combined effect of a highly bone-friendly surface coupled with these osteopromotive moieties results in enhanced osteogenic activity. A growing body of evidence demonstrates that Si_3N_4 is a promising therapeutic bioceramic with the potential to revolutionize the treatment of a variety of osseous diseases and conditions.

References

1. Petzow, G.; Herrmann, M. Silicon Nitride Ceramics. In *High Performance Non-Oxide Ceramics II SE - 2*; Jansen, M., Ed.; Structure and Bonding; Springer Berlin Heidelberg, 2002; Vol. 102, pp. 47–167 ISBN 978-3-540-43132-9.
2. Katz, R.N. Commercial Applications of Silicon Nitride Based Ceramics. *Mater. Technol.* **1993**, *8*, 142–148, https://doi.org/10.1080/10667857.1993.11784967.
3. Nicholson, K.C. Manufacture of Silicon Nitride-Bonded Articles. *US Pat. 2618,565* **1952**.
4. Nicholson, K.C. Silicon Nitride-Bonded Refractory Oxide Bodies and Method of Making. *US Pat. 2,636,828* **1953**.
5. G R Terwilliger; F F Lange Hot-Pressing Behavior of Si_3N_4. *J. Am. Ceram. Soc.* **1974**, *57*, 25–29.
6. Becher, P.F. Microstructural Design of Toughened Ceramics. *J. Am. Ceram. Soc.* **1991**, *74*, 255–269, https://doi.org/10.1111/j.1151-2916.1991.tb06872.x.
7. Becher, P.F.; Sun, E.Y.; Plucknett, K.P.; Alexander, K.B.; Hsueh, C.; Lin, H.; Waters, S.B.; Westmoreland, C.G. Microstructural Design of Silicon Nitride with Improved Fracture Toughness: I, Effects of Grain Shape and Size. *J. Am. Ceram. Soc.* **1998**, *81*, 2821–2830.
8. Ohji, T.; Hirao, K.; Kanzaki, S. Fracture Resistance Behavior of Highly Anisotropic Silicon Nitride. *J. Am. Ceram. Soc.* **1995**, *78*, 3125–3128.
9. Emoto, H.; Mitomo, M. Control and Characterization of Abnormally Grown Grains in Silicon Nitride Ceramics. *J. Eur. Ceram. Soc.* **1997**, *17*, 797–804, https://doi.org/10.1016/S0955-2219(96)00139-2.
10. Hirosaki, N.; Akimune, Y.; Mitomo, M. Effect of Grain Growth of ß-Silicon Nitride on Strength, Weibull Modulus, and Fracture Toughness. *J. Am. Ceram. Soc.* **1993**, *76*, 1892–1894, https://doi.org/10.1111/j.1151-2916.1993.tb06670.x.
11. Wu, F.; Zhuang, H.; Ma, L.; Fu, X. Self-Reinforced Silicon Nitride by Gas Pressure Sintering. In *A Collection of Papers Presented at the 94th Annual Meeting and the 1992 Fall Meeting of the Materials & Equipment/Whitewares Manufacturing: Ceramic Engineering and Science Proceedings*; Ceramic Engineering and Science Proceedings; 1993; pp. 321–332 ISBN 9780470314050.
12. Kawashima, T.; Okamoto, H.; Yamamoto, H.; Kitamura, A. Grain Size Dependence of the Fracture Toughness of Silicon Nitride Ceramics. *J. Ceram. Soc. Japan* **1991**, *99*, 320–323.
13. Li, C.-W.; Yamanis, J. Super-Tough Silicon Nitride with R-Curve Behavior. *Ceram. Eng. Sci. Proc.* **1989**, *10*, 632–645.
14. Tani, E.; Umebayashi, S.; Kishi, K.; Kobayashi, K.; Nishijima, M. Gas-Pressure Sintering of Si_3N_4 with Concurrent Addition of Al_2O_3 and 5 wt% Rare Earth Oxide: High Fracture Toughness Si_3N_4 With Fiber-Like Structure. *Bull. Am. Ceram. Soc.* **1986**, *65*, 1311–1315.
15. Pezzotti, G. Si_3N_4/SiC-Platelet Composite without Sintering Aids: A Candidate for Gas Turbine Engines. *J. Am. Ceram. Soc.* **1993**, *76*, 1313–1320, https://doi.org/10.1111/j.1151-2916.1993.tb03757.x.
16. Richerson, D.W. Historical Review of Addressing the Challenges of Use of Ceramic Components in Gas Turbine Engines 2006, 241–254.
17. Takahashi, J.; Mizutani, T.; Sugihara, H.Y.; Nagata, S.; Kato, S.; Hiraguri, Y.; Takeoka, S.; Tsuchiya, M.; Kuno, R.; Kakinuma, S. Suspension Culture in a Rotating Bioreactor for Efficient Generation of Human Intestinal Organoids. *Cell Reports Methods* **2022**, *2*, e1–e5.

18. Takasato, M.; Er, P.X.; Becroft, M.; Vanslambrouck, J.M.; Stanley, E.G.; Elefanty, A.G.; Little, M.H. Directing Human Embryonic Stem Cell Differentiation Towards a Renal Lineage Generates a Self-Organizing Kidney. *Nat. Cell Biol.* **2014**, *16*, 118–126, https://doi.org/10.1038/ncb2894.
19. Osakada, F.; Jin, Z.-B.; Hirami, Y.; Ikeda, H.; Danjyo, T.; Watanabe, K.; Sasai, Y.; Takahashi, M. In vitro Differentiation of Retinal Cells from Human Pluripotent Stem Cells by Small-Molecule Induction. *J. Cell Sci.* **2009**, *122*, 3169–3179, https://doi.org/10.1242/jcs.050393.
20. Mansouri, M.; Fussenegger, M. Therapeutic Cell Engineering: Designing Programmable Synthetic Genetic Circuits in Mammalian Cells. *Protein Cell* **2022**, *13*, 476–489, https://doi.org/10.1007/s13238-021-00876-1.
21. Khalil, A.S.; Collins, J.J. Synthetic Biology: Applications Come of Age. *Nat. Rev. Genet.* **2010**, *11*, 367–379, https://doi.org/10.1038/nrg2775.
22. Mauger, D.M.; Cabral, B.J.; Presnyak, V.; Su, S. V; Reid, D.W.; Goodman, B.; Link, K.; Khatwani, N.; Reynders, J.; Moore, M.J. mRNA Structure Regulates Protein Expression through Changes in Functional Half-Life. *Proc. Natl. Acad. Sci.* **2019**, *116*, 24075–24083.
23. Cazzola, M.; Skoda, R.C. Translational Pathophysiology: A Novel Molecular Mechanism of Human Disease. *Blood* **2000**, *95*, 3280–3288, https://doi.org/10.1182/blood.V95.11.3280.
24. Pezzotti, G.; Zhu, W.; Terai, Y.; Marin, E.; Boschetto, F.; Kawamoto, K.; Itaka, K. Raman Spectroscopic Insight into Osteoarthritic Cartilage Regeneration by mRNA Therapeutics Encoding Cartilage-Anabolic Transcription Factor Runx1. *Mater. Today Bio* **2022**, *13*, 100210, https://doi.org/10.1016/j.mtbio.2022.100210.
25. Shastri, V.P. In vivo Engineering of Tissues: Biological Considerations, Challenges, Strategies, and Future Directions. *Adv. Mater.* **2009**, *21*, 3246–3254, https://doi.org/10.1002/adma.200900608.
26. Nussinov, R. Advancements and Challenges in Computational Biology. *PLOS Comput. Biol.* **2015**, *11*, e1004053.
27. Szczęsny, G.; Kopec, M.; Politis, D.J.; Kowalewski, Z.L.; Łazarski, A.; Szolc, T. A Review on Biomaterials for Orthopaedic Surgery and Traumatology: From Past to Present. *Materials (Basel).* **2022**, *15*, 3622.
28. Florea, D.A.; Albuleț, D.; Grumezescu, A.M.; Andronescu, E. Surface Modification – A Step Forward to Overcome the Current Challenges in Orthopedic Industry and to Obtain an Improved Osseointegration and Antimicrobial Properties. *Mater. Chem. Phys.* **2020**, *243*, 122579, https://doi.org/10.1016/j.matchemphys.2019.122579.
29. Martin, V.; Bettencourt, A. Bone Regeneration: Biomaterials as Local Delivery Systems with Improved Osteoinductive Properties. *Mater. Sci. Eng. C* **2018**, *82*, 363–371, https://doi.org/10.1016/j.msec.2017.04.038.
30. Mazzocchi, M.; Bellosi, A. On the Possibility of Silicon Nitride as a Ceramic for Structural Orthopaedic Implants. Part I: Processing, Microstructure, Mechanical Properties, Cytotoxicity. *J. Mater. Sci. Mater. Med.* **2008**, *19*, 2881–2887, https://doi.org/10.1007/s10856-008-3417-2.
31. Mazzocchi, M.; Gardini, D.; Traverso, P.L.; Faga, M.G.; Bellosi, A. On the Possibility of Silicon Nitride as a Ceramic for Structural Orthopaedic Implants. Part II: Chemical Stability and Wear Resistance in Body Environment. *J. Mater. Sci. Mater. Med.* **2008**, *19*, 2889–2901, https://doi.org/10.1007/s10856-008-3437-y.
32. Bal, B.S.; Rahaman, M.N. Orthopedic Applications of Silicon Nitride Ceramics. *Acta Biomater.* **2012**, *8*, 2889–2898, https://doi.org/10.1016/j.actbio.2012.04.031.
33. Pezzotti, G. Silicon Nitride: A Bioceramic with a Gift. *ACS Appl. Mater. Interfaces* **2019**, *11*, 26619–26636, https://doi.org/10.1021/acsami.9b07997.
34. Howlett, C.R.; McCartney, E.; Ching, W. The Effect of Silicon Nitride Ceramic on Rabbit Skeletal Cells and Tissue. *Clin. Orthop. Relat. Res.* **1989**, *244*, 293–304.
35. Sorrell, C.C.; Hardcastle, P.H.; Druitt, R.K.; Howlett, C.R.; McCartney, E.R. Results of 15-Year Clinical Study of Reaction Bonded Silicon Nitride Intervertebral Spacers. *Proc. 7th World Biomater. Conf.* **2004**, 1872.

36. Mobbs, R.J.; Rao, P.J.; Phan, K.; Hardcastle, P.; Choy, W.J.; McCartney, E.R.; Druitt, R.K.; Mouatt, C.A.L.; Sorrell, C.C. Anterior Lumbar Interbody Fusion Using Reaction Bonded Silicon Nitride Implants: Long Term Case Series of the First Synthetic ALIF Spacer Implanted in Humans. *World Neurosurg.* **2018**, *120*, 256–264, https://doi.org/10.1016/j.wneu.2018.08.237.
37. Albrektsson, T.; Johansson, C. Osteoinduction, Osteoconduction and Osseointegration. *Eur. Spine J.* **2001**, *10*, 96–101, https://doi.org/10.1007/s005860100282.
38. Daculsi, G.; Fellah, B.H.; Miramond, T.; Durand, M. Osteoconduction, Osteogenicity, Osteoinduction, What are the Fundamental Properties for a Smart Bone Substitutes. *IRBM* **2013**, *34*, 346–348, https://doi.org/10.1016/j.irbm.2013.07.001.
39. Sun, L.; Berndt, C.C.; Gross, K.A.; Kucuk, A. Material Fundamentals and Clinical Performance of Plasma-Sprayed Hydroxyapatite Coatings: A Review. *J. Biomed. Mater. Res. Appl. Biomater.* **2001**, *58*, 570–592, https://doi.org/10.1002/jbm.xxxx.
40. Hing, K.A.; Best, S.M.; Tanner, K.E.; Bonfield, W.; Revell, P.A. Mediation of Bone Ingrowth in Porous Hydroxyapatite Bone Graft Substitute. *J. Biomed. Mater. Res. Part A* **2004**, *68A*, 187–200, https://doi.org/10.1002/jbm.a.10050.
41. Lee, K.-Y.; Park, M.; Kim, H.-M.; Lim, Y.-J.; Chun, H.-J.; Kim, H.; Moon, S.-H. Ceramic Bioactivity: Progresses, Challenges and Perspectives. *Biomed. Mater.* **2006**, *1*, R31.
42. Dorozhkin, S. V. Calcium Orthophosphate-Based Biocomposites and Hybrid Biomaterials. *J. Mater. Sci.* **2009**, *44*, 2343–2387, https://doi.org/10.1007/s10853-008-3124-x.
43. Bodhak, S.; Bose, S.; Bandyopadhyay, A. Role of Surface Charge and Wettability on Early Stage Mineralization and Bone Cell-Materials Interactions of Polarized Hydroxyapatite. *Acta Biomater.* **2009**, *5*, 2178–88, https://doi.org/10.1016/j.actbio.2009.02.023.
44. Lang, S.B.; Tofail, S.A.M.; Kholkin, A.L.; Wojtaś, M.; Gregor, M.; Gandhi, A.A.; Wang, Y.; Bauer, S.; Krause, M.; Plecenik, A. Ferroelectric Polarization in Nanocrystalline Hydroxyapatite Thin Films on Silicon. *Sci. Rep.* **2013**, *3*, 2215, https://doi.org/10.1038/srep02215.
45. Kumar, D.; Gittings, J.P.; Turner, I.G.; Bowen, C.R.; Bastida-Hidalgo, A.; Cartmell, S.H.; Hidalgo-Bastida, L. a; Bastida-Hidalgo, A.; Cartmell, S.H. Polarization of Hydroxyapatite: Influence on Osteoblast cell Proliferation. *Acta Biomater.* **2010**, *6*, 1549–1554, https://doi.org/10.1016/j.actbio.2009.11.008.
46. Zhu, W.; McEntire, B.J.; Enomoto, Y.; Boffelli, M.; Pezzotti, G. Point-Defect Populations as Induced by Cation/Anion Substitution in β-Si_3N_4 Lattice – A Cathodoluminescence Study. *J. Phys. Chem. C* **2015**, *119*, 3279–3287, https://doi.org/10.1021/jp509995c.
47. Uskokovic, V. The Role of Hydroxyl Channel in Defining Selected Physicochemical Peculiarities Exhibited by Hydroxyapatite. *RSC Adv.* **2015**, *5*, 36614–36633, https://doi.org/10.1039/C4RA17180B.
48. Pezzotti, G.; McEntire, B.J.; Bock, R.; Boffelli, M.; Zhu, W.; Vitale, E.; Puppulin, L.; Adachi, T.; Yamamoto, T.; Kanamura, N.; et al. Silicon Nitride: A Synthetic Mineral for Vertebrate Biology. *Sci. Rep.* **2016**, *6*, 31717, https://doi.org/10.1038/srep11384.
49. Pezzotti, G.; McEntire, B.J.; Bock, R.M.; Zhu, W.; Boschetto, F.; Rondinella, A.; Marin, E.; Marunaka, Y.; Adachi, T.; Yamamoto, T.; et al. In situ Spectroscopic Screening of Osteosarcoma Living Cells on Stoichiometry-Modulated Silicon Nitride Bioceramic Surfaces. *ACS Biomater. Sci. Eng.* **2016**, *2*, 1121–1134, https://doi.org/10.1021/acsbiomaterials.6b00126.
50. Bock, R.M.; McEntire, B.J.; Bal, B.S.; Rahaman, M.N.; Boffelli, M.; Pezzotti, G. Surface Modulation of Silicon Nitride Ceramics for Orthopaedic Applications. *Acta Biomater.* **2015**, *26*, 318–330.
51. Cabral-Pacheco, G.A.; Garza-Veloz, I.; Castruita-De la Rosa, C.; Ramirez-Acuña, J.M.; Perez-Romero, B.A.; Guerrero-Rodriguez, J.F.; Martinez-Avila, N.; Martinez-Fierro, M.L. The Roles of Matrix Metalloproteinases and Their Inhibitors in Human Diseases. *Int. J. Mol. Sci.* 2020, *21*.
52. Guntur, A.R.; Rosen, C.J. IGF-1 Regulation of Key Signaling Pathways in Bone. *Bonekey Rep.* **2013**, *2*, 1–6.

53. Kawasaki, T. Theory of Chromatography of Rigid Molecules on Hydroxyapatite Columns with Small Loads. *J. Chromatogr. A* **1978**, *151*, 95–112, https://doi.org/10.1016/S0021-9673(00)85374-1.
54. Kawasaki, T. Theory of Chromatography on Hydroxyapatite Columns with Small Loads. *J. Chromatogr. A* **1978**, *157*, 7–42, https://doi.org/10.1016/S0021-9673(00)92319-7.
55. Kawasaki, T.; Takahashi, S.; Ideda, K. Hydroxyapatite High-Performance Liquid Chromatography: Column Performance for Proteins. *Eur. J. Biochem.* **1985**, *152*, 361–371, https://doi.org/10.1111/j.1432-1033.1985.tb09206.x.
56. Gorbunoff, M.J. The Interaction of Proteins with Hydroxyapatite. I. Role of Protein Charge and Structure. *Anal. Biochem.* **1984**, *136*, 425–432, https://doi.org/10.1016/0003-2697(84)90239-2.
57. Gorbunoff, M.J. The Interaction of Proteins with Hydroxyapatite. II. Role of Acidic and Basic Groups. *Anal. Biochem.* **1984**, *136*, 433–439, https://doi.org/10.1016/0003-2697(84)90240-9.
58. Gorbunoff, M.J.; Timasheff, S.N. The Interaction of Proteins with Hydroxyapatite. III. Mechanism. *Anal. Biochem.* **1984**, *136*, 440–445, https://doi.org/10.1016/0003-2697(84)90241-0.
59. Schwarz, K. A Bound Form of Silicon in Glycosaminoglycans and Polyuronides. *Proc. Natl. Acad. Sci. U. S. A.* **1973**, *70*, 1608–1612, https://doi.org/10.1073/pnas.70.5.1608.
60. Shepherd, J.H.; Shepherd, D. V.; Best, S.M. Substituted Hydroxyapatites for Bone Repair. *J. Mater. Sci. Mater. Med.* **2012**, *23*, 2335–2347, https://doi.org/10.1007/s10856-012-4598-2.
61. Jugdaohsingh, R. Silicon and Bone Health. *J Nutr Heal. Aging* **2007**, *11*, 99–110.
62. Jurkić, L.M.; Cepanec, I.; Pavelić, S.K.; Pavelić, K. Biological and Therapeutic Effects of Ortho-Silicic Acid and some Ortho-Silicic Acid-Releasing Compounds: New Perspectives for Therapy. *Nutr. Metab. (Lond).* **2013**, *10*, 2, https://doi.org/10.1186/1743-7075-10-2.
63. Mladenovic, Z.; Johansson, A.; Willman, B.; Shahabi, K.; Bjorn, E.; Ransjo, M. Soluble Silica Inhibits Osteoclast Formation and Bone Resorption in vitro. *Acta Biomater.* **2014**, *10*, 406–418, https://doi.org/10.1016/j.actbio.2013.08.039.
64. Reffitt, D.M.; Ogston, N.; Jugdaohsingh, R.; Cheung, H.F.J.; Evans, B.A.J.; Thompson, R.P.H.; Powell, J.J.; Hampson, G.N. Orthosilicic Acid Stimulates Collagen Type 1 Synthesis and Osteoblastic Differentiation in Human Osteoblast-Like Cells in vitro. *Bone* **2003**, *32*, 127–135, https://doi.org/10.1016/S8756-3282(02)00950-X.
65. Costa-Rodrigues, J.; Reis, S.; Castro, A. Bone Anabolic Effects of Soluble Si: In Vitro Studies with Human Mesenchymal Stem Cells and CD14. *Stem Cells* **2015**, *2016*.
66. Thian, E.S.; Huang, J.; Best, S.M.; Barber, Z.H.; Brooks, R.A.; Rushton, N.; Bonfield, W. The Response of Osteoblasts to Nanocrystalline Silicon-Substituted Hydroxyapatite Thin Films. *Biomaterials* **2006**, *27*, 2692–2698, https://doi.org/10.1016/j.biomaterials.2005.12.019.
67. Thian, E.S.; Huang, J.; Best, S.M.; Barber, Z.H.; Bonfield, W. Silicon-Substituted Hydroxyapatite: The Next Generation of Bioactive Coatings. *Mater. Sci. Eng. C* **2007**, *27*, 251–256.
68. Beck, G.R.; Ha, S.-W.W.; Camalier, C.E.; Yamaguchi, M.; Li, Y.; Lee, J.-K.K.; Weitzmann, M.N. Bioactive Silica-Based Nanoparticles Stimulate Bone-Forming Osteoblasts, Suppress Bone-Resorbing Osteoclasts, and Enhance Bone Mineral Density in vivo. *Nanomedicine Nanotechnology, Biol. Med.* **2012**, *8*, 793–803, https://doi.org/10.1016/j.nano.2011.11.003.
69. Ha, S.W.; Neale Weitzmann, M.; Beck, G.R. Bioactive Silica Nanoparticles Promote Osteoblast Differentiation Through Stimulation of Autophagy and Direct Association with LC3 and p62. *ACS Nano* **2014**, *8*, 5898–5910, https://doi.org/10.1021/nn5009879.
70. Calvo-Guirado, J.L.; Garces, M.; Delgado-Ruiz, R.A.; Ramirez Fernandez, M.P.; Ferres-Amat, E.; Romanos, G.E. Biphasic β-TCP Mixed with Silicon Increases Bone Formation in Critical Site Defects in Rabbit Calvaria. *Clin. Oral Implants Res.* **2015**, *26*, 891–897, https://doi.org/10.1111/clr.12413.
71. Escott-Stump, S. *Nutrition and Diagnosis-Related Care*; 5th Ed.; Lippincott Williams & Wilkins: Philadelphia, PA USA, 2008; ISBN 0781798450.
72. Rizzo, D. *Fundamentals of Anatomy and Physiology*; 4th Ed.; Cengage Learning: Boston, MA USA, 2015; ISBN 130546477X.

73. Brandi, M.L.; Hukkanen, M.; Umeda, T.; Moradi-Bidhendi, N.; Bianchi, S.; Gross, S.S.; Polak, J.M.; MacIntyre, I. Bidirectional Regulation of Osteoclast Function by Nitric Oxide Synthase Isoforms. *Proc. Natl. Acad. Sci. U. S. A.* **1995**, *92*, 2954–2958, https://doi.org/10.1073/pnas.92.7.2954.
74. Riancho, J.A.; Salas, E.; Zarrabeitia, M.T.; Olmos, J.M.; Amado, J.A.; Fernandez-Luna, J.L.; Gonzalez-Macias, J. Expression and Functional Role of Nitric Oxide Synthase in Osteoblast-Like Cells. *J. Bone Miner. Res.* **1995**, *10*, 439–446, https://doi.org/10.1002/jbmr.5650100315.
75. Evans, D.M.; Ralston, S.H. Nitric Oxide and Bone. *J. Bone Miner. Res.* **1996**, *11*, 300–305, https://doi.org/10.1002/jbmr.5650110303.
76. Ralston, S.H. Nitric Oxide and Bone: What a Gas! *Revmatologiia* **1997**, *36*, 831–838.
77. Wimalawansa, S.J. Nitric Oxide and Bone. *Ann. N. Y. Acad. Sci.* **2010**, *1192*, 391–403, https://doi.org/10.1111/j.1749-6632.2009.05230.x.
78. Saura, M.; Tarin, C.; Zaragoza, C. Recent Insights into the Implication of Nitric Oxide in Osteoblast Differentiation and Proliferation During Bone Development. *Sci. World J.* **2010**, *10*, 624–632, https://doi.org/10.1100/tsw.2010.58.
79. MacPherson, H.; Noble, B.S.; Ralston, S.H. Expression and Functional Role of Nitric Oxide Synthase Isoforms in Human Osteoblast-Like Cells. *Bone* **2016**, *24*, 179–185, https://doi.org/10.1016/S8756-3282(98)00173-2.
80. Pezzotti, G.; Bock, R.M.; McEntire, B.J.; Jones, E.; Boffelli, M.; Zhu, W.; Baggio, G.; Boschetto, F.; Puppulin, L.; Adachi, T.; et al. Silicon Nitride Bioceramics Induce Chemically Driven Lysis in Porphyromonas Gingivalis. *Langmuir* **2016**, *32*, 3024–3035, https://doi.org/10.1021/acs.langmuir.6b00393.
81. Rahal, A.; Kumar, A.; Singh, V.; Yadav, B.; Tiwari, R.; Chakraborty, S.; Dhama, K. Oxidative Stress, Prooxidants, and Antioxidants: The Interplay. *Biomed Res. Int.* **2014**, *2014*, 761264, https://doi.org/10.1155/2014/761264.
82. Parker, D.R.; Norvell, W.A.; Chaney, R.L. GEOCHEM-PC— A Chemical Speciation Program for IBM and Compatible Personal Computers. In *Chemical Equilibrium and Reaction Models*; Loeppert, R.H., Schwab, A.D., Goldberg, S., Eds.; Soil Science Society of America and American Society of Agronomy: Madison, SI, USA, 1995; pp. 253–269.
83. Pezzotti, G.; McEntire, B.; Zhu, W.; Puppulin, L.; Boffelli, M.; Rahaman, M.N.; Bal, B.S. The Scientific Rationale for Using Silicon Nitride in Biomedical Implants. In *Biomedical Nanomaterials: From Design to Implementation,*; Webster, T.J., Yazici, H., Eds.; IET Press: London, UK, 2016; p. 281.
84. Pezzotti, G. Silicon Nitride as a Biomaterial. *J. Ceram. Soc. Japan* **2023**, *131*, 398–428, https://doi.org/10.2109/jcersj2.23057.
85. Kajdas, C. General Approach to Mechanochemistry and Its Relation to Tribochemistry. In *Tribology in Engineering*; INTEC, 2013; pp. 209–240 ISBN 4043851235.
86. Mezzasalma, S. Characterization of Silicon Nitride Surface in Water and Acid Environment: A General Approach to the Colloidal Suspensions. *J. Colloid Interface Sci.* **1996**, *180*, 413–420, https://doi.org/10.1006/jcis.1996.0320.
87. Sonnefeld, J. Determination of Surface Charge Density Parameters of Silicon Nitride. *Colloids Surfaces A Physicochem. Eng. Asp.* **1996**, *108*, 27–31, https://doi.org/10.1016/0927-7757(95)03356-4.
88. Sulpizi, M.; Gaigeot, M.-P.; Sprik, M. The Silica−Water Interface: How The Silanols Determine The Surface Acidity And Modulate The Water Properties. *J. Chem. Theory Comput.* **2012**, *8*, 1037–1047, https://doi.org/10.1021/ct2007154.
89. Shapovalov, V.; Truong, T.N. Ab Initio Study of Water Adsorption on α-Al_2O_3 (0001) Crystal Surface. *J. Phys. Chem. B* **2000**, *104*, 9859–9863, https://doi.org/10.1021/jp001399g.
90. Fernández, E.M.; Eglitis, R.I.; Borstel, G.; Balbás, L.C. Ab initio Calculations of H2O and O2 Adsorption on Al_2O_3 Substrates. *Comput. Mater. Sci.* **2007**, *39*, 587–592, https://doi.org/10.1016/j.commatsci.2006.08.010.
91. Lu, Y.-H.; Chen, H.-T. Hydrogen Generation by the Reaction of H2O with Al_2O_3-based Materials: A Computational Analysis. *Phys. Chem. Chem. Phys.* **2015**, *17*, 6834–6843.

92. Kisnieriené, V.; Lapeikaitė, I. When Chemistry Meets Biology: The Case of Aluminium–A Review. *Chemija* **2015**, *26*, 148–158.
93. Zanocco, M.; Marin, E.; Rondinella, A.; Boschetto, F.; Horiguchi, S.; Zhu, W.; McEntire, B.J.; Bock, R.M.; Bal, B.S.; Pezzotti, G. The Role of Nitrogen Off-Stoichiometry in the Osteogenic Behavior of Silicon Nitride Bioceramics. *Mater. Sci. Eng. C* **2019**, *105*, 110053, https://doi.org/10.1016/j.msec.2019.110053.
94. Pezzotti, G.; Oba, N.; Zhu, W.; Marin, E.; Rondinella, A.; Boschetto, F.; McEntire, B.J.; Yamamoto, K.; Bal, B.S. Human Osteoblasts Grow Transitional Si/N Apatite in Quickly Osteointegrated Si_3N_4 Cervical Insert. *Acta Biomater.* **2017**, *64*, 411–420.
95. Pezzotti, G.; Marin, E.; Adachi, T.; Rondinella, A.; Boschetto, F.; Zhu, W.-L.; Sugano, N.; Bock, R.M.; McEntire, B.J.; Bal, B.S. Bioactive Silicon Nitride: A New Therapeutic Material for Osteoarthropathy. *Sci. Rep.* **2017**, *7*, 44848.
96. Notingher, I.; Bisson, I.; Polak, J.M.; Hench, L.L. In situ Spectroscopic Study of Nucleic Acids in Differentiating Embryonic Stem Cells. *Vib. Spectrosc.* **2004**, *35*, 199–203, https://doi.org/10.1016/j.vibspec.2004.01.014.
97. Notingher, I.; Bisson, I.; Bishop, A.E.; Randle, W.L.; Polak, J.M.P.; Hench, L.L. In situ Spectral Monitoring of mRNA Translation in Embryonic Stem Cells during Differentiation in Vitro. *Anal. Chem.* **2004**, *76*, 3185–3193, https://doi.org/10.1021/ac0498720.
98. Lovrić, J.; Cho, S.J.; Winnik, F.M.; Maysinger, D. Unmodified Cadmium Telluride Quantum dots Induce Reactive Oxygen Species Formation Leading to Multiple Organelle Damage and Cell Death. *Chem. Biol.* **2005**, *12*, 1227–1234, https://doi.org/10.1016/j.chembiol.2005.09.008.
99. Chen, L. Manufactered Aluminum Oxide Nanoparticles Decrease Expression of Tight Junction Proteins in Bain Vasculature. *J. Neuroimmune Pharmacol.* **2008**, *3*, 286–295, https://doi.org/10.1007/s11481-008-9131-5.Manufactured.
100. Freyre-Fonseca, V.; Delgado-Buenrostro, N.L.; Gutierrez-Cirlos, E.B.; Calderon-Torres, C. M.; Cabellos-Avelar, T.; Sanchez-Perez, Y.; Pinzon, E.; Torres, I.; Molina-Jijon, E.; Zazueta, C.; et al. Titanium Dioxide Nanoparticles Impair Lung Mitochondrial Function. *Toxicol. Lett.* **2011**, *202*, 111–119, https://doi.org/10.1016/j.toxlet.2011.01.025.
101. Teodoro, J.S.; Simões, A.M.; Duarte, F. V.; Rolo, A.P.; Murdoch, R.C.; Hussain, S.M.; Palmeira, C.M. Assessment of the Toxicity of Silver Nanoparticles in vitro: A Mitochondrial Perspective. *Toxicol. Vitr.* **2011**, *25*, 664–670, https://doi.org/10.1016/j.tiv.2011.01.004.
102. Zou, S.; Huang, J.; Best, S.; Bonfield, W. Crystal Imperfection Studies of Pure and Silicon Substituted Hydroxyapatite using Raman and XRD. *J. Mater. Sci. Mater. Med.* **2005**, *16*, 1143–1148, https://doi.org/10.1007/s10856-005-4721-8.
103. Pezzotti, G.; Adachi, T.; Boschetto, F.; Zhu, W.; Zanocco, M.; Marin, E.; Bal, B.S.S.; McEntire, B.J.B.J. Off-Stoichiometric Reactions at the Cell-Substrate Biomolecular interface of Biomaterials: In situ and Ex situ Monitoring of Cell Proliferation, Differentiation, and Bone Tissue Formation. *Int. J. Mol. Sci.* **2019**, *20*, 1–23, https://doi.org/10.3390/ijms20174080.
104. Pezzotti, G.; Bock, R.M.; McEntire, B.J.; Adachi, T.; Marin, E.; Boschetto, F.; Zhu, W.; Mazda, O.; Bal, B.S. In vitro Antibacterial Activity of Oxide and Non-Oxide Bioceramics for Arthroplastic Devices: I. In situ Time-Lapse Raman Spectroscopy. *Analyst* **2018**, *143*, 3708–3721, https://doi.org/10.1039/C8AN00233A.
105. Lundberg, J.O.; Weitzberg, E.; Gladwin, M.T. The Nitrate-Nitrite-Nitric Oxide Pathway in Physiology and Therapeutics. *Nat. Rev. Drug Discov.* **2008**, *7*, 156–167.
106. Zweier, J.L.; Wang, P.; Samouilov, A.; Kuppusamy, P. Enzyme-independent formation of nitric oxide in biological tissues. *Nat. Med.* **1995**, *1*, 804–809, https://doi.org/10.1038/nm0895-804.
107. Godber, B.L.J.; Doel, J.J.; Sapkota, G.P.; Blake, D.R.; Stevens, C.R.; Eisenthal, R.; Harrison, R. Reduction of Nitrite to Nitric Oxide Catalyzed by Xanthine Oxidoreductase. *J. Biol. Chem.* **2000**, *275*, 7757–7763, https://doi.org/10.1074/jbc.275.11.7757.
108. Schneider, M. The Importance of Ammonia in Mammalian Cell Culture. *J. Biotechnol.* **1996**, *46*, 161–185, https://doi.org/10.1016/0168-1656(95)00196-4.

109. Williamson, G.; Tamburrino, G.; Bizior, A.; Boeckstaens, M.; Dias Mirandela, G.; Bage, M. G.; Pisliakov, A.; Ives, C.M.; Terras, E.; Hoskisson, P.A.; et al. A Two-Lane Mechanism for Selective Biological Ammonium Transport. *Elife* **2020**, *9*, e57183, https://doi.org/10.7554/eLife.57183.
110. Stein, L.Y.; Klotz, M.G. The Nitrogen Cycle. *Curr. Biol.* **2016**, *26*, R94–R98.
111. Klein-Nulend, J.; Van Oers, R.F.M.; Bakker, A.D.; Bacabac, R.G. Nitric Oxide Signaling in Mechanical Adaptation of Bone. *Osteoporos. Int.* **2014**, *25*, 1427–1437, https://doi.org/10.1007/s00198-013-2590-4.
112. Abnosi, M.H.; Pari, S. Exogenous Nitric Oxide Induced Early Mineralization in Rat Bone Marrow Mesenchymal Stem Cells via Activation of Alkaline Phosphatase. *Iran. Biomed. J.* **2019**, *23*, 142.
113. Boyce, B.F.; Yao, Z.; Xing, L. Functions of Nuclear Factor κB in Bone. *Ann. N. Y. Acad. Sci.* **2010**, *1192*, 367–375, https://doi.org/10.1111/j.1749-6632.2009.05315.x.
114. Veeriah, V.; Zanniti, A.; Paone, R.; Chatterjee, S.; Rucci, N.; Teti, A.; Capulli, M. Interleukin-1β, Lipocalin 2 and Nitric Oxide Synthase 2 are Mechano-Responsive Mediators of Mouse and Human Endothelial Cell-Osteoblast Crosstalk. *Sci. Rep.* **2016**, *6*, 29880, https://doi.org/10.1038/srep29880.
115. Nakamura, I.; Jimi, E.B.T.-V.& H. Regulation of Osteoclast Differentiation and Function by Interleukin-1. In *Interleukins*; Academic Press, 2006; Vol. 74, pp. 357–370 ISBN 0083-6729.
116. McEntire, B.J.; Bal, B.S.; Rahaman, M.N.; Chevalier, J.; Pezzotti, G. Ceramics and Ceramic Coatings in Orthopaedics. *J. Eur. Ceram. Soc.* **2015**, *35*, 4327–4369, https://doi.org/10.1016/j.jeurceramsoc.2015.07.034.

The Antipathogenic Properties of Silicon Nitride

Giuseppe Pezzotti and Bryan J. McEntire

1 Introduction

Antimicrobials stand as one of the most remarkable triumphs in the annals of medical science. Their impact on human health is undeniable, having saved countless lives and dramatically curbed the scourge of infectious diseases that plagued humanity. Although a scientific understanding of microbial pathogenicity has only occurred in the modern era, prior human societies turned to naturopathic materials to inhibit or cure infectious disease [1]. For instance, tetracycline was unknowingly ingested as part of the diet of ancient African agriculturalists (AD 350). It was postulated that grain stored in mud bins served as a cultivation medium for *Streptomycetes*, which is a mold-like bacteria that naturally produces tetracyclines [2]. Similarly, the red soils from Jordan have a long history of use as an inexpensive alternative to modern pharmaceuticals for treating skin infections, a practice that continues today. This traditional treatment led to the discovery of antibiotic-producing bacteria within these soils. Microorganisms, which are abundantly found in rich soil ecosystems, possess remarkable biosynthetic capabilities, producing a vast array of antimicrobial compounds; soil has played a pivotal role in the discovery and development of numerous modern antibiotics, including streptomycin, actinomycin, erythromycin, and vancomycin [3]. Finally, Chinese herbalists have long recognized the effectiveness of various plant-based compounds, including *qinghaosu*, a herb extracted from *Artemisia* plants and used in the treatment of malaria and other illnesses [4].

G. Pezzotti (✉)
Ceramic Physics Laboratory, Kyoto Institute of Technology, Sakyo-ku, Matsugasaki, Kyoto, Japan
e-mail: pezzotti@kit.ac.jp

B. J. McEntire
SINTX Technologies, Inc., Salt Lake City, UT, USA

B. S. Bal et al. (eds.), *Silicon Nitride Bioceramics*,
https://doi.org/10.1007/978-3-031-67047-3_6

The advent of the modern antibiotic era can be attributed to the groundbreaking work of Paul Ehrlich. His visionary concept of the "magic bullet"—a therapeutic agent that selectively targets and eliminates pathogens without harming host cells—revolutionized the approach to infectious diseases. Ehrlich's primary objective was to develop an effective treatment for syphilis, a devastating and widespread disease at the time. However, his enduring legacy lies in the establishment of a systematic methodology for screening potential drug candidates, a process that continues to underpin modern drug discovery, albeit in a highly automated form [1, 5]. In line with Ehrlich's concept of the "magic bullet," silicon nitride (Si_3N_4) is a crystalline compound that seemingly fulfills his criteria. In many respects, it can be regarded as a solid-state analog of penicillin. For Si_3N_4, Ehrlich's notion of high chemical selectivity could be extended beyond mere safety to encompass the active support of mammalian cells while eradicating pathogens. It is a unique solid-state compound that is both metabolically active and selectively toxic.

While the previous chapter explored Si_3N_4's favorable interactions with mammalian cells, this chapter delves into the molecular mechanisms underlying its remarkable antipathogenic properties. Initially considered solely as a structural ceramic, Si_3N_4 also possesses antibacterial efficacy. This property was first reported by the phenomenological studies of Gorth et al. [6] and Webster et al. [7] in 2012, and later rigorously linked to Si_3N_4's unique surface chemistry [8, 9]. Since then, over 50 peer-reviewed publications have documented its success in lysing or inactivating a wide spectrum of pathogenic bacteria, fungi, and viruses, including variants of SARS-CoV-2. This chapter will review a portion of this literature beginning with bacteria, then viruses, and finally fungi.

2 Antibacterial Properties of Silicon Nitride

This section unfolds in two parts. First, it presents two illustrative phenomenological studies showcasing Si_3N_4's efficacy against common pathogenic bacteria. Subsequently, a detailed analysis delves into the underlying lytic mechanism of Si_3N_4, drawing parallels and contrasts with its osteopromotive capabilities.

2.1 Si_3N_4's Bacteriolytic Effectiveness

Figure 1a, b illustrates the in vitro metabolic responses of gram-positive *Staphylococcus epidermidis* (*S. epidermidis*) and gram-negative *Escherichia coli* (*E. coli*) bacteria, respectively, exposed to Si_3N_4 at various times. 4′,6-diamidino-2-phenylindole (DAPI, blue), carboxyfluorescein diacetate CFDA (green), and propidium iodide (PI) (red) fluorescence micrographs on the right side of each figure quantify the total number and fractions of live and dead bacteria, respectively. All data were obtained at 48 h of exposure. Distinct differences in the behavior of the

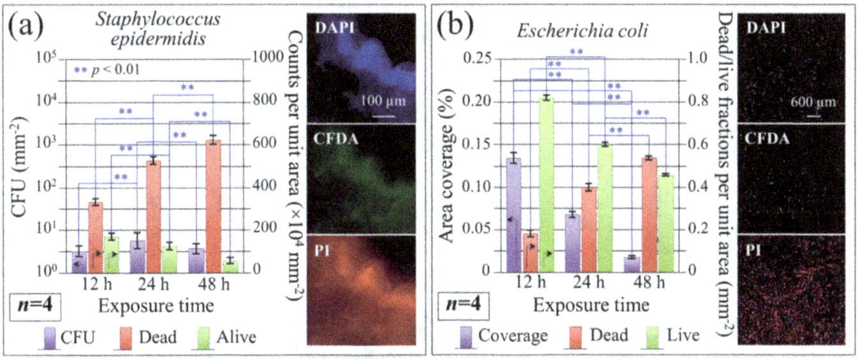

Fig. 1 Time-lapse results of biological assays on (**a**) *S. epidermidis* and (**b**) *E. coli* exposed to Si_3N_4 substrates; in the inset, DAPI blue (nuclei), CFDA green (live), and PI red (dead)-stained fluorescence micrographs of the two strains at 48-h exposure. In the micrographs on the right side of each section (cf. labels in the insets), note the formation of biofilm and the isolated colonies in *S. epidermidis* and *E. coli*, respectively. (Adapted with permission from Ref. [9])

two pathogens were observed. Unlike *E. coli*, *S. epidermidis* exhibited biofilm formation on the ceramic surface. This is vividly evident in the distribution of colored pixels in the respective micrographs [9].

S. epidermidis exhibited a relatively uniform and continuous distribution, whereas *E. coli* formed isolated spots. This distinction influenced the method of counting colony-forming units (CFU) on the substrates. For *S. epidermidis*, CFUs were readily countable and showed a slight increase at 24 h followed by a decrease at 48 h (cf., Fig. 1a). In contrast, only area coverage could be automatically computed for *E. coli*, revealing a clear monotonic decline from 12 to 48 h of exposure (cf., Fig. 1b). Independent of bacterial proliferation, a common feature in the CFU counts and area coverage graphs was the impact of Si_3N_4 on bacterial growth. As evident in the CFDA and PI fluorescence images, the number of green pixels (live bacteria) distinctly decreased, either as a diffuse CFDA green-stained pattern (*S. epidermidis*) or as isolated CFDA green-stained dots (*E. coli*). Consistent with the declining CFU counts and area coverage, the number of dead bacteria (red pixels) steadily increased over time for both strains. This suggests a pronounced decrease in bacterial proliferation with exposure to Si_3N_4. While the intrinsic metabolisms of the two strains differed, a common bactericidal trend emerged under identical testing conditions.

To validate the antibacterial efficacy of Si_3N_4 in practical orthopedic applications, the behavior of uncoated and Si_3N_4-coated acetabular shells crafted from a Ti alloy was evaluated. An acetabular shell serves as a critical component in hip arthroplasty, anchoring the artificial hip joint to the acetabulum. Commercially available uncoated Ti-alloy shells feature a rough surface mimicking the porous structure of trabecular bone. In a simple demonstrative experiment, one Ti-alloy shell was coated with a thin Si_3N_4 layer using laser cladding [10]. To enhance bone integration and expedite post-surgery recovery, the manufacturer of these shells also offers them in an optional plasma-sprayed hydroxyapatite coating. Figure 2a–c depicts uncoated,

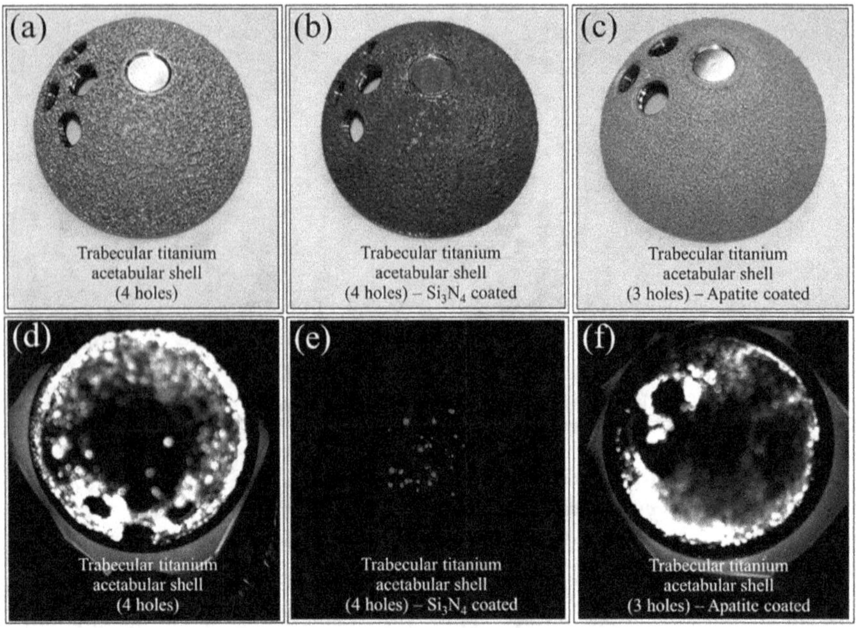

Fig. 2 (**a**) Uncoated, (**b**) Si_3N_4-coated, and (**c**) hydroxyapatite-coated Ti-alloy acetabular shells used in artificial hip joints; (**d–f**) results of luciferin "glowing" experiments at 48 h conducted on the above Ti-alloy acetabular shells, respectively, equally contaminated with gram-negative *E. coli* bacteria. (Adapted with permission from Ref. [9])

Si_3N_4-coated, and hydroxyapatite-coated Ti-alloy acetabular shells, respectively. Figure 2d–f presents the results of luciferin "glowing" experiments using gram-negative *E. coli* as a contaminant on the three acetabular shells (cf., labels in the insets). The chemiluminescent oxyluciferin luciferase signal, recorded at 48 h, illustrates the disparity in bacterial contamination levels among the acetabular shells. The uncoated shell exhibited a bright glow due to fluorescent bacteria, while the Si_3N_4-coated shell remained essentially dark, with only a limited number of fluorescent spots. These spots corresponded to microscopic regions where the Si_3N_4 coating had not fully covered the porous surface. Notably, the fluorescent patterns of a poorly coated microscopic region (Fig. 2b) bear a resemblance to the microscopic bacterial enclaves in Fig. 2e. In contrast to the uncoated shell, the hydroxyapatite-coated shell showed no discernable difference in bacterial contamination (cf., Fig. 2d–f). This suggests that while hydroxyapatite may be beneficial for bone formation, it lacks antibacterial efficacy. In light of the osteopromotive properties of Si_3N_4 discussed in the previous chapter, it is evident that Si_3N_4's dual osteogenic and antibacterial capabilities make it a remarkably unique biomaterial for hip joint applications.

These experimental results demonstrated the ability of Si_3N_4 to eradicate both gram-positive and gram-negative bacteria and also the practicality of applying laser-cladded Si_3N_4 coatings on actual metal arthroplasty components. Based on these observations, it was hypothesized that nitrogen moieties released from the Si_3N_4 surface perform two important roles. They can penetrate both prokaryotic and eukaryotic cells, inducing lysis in bacteria and upregulating cellular processes in mammalian cells. While this chapter focuses principally on the antipathogenic properties of Si_3N_4, it is important and informative to describe this dualism in parallel.

2.2 Si_3N_4's Mediated Mammalian Metabolism and Selective Bacterial Toxicity

Biocompatibility is defined as the ability of a synthetic material to interact with living tissue without causing a toxic or immunological reaction [11]. Biocompatible materials can be further classified into three main categories: bioinert (i.e., non-reactive), bioresorbable (i.e., can dissolve in the human body), and bioactive (i.e., can integrate with living tissue) [12, 13]. Bioglasses and synthetic hydroxyapatite are examples of bioactive inorganic compounds, while ß-tricalcium phosphate is an example of a bioresorbable ceramic. Alumina and silicon nitride are generally considered to be bioinert [14–16]. However, the classification of a ceramic as bioinert is based on the misconception that strong ionic or covalent bonds cannot be affected by the biological environment. This assumption is unrealistic, as addressed in the previous chapter. There will always be chemical interactions at the bioceramic interface with cells or pathogens. Indeed, from a phenomenological viewpoint, the results reviewed thus far show that ionic species eluted from ceramic surfaces affect cellular metabolism. These chemical interactions determine the fate of both eukaryotic and prokaryotic cells. The steep chemical concentration and pH gradients at the bioceramic surfaces (i.e., local vs far-field) significantly affect the immunochemical responses of mesenchymal cells and can either favor or delay cell differentiation, proliferation, and osteogenesis. Similarly, these same gradients also affect the ability of prokaryotes to attach to surfaces, multiply, and form protective biofilms, leading to periprosthetic infections. These findings alone demonstrate that bioceramics are not bioinert. Significant questions remain about the specific roles of trace amounts of eluted and hydrated anion/cation species, such as N/Si and O/Al, in eukaryotic and prokaryotic cell metabolism. To date, relatively few studies have been published on the structure of new bone synthesized by cells on different bioceramics, primarily due to the misconception that bioceramic surfaces are inert. Conducting such research could help dispel the misleading "bioinert paradigm" for bioceramics. Clarifying these roles is important for the development of new bioceramics and opens new avenues of biochemical research.

As described in Chap. 4, the surface chemistry of Si_3N_4 is characterized by the elution of two nitrogen species (NH_3 and NH_4^+) and the formation of surface silanols. The effects of these eluted species on cell metabolism are provided in

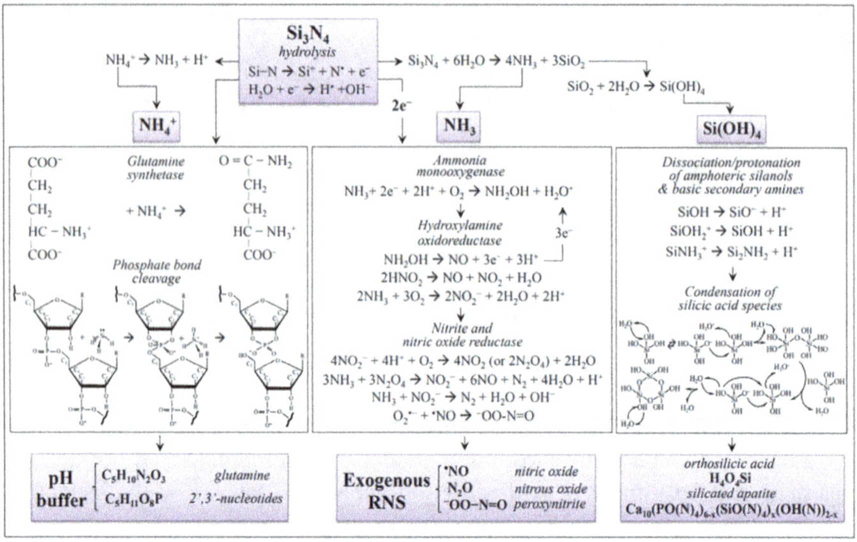

Fig. 3 Main effects of Si_3N_4-eluted moieties on cell metabolism. (Left panel) NH_4^+ ammonium species enter the cytoplasm through specific channels and contribute to the cellular nutrient cycle. (Center panel) NH_3-eluted species freely diffuse across the cell membrane and require oxygen radicals to be enzymatically oxidized into hydroxylamine (NH_2OH) by ammonia monooxygenase. (Right panel) Dissociation of surface silanols leads to the formation of silicic acid chains. (Reprinted with permission from Ref. [17])

Fig. 3 [17]. Ammonium ions (NH_4^+) can only enter the cytoplasm through specific transport channels. These channels are used by cells in their nutrient cycle (cf., left panel of Fig. 3). Specifically, NH_4^+ is used by glutamine synthetase, which catalyzes the reaction of NH_4^+ with glutamate to form glutamine. This nutrient cycle is further enhanced by the pH buffering induced by Si_3N_4 in the extracellular space, and the process increases alkaline phosphatase (ALP). Calcium concentration also increases, which enhances the binding of Ca^{2+} to the COO^- terminus of the deprotonated serine residue to form Ca^{2+} complexes [18]. Conversely, NH_3 is a small, highly volatile molecule that freely diffuses across the cell membrane. It requires oxygen radicals to be enzymatically oxidized into hydroxylamine (NH_2OH) by ammonia monooxygenase (cf., central panel of Fig. 3). Unpaired electrons can react with adsorbed O_2 to form $O_2^{-\bullet}$ radical anions and other highly oxidative protonated radicals. This may be partly due to the dissociation of silanols and the formation of superoxide radicals ($\equiv SiO^\bullet$ and $\equiv SiO_2^\bullet$). However, the main source of free electrons is from cellular mitochondria. Mitochondria produce oxygen radicals as byproducts of oxidative phosphorylation, which occurs when electrons leak from the electron transport chain [19]. The enzymatic catalysis of cytochrome c oxidase facilitates the transfer of four electrons to an O_2 molecule. This process can generate free electrons that are capable of oxidizing small concentrations of exogenous NH_3 to NH_2OH (cf., chemical reactions listed in the central panel of Fig. 3). Successive reactions catalyzed by hydroxylamine oxidoreductase lead to the formation of nitric oxide (NO) radicals. NO is a signaling molecule that plays a fundamental role in many cellular processes [18]. The isoenzymes endothelial nitric oxide synthase (eNOS) and inducible nitric oxide synthase (iNOS) regulate the production of nitric oxide (NO) and mesenchymal cell fate.

NO molecules freely diffuse through the intercellular space and act as messengers through the lyase enzyme guanylate cyclase, which regulates intracellular calcium levels [20]. Nitric oxide has a dual role in regulating cell proliferation and death. At low concentrations, NO promotes cell proliferation, while at high concentrations, it induces apoptosis. NO has also been shown to promote osteogenic differentiation of mesenchymal stem cells in vitro [21]. At exogenous NO concentrations (i.e., $<1.2\ \mathrm{nmol{\cdot}dm^{-3}}$), ALP is upregulated, and the ability of mesenchymal cells to proliferate and differentiate into osteoblasts is greatly enhanced. Conversely, concentrations of about $1\ \mathrm{mmol{\cdot}dm^{-3}}$ significantly reduce cell viability and lead to cell caspase-mediated cell death. The threshold concentration of NO required to induce cell apoptosis is three orders of magnitude greater than the amount of ammonia locally released by Si_3N_4. This suggests that Si_3N_4 has a favorable dissolution kinetic profile for human biocompatibility.

Another important aspect of Si_3N_4's surface chemistry is the dissociation of surface silanols to form silicic acid chains (cf., right panel of Fig. 3). The elution of silicic acid is beneficial to osteogenesis, which is consistent with the well-established understanding that silicon is a potent promoter of bone mineralization [22, 23]. Analysis of a Si_3N_4 ex vivo spinal implant revealed significant amounts of silicon incorporated into the apatite structure at the bone-implant interface [24]. Another study reported the presence of $(SiO_4)^{4-}$ tetrahedra and Si nanodots at in vitro-grown bone near Si_3N_4 surfaces [25]. These observations demonstrate that osteoblasts actively scavenge Si from Si_3N_4 and incorporate it into new bone.

How do the eluted moieties from Si_3N_4, which are beneficial to eukaryotes, affect prokaryotes? Figure 4a, b shows, respectively, an optical micrograph and a fluorescence image of NO molecules from eukaryotic cells after 24 h of incubation on Si_3N_4. NO was imaged using diaminofluorescein-2 diacetate (DAF-2(NO)). The fluorescence micrograph in (b) provides experimental evidence for the formation of NO in the cytoplasm, supporting the hypothesis that mitochondria metabolize ammonia. Free electrons from ammonia oxidation reduce oxygen atoms to produce hydroxylamine oxidoreductase, which then catalyzes the oxidation of hydroxylamine (NH_2OH) to nitrite ions (NO_2^-). This process releases four electrons, two of which can sustain ammonia oxidation (cf., central panel in Fig. 3). These off-stoichiometric reactions are unique to Si_3N_4 and ultimately lead to the formation of NO radicals and other highly reactive nitrogen species (RNS), such as peroxynitrite ($OONO^-$). Figure 4c shows time-lapse plots of NO production and cell proliferation. Both NO formation and cell proliferation increase monotonically over time. By comparing optical and fluorescence microscopy images (see arrowed locations), NO fluorescent moieties were observed within the osteoblasts, supporting the thesis of an activated endothelial nitric oxide synthase (eNOS) pathway.

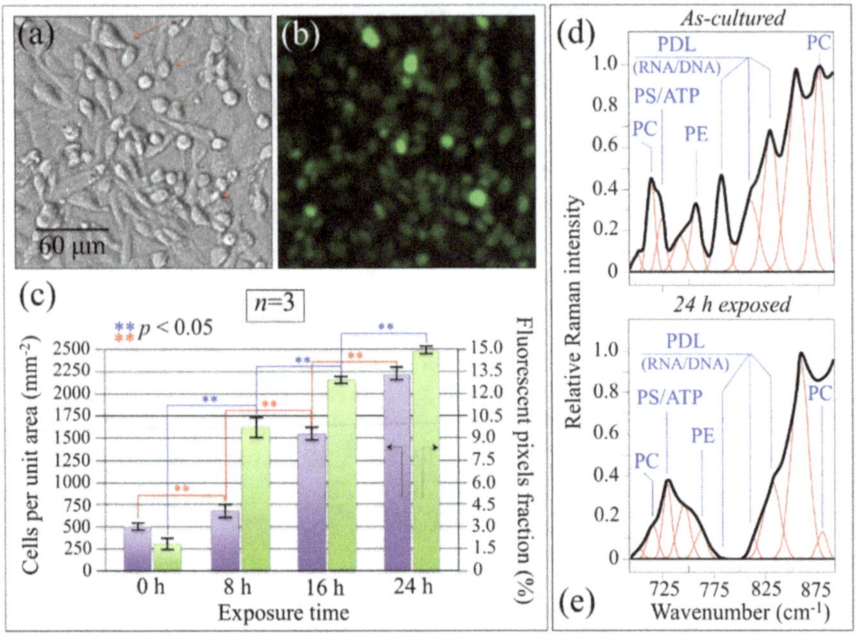

Fig. 4 (**a**) Optical image of human osteoblasts cultured on a Si_3N_4 substrate for 24 h. (**b**) Relative location of NO molecules upon staining with DAF green. (**c**) Time dependence of cell population and NO concentration as a function of exposure time to Si_3N_4. (**d** and **e**) Raman spectra in the wavenumber range of 700–900 cm^{-1} collected on cells as cultured and after 24 h of exposure to Si_3N_4. (Adapted with permission from Ref. [9])

Figure 5 shows the same DAF-2(NO) experiment conducted on a culture of *S. epidermidis*. Unlike the larger, optically visible cells in Fig. 4a, bacteria were only visible after staining their nuclei with 4′,6-diamidino-2-phenylindole (DAPI blue) (Fig. 5a). Figure 5b shows a fluorescence image of NO production at 24 h, while the time-lapse dependencies of NO production and bacterial proliferation are given in Fig. 5c. Notably, the distribution of NO in the bacterial culture appears to be much more dispersed throughout the extracellular environment. This suggests an inducible nitric oxide synthase (iNOS)-like pathway. However, the most striking features in Fig. 5c are that both the bacterial population and NO concentration were highest at ~24 h of exposure. The peak in NO production and the abrupt end of bacterial proliferation suggest that bacteriolysis begins at 24 h.

The difference in cell proliferation and NO production trends between Figs. 4c and 5c is a key to understanding the effects of Si_3N_4 surface chemistry on cells and bacteria [26–28]. To elucidate the chemistry underlying the observed behaviors, in situ Raman spectroscopy was performed on living cells and bacteria in culture experiments. Figure 4d, e shows Raman spectra for osteoblasts as cultured and after 48 h of exposure to Si_3N_4, respectively. The low-frequency spectral region at 700–900 cm^{-1} contains important metabolic signatures for both lipids and

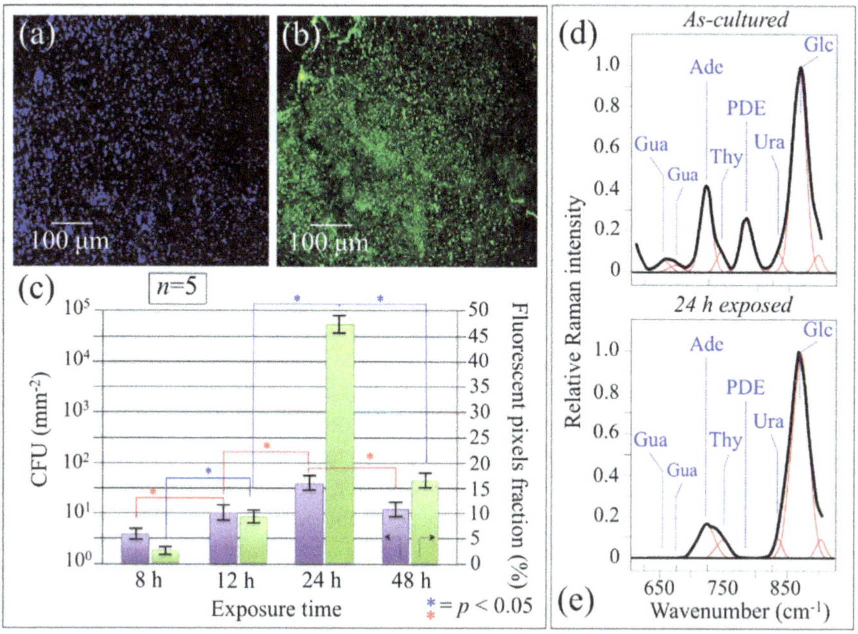

Fig. 5 (**a**) DAPI blue image of *S. epidermidis* cultured on a Si_3N_4 substrate for 24 h. (**b**) Relative location of NO molecules upon staining with DAF green. (**c**) Time dependency of bacterial population and NO concentration as a function of exposure time to Si_3N_4. (**d** and **e**) Raman spectra in the wavenumber range of 700–900 cm^{-1} obtained on bacteria as cultured and after 24 h of exposure to Si_3N_4. (Adapted with permission from Ref. [9])

DNA/RNA molecules [29]. The Raman spectra acquired at 48 h showed significant differences in bands related to both lipids (i.e., phosphatidylcholine, PC; phosphatidylserine, PS; and phosphatidylethanolamine, PE) and RNA/DNA (i.e., phosphodiester linkage PDL) [9]. Note that the increase in phosphatidylserine (PS) band intensity coincides with the reductions in phosphatidylcholine (PC) and phosphatidylethanolamine (PE) band intensities. These changes suggest that the collapse of phospholipid asymmetry is occurring, which is a hallmark of osteoblast differentiation and bone formation [30]. Note also that the low-frequency phosphatidylserine (PS) band is also contributed to by adenosine triphosphate (ATP) [31, 32]. Osteocytes use ATP as a chemical messenger to signal to other bone cells [33]. The observed increase in ATP is accompanied by a noticeable disappearance of RNA/DNA phosphodiester linkage bands (*C5′*-O-P-O-*C3′* bonds), as evidenced by the decrease in the PDL band in Fig. 4d, e [34, 35]. The disappearance of these bands indicates that the phosphodiester linkage is being cleaved, suggesting a significant decrease in the concentration of nucleic acids due to osteoblast activation on Si_3N_4. Non-translated RNA is then reactivated via polyadenylation to synthesize the specific proteins required for cell differentiation [34, 36]. To put it another way, the cells use their pool of dormant mRNA and some of their DNA to produce new proteins for proliferation. In their activated state, monovalent ammonium ions (NH_4^+) also play an important role in mRNA self-cleavage, including the formation of a phosphorane intermediate via proton transfer [37]. Thaplyal et al. [38] showed that NH_4^+ ions can substitute for N4 amines. They first accept a proton from an exocyclic amine and then they donate one.

The DNA bands belonging to guanine and adenine (labeled as Gua and Ade, respectively) are the main variations in the Raman spectrum of *S. epidermidis* after being exposed to Si_3N_4 for 0 to 24 h. These variations are shown in Fig. 5d, e, respectively [9]. The doublet low-frequency DNA Gua (ring-breathing) bands disappear in the spectrum of this bacterium, which is a striking feature [39, 40]. The ring-breathing band of DNA Ade showed a significant decrease in intensity and broadening [39, 41]. In contrast, the two bands assigned to thymine (Thy) and uracil (Ura) remained unchanged [39, 42]. The results can be interpreted by considering that DNA bases containing the less stable C=N double bond (i.e., Ade and Gua) are more prone to disruption in the presence of NH_3 than Thy and Ura, since the latter molecules only contain more stable C-N single bonds. The predominant vibration of the COC 1,4 glycosidic link in carbohydrates (labeled as Glc) remains unchanged, while the PDE band, which embodies the symmetric stretching of the phosphodiester *C5′*-O-P-O-*C3′* bonds in DNA, fully disappears by a mechanism similar to that described above for osteoblasts.

Comparative fluorescence analyses, time-lapse biological data, and in situ Raman spectroscopy on osteoblasts and bacteria provide a consistent depiction of the molecular processes taking place on Si_3N_4. The primary characteristics revealed by this set of analyses are summarized in Fig. 6a, b for cells and bacteria, respectively. In responding to Si_3N_4, the physiological difference between cells and bacteria is their ability and inability, respectively, to metabolize NH_3 through ammonia monooxygenase. In eukaryotic cells, the mitochondria are the primary consumers of oxygen and the primary source of oxygen radicals. Within mitochondria, the transfer of four electrons to an O_2 molecule occurs directly through cytochrome c oxidase enzymatic catalysis, resulting in the formation of byproducts. These byproducts are partially reduced species due to the leakage of electrons from transport chains involved in the transfer [19].

If the NO byproduct remains below a threshold concentration (as is the case for Si_3N_4 exposure), it contributes to cell signaling, differentiation, and proliferation. The lower part in Fig. 6a schematically shows that the glutamine synthetase pathway is an important source of energy and C/N elements for the synthesis of amino acids, nucleotides, glutathione, and hexosamine. This explains why the populations of both cells and bacteria initially increase with time due to their interaction with Si_3N_4. Cell populations can grow continuously because equilibrium is maintained between eluted and metabolized nitrogen. This includes a proper response by ammonia monooxygenase to the local increase in pH. In contrast, in prokaryotic cells lacking mitochondria, the combination of eluted NH_4^+ ions and NH_3 molecules exceeds the rate of population growth, resulting in an unmetabolized surplus of nitrogen moieties in the intracellular space. Ammonia's direct interaction with other macromolecules in the cytoplasmic enclave triggers a cascading reaction that culminates in the

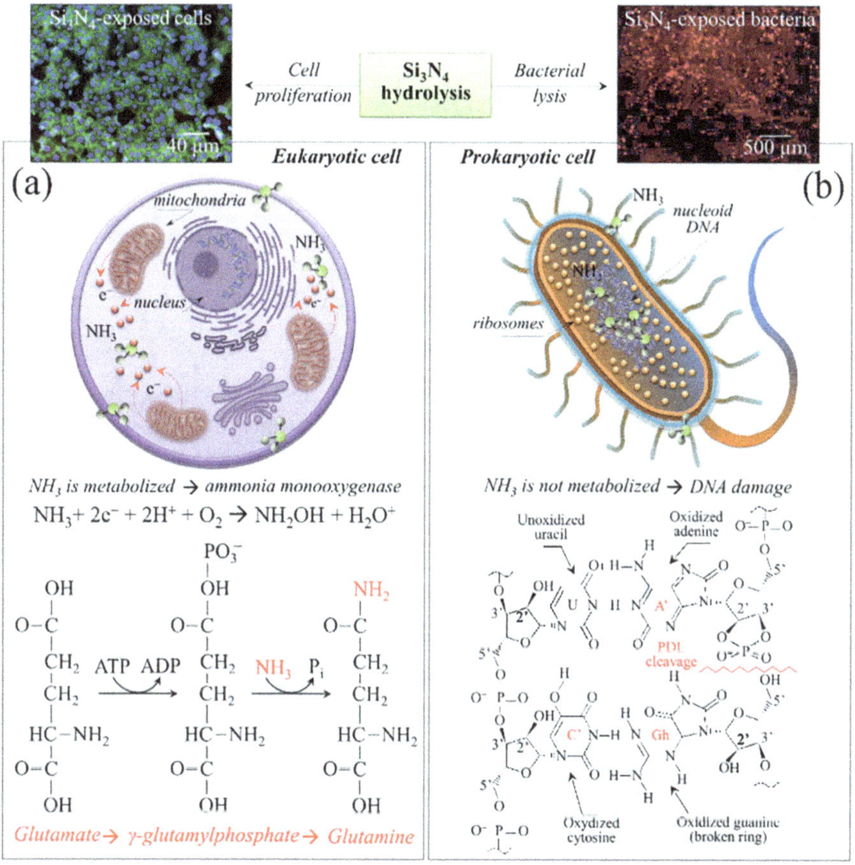

Fig. 6 Top panel: Comparative fluorescence analyses summarizing the effect of Si_3N_4 hydrolysis on cellular proliferation and bacterial lysis. The molecular machinery occurring on Si_3N_4 in an aqueous environment is shown in (**a**) for eukaryotic cells and (**b**) for bacteria. Eukaryotic cells metabolize ammonia by exploiting electron transfer from mitochondria, leading to the formation of RNA byproducts that promote cell signaling, differentiation, and proliferation (e.g., the glutamine synthetase pathway as an important source of energy and amino acid synthesis). In contrast, bacteria lack mitochondria and are unable to metabolize ammonia, which leads to DNA damage (prevalently oxidation of purine bases and PDL cleavage) (Used with permission from Ref. [50])

formation of toxic concentrations of reactive nitrogen species (RNS), as illustrated in the middle panel of Fig. 6. The RNS formed include nitric oxide (NO) and peroxynitrite ($ONOO^-$) [43]. Peroxynitrite has a strong oxidizing effect that opens the guanine ring (G → Gh) and causes the disappearance of its ring vibration bands. Additionally, $ONOO^-$ induces a strong structural modification by oxidizing adenine (A → A′), as shown in the lower panel of Fig. 6b. Ammonia also directly cuts phosphodiester (PDE) linkages (cf., Fig. 5d, e).

Finally, it is important to note that ammonia may act as an indirect nitrogen source to support osteoblast proliferation. This has also been observed for other types of cells [44–49]. However, some cells are unable to use ammonia to support their growth. In other words, different cells vary widely in their ability to exploit ammonia as an alternative nitrogen source [44]. Therefore, when using Si_3N_4 for different human tissues, its surface chemistry should be optimized to match its nitrogen release kinetics for each cell type.

In summary, the surface chemistry that underpins Si_3N_4's osteopromotive properties also contributes to its ability to resist bacterial adhesion and proliferation. The microscopic dissolution of its surface releases nitrogenous moieties, including ammonia (NH_3) and ammonium (NH_4^+), into the surrounding environment. These compounds are enzymatically converted into reactive nitrogen species (RNS), such as nitric oxide (NO) and peroxynitrite (^-OONO), which disrupt prokaryotic membranes and damage cellular DNA [51, 52]. Although the full extent of Si_3N_4's antibacterial efficacy remains to be fully elucidated, research in this area is rapidly growing. Recent studies have demonstrated its effectiveness against various nosocomial strains, problematic oral strains, and at least one virulent antibiotic-resistant strain (i.e., MRSA). These pathogens contribute significantly to various illnesses, including an implant-related condition known as periprosthetic joint infection (PJI). PJIs are a serious complication after joint replacement surgery and can cause significant patient morbidity. Treating PJIs often involves a two-stage procedure: removing the implant, cleaning the infected area, applying systemic antibiotics, and inserting a temporary antibiotic-loaded implant. Then, once the infection is eradicated, the temporary implant is replaced with a permanent new device [53]. However, this approach also raises concerns about antibiotic resistance. Si_3N_4 could play an effective role in reducing the prevalence of this iatrogenic complication and mitigating the effects of antibiotic resistance. Si_3N_4 will not only prevent the initial formation of a biofilm, which leads to an infection, but because of its chemical-based lytic mechanism, it will also reduce the chance of bacterial resistance.

3 Antiviral Properties of Silicon Nitride

Ethanol, in combination with sodium hypochlorite, hydrogen peroxide, or ultraviolet light, is commonly used to inactivate viruses. However, the use of these and other methods is limited by their potential toxicity to human cells. Silicon nitride has been proposed as an alternative surface disinfectant against COVID-19 due to both its biocompatibility and antibacterial properties [54]. In light of this perspective, Si_3N_4 possesses the inherent ability to suppress viral transmission on surfaces, fabrics or masks, and other fomites, while simultaneously posing no safety concerns for humans. This hypothesis was corroborated in a previous study that demonstrated the complete inactivation of influenza A and enterovirus within a mere minute of exposure to a dilute aqueous suspension of Si_3N_4 powder [55].

Building upon its demonstrated efficacy against these prevalent viruses, a subsequent investigation was conducted to evaluate the effectiveness of Si_3N_4 against the highly contagious SARS-CoV-2 virus [54]. The results of this study are shown in

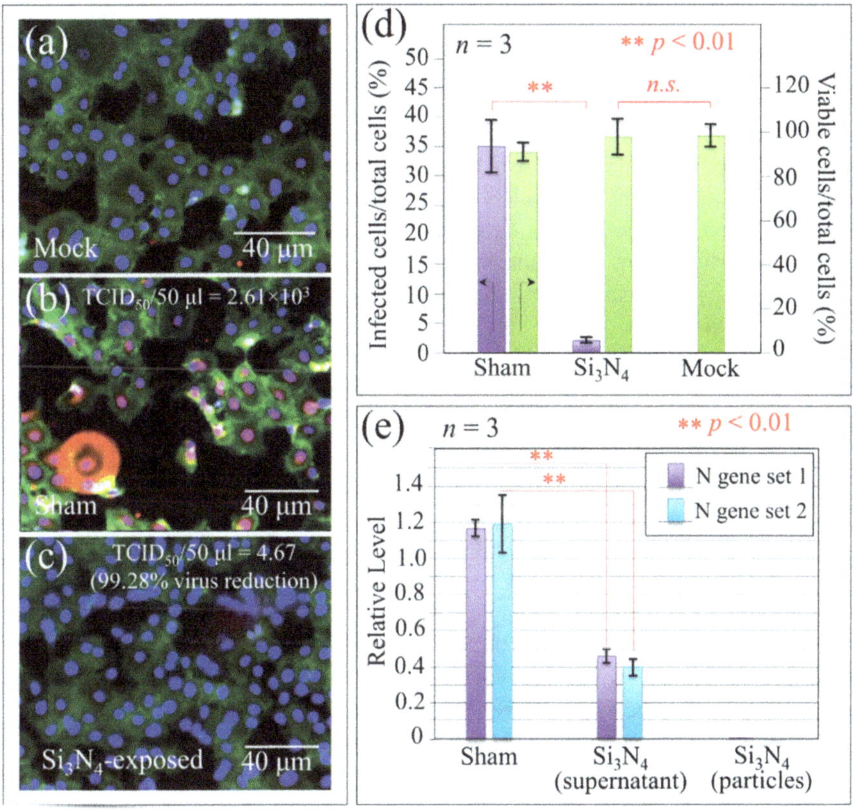

Fig. 7 Immunofluorescence microscopy images of VeroE6/TMPRSS2 cells: (**a**) Mock (unexposed to SARS-CoV-2 virions), (**b**) Sham (exposed to virions without Si_3N_4), and (**c**) exposed to virions suspended for 1 min in an aqueous solution containing 5 vol.% Si_3N_4 powder. Inoculated cells were stained with anti-SARS coronavirus envelope antibody (red), F-actin (green), and cell nuclei (DAPI, blue). (**d**) TCID50 assay results for virions exposed to Si_3N_4 (5 vol.%) at 1-min inactivation time, quantifying the immunofluorescence imaging. (**e**) The relative level of viral RNA fragmentation upon 1-min contact with Si_3N_4 powder, as detected by RT-PCR experiments on virions N-gene sequence. (Adapted and used with permission under Creative Commons License [54])

Fig. 7a–e. Figure 7a–c shows immunofluorescence images of VeroE6/TMPRSS2 cells: (a) Mock (unexposed to SARS-CoV-2 virions), (b) Sham (cells inoculated with SARS-CoV-2 virions unexposed to Si_3N_4 powder), and (c) Si_3N_4-exposed (cells inoculated with the same virions suspended for 1 min in a hydrosol containing 5 vol.% Si_3N_4 powder) [54]. The inoculated cells were stained with anti-SARS coronavirus envelope antibody (red), viable cell F-actin (Phalloidin-stained, green), and cell nuclei (DAPI-stained, blue). As expected, VeroE6/TMPRSS2 cells unexposed to virions (mock sample) showed no red staining (cf., Fig. 7a). Conversely, the sham cells were heavily infected with the virus, as evidenced by the abundance of red pixels in Fig. 7b. In stark contrast, the Si_3N_4 powder suspension effectively inactivated the SARS-CoV-2 virions, resulting in a near absence of red pixels in Fig. 7c. Figure 7d illustrates the results of $TCID_{50}$ assays for virions exposed to Si_3N_4 at 1 min, demonstrating an impressive >99% reduction in viral load. Additionally, Fig. 7e depicts the outcome of an RT-PCR examination of viral fragmentation following 1 min of contact with the Si_3N_4 powder suspension. Unlike the sham sample, the viral RNA exhibited substantial fragmentation upon contact with Si_3N_4 powder. Notably, no RNA was detected from a centrifuged sample after 1 min of exposure to Si_3N_4. The data presented in Fig. 7 underscore the critical role of Si_3N_4's surface chemistry in inactivating SARS-CoV-2. The virucidal mechanism of Si_3N_4 appears to share similarities with its bacteriolytic action, with eluted ammonia playing a crucial role in disrupting viral structure. Additionally, the proteins on the viral envelope exhibit a strong affinity for Si_3N_4 particles, suggesting a highly effective "competitive binding" mechanism. Once the virions come into contact with the Si_3N_4 particles, eluted ammonia penetrates the envelope and fragments the viral RNA backbone, effectively inactivating the virus [56]. These findings highlight the potential of Si_3N_4 as a potent antiviral agent against SARS-CoV-2 and other viral pathogens [55].

The combined evidence from RT-PCR and fluorescence microscopy suggests that SARS-CoV-2 inactivation occurs through a multistep mechanism: (1) Electrical Trapping and "competitive binding": The positively charged Si_3N_4 particles electrostatically attract and bind to the negatively charged viral envelope proteins, effectively trapping the virions; and (2) "Ammonia poisoning": Eluted ammonia from the Si_3N_4 particles penetrates the trapped virions, disrupting their internal structure and fragmenting the viral RNA backbone. This disruption leads to complete viral inactivation. This two-pronged mechanism, akin to a "catch and kill" strategy, highlights the remarkable efficacy of Si_3N_4 in neutralizing SARS-CoV-2.

3.1 Si_3N_4's Inhibition of Virion Attachment and Replication

To elucidate the underlying mechanisms responsible for SARS-CoV-2 inactivation by Si_3N_4, Raman spectroscopy was employed to analyze virions before and after 1-min exposure to Si_3N_4 particles in an aqueous suspension. This analysis revealed significant RNA and spike receptor mutations, along with other profound molecular structure and symmetry alterations. These alterations included changes in sulfur-containing amino acids, disrupted hydrophobic interactions of tyrosine residues, and modifications to protein secondary structures [56]. Furthermore, alterations in the vibrational frequencies of RNA purines and pyrimidines, along with dissimilarities in protein secondary structures, were also observed. Figure 8a presents a schematic representation of the interaction between a virion and Si_3N_4, resulting in ammonia elution and the subsequent formation of RNS (encompassing NO and $O{=}N\text{-}O\text{-}O^-$ moieties) via previously discussed off-stoichiometric reactions. Figure 8b additionally illustrates a schematic of the oxidation of RNA purines to yield formamidopyrimidine (Fapy). This depiction incorporates vibrational spectra of

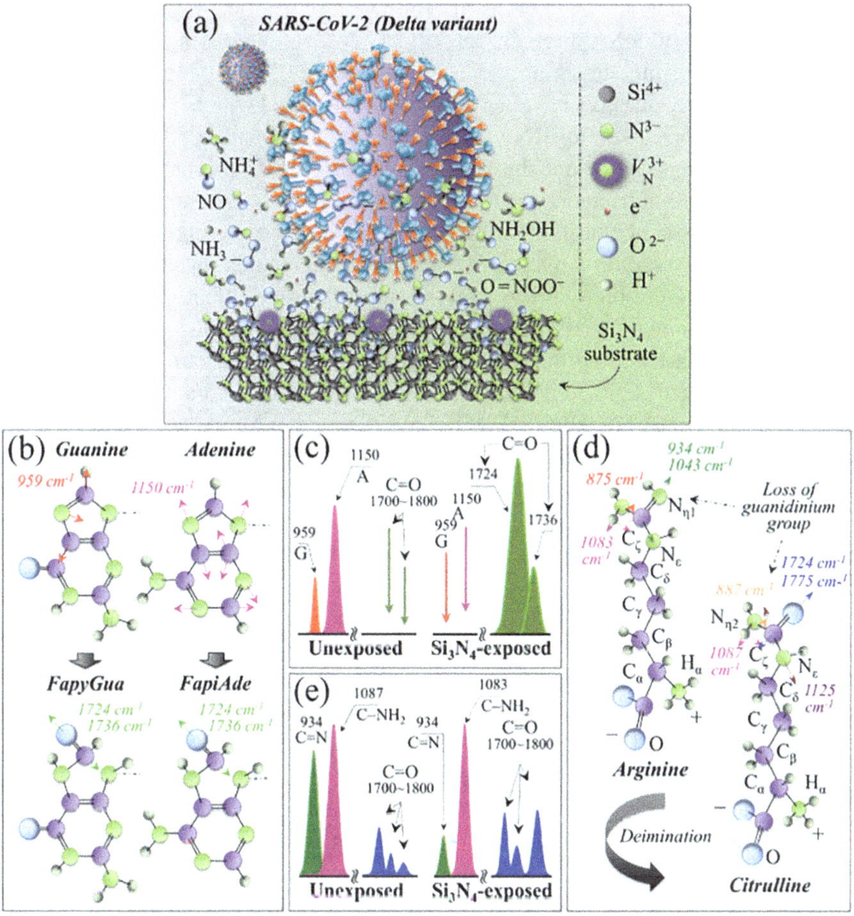

Fig. 8 (**a**) Schematic representation of the interaction between a SARS-CoV-2 virion and the Si_3N_4 free surface in an aqueous medium, leading to ammonia elution and subsequent formation of NO and O=N-O-O$^-$ RNS. (**b**) Schematic representation of the oxidation of RNA guanine (G) and adenine (A) purines to formamidopyrimidines (Fapy). Vibrational spectra of both purine and Fapy species are included. In (**c**), in-plane ring deformation of C-N-C in G and C-N ring stretching mode in A (at 959 and 1150 cm^{-1}, respectively) disappears after instantaneous contact with Si_3N_4 powder, while two prominent C=O stretching signals appear at 1721 and 1736 cm^{-1} (data collected on JPN/TY/WK-521, also referred to as the ancestral Japanese variant). Panel (**d**) presents a schematic representation of the post-translational deimination of arginine spike residues, resulting in the loss of the guanidinium group and the formation of citrulline. Vibrational spectra for both configurations are provided in the inset. In panel (**e**), the arginine Raman marker at 934 cm^{-1} (N-C-N symmetric stretching) tends to noticeably disappear upon exposure to Si_3N_4, while the C-NH_2 antisymmetric bending vibration at 1087 cm^{-1} maintains signal intensity (data collected on TY11–927, also referred to as the Delta variant). (Reprinted and adapted with permission from Ref. [61])

guanine (G) and adenine (A) before and following oxidation. As shown in Fig. 8c, Raman markers of RNA purines, namely, C-N-C in-plane ring deformation in guanine and C-N ring stretching mode in adenine (at 959 and 1150 cm^{-1}, respectively), completely disappeared after instantaneous contact with Si_3N_4 [56]. Notably, the spectroscopic markers of RNA pyrimidines (at 1054 and 1038 cm^{-1} for uracil and cytosine, respectively; not shown) remained unaltered. This same behavior was previously noticed in bacteriolysis (cf. Figs. 5d, e and 6b) of *S. epidermidis* exposed to Si_3N_4. The resilience of pyrimidine structures when interacting with RNA stems from the enhanced stability of the C-N single bonds in their exocyclic ring compared to the C=N double bonds in the pentose ring of purines. This inherent stability renders pyrimidine structures less susceptible to alterations induced by RNA interactions. In contrast, the prominent intensity of the two C=O stretching signals at 1721 and 1736 cm^{-1} (cf., Fig. 8c) implies the formation of Fapy molecules (cf., C=O bonds in open Fapy rings in Fig. 8b). As reported by Douki and Cadet [57], the oxidative action of peroxynitrite anions on RNA purines results in the formation of 8-oxo-7,8-dihydro-2′B-deoxyguanosine (8-oxo-dGuo) at a relatively low yield. This formation is subsequently linked to the development of 2,6-diamino-4-hydroxy-5-formamidopyrimidine (FapyGua). Similar structural alterations at the molecular level occur in FapyAde (i.e., 4,6-diamino-5-formamidopyrimidine), an oxidation product of the adenine base [58]. Figure 8d illustrates a schematic representation of the post-translational deimination of arginine spike residues. One of the defining features of the SARS-CoV-2 Delta variant is the L452R spike receptor-binding motif substitution, which enhances infectivity by augmenting the binding affinity between the spike protein and the human cell ACE2 receptor [59]. The L452R mutation, which replaces leucine with arginine, introduces a positive charge at a critical position in the receptor-binding domain of the spike protein. Since the ACE2 receptor charges are negative, the L452R mutation significantly enhances the binding affinity of the Delta variant spike protein. A polar amino acid, arginine, is one of the most hydrophilic natural amino acids, with its large side-chain guanidinium ($CH_6N_3^+$) group capable of storing up to six hydrogen bonds (cf., Fig. 8d). Due to its pKa exceeding 12, arginine retains its positive charge in essentially all biological environments. The compatibility of the highly hydrophilic arginine residue within the hydrophobic environment of Delta virions has been previously described [60]. The guanidinium group possesses five dipolar N-H protons that can act as hydrogen bond donors, while one electron pair can serve as a hydrogen bond acceptor. Upon interaction with cell membranes, phospholipids primarily offer hydrogen bond acceptor groups through both esters and phosphate oxygens. Consequently, Delta virions exhibit a strong affinity for the cell membrane, enhancing the likelihood of their attachment to the negatively charged ACE2 receptor.

Within the arginine markers observed in the Delta variant's infrared spectrum, the band at 934 cm^{-1}, corresponding to the N-C-N symmetric stretching vibration, can be considered a unique vibrational fingerprint of the guanidinium group [56]. The relative intensity of this band serves as a marker for strain infectivity. Interestingly, this band tends to disappear upon immediate exposure to Si_3N_4 (cf., Fig. 8e), likely due to the chemical environment established at the solid/virion interface.

In contrast, exposure to Si_3N_4 does not appear to influence the C-NH_2 antisymmetric bending vibration at 1087 cm^{-1} [62, 63], as its signal intensity remains unchanged (cf., Fig. 8e) [56]. One plausible explanation for this spectroscopic observation lies in the post-translational modification of arginine to citrulline through a deimination process, as schematically illustrated in Fig. 8d. During deimination, one terminal nitrogen of the arginine side chain is replaced by a double-bonded oxygen, resulting in the loss of positive charge and alterations to the protein's secondary structure. This substantial structural transformation results in the elimination of the N-C-N symmetric stretching vibrations associated with the arginine guanidinium group (i.e., the 934 cm^{-1} band), while preserving the C-NH_2 asymmetric bending mode, which shifts to a slightly lower frequency of 1083 cm^{-1} (cf., Fig. 8e). Upon exposure to Si_3N_4, the bands at 1125 cm^{-1} (predominantly C-N stretching) and 1655 cm^{-1} (C-NH_2 symmetric bending) [62, 63] exhibit constant or marginally increased intensities [56]. These alterations arise from an altered molecular structure devoid of the guanidinium group.

Consistent with the structural transformation into citrulline, there is a notable increase in the relative intensity of C=O bands within the 1700~1800 cm^{-1} range (cf., the labels in the inset of Fig. 8d and their corresponding Raman spectral bands in Fig. 8e). This observation aligns with the replacement of the N-C-N symmetric stretching vibrations associated with the arginine guanidinium group (i.e., the 934 cm^{-1} band) upon exposure to Si_3N_4. The guanidinium group of arginine, introduced through the L452R spike substitution, plays a crucial role in strengthening the electrochemical grip on cells. This group possesses two proton donor groups (OH and NH), six proton acceptor sites (N and O), and six or seven bonds that enable molecular rotations, contributing to its ability to form strong electrostatic interactions with cell membranes. Accordingly, a range of different tautomers and conformers, encompassing hydrated [64, 65], protonated [66], and alkali-cationized structures [67, 68], can exist due to stabilization by various intermolecular hydrogen bonds. These Raman experiments elucidate that brief exposure of Delta virions to a 5 vol.% hydrosol of Si_3N_4 particles simultaneously triggers RNA purine oxidation and arginine deimination in the viral spike protein. These post-translational modifications of the spike protein hinder the ability of SARS-CoV-2 virions to infect cells, while RNA damage at purine sites impedes viral replication.

A doctoral thesis from Uppsala University highlighted the efficacy of Si_3N_4 against the SARS-CoV-2 virus and the persistent human adenovirus type 5 [69, 70]. In their investigation into the efficacy of Si_3N_4 against SARS-CoV-2, the researchers hypothesized that a surface-modified Si_3N_4 material with enhanced hydrophilicity, achieved through heat treatment, could lead to the effective attachment and inactivation of the virions. To test this hypothesis, Si_3N_4 powders were oxidized and characterized, and their ability to inactivate SARS-CoV-2 was evaluated. Their findings revealed that oxidized Si_3N_4 exhibited remarkable effectiveness in binding and inactivating the virus after as little as 1 min of contact, suggesting its potential as a viable agent for inhibiting the spread of COVID-19 under specific conditions. For the human adenovirus, the researchers also found that Si_3N_4 remarkably reduced adenoviral infectivity in both powder and bulk forms. Si_3N_4 powder inactivated 98% of infectious adenoviruses within 1 min of contact, while bulk silicon surfaces achieved a 73% reduction in infectivity within 10 min. The findings of both experiments demonstrate the potential of silicon nitride as a broad-spectrum antiviral agent capable of combating both DNA and RNA viruses.

3.2 Inactivation of the Herpes Simplex Virus by Si_3N_4

The herpes simplex virus (HSV-1) was also found to be instantaneously inactivated by contact with Si_3N_4 [71]. Infection by this virus is particularly problematic because it can cycle through periods of dormancy and activity in the body. Current therapeutics manage recurrent symptoms associated with break-outs but are essentially ineffective in eliminating the virus. Mechanistically, HSV-1 is inactivated similar to other virions. The hydrolytic reaction of Si_3N_4 releases ammonia, which is enzymatically converted to various reactive nitrogen species including nitric oxide and peroxynitrite. These RNS moieties penetrate the viral capsule and cleave the virus' phosphodiester DNA backbone leading to instantaneous and irreversible damage [43]. However, this HSV-1 research also identified the importance of silanols in this surface reaction. Hydrolysis of the Si_3N_4 generates an unstable intermediate molecule featuring a penta-coordinated silicon complex as follows [71]:

$$\mathrm{Si(OH)_4 + (OH)^- \rightarrow [Si(OH)_5]^- \rightarrow [(OH)_3SiO]^- \cdot H_2O} \tag{1}$$

The activation energy of this reaction exhibits pH dependence, with a significant decrease observed as surface silanols at Si–O–Si bridges transitioned from the protonated state ($Si–OH_2^+$) to the neutral state (Si–OH) and subsequently to the deprotonated state ($Si–O^-$) as pH increases. Consequently, an increase in alkaline pH values promotes the elution of silanols by enhancing their solubility. Within the biological milieu, the periodic renewal of the surface through ammonia and silicic acid elution is a unique characteristic. It sustains the effectiveness of bioactive reactions with tissue and microorganisms. This research indicates that complete and immediate inactivation of HSV-1 can be achieved using a minor proportion of Si_3N_4 particles contained within creams, gels, and oral rinses. Localized delivery via a sustained hydrolytic release in polymeric matrix-based antiherpetic devices can effectively suppress viral reactivation and provide sustained high local efficacy for extended durations. As an added benefit, any retained Si_3N_4 particles are expected to be completely resorbed within a short period without affecting human tissues [72, 73].

3.3 *Demonstrating the Antiviral Potential of Si_3N_4: Prototype N95 Face Masks with Embedded Powder*

Finally, a practical antiviral application for Si_3N_4 was demonstrated by Simpson et al. [74]. In this instance, Si_3N_4 powder was embedded into nonwoven fabric and assembled into N95 prototype face masks. Four- or five-layer masks and representative mask swatches were fabricated for standard industrial mask tests including ASTM F2299 and EN 13274-7 for filtration efficiency, EN 14683 for differential pressure drop, and ISO 18562-2 for particle shedding. The prototype masks and swatches consisted of outer spunbond polypropylene (PP) fabric layers (45 g/m^2), inner melt-blown layers (~ 50 g/m^2), and a central spunbond layer with embedded Si_3N_4 powder (20–30 wt%) sandwiched between the other layers. The fabric layers were joined using ultrasonic welding. Antiviral testing of assembled swatches was performed as per ISO 18184 using several viruses including SARS-CoV-2. Swatches were autoclaved and then exposed to viral titers for 5–120 min. A picture of a prototype face mask is provided in Fig. 9, and SEM photographs of the Si_3N_4-infused fibers at different magnifications are shown in Fig. 10a–c.

Two independent laboratories evaluated the filtration efficacy of the masks. Test results were similar, with filtration efficiencies reaching 99.18% and 99.97%, demonstrating that the masks surpassed the N95 particle filtration standard. The risk of Si_3N_4 particle release was assessed using measured values for PM2.5 and PM10 particles. Both were measured as $\leq 1.25 \pm 0.5$ μg/m^3, within the permissible limits of 12 and 150 μg/m^3, respectively. The breathability of the masks was determined using differential pressure drops across the fabric. The result was 53 ± 2 Pa/cm^2, which aligns with EU and US specifications for Type IIR medical masks.

Fig. 9 Prototype N-95 type face mask with embedded Si_3N_4 powder. (Used with permission under Creative Commons License [74])

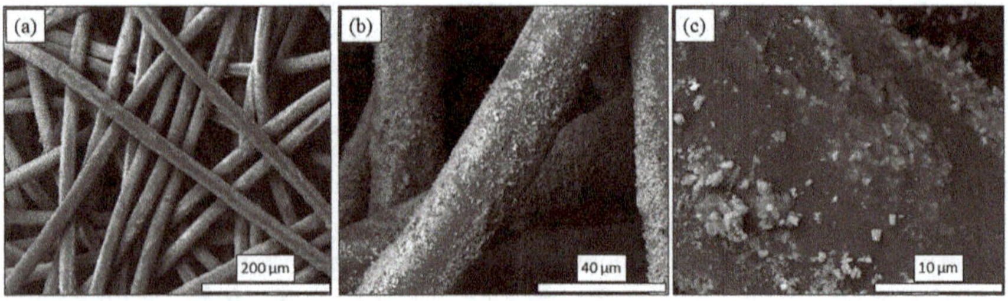

Fig. 10 SEM photomicrographs of spunbond polypropylene fibers coated with AP^2 Si_3N_4 particles: (**a**) ×500, (**b**) ×2,500, and (**c**) ×10,000 magnification. (Used with permission under Creative Commons License [74])

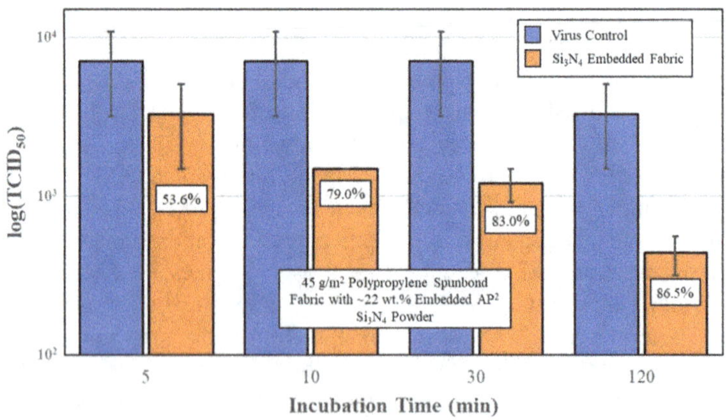

Fig. 11 Virus titers and % SARS-CoV-2 virus inactivation, after incubation with polypropylene fabric embedded with AP^2 Si_3N_4 powder. Error bars represent the standard error of the mean values. (Used with permission under Creative Commons License [74])

Inactivation results for the SARS-CoV-2 virus as a function of exposure time are given in Fig. 11. The Si_3N_4-embedded fabric exhibited a 50% reduction in viral load after just 5 mins of incubation. Viral inactivation continued at longer exposure times, with reductions of 79%, 83%, and ~87% at 10, 30, and 120 mins, respectively. While greater than a 2-log reduction was achieved at 120 mins, the antiviral efficacy of the fabric samples was lower compared to simple exposure to powders in a sol. This observation was probably due to the hydrophobic nature of the spunbond PP fibers, which likely hindered direct contact between the embedded powder particles and the virogenic solution. Based on the findings of this initial prototype study, Si_3N_4-embedded nonwoven fabric holds promise as a significant advancement in combating respiratory diseases. Integrating this material into facial masks could transform personal protective equipment from mere "capture and retain" devices to effective "capture and kill" barriers against viral pathogens [74].

In summary, while other inorganic antiviral compounds, such as copper, can also instantaneously inactivate viruses [54], only Si_3N_4 has demonstrated biocompatibility with host cells. This unique property makes Si_3N_4 a promising candidate for replacing toxic and allergenic compounds in long-term environmental sanitation applications, such as disinfectant sprays, face masks, surgical drapes, and other garments in hospitals. Its measured use could significantly reduce viral transmission among healthcare workers and patients alike. The biocompatibility of Si_3N_4, unaffected by either its anionic or cationic surface chemistry, highlights its potential as a vital tool in curbing the spread of evolving pathogenic viruses.

4 Antifungal Properties of Silicon Nitride

4.1 *Antifungal Potential of Si_3N_4 in PMMA Composites: A Strategy for Oral Candidiasis Management*

While *Candida albicans* (*C. albicans*) is typically a benign resident of the human gut microbiome, uncontrolled growth and a shift from yeast to hyphal form can lead to severe systemic infections when the immune system is compromised, or the gut microbiome is disrupted [75]. The yeast-to-hypha transition serves as a key pathogenetic mechanism for *C. albicans*, while phenotypic switching, biofilm formation, and the secretion of hydrolytic enzymes contribute to its diverse virulence repertoire [76]. The ability of this polymorphic fungus to exist as yeast cells and transform into a pathogenic form presents a significant challenge in modern medical mycology. The alarming rise in Candida-associated infections highlights the urgent need for novel antifungal therapies, as the efficacy of existing drugs is waning due to increasing drug resistance. This underscores the importance of elucidating the mechanisms underlying the dysbiotic transition of *C. albicans*. In the elderly population, excessive polypharmacy and the use of certain drug classes are associated with reduced salivary flow (hyposalivation) and dry mouth (xerostomia) [77]. Both of these conditions are strongly linked to an increased risk of oral candidiasis [78]. Additionally, this patient population is typically also comprised of individuals who have partial or complete edentulism.

Polymethylmethacrylate (PMMA) dentures have been a mainstay of prosthodontic treatment for decades due to their affordability, biocompatibility, and ease of fabrication. However, several challenges are associated with PMMA dentures, including poor adhesion to mucosal tissues, and harboring of *Candida albicans*, leading to oral thrush [79]. Studies have demonstrated that yeast cells are present in the oral cavity of 25 and 50% of healthy adults and children, respectively, but this prevalence notably increases to 70% among denture wearers, ranging from 60% to 100% [80]. In general, dental implants act as reservoirs and incubators for oral infections, while PMMA presents a more favorable environment for the proliferation of *C. albicans* due to its porous nature. Despite efforts to develop functional prostheses (e.g., glass, metals, sealed resins) [81–86] and antibiotic-infused materials [87–90] to combat *C. albicans* colonization, none have demonstrated significant efficacy in preventing fungal overgrowth. Consequently, clinical strategies have primarily focused on maintaining a healthy oral environment as a preventive measure rather than relying solely on novel candidacidal materials [91].

From prior work on oomycetes [92], it was posited that Si_3N_4 might be a promising therapy for controlling the proliferation of *C. albicans* and preventing subsequent candidiasis [83]. Moreover, Si_3N_4 exhibits the potential to counteract the resistance of *C. albicans* to conventional antifungal agents [87, 93]. Consequently, a study was designed and conducted to test whether the addition of Si_3N_4 particles to PMMA to form a composite dental device would be effective against *C. albicans* [94]. To maintain clinical relevance, Si_3N_4 particle dispersions ranging from 2 to 30 vol.% were incorporated into a PMMA resin matrix. The underlying hypothesis was that Si_3N_4-loaded PMMA would exhibit a statistically significant difference in candidacidal activity compared to pure PMMA.

Test batches were prepared by intricately mixing PMMA resin with ß-Si_3N_4 powder at concentrations of ~2, 6, 8, 10, 15, and 30 vol.% (i.e., 5.4–54.3 wt.% Si_3N_4). Each batch was then mixed with benzoyl peroxide, methyl methacrylate monomer, N,N-dimethyl-p-toluidine, and ethylene glycol dimethacrylate following the manufacturer's instructions. To determine the maximum volume fraction of Si_3N_4 that could be incorporated without compromising PMMA mechanical properties, rectangular test specimens ($15 \times 3 \times 3$ mm) were fabricated from each batch and subjected to mechanical properties testing, including maximum tensile stress, Young's modulus, and work of fracture. It was found that batches with Si_3N_4 fractions >10 vol.% exhibited brittle behavior, characterized by reduced tensile strength and diminished deformation energy. *C. albicans* cultures were prepared at a concentration of 1×10^6 CFU and allowed to proliferate on pure PMMA and the remaining PMMA/Si_3N_4 test specimens for 24 h. Metabolic activity was then assessed using an optical density (OD) reader and a microbial viability assay kit.

The results of these tests are provided in Fig. 12. While candidacidal efficacy in PMMA/Si_3N_4 composites was achievable only at Si_3N_4 volume fractions exceeding 6%, maintaining adequate mechanical properties necessitated restricting the Si_3N_4 content to a narrow range between 6 and 10 vol.%, (i.e., 15.0–23.5 wt% Si_3N_4). An intermediate fraction of Si_3N_4 at 8 vol.% was selected for further testing.

Fluorescence microscopy and CFU counting were subsequently utilized for cells exposed to selected substrates, then fixed with 95% ethanol and stained using DAPI (blue; targeting nuclei) and a specific fluorescent stain kit for fungi, Fungiflora Y® (green; targeting fungal cell wall chitin, glucans, and polysaccharides). Representative micrographs for cell/CFU counting and fluorescence microscopy assessments are presented in Fig. 13a–e. The micrographs in Fig. 13a–c were obtained following violet staining. They depict as-cultured *C. albicans* (a) and their proliferation after 24 h of exposure to PMMA and PMMA/8 vol.% Si_3N_4 substrates (b) and (c), respectively. Enlarged images of stained cells are provided in the insets of panels (b) and (c). In (d), the Candida detector kit visually illustrates the change in

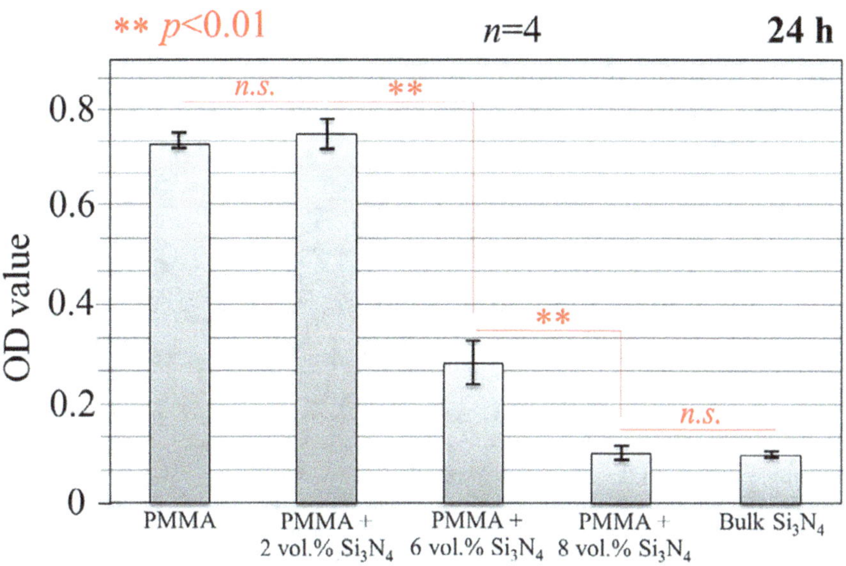

Fig. 12 *C. albicans* cell proliferation as a function of Si_3N_4 volume fraction, determined using optical density microbial viability assays. Data are presented as mean values ± standard deviation ($n = 4$). Statistical significance ($p < 0.01$) is indicated by two asterisks. (Used with permission from Ref. [94])

concentration (CFU/mL) for cells exposed to pure PMMA (negative control), PMMA/8 vol.% Si_3N_4, and bulk Si_3N_4 (positive control).

The results of this assay, where the persistence of the red color signifies the absence of yeast cell proliferation, underscore the candidacidal efficacy of Si_3N_4, both as a powder embedded within PMMA and as a bulk substrate. Fig. 13c presents a plot depicting the direct cell counting of cells exposed to PMMA and PMMA/8 vol.% Si_3N_4 substrates. Inoculum concentrations on the substrates were 10^6 cells/mL. Comparisons were made using control samples (cultured cells) and cells exposed to bulk Si_3N_4 substrates under the same conditions. These comparisons corroborated the ease of cell proliferation on the PMMA monolithic substrate. Additionally, they demonstrated that a relatively small fraction of Si_3N_4 in PMMA resulted in a substantial reduction in proliferation, comparable to that observed on bulk Si_3N_4 substrates. Notably, the direct cell counting method revealed a 3–4 $\log_{10}$ reduction in pathogen viability (99.99%).

In situ Raman spectroscopy was utilized to evaluate the mechanism behind Si_3N_4's antifungal effectiveness. Similar to bacteria and viruses, hydrolysis of Si_3N_4 effectively inhibited the proliferation of *C. albicans*. Nevertheless, the underlying lytic mechanisms differed somewhat from those observed in bacteria and viruses. Figure 14 presents a schematic view of these lytic mechanisms.

In brief, the homolytic cleavage of Si-N bonds, triggered by hydrolysis at the Si_3N_4 surface, simultaneously releases nitrogen anions and unpaired electrons.

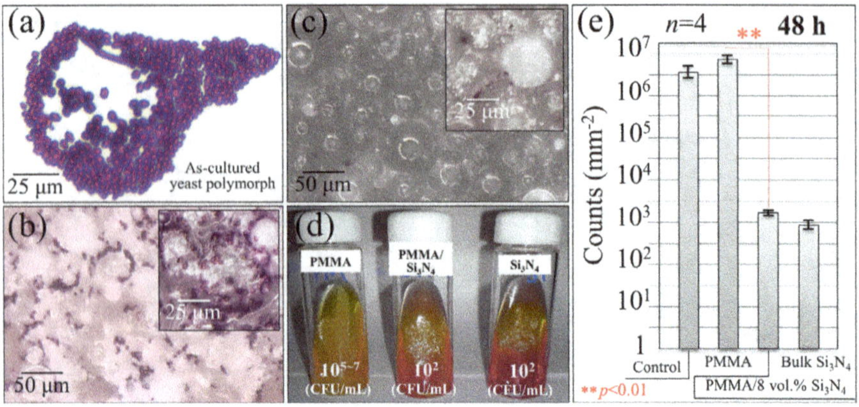

Fig. 13 (**a**) As-cultured *C. albicans* after violet PAS staining. Proliferation of *C. albicans* after 48-h exposure to PMMA (**b**) and PMMA/8 vol.% Si_3N_4 (**c**) substrates. Enlarged views are provided in the insets. (**d**) CFU assessments using the Candida detector kit in a selective medium. CFU concentrations in the insets are in CFU/mL according to the color scale provided by the kit manufacturer. Yeast cells were exposed to PMMA (negative control), PMMA/8 vol.% Si_3N_4, and bulk Si_3N_4 (positive control) substrates. The preservation of the original red color of the medium indicates no yeast cell proliferation. (**e**) Direct cell counting of violet PAS-stained cells exposed to PMMA and PMMA/8 vol.% Si_3N_4 composite, averaging over 20 micrographs. A comparison is made with the control samples (cultured yeast cells) and cells exposed to bulk Si_3N_4 substrates. (Used with permission from Ref. [94])

These free electrons disassociate water molecules in the vicinity, generating oxygen radicals. In turn, these radicals catalyze the oxidation of NH_3 into hydroxylamine, initiating a series of off-stoichiometric reactions that culminate in the formation of reactive nitrogen species (RNS), including NO, NO_2, and peroxynitrite $ONOO^-$ [9]. Localized accumulation of $ONOO^-$ radicals was observed in the mitochondria of *Candida* cells exposed to the PMMA/Si_3N_4 composite, whereas these radicals were absent in the mitochondria of control cells exposed only to PMMA. These in situ findings suggest that the material properties of Si_3N_4 resemble the antifungal activity of macrophages, which simultaneously generate NO radicals and superoxide anions ($O_2{\bullet}^-$) leading to the formation of antifungal $ONOO^-$.

In summary, a small proportion (8 vol.%) of Si_3N_4 was demonstrated to be adequate to impart antifungal properties to a PMMA-based composite. Although evidence suggests that the mechanism is analogous to both antiviral and antibacterial processes, it was newly revealed that the exogenous antifungal activity of Si_3N_4 mimics that of the endogenous human immune system. The looming threat of fungal infections to global health, with projected increases in morbidity and mortality [95], underscores the potential of Si_3N_4 to mitigate the impact of candidiasis in dentistry and orthopedics. This study showed that its effectiveness in inhibiting fungal growth can be translated to medical devices, thereby reducing the risk of infections among vulnerable populations, such as the elderly and immunocompromised individuals.

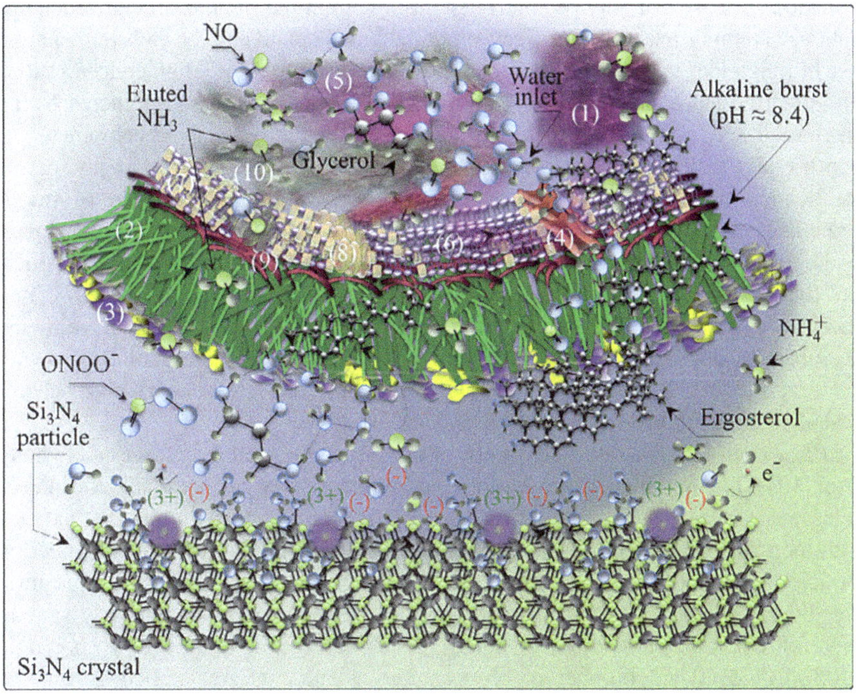

Fig. 14 Schematic diagram illustrating the candidacidal activity of Si_3N_4 due to both reactive nitrogen species (RNS) formation and the direct effect of ammonia (NH_3) in cytoplasmic space. NH_3 freely penetrates the cell membrane, inducing water influx and disrupting the endocytotic pH balance, leading to alkalization and NH_4^+ formation. In response to the osmotic stress caused by increased water influx and the associated alteration in cell volume, the cell produces and expels glycerol-water hydrates, serving as osmolytes. Simultaneously, the cell interprets the high concentration of RNS at the Si_3N_4 interface as a host immune response and triggers an overproduction of ergosterol. The inset labels the following features: (1) mitochondrion, (2) β-1,3-glucans and β-1,6-glucans, (3) anchored proteins and mannans, (4) ion channel, (5) vacuole, (6) plasma membrane, (7) lipid droplets, (8) lipid raft, (9) chitin, and (10) secretory vesicle. (Used with permission from Ref. [94])

4.2 *Si_3N_4 Powder: A Potential Replacement for Current Fungicides in Grapevine Protection*

In agrochemistry, the application of Si_3N_4 powder could potentially replace current fungicides in preventing grapevine infections and improving harvest yields [96]. The treatment of downy mildew, which is caused by the oomycete *Plasmopara viticola*, requires frequent application of antifungal agents to control the infection [97]. Nearly two-thirds of all fungicides in the EU are used for this purpose [98]. However, after fulfilling their function, these products pollute the environment. Although *Plasmopara viticola* is native to North America, it was accidentally introduced into Europe at the end of the nineteenth century, and into Japan in 2009. This voracious disease can only be controlled through multiple weather-modulated applications. To minimize adverse effects on both human health and the environment, the use of fungicides is limited to specific types and concentrations. High concentrations are effective but environmentally prohibited. Yet, inadequate dosage increases the likelihood of pathogen resistance [99]. Consequently, the development of eco-friendly antifungal products has received significant attention as an alternative to chemical treatments [100]. Some of the new concepts include the use of microorganisms as biocontrol agents or as systemic resistance inducers against plant pathogens [101]. An important step forward is the breeding of resistant varieties, but this strategy does not always meet the market's quality expectations [97]. Yet, recent advancements in genomic and transcriptomic analytical methods have fueled the growing popularity of this approach [102]. Furthermore, the public's growing concern over health, safety, and the environment has made the search for agrochemical alternatives a priority in today's society [103].

Plasmopara viticola attacks all parts of the grapevine, including leaves and young fruit [104]. Favored by temperate and humid weather, asexual sporangia (zoospores) release spores that eventually attach to stomata, forming a germ tube that extends into the substomatal cavity. The germ tube eventually transforms into an infection vesicle. A primary hypha emerges and rapidly develops branches, with haustoria (feeding structures) penetrating plant cells to obtain nutrients. Following several days of infective incubation, sporangiophores emerge to release new sporangia [105]. A significant number of phytosanitary treatments are applied during the growing season, with the type, amount, and timing determined by the disease stage, grapevine variety, and environmental conditions. In most regions, managing downy mildew requires multiple applications, starting early in the growth cycle [98]. Broad-spectrum preventive or systemic fungicides are typically used, but their frequent use, high cost, and detrimental long-term effects on the environment necessitate more effective, sustainable, and eco-friendly alternatives. Traditional fungicidal compounds should be reserved for situations where they are necessary [106].

Silicon nitride has proven to be effective in its ability to knock down *Plasmopara viticola* early in its infection cycle [92]. The antifungal mechanisms of Si_3N_4 mirror those previously demonstrated for its antibacterial and antiviral properties. Moreover, Si_3N_4 is compatible with plant cells and poses no harm to soil. To illustrate the potential of Si_3N_4, Fig. 15a–c shows time-lapse micrographs of sporangia in an aqueous environment. These images depict sporangia at different stages: (a) prepared for germination, (b) discharging zoospores, and (c) an empty sporangium approximately 3 h later (also shown in the inset micrograph). Similarly, Fig. 15d–f showcases time-lapse micrographs of sporangia interacting with a 1.5 vol.% hydrosol of Si_3N_4 powder. The Si_3N_4 particles exhibited a strong electrostatic attraction to the external wall of the sporangium. Contact between the particles and the sporangia occurred almost instantaneously upon introduction into the suspension. This contact gradually increased until the entire sporangium surface was covered by Si_3N_4 granules. This resulted in the complete blockage of germination, as no mobile zoospores were detected in the treated samples. The observation of

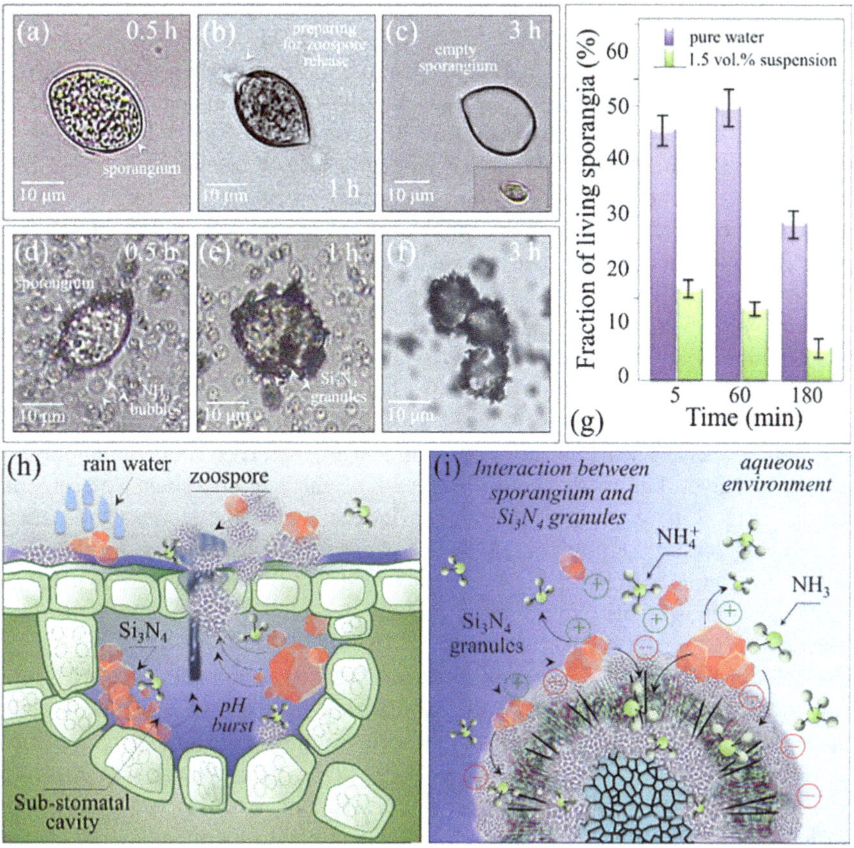

Fig. 15 Time-lapse micrographs of sporangia in pure water and 1.5 vol.% Si_3N_4 water suspension. (**a–c**) Sporangia in pure water at 0.5 h, 1 h, and 3 h, respectively. (**d–f**) Sporangia embedded in 1.5 vol.% Si_3N_4 water suspension at the same corresponding times. Arrows in the insets indicate ammonia gas bubbles. (**g**) Fractional plots of living sporangia as a function of time, comparing viability in pure water and 1.5 vol.% Si_3N_4 hydrosol. (**h**) Schematic representation of pH buffering, leading to ammonia elution in stomata entrapping Si_3N_4 granules after a rain event. (**i**) Schematic illustration of charge attraction between Si_3N_4 granules and sporangia, resulting in the formation of ammonium and ammonia, which act as antimicrobial agents at the pathogen/Si_3N_4 interface. (Adapted and used with permission under Creative Commons License [92])

bubbles and an increase in pH, as evident in Fig. 15d (refer to inset labels), indicate the production of gaseous ammonia (NH_3). This volatile molecule readily traverses the sporangia's cell wall. Additionally, ammonia freely diffuses across the fungal cell membranes, allowing it to enter the cells [107]. Fig. 15g presents fractional plots as a function of time, comparing the viability of sporangia suspended in pure water and in a 1.5 vol.% Si_3N_4 hydrosol. These plots were generated through direct counting of sporangia stained with fluorescein diacetate [92]. The viability of sporangia exposed to the Si_3N_4 hydrosol exhibited a clear downward trend, with a fivefold reduction compared to pure water after 180 min. This indicates that the interaction between sporangia and Si_3N_4 follows the "catch and kill" mechanism described earlier in this chapter. Findings from experiments using leaf disks from two distinct grapevine cultivars, *Cabernet Sauvignon* and *Cannonau* (not shown), corroborate the hypothesis of sporangia inactivation [92]. Control leaf samples from both cultivars inoculated with pathogens that had not been exposed to Si_3N_4 consistently exhibited severe infection 5 days after inoculation, leading to necrosis at the infection sites. Conversely, no signs of phytotoxicity were observed in leaves from either cultivar after the same exposure when inoculated with sporangia that had been exposed to Si_3N_4.

Plasmopara viticola gains entry into the host leaf tissue through the stomata, establishing itself within the substomatal air spaces where it undergoes a slow developmental process for approximately 12–15 h before forming the initial haustorium [108]. This stage is recognized as the most critical phase of the overall infection process. As a preventive measure, Si_3N_4 particles could be applied to grapevine leaves before the commencement of this period. Upon entering the stomata (Fig. 15h), these particles remain lodged within the stomata throughout the growing season. During rainfall events, water repeatedly triggers the elution of ammonium moieties from the Si_3N_4 particles, leading to a rapid increase in pH. This creates an inhospitable environment for the sporangia, effectively preventing infection.

In contrast, the use of currently available systemic fungicides is increasingly restricted due to environmental concerns. Additionally, preventive products based on copper hydroxide and copper sulfate offer limited protection to grapevines, providing only short-term efficacy. Their nonsystemic nature limits their effectiveness in controlling established infections. Furthermore, their susceptibility to rain washout diminishes their protective capabilities [109]. Consequently, a solid-state fungicide like Si_3N_4 offers several advantages over conventional fungicides. Its repeated activation during rain events provides sustained protection against the pathogen, even after the infection has entered leaf tissue. Future studies should explore Si_3N_4's potential to inhibit sporangia release from the stroma during secondary infection cycles, potentially slowing disease progression and reducing the need for repeated applications.

The multi-mechanistic antifungal activity of Si_3N_4 shares similarities with its antibacterial mechanism. Experimental observations in Fig. 15d–f demonstrate the initial electrostatic attraction of Si_3N_4 granules to sporangia. Subsequently, nitrogen species generated on the Si_3N_4 surface induce protein alterations in the pathogens, trigger DNA nitrosative damage, and activate metabolic enzymes that modify pathogen membrane structures [92]. The cytoplasmic lysis induced by the diffusion of NH_3 across pathogen cell walls, as illustrated in Fig. 15i, diminishes the likelihood of adaptive mutations arising within the pathogen population.

The use of Si_3N_4 as a fungicide holds promise for a wide range of plant species, as it is composed solely of environmentally friendly elements that have been present on Earth since its formation [110]. Silicon is a well-established fertilizer, particularly effective in mitigating biotic and abiotic stresses [111, 112]. Ammonium, the primary inorganic nitrogen source for organic nitrogen synthesis, is readily absorbed and utilized by plants through root-specific transporters [113]. In grapevines, ammonium accounts for up to 80% of total nitrogen before veraison, decreasing to 5–10% after maturation and even lower after fermentation [114]. The nitrogen eluted from Si_3N_4 could potentially contribute to improved berry quality and fermentation conditions.

5 Conclusion

Silicon nitride, an advanced ceramic material engineered for exceptional performance under extreme conditions, was initially envisioned for high-temperature engine applications but it eventually found its niche in bearing and wear components due to its remarkable strength, toughness, hardness, and wear resistance. However, its biological potential remained largely unexplored until the early twenty-first century, when it emerged as a promising candidate for combating pathogenic microbes. Remarkably, silicon nitride aligns seamlessly with Ehrlich's concept of a "magic bullet," exhibiting exceptional biocompatibility while effectively eradicating bacteria, viruses, and fungi. Its unique combination of biocompatibility, surface and bulk properties, and antipathogenicity has garnered significant attention in the medical field, particularly in the development of orthopedic implants and dental applications.

Si_3N_4 exhibits potent antibacterial activity against a wide range of bacterial strains, including both gram-positive and gram-negative species. Its antibacterial effect is attributed to a simple surface chemical reaction that releases nitrogen and moieties, which are converted to reactive nitrogen species (RNS). These RNS are known to disrupt prokaryotic cell membranes and interfere with their metabolic processes, leading to cell lysis. Studies reviewed in this chapter have also demonstrated the antiviral efficacy of Si_3N_4 against some of the most prevalent and virulent viruses. The antiviral mechanism of Si_3N_4 shares similarities with its antibacterial effect. An electrostatic attraction between virions and Si_3N_4 particles accelerates their inactivation, a phenomenon referred to as a "catch-and-kill mechanism." In vitro studies demonstrated essentially instantaneous inactivation upon contact. Si_3N_4 shows promise in combating fungal infections as well, particularly against *Candida albicans*, a common fungal pathogen associated with dental prostheses. Mechanistically, the antifungal activity of Si_3N_4 is attributed to its ability to release RNS that disrupt fungal cell membranes and inhibit biofilm formation. Finally, it was shown that Si_3N_4 prevents the growth of a persistent agricultural fungus, *Plasmopara viticola*, and therefore has the potential to increase crop yields and fruit quality.

Silicon nitride's remarkable antipathogenic properties position it as a promising material for the development of next-generation antimicrobial medical devices and other nonmedical uses. Its ability to combat a broad spectrum of microbes, including bacteria, viruses, and fungi, makes it a versatile tool for infection prevention and control. As research continues to explore the full potential of Si_3N_4, it is anticipated to play an increasingly significant role in safeguarding human health from the threat of microbial diseases.

References

1. Aminov, R.I. A Brief History of the Antibiotic Era: Lessons Learned and Challenges for the Future. *Front. Microbiol.* **2010**, *1*, 134, https://doi.org/10.3389/fmicb.2010.00134.
2. Bassett, E.J.; Keith, M.S.; Armelagos, G.J.; Martin, D.L.; Villanueva, A.R. Tetracycline-Labeled Human Bone from Ancient Sudanese Nubia (A.D. 350). *Science (80-.).* **1980**, *209*, 1532–1534, https://doi.org/10.1126/science.7001623.
3. Joseph, I.O.F.; Wall, T.E.; Tanner, J.R.; Tawaha, K.; Alali, F.Q.; Li, C.; Oberlies, N.H. Proliferation of Antibiotic-Producing Bacteria and Concomitant Antibiotic Production as the Basis for the Antibiotic Activity of Jordan's Red Soils. *Appl. Environ. Microbiol.* **2009**, *75*, 2735–2741, https://doi.org/10.1128/AEM.00104-09.
4. Cui, L.; Su, X. Discovery, Mechanisms of Action and Combination Therapy of Artemisinin. *Expert Rev. Anti. Infect. Ther.* **2009**, *7*, 999–1013, https://doi.org/10.1586/eri.09.68.
5. Schwartz, R.S. Paul Ehrlich's Magic Bullets. *N. Engl. J. Med.* **2004**, *350*, 1079–1080, https://doi.org/10.1056/NEJMp048021.
6. Gorth, D.J.; Puckett, S.; Ercan, B.; Webster, T.J.; Rahaman, M.; Bal, B.S. Decreased Bacteria Activity on Si_3N_4 Surfaces Compared with PEEK or Titanium. *Int. J. Nanomedicine* **2012**, *7*, 4829–4840, https://doi.org/10.2147/IJN.S35190.
7. Webster, T.J.; Patel, A.A.; Rahaman, M.N.; Bal, B.S. Anti-Infective and Osteointegration Properties of Silicon Nitride, Poly (Ether Ether Ketone), and Titanium Implants. *Acta Biomater.* **2012**, *8*, 4447–4454.
8. Pezzotti, G. A Spontaneous Solid-State NO Donor to Fight Antibiotic Resistant Bacteria. *Mater. Today Chem.* **2018**, *9*, 80–90, https://doi.org/10.1016/j.mtchem.2018.05.004.
9. Pezzotti, G. Silicon Nitride: A Bioceramic with a Gift. *ACS Appl. Mater. Interfaces* **2019**, *11*, 26619–26636, https://doi.org/10.1021/acsami.9b07997.
10. Zanocco, M.; Marin, E.; Rondinella, A.; Boschetto, F.; Horiguchi, S.; Zhu, W.; McEntire, B.J.; Bock, R.M.; Bal, B.S.; Pezzotti, G. The Role of Nitrogen Off-Stoichiometry in the Osteogenic Behavior of Silicon Nitride Bioceramics. *Mater. Sci. Eng. C* **2019**, *105*, 110053, https://doi.org/10.1016/j.msec.2019.110053.
11. Williams, D. Revisiting the Definition of Biocompatibility. *Med. Device Technol.* **2003**, *14*, 10–13.
12. Hench, L.L.; West, J.K. The Sol-Gel Process. *Chem. Rev.* **1990**, *90*, 33–72, https://doi.org/10.1021/cr00099a003.
13. Bhatia, S.K. *Biomaterials for Clinical Applications*; Springer Science & Business Media: New York City, NY USA, 2010; ISBN 1441969209.
14. Hench, L.L. Bioceramics: From Concept to Clinic. *J. Am. Ceram. Soc.* **1991**, *74*, 1487–1510, https://doi.org/10.1111/j.1151-2916.1991.tb07132.x.
15. Salinas, A.J.; Esbrit, P.; Vallet-Regí, M.; Vallet-Regi, M. A Tissue Engineering Approach Based on the Use of Bioceramics for Bone Repair. *Biomater. Sci.* **2013**, *1*, 40–51, https://doi.org/10.1039/c2bm00071g.
16. Piconi, C. Bioinert Ceramics: State-of-the-Art. *Key Eng. Mater.* **2017**, *758 KEM*, 3–13, https://doi.org/10.4028/www.scientific.net/KEM.758.3.
17. Pezzotti, G. Surface Chemistry of Bioceramics: The Missing Key. In *Bioceramics – From Macro to Nano Scale*; Osaka, A., Narayan, R., Eds.; Elsevier B.V.: New York City, NY USA, 2021; pp. 297–324.

18. Schneider, M. The Importance of Ammonia in Mammalian Cell Culture. *J. Biotechnol.* **1996**, *46*, 161–185, https://doi.org/10.1016/0168-1656(95)00196-4.
19. Vatansever, F.; de Melo, W.C.M.A.; Avci, P.; Vecchio, D.; Sadasivam, M.; Gupta, A.; Chandran, R.; Karimi, M.; Parizotto, N.A.; Yin, R.; et al. Antimicrobial Strategies Centered Around Reactive Oxygen Species – Bactericidal Antibiotics, Photodynamic Therapy, and Beyond. *FEMS Microbiol. Rev.* **2013**, *37*, 955–989, https://doi.org/10.1111/1574-6976.12026.
20. Wimalawansa, S.J. Nitric Oxide: Novel Therapy for Osteoporosis. *Expert Opin. Pharmacother.* **2008**, *9*, 3025–3044, https://doi.org/10.1517/14656560802197162.
21. Abnosi, M.H.; Pari, S. Exogenous Nitric Oxide Induced Early Mineralization in Rat Bone Marrow Mesenchymal Stem Cells via Activation of Alkaline Phosphatase. *Iran. Biomed. J.* **2019**, *23*, 142.
22. Henstock, J.R.; Canham, L.T.; Anderson, S.I. Silicon: The Evolution of its Use in Biomaterials. *Acta Biomater.* **2015**, *11*, 17–26, https://doi.org/10.1016/j.actbio.2014.09.025.
23. Shie, M.-Y.; Ding, S.-J.; Chang, H.-C. The Role of Silicon in Osteoblast-Like Cell Proliferation and Apoptosis. *Acta Biomater.* **2011**, *7*, 2604–2614, https://doi.org/10.1016/j.actbio.2011.02.023.
24. Pezzotti, G.; Oba, N.; Zhu, W.; Marin, E.; Rondinella, A.; Boschetto, F.; McEntire, B.J.; Yamamoto, K.; Bal, B.S. Human Osteoblasts Grow Transitional Si/N Apatite in Quickly Osteointegrated Si_3N_4 Cervical Insert. *Acta Biomater.* **2017**, *64*, 411–420.
25. Pezzotti, G.; Marin, E.; Adachi, T.; Rondinella, A.; Boschetto, F.; Zhu, W.-L.; Sugano, N.; Bock, R.M.; McEntire, B.J.; Bal, B.S. Bioactive Silicon Nitride: A New Therapeutic Material for Osteoarthropathy. *Sci. Rep.* **2017**, *7*, 44848.
26. Pacher, P.; Beckman, J.S.; Liaudet, L. Nitric Oxide and Peroxynitrite in Health and Disease. *Physiol. Rev.* **2007**, *87*, 315–424.
27. Mungrue, I.N.; Husain, M.; Stewart, D.J. The Role of NOS in Heart Failure: Lessons from Murine Genetic Models. *Heart Fail. Rev.* **2002**, *7*, 407–422, https://doi.org/10.1023/A:1020762401408.
28. Van'T Hof, R.J.; Ralston, S.H. Cytokine-Induced Nitric Oxide Inhibits Bone Resorption by Inducing Apoptosis of Osteoclast Progenitors and Suppressing Osteoclast Activity. *J. Bone Miner. Res.* **1997**, *12*, 1797–1804, https://doi.org/10.1359/jbmr.1997.12.11.1797.
29. Pezzotti, G.; Horiguchi, S.; Boschetto, F.; Adachi, T.; Marin, E.; Zhu, W.; Yamamoto, T.; Kanamura, N.; Ohgitani, E.; Mazda, O. Raman Imaging of Individual Membrane Lipids and Deoxynucleoside Triphosphates in Living Neuronal Cells during Neurite Outgrowth. *ACS Chem. Neurosci.* **2018**, *9*, 3038–3048, https://doi.org/10.1021/acschemneuro.8b00235.
30. Kodigepalli, K.M.; Bowers, K.; Sharp, A.; Nanjundan, M. Roles and Regulation of Phospholipid Scramblases. *FEBS Lett.* **2015**, *589*, 3–14, https://doi.org/10.1016/j.febslet.2014.11.036.
31. Rimai, L.; Cole, T.; Parsons, J.L.; Hickmott, J.T.; Carew, E.B. Studies of Raman Spectra of Water Solutions of Adenosine Tri-, Di-, and Monophosphate and Some Related Compounds. *Biophys. J.* **1969**, *9*, 320–329, https://doi.org/10.1016/S0006-3495(69)86389-7.
32. Bhaumik, A.; Shearin, A.M.; Delong, R.; Wanekaya, A.; Ghosh, K. Probing the Interaction at the Nano–Bio Interface Using Raman Spectroscopy: ZnO Nanoparticles and Adenosine Triphosphate Biomolecules. *J. Phys. Chem. C* **2014**, *118*, 18631–18639, https://doi.org/10.1021/jp506200a.
33. Kringelbach, T.M.; Aslan, D.; Novak, I.; Schwarz, P.; Jørgensen, N.R. UTP-Induced ATP Release is a Fine-Tuned Signalling Pathway in Osteocytes. *Purinergic Signal.* **2014**, *10*, 337–347, https://doi.org/10.1007/s11302-013-9404-1.
34. Lopes, R.P.; Marques, M.P.M.; Valero, R.; Tomkinson, J.; Batista De Carvalho, L.A.E. Guanine-A Combined Study Using Vibrational Spectroscopy and Theoretical Methods. *Spectrosc. Int.* **2012**, *27*, 273–292, https://doi.org/10.3233/978-1-61499-184-7-127.
35. Notingher, I.; Bisson, I.; Bishop, A.E.; Randle, W.L.; Polak, J.M.P.; Hench, L.L. *In situ* Spectral Monitoring of mRNA Translation in Embryonic Stem Cells during Differentiation *in vitro*. *Anal. Chem.* **2004**, *76*, 3185–3193, https://doi.org/10.1021/ac0498720.

36. Lopes, R.P.; Valero, R.; Tomkinson, J.; Marques, M.P.M.; Batista De Carvalho, L.A.E. Applying Vibrational Spectroscopy to the Study of Nucleobases-Adenine as a Case-Study. *New J. Chem.* **2013**, *37*, 2691–2699, https://doi.org/10.1039/c3nj00445g.
37. Fedor, M.J.; Williamson, J.R. The Catalytic Diversity of RNAs. *Nat. Rev. Mol. Cell Biol.* **2005**, *6*, 399–412, https://doi.org/10.1038/nrm1647.
38. Thaplyal, P.; Ganguly, A.; Hammes-Schiffer, S.; Bevilacqua, P.C. Inverse Thio Effects in the Hepatitis Delta Virus Ribozyme Reveal that the Reaction Pathway is Controlled by Metal Ion Charge Density. *Biochemistry* **2015**, *54*, 2160–2175, https://doi.org/10.1021/acs.biochem.5b00190.
39. Arp, D.J.; Stein, L.Y. Metabolism of Inorganic N Compounds by Ammonia-Oxidizing Bacteria. *Crit. Rev. Biochem. Mol. Biol.* **2003**, *38*, 471–495, https://doi.org/10.1080/10409230390267446.
40. Wimalawansa, S.J. Nitric Oxide and Bone. *Ann. N. Y. Acad. Sci.* **2010**, *1192*, 391–403, https://doi.org/10.1111/j.1749-6632.2009.05230.x.
41. Riancho, J.A.; Salas, E.; Zarrabeitia, M.T.; Olmos, J.M.; Amado, J.A.; Fernandez-Luna, J.L.; Gonzalez-Macias, J. Expression and Functional Role of Nitric Oxide Synthase in Osteoblast-Like Cells. *J. Bone Miner. Res.* **1995**, *10*, 439–446, https://doi.org/10.1002/jbmr.5650100315.
42. MacPherson, H.; Noble, B.S.; Ralston, S.H. Expression and Functional Role of Nitric Oxide Synthase Isoforms in Human Osteoblast-Like Cells. *Bone* **2016**, *24*, 179–185, https://doi.org/10.1016/S8756-3282(98)00173-2.
43. Burrows, C.J.; Muller, J.G. Oxidative Nucleobase Modifications Leading to Strand Scission. *Chem. Rev.* **1998**, 1109–1151, https://doi.org/10.1021/cr960421s.
44. Spinelli, J.B.; Yoon, H.; Ringel, A.E.; Jeanfavre, S.; Clish, C.B.; Haigis, M.C. Metabolic Recycling of Ammonia via Glutamate Dehydrogenase Supports Breast Cancer Biomass. *Science (80-.).* **2017**, *358*, 941–946, https://doi.org/10.1126/science.aam9305.
45. Takeuchi, Y.; Nakayama, Y.; Fukusaki, E.; Irino, Y. Glutamate Production from Ammonia via Glutamate Dehydrogenase 2 Activity Supports Cancer Cell Proliferation Under Glutamine Depletion. *Biochem. Biophys. Res. Commun.* **2018**, *495*, 761–767, https://doi.org/10.1016/j.bbrc.2017.11.088.
46. Meng, M.; Chen, S.; Lao, T.; Liang, D.; Sang, N. Nitrogen Anabolism Underlies the Importance of Glutaminolysis in Proliferating Cells. *Cell Cycle* **2010**, *9*, 3921–3932, https://doi.org/10.4161/cc.9.19.13139.
47. Merhi, A.; Delrée, P.; Marini, A.M. The Metabolic Waste Ammonium Regulates mTORC2 and mTORC1 Signaling. *Sci. Rep.* **2017**, *7*, 44602, https://doi.org/10.1038/srep44602.
48. Davie, E.; Forte, G.M.A.; Petersen, J. Nitrogen Regulates AMPK to Control TORC1 Signaling. *Curr. Biol.* **2015**, *25*, 445–454.
49. Lie, S.; Wang, T.; Forbes, B.; Proud, C.G.; Petersen, J. The Ability to Utilise Ammonia as Nitrogen Source is Cell Type Specific and Intricately Linked to GDH, AMPK and mTORC1. *Sci. Rep.* **2019**, *9*, 1461, https://doi.org/10.1038/s41598-018-37509-3.
50. Pezzotti, G. Silicon Nitride as a Biomaterial. *J. Ceram. Soc. Japan* **2023**, *131*, 398–428, https://doi.org/10.2109/jcersj2.23057.
51. Pezzotti, G.; Bock, R.M.; McEntire, B.J.; Adachi, T.; Marin, E.; Boschetto, F.; Zhu, W.; Mazda, O.; Bal, B.S. *In vitro* Antibacterial Activity of Oxide and Non-Oxide Bioceramics for Arthroplastic Devices: I. *In situ* Time-Lapse Raman Spectroscopy. *Analyst* **2018**, *143*, 3708–3721, https://doi.org/10.1039/C8AN00233A.
52. Boschetto, F.; Toyama, N.; Horiguchi, S.; Bock, R.M.; McEntire, B.J.; Adachi, T.; Marin, E.; Zhu, W.; Mazda, O.; Bal, B.S.; et al. *In vitro* Antibacterial Activity of Oxide and Non-Oxide Bioceramics for Arthroplastic Devices: II. In situ Time-Lapse Fourier Transform Infrared Spectroscopy. *Analyst* **2018**, *143*, 2128–2140, https://doi.org/10.1039/C8AN00234G.
53. Mian, H.M.; Lyons, J.G.; Perrin, J.; Froehle, A.W.; Krishnamurthy, A.B. A Review of Current Practices in Periprosthetic Joint Infection Debridement and Revision Arthroplasty. *Arthroplasty* **2022**, *4*, 1–12, https://doi.org/10.1186/s42836-022-00136-5.

54. Pezzotti, G.; Ohigitani, E.; Shin-Ya, M.; Adachi, T.; Marin, E.; Boshetto, F.; Zhu, W.; Mazda, O. Instantaneous "Catch-and-Kill" Inactivation of SARS-CoV-2 by Nitride Ceramics. *Clin. Transl. Med.* **2020**, 1–4, https://doi.org/10.1002/ctm2.212.
55. Pezzotti, G.; Boschetto, F.; Ogitani, E.; Fujita, Y.; Zhu, W.; Marin, E.; McEntire, B.J.; Bal, B. S.; Mazda, O. Silicon Nitride: A Potent Solid-State Bioceramic Inactivator of ssRNA Viruses. *Sci. Rep.* **2021**, *11*, 2977, https://doi.org/10.1038/s41598-021-82608-3.
56. Pezzotti, G.; Boschetto, F.; Ohgitani, E.; Fujita, Y.; Shin-Ya, M.; Adachi, T.; Yamamoto, T.; Kanamura, N.; Marin, E.; Zhu, W.; et al. Mechanisms of Instantaneous Inactivation of SARS-CoV-2 by Silicon Nitride Bioceramic. *Mater. Today Bio* **2021**, 100144, https://doi.org/10.1016/j.mtbio.2021.100144.
57. Douki, T.; Cadet, J. Peroxynitrite Mediated Oxidation of Purine Bases of Nucleosides and Isolated DNA. *Free Radic. Res.* **1996**, *24*, 369–380, https://doi.org/10.3109/10715769609088035.
58. Douki, T.; Martini, R.; Ravanat, J.L.; Turesky, R.J.; Cadet, J. Measurement of 2,6-diamino-4-hydroxy-5-formamidopyrimidine and 8-oxo-7,8-dihydroguanine in Isolated DNA Exposed to Gamma Radiation in Aqueous Solution. *Carcinogenesis* **1997**, *18*, 2385–2391, https://doi.org/10.1093/carcin/18.12.2385.
59. Mlcochova, P.; Kemp, S.A.; Dhar, M.S.; Papa, G.; Meng, B.; Ferreira, I.A.T.M.; Datir, R.; Collier, D.A.; Albecka, A.; Singh, S.; et al. SARS-CoV-2 B.1.617.2 Delta Variant Replication and Immune Evasion. *Nature* **2021**, *599*, 114–119, https://doi.org/10.1038/s41586-021-03944-y.
60. Hristova, K.; Wimley, W.C. A Look at Arginine in Membranes. *J. Membr. Biol.* **2011**, *239*, 49–56, https://doi.org/10.1007/s00232-010-9323-9.
61. Pezzotti, G.; Ohgitani, E.; Fujita, Y.; Imamura, H.; Shin-Ya, M.; Adachi, T.; Yamamoto, T.; Kanamura, N.; Marin, E.; Zhu, W. Raman Fingerprints of the SARS-CoV-2 Delta Variant and Mechanisms of Its Instantaneous Inactivation by Silicon Nitride Bioceramics. *ACS Infect. Dis.* **2022**, https://doi.org/10.1021/acsinfecdis.2c00200.
62. Hernández, B.; Pflüger, F.; Derbel, N.; De Coninck, J.; Ghomi, M. Vibrational Analysis of Amino Acids and Short Peptides in Hydrated Media. VI. Amino Acids with Positively Charged Side Chains: l-Lysine and l-Arginine. *J. Phys. Chem. B* **2010**, *114*, 1077–1088, https://doi.org/10.1021/jp909517r.
63. Kumar, S.; Rai, S.B. Spectroscopic Studies of L-Arginine Molecule. *Indian J. Pure Appl. Phys.* **2010**, *48*, 251–255.
64. Overman, S.A.; Thomas, G.J. Amide Modes of the α-Helix: Raman Spectroscopy of Filamentous Virus *fd* Containing Peptide ^{13}C and ^{2}H Labels in Coat Protein Subunits. *Biochemistry* **1998**, *37*, 5654–5665, https://doi.org/10.1021/bi972339c.
65. Nagy, P.I.; Noszál, B. Theoretical Study of the Tautomeric/Conformational Equilibrium of Aspartic Acid Zwitterions in Aqueous Solution. *J. Phys. Chem. A* **2000**, *104*, 6834–6843, https://doi.org/10.1021/jp0006227.
66. Wyttenbach, T.; Witt, M.; Bowers, M.T. On the Stability of Amino Acid Zwitterions in the Gas Phase: The Influence of Derivatization, Proton Affinity, and Alkali Ion Addition. *J. Am. Chem. Soc.* **2000**, *122*, 3458–3464, https://doi.org/10.1021/ja992546v.
67. Wyttenbach, T.; Witt, M.; Bowers, M.T. On the Question of Salt Bridges of Cationized Amino Acids in the Gas Phase: Glycine and Arginine. *Int. J. Mass Spectrom.* **1999**, *182–183*, 243–252, https://doi.org/10.1016/S1387-3806(98)14255-0.
68. Bhunia, S.; Srivastava, S.K.; Materny, A.; Ojha, A.K. A Vibrational and Conformational Characterization of Arginine at Different pH Values Investigated Using Raman Spectroscopy Combined with DFT Calculations. *J. Raman Spectrosc.* **2016**, *47*, 1073–1085, https://doi.org/10.1002/jrs.4918.
69. Katsaros, I.; Ling, J.; Donzel-Gargand, O.; Akusjärvi, G.; Lundkvist, Å.; Persson, C.; Xia, W.; Engqvist, H. Antiviral Properties of Oxidized Silicon Nitride Against SARS-CoV-2, Uppsala University: Uppsala, Sweden, 2023.

70. Katsaros, I.; Carlsson, A.; Persson, C.; Akusjärvi, G.; Xia, W.; Engqvist, H. Contact Inactivation of Human Adenovirus Type 5 by Silicon Nitride Ceramics, Uppsala University: Uppsala, Sweden, 2023.
71. Pezzotti, G.; Ohgitani, E.; Ikegami, S.; Shin-ya, M.; Adachi, T. Instantaneous Inactivation of Herpes Simplex Virus by Silicon Nitride Bioceramics. *J. Mol. Sci.* **2023**, *24*, 12657, https://doi.org/10.3390/ijms241612657.
72. Olofsson, J.; Grehk, T.; Berlind, T. Evaluation of Silicon Nitride as a Wear Resistant and Resorbable Alternative for Total Hip Joint Replacement. *Biomatter* **2012**, *2*, 94–102.
73. Pettersson, M.; Skjoldebrand, C.; Filho, L.; Engqvist, H.; Persson, C. Morphology and Dissolution Rate of Wear Debris from Silicon Nitride Coatings. *ACS Biomater. Sci. Eng.* **2016**, 1–28.
74. Simpson, S.; Mcminn, C.; Mondfrans, S.M. Van; Hendry, J.; Ronayne, S.; Dewhurst, S.; Feng, C.; Bal, B.S.; Bock, R.M.; McEntire, B.J. A Novel Antipathogenic Agent for Nonwoven Fabric. *Biomed. Mater. Devices* **2022**, https://doi.org/10.1007/s44174-022-00001-8.
75. Talapko, J.; Juzbašić, M.; Matijević, T.; Pustijanac, E.; Bekić, S.; Kotris, I.; Škrlec, I. Candida Albicans-The Virulence Factors and Clinical Manifestations of Infection. *J. Fungi* **2021**, *7*, 1–19, https://doi.org/10.3390/jof7020079.
76. R., N.J.; J., C.S.; Bernhard, H. Candida Albicans Secreted Aspartyl Proteinases in Virulence and Pathogenesis. *Microbiol. Mol. Biol. Rev.* **2003**, *67*, 400–428, https://doi.org/10.1128/mmbr.67.3.400-428.2003.
77. Viljakainen, S.; Nykänen, I.; Ahonen, R.; Komulainen, K.; Suominen, A.L.; Hartikainen, S.; Tiihonen, M. Xerostomia Among Older Home Care Clients. *Community Dent. Oral Epidemiol.* **2016**, *44*, 232–238, https://doi.org/10.1111/cdoe.12210.
78. Shinozaki, S.; Moriyama, M.; Hayashida, J.-N.; Tanaka, A.; Maehara, T.; Ieda, S.; Nakamura, S. Close Association Between Oral Candida Species and Oral Mucosal Disorders in Patients with Xerostomia. *Oral Dis.* **2012**, *18*, 667–672, https://doi.org/10.1111/j.1601-0825.2012.01923.x.
79. Baumgardner, D.J. Oral Fungal Microbiota: To Thrush and Beyond. *J. Patient-Centered Res. Rev.* **2019**, *6*, 252–261, https://doi.org/10.17294/2330-0698.1705.
80. Berdicevsky, I.; Ben-Aryeh, H.; Szargel, R.; Gutman, D. Oral Candida in Asymptomatic Denture Wearers. *Int. J. Oral Surg.* **1980**, *9*, 113–115, https://doi.org/10.1016/S0300-9785(80)80047-0.
81. Beldüz, N.; Kamburoğlu, A.; Yilmaz, Y.; Tosun, I.; Beldüz, M.; Kara, C. Evaluation of *Candida albicans* Biofilm Formation on Various Dental Restorative Material Surfaces. *Niger. J. Clin. Pract.* **2017**, *20*, 355–360, https://doi.org/10.4103/1119-3077.198388.
82. Bürgers, R.; Hahnel, S.; Reichert, T.E.; Rosentritt, M.; Behr, M.; Gerlach, T.; Handel, G.; Gosau, M. Adhesion of *Candida albicans* to Various Dental Implant Surfaces and the Influence of Salivary Pellicle Proteins. *Acta Biomater.* **2010**, *6*, 2307–2313, https://doi.org/10.1016/j.actbio.2009.11.003.
83. Lamfon, H.; Porter, S.R.; McCullough, M.; Pratten, J. Formation of *Candida albicans* Biofilms on Non-Shedding Oral Surfaces. *Eur. J. Oral Sci.* **2003**, *111*, 465–471, https://doi.org/10.1111/j.0909-8836.2003.00084.x.
84. Geurtsen, W. Substances Released from Dental Resin Composites and Glass Ionomer Cements. *Eur. J. Oral Sci.* **1998**, *106*, 687–695, https://doi.org/10.1046/j.0909-8836.1998.eos10602ii04.x.
85. Friedl, K.-H.; Schmalz, G.; Miller, K.-A.; Shams, M. Resin-Modified Glass Ionomer Cements: Fluoride Release and Influence on Streptococcus mutans Growth. *Eur. J. Oral Sci.* **1997**, *105*, 81–85, https://doi.org/10.1111/j.1600-0722.1997.tb00184.x.
86. Park, S.E.; Periathamby, A.R.; Loza, J.C. Effect of Surface-Charged Poly(Methyl Methacrylate) on the Adhesion of *Candida albicans*. *J. Prosthodont.* **2003**, *12*, 249–254, https://doi.org/10.1016/S1059-941X(03)00107-4.
87. Richardson, M.D. Changing Patterns and Trends in Systemic Fungal Infections. *J. Antimicrob. Chemother.* **2005**, *56*, i5–i11, https://doi.org/10.1093/jac/dki218.

88. Pizzo, G.; Giuliana, G.; Milici, M.E.; D'Angelo, M. Effect of Antimicrobial Mouthrinses on the *in vitro* Adhesion of *Candida albicans* to Human Buccal Epithelial Cells. *Clin. Oral Investig.* **2001**, *5*, 172–176, https://doi.org/10.1007/s007840100121.
89. Salim, N.; Moore, C.; Silikas, N.; Satterthwaite, J.; Rautemaa, R. Candidacidal Effect of Fluconazole and Chlorhexidine Released from Acrylic Polymer. *J. Antimicrob. Chemother.* **2013**, *68*, 587–592, https://doi.org/10.1093/jac/dks452.
90. Marra, J.; Paleari, A.G.; Rodriguez, L.S.; Leite, A.R.P.; Pero, A.C.; Compagnoni, M.A. Effect of an Acrylic Resin Combined with an Antimicrobial Polymer on Biofilm Formation. *J. Appl. Oral Sci.* **2012**, *20*, 643–648, https://doi.org/10.1590/S1678-77572012000600009.
91. Metwalli, K.H.; Khan, S.A.; Krom, B.P.; Jabra-Rizk, M.A. *Streptococcus mutans*, *Candida albicans*, and the Human Mouth: A Sticky Situation. *PLoS Pathog.* **2013**, *9*, e1003616, https://doi.org/10.1371/journal.ppat.1003616.
92. Pezzotti, G.; Fujita, Y.; Boschetto, F.; Zhu, W.; Marin, E.; Vandelle, E.; McEntire, B.J.; Bal, B.S.; Giarola, M.; Makimura, K.; et al. Activity and Mechanism of Action of the Bioceramic Silicon Nitride as an Environmentally Friendly Alternative for the Control of the Grapevine Downy Mildew Pathogen *Plasmopara viticola*. *Front. Microbiol.* **2020**, *11*, 1–17, https://doi.org/10.3389/fmicb.2020.610211.
93. Cowen, L.E.; Anderson, J.B.; Kohn, L.M. Evolution of Drug Resistance in *Candida albicans*. *Annu. Rev. Microbiol.* **2002**, *56*, 139–165, https://doi.org/10.1146/annurev.micro.56.012302.160907.
94. Pezzotti, G.; Asai, T.; Adachi, T.; Ogitani, E.; Yamamoto, T.; Kanamura, N.; Boschetto, F.; Zhu, W.; Zanocco, M.; Marin, E.; et al. Antifungal Activity of Polymethyl Methacrylate/Si_3N_4 Composites against *Candida albicans*. *Acta Biomater.* **2021**, *126*, 259–276, https://doi.org/10.1016/j.actbio.2021.03.023.
95. Kojic, E.M.; Darouiche, R.O. Candida Infections of Medical Devices. *Clin. Microbiol. Rev.* **2004**, *17*, 255–267, https://doi.org/10.1128/CMR.17.2.255-267.2004.
96. Armijo, G.; Schlechter, R.; Agurto, M.; Muñoz, D.; Nuñez, C.; Arce-Johnson, P. Grapevine Pathogenic Microorganisms: Understanding Infection Strategies and Host Response Scenarios. *Front. Plant Sci.* **2016**, *7*, 1–18, https://doi.org/10.3389/fpls.2016.00382.
97. Gessler, C.; Pertot, I.; Perazzolli, M. *Plasmopara viticola*: A Review of Knowledge on Down Mildew of Grapevine and Effective Disease Management. *Phytopathol. Mediterr.* **2011**, *50*, 3–44.
98. Muthmann, R.; Nardin, P. The Use of Plant Protection Products in the European Union: Data 1992–2003. *Off. Off. Publ. Eur. Communities, Luxemb.* **2007**.
99. Delmas, C.E.L.; Dussert, Y.; Delière, L.; Couture, C.; Mazet, I.D.; Richart Cervera, S.; Delmotte, F. Soft Selective Sweeps in Fungicide Resistance Evolution: Recurrent Mutations Without Fitness Costs in Grapevine Downy Mildew. *Mol. Ecol.* **2017**, *26*, 1936–1951, https://doi.org/10.1111/mec.14006.
100. Reganold, J.P.; Wachter, J.M. Organic Agriculture in the Twenty-First Century. *Nat. Plants* **2016**, *2*, 15221, https://doi.org/10.1038/nplants.2015.221.
101. Perazzolli, M.; Dagostin, S.; Ferrari, A.; Elad, Y.; Pertot, I. Induction of Systemic Resistance Against *Plasmopara viticola* in Grapevine by Trichoderma Harzianum T39 and Benzothiadiazole. *Biol. Control* **2008**, *47*, 228–234, https://doi.org/10.1016/j.biocontrol.2008.08.008.
102. Wang, B.; Kumar, V.; Olson, A.; Ware, D. Reviving the Transcriptome Studies: An Insight into the Emergence of Single-Molecule Transcriptome Sequencing. *Front. Genet.* **2019**, *10*, 1–11, https://doi.org/10.3389/fgene.2019.00384.
103. German, R.N.; Thompson, C.E.; Benton, T.G. Relationships Among Multiple Aspects of Agriculture's Environmental Impact and Productivity: A Meta-Analysis to Guide Sustainable Agriculture. *Biol. Rev.* **2017**, *92*, 716–738, https://doi.org/10.1111/brv.12251.
104. Burruano, S. The Life-Cycle of *Plasmopara viticola*, Cause of Downy Mildew of Vine. *Mycologist* **2000**, *14*, 179–182, https://doi.org/10.1016/S0269-915X(00)80040-3.

105. Kiefer, B.; Riemann, M.; Büche, C.; Kassemeyer, H.-H.; Nick, P. The Host Guides Morphogenesis and Stomatal Targeting in the Grapevine Pathogen *Plasmopara viticola*. *Planta* **2002**, *215*, 387–393, https://doi.org/10.1007/s00425-002-0760-2.
106. Caffi, T.; Rossi, V.; Bugiani, R. Evaluation of a Warning System for Controlling Primary Infections of Grapevine Downy Mildew. *Plant Dis.* **2010**, *94*, 709–716, https://doi.org/10.1094/PDIS-94-6-0709.
107. MacMillan, A. The Entry of Ammonia into Fungal Cells. *J. Exp. Bot.* **1956**, *7*, 113–126, https://doi.org/10.1093/jxb/7.1.113.
108. Farina, G.; Barbieri, N.; Bassi, M.; Betto, E. *Plasmopara Viticola* in Leaves of Vitis Vinifera. An Electron Microscopic Study. *Riv. di Patol. Veg.* **1976**, *12*, 43–51.
109. Agrios, G.N. *Plant Pathology*; Academic Press, Inc: San Diego, CA, USA, 1997; ISBN 0080473784.
110. Riley, F.L. Silicon Nitride and Related Materials. *J. Am. Ceram. Soc.* **2000**, *83*, 245–265.
111. Fauteux, F.; Chain, F.; Belzile, F.; Menzies, J.G.; Bélanger, R.R. The Protective Role of Silicon in the Arabidopsis–Powdery Mildew Pathosystem. *Proc. Natl. Acad. Sci.* **2006**, *103*, 17554–17559, https://doi.org/10.1073/pnas.0606330103.
112. Fauteux, F.; Rémus-Borel, W.; Menzies, J.G.; Bélanger, R.R. Silicon and Plant Disease Resistance Against Pathogenic Fungi. *FEMS Microbiol. Lett.* **2005**, *249*, 1–6, https://doi.org/10.1016/j.femsle.2005.06.034.
113. Glass, A.D.M.; Britto, D.T.; Kaiser, B.N.; Kinghorn, J.R.; Kronzucker, H.J.; Kumar, A.; Okamoto, M.; Rawat, S.; Siddiqi, M.Y.; Unkles, S.E.; et al. The Regulation of Nitrate and Ammonium Transport Systems in Plants. *J. Exp. Bot.* **2002**, *53*, 855–864, https://doi.org/10.1093/jexbot/53.370.855.
114. Martins, V.; Cunha, A.; Geros, H.; Hanana, M.; Blumwald, E. Mineral Compounds in the Grape Berry. In *The Biochemistry of Grape Berry*; Geros, H., Chaves, M., Delrot, S., Eds.; Bentham Science Publishers: Sharjah, U.A.E., 2012; pp. 23–43.

Silicon Nitride Biocomposites

Elia Marin, Giuseppe Pezzotti, and Bryan J. McEntire

1 Introduction

Beyond its widespread application as a bulk ceramic in such diverse markets as aerospace, industrial, and biomedical, silicon nitride (Si_3N_4) has also garnered attention as an additive for producing particulate-based composites in various matrices [1]. Si_3N_4's relatively favorable thermal stability allows for successful combination with various low-melting-point metals, particularly alloys of magnesium [2] and aluminum [3], and even some higher-melting metals such as copper, albeit requiring advanced powder metallurgy techniques [4]. However, as the matrix's melting point increases, the oxidation of Si_3N_4 particles accelerates, resulting in nitrogen release. This released nitrogen readily bonds with metals exhibiting higher affinity, such as iron [5], chromium [6], and titanium [7]. Consequently, Si_3N_4 has not been extensively used as an additive to metal-matrix composites. Si_3N_4 has also found limited use in ceramic matrix composites due primarily to its high-temperature oxidation and thermal degradation. Accordingly, publications on ceramic matrix composites containing Si_3N_4 have been relatively rare, with notable exceptions that include carbon fiber nanotube [8], graphene [9], silicon carbide [10], and boron nitride [11].

Fabrication of fully dense metal- or ceramic-matrix composites using Si_3N_4 requires careful control over chemical composition and necessitates the use of

E. Marin
Ceramic Physics Laboratory, Kyoto Institute of Technology, Kyoto, Japan

G. Pezzotti
Ceramic Physics Laboratory, Kyoto Institute of Technology, Sakyo-ku, Matsugasaki, Kyoto, Japan

B. J. McEntire (✉)
SINTX Technologies, Inc., Salt Lake City, UT, USA
e-mail: bmcentire@sintx.com

B. S. Bal et al. (eds.), *Silicon Nitride Bioceramics*,
https://doi.org/10.1007/978-3-031-67047-3_7

high partial gas pressures of nitrogen (N_2) [12], typically in a hot-isostatic press (HIP). Mitigating the unavoidable nitrogen loss during high-temperature composite processing has been exceptionally challenging. It requires the adoption of more sophisticated and experimental production techniques such as spark plasma sintering [13] or laser cladding [14]. However, these rapid and intricate processing techniques pose challenges in adapting them to produce Si_3N_4 composite materials without compromising stoichiometry, homogeneity, and density.

Si_3N_4 has demonstrated remarkable versatility as an additive for a wide range of polymeric matrices, encompassing both thermoplastics and rubbers. Its compatibility extends to commonly used polymers like polyethylene [15], polyaryletherketones (e.g., polyetheretherketone (PEEK) [16] or poly(etherketoneketone) (PEKK) [17]), polymethylmethacrylate (PMMA) [18], polypropylenes [19], nylon [20], and polystyrene [21]. Moreover, Si_3N_4 has been successfully incorporated into styrene butadiene [22], acrylic [23], and silicone rubbers [24]. The wide range of available Si_3N_4 particle shapes and sizes provides considerable versatility to polymer-based composite manufacturing, including 3D printing technologies [25]. Indeed, this technique has been successfully applied to polypropylene, polyethylene [25], and PMMA [26]. While 3D printing incurs higher costs compared to conventional polymer production methods like injection molding, the value-added benefits of incorporating Si_3N_4 powders into these relatively inexpensive materials outweigh the initial investment.

Bulk silicon nitride is a structural ceramic with exceptional mechanical, wear, thermal, electrical, and biological properties. It has a compressive strength of about 4 GPa, flexural strength of approximately 800–1000 MPa, bulk modulus of 240–250 GPa, fracture toughness between 4.4 and 15 $MPa{\cdot}m^{1/2}$, hardness of 15 GPa, thermal conductivity of 30–50 W/m·K [27], and resistivity of 10^9–10^{11} $\Omega{\cdot}m$ [28]. Si_3N_4 particles are incorporated into metal, ceramic, or polymer matrices to enhance specific properties of the composite. For instance, the addition of a small fraction of hard reinforcing particles can modestly improve the compressive strength, elastic modulus, and wear performance of polymers [29]; but it may also negatively impact tensile and bending strength [30] as well as fatigue resistance [31]. This is particularly true if a strong interfacial bond between the two phases is not well-established. While reinforcement particles of a few microns or less have been shown to enhance the initial tensile strength of polymeric matrices, adding ceramic particulates generally leads to a decline in their long-term performance, especially under complex stress fields or fatigue loading. As an example, the effect of Si_3N_4 particle loading on the mechanical properties of an advanced PMMA/Si_3N_4 biocomposite is provided in Fig. 1. Note that the maximum tensile strength, elastic modulus, and work of fracture in Fig. 1a–c, respectively, increase up to about 10 ~ 15 vol.% Si_3N_4, and precipitously decline thereafter.

Only hardness and wear resistance consistently benefit from any size fraction of hard Si_3N_4 particles, particularly under erosive and abrasive wear conditions [32–34]. Like mechanical properties, wear resistance is optimized with micron-sized and nanometric Si_3N_4 particles. As shown in Fig. 2, Shi et al. found that even at low fractions, Si_3N_4 nanoparticles created strong interfacial adhesion with epoxy-based

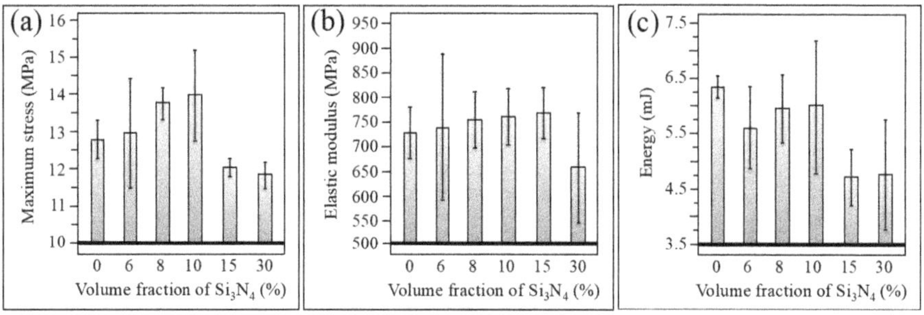

Fig. 1 The effect of Si_3N_4 particle concentration on the mechanical properties of PMMA/Si_3N_4 composites. (**a**) Maximum tensile stress; (**b**) elastic modulus; and (**c**) work of fracture. (Reprinted with permission from Ref. [18], supporting information, Fig. S-1)

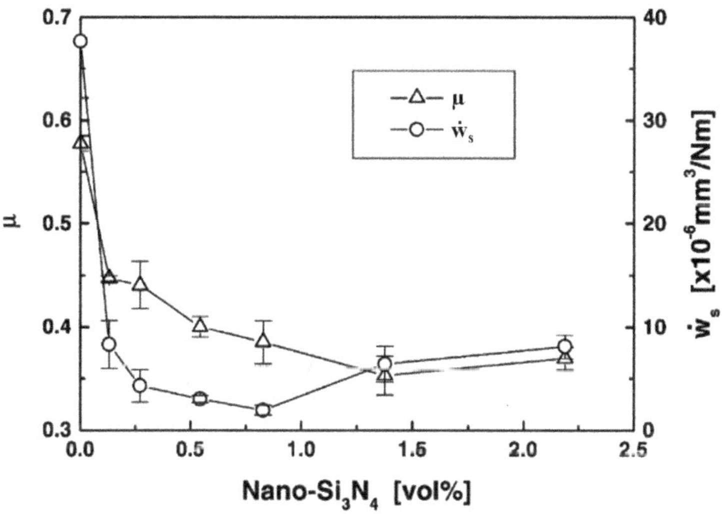

Fig. 2 Frictional coefficient, μ, and specific wear rate, w_s, of nano- Si_3N_4/epoxy composites as a function of filler content. (Reprinted with permission [35])

resins, leading to lower wear, coefficient of friction, and resistance to thermal distortion [35]. Beyond serving as matrices for conventional reinforcements, polymers can also be infiltrated into porous Si_3N_4 to enhance mechanical strength, particularly in biomedical applications where successful osteointegration relies heavily on optimized pore size and distribution [36].

Although increasing the mechanical strength of composites with Si_3N_4 particles can have drawbacks and limitations, other technological properties can benefit, including thermal [15, 37] and electrical conductivity [9, 19, 21]. Silicon nitride has been demonstrably effective in enhancing the thermal conductivity of diverse polymers like polyethylene [15], epoxy resins [38], and acrylates [39], showcasing its potential for industrial applications in electromagnetic absorption materials

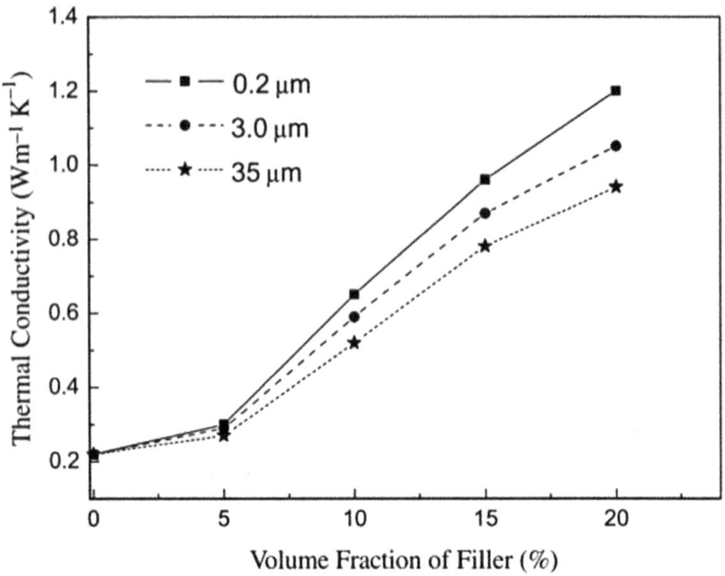

Fig. 3 Thermal conductivity of fiber-reinforced polyethylene composite with added Si_3N_4 particles as a function of volume fraction of filler and average particle size. (Reprinted with permission [15])

[40]. An example of how Si_3N_4 particles affect the thermal conductivity of polyethylene is provided in Fig. 3. Likewise, Si_3N_4 particles have proven to be successful in augmenting the dielectric properties of polymers [41], particularly in polyimides [42], polystyrene [21], and polytetrafluoroethylene (PTFE) [43]. The concentration of Si_3N_4 particles in these various applications is carefully tailored to balance the desired additive properties without compromising the material's inherent characteristics. This chapter delves into the incorporation of Si_3N_4 particles into various materials with an emphasis on biomedical applications, inevitably addressing the impact of these additions on the matrix's physical, mechanical, chemical, thermal, and biological properties.

2 Silicon Nitride Metal Matrix Composites

2.1 *Nonbiological Applications of Si_3N_4 Metal Matrix Composites*

While a limited number of Si_3N_4-based metal matrix composites are commercially available or under active research, their dèvelopment faces a significant challenge. It is difficult to achieve homogeneous mixing of the two materials at sufficiently low temperatures to prevent undesirable Si_3N_4 oxidation. Magnesium and aluminum composites are exceptions due to their low melting points (650 ~ 660 °C). One of the most common techniques for fabricating such composites involves powder metallurgy with conventional heating under vacuum [44, 45] or in an argon atmosphere [46]. However, other techniques have also been explored, including disintegrated melt deposition [47], ultrasonic-assisted stir-casting [48], and microwave sintering [49]. Most techniques employed in the fabrication of magnesium matrix composites are readily applicable to aluminum, achieving comparable results. In both instances, the introduction of Si_3N_4 reinforcements demonstrably enhances hardness, compressive strength, and Young's modulus [50]. However, the two metallic matrices exhibit seemingly disparate affinities for Si_3N_4. Notably, magnesium matrices accommodate relatively low Si_3N_4 content without incurring porosity concerns; but poorly controlled temperature and atmosphere during processing can lead to the formation of undesirable phases, including MgO, Mg_2Si, Mg_3N_2, and elemental Si [49].

Several stainless-steel matrix Si_3N_4 composites have been reported, primarily fabricated via spark plasma sintering [51], stir-casting [52], powder injection molding [53], or selective laser melting [54]. During high-temperature sintering, Si_3N_4 undergoes solid-solid reactions with the metal matrix, leading to diffusive intermixing of silicon with the constituent metallic elements, including some intermetallic compounds: Cr_3Si, Cr_5Si_3, CrSi, and $CrSi_2$ [55, 56]. Chromium nitride (Cr_2N) is formed above about 1300 °C in a two-step reaction that releases gaseous nitrogen [56]:

$$Si_3N_4(s) + 9Cr\,(s) \rightarrow 3Cr_3Si\,(s) + 2\,N_2(g); \Delta G^\circ = 405.0\text{–}0.3(T)\ \mathrm{kJ/mol \cdot K} \quad (1)$$

$$4Cr(s) + N_2(g) \rightarrow 2Cr_2N(s); \Delta G^\circ = -99.20 + 0.05\ \mathrm{kJ/mol \cdot K} \quad (2)$$

The presence of chromium nitride increases both the hardness and Young's modulus of the composite [57]. Silicon also reacts with iron to form an intermetallic phase, leading to localized improvements in hardness, strength, and Young's modulus. However, for optimal property enhancement, direct silicon addition is always preferred over the use of Si_3N_4 as an additive [58].

Given the inherent challenges associated with steel matrix composites, the relative scarcity of titanium-matrix Si_3N_4 composites reported in the literature is unsurprising. The synthesis of these composites occurs at temperatures over 1300 °C and frequently leads to the formation of intermetallic compounds such as Ti_5Si_3 or Ti_3Si [59], or the formation of titanium nitride (TiN) [60].

$$Si_3N_4(s) + 4Ti(s) \rightarrow 4TiN(s) + 3Si(s); \Delta G^\circ = -613 + .041(T)\ \mathrm{kJ/mol \cdot K} \quad (3)$$

Similar to Cr_2N, the presence of TiN precipitates within the composite's microstructure results in increased hardness and elastic modulus. Alternatively, due to titanium's strong affinity for nitrogen, mixed powders can be nitrided in a nitrogen atmosphere to produce TiN/Si_3N_4 composites [61].

Analogous to steel and titanium, copper-based Si_3N_4 composites can be fabricated through various powder metallurgy techniques, with a similar impact on their chemical composition [4, 62]. Alternatively, electrodeposition presents a viable route for synthesizing Cu/Si_3N_4 composites. This technique retains the inherent properties of the bulk material while enabling the incorporation of Si_3N_4. This targeted enhancement leads to a significant improvement in tribological performance, characterized by increased surface hardness and reduced friction coefficient [63, 64]. Silicon nitride composite coatings can also be electrodeposited onto different substrates using zinc [65], cobalt [66], and nickel [67] to improve their tribological performance and corrosion resistance.

Of the aforementioned matrix composites, the rationale for including Si_3N_4 revolved around enhancing specific physical or mechanical properties. Si_3N_4 primarily increased hardness, reduced friction, and improved wear and corrosion resistance. It also increased strength through particulate dispersion or diffusion of nitrogen into the matrix. With electrodeposited composites, the dispersed phase not only facilitated dispersion strengthening but also served as a nucleation site for finer grains, leading to grain refinement strengthening and preferential grain orientation [63]. This microstructural refinement further improved corrosion resistance by reducing the size of inter-columnar pores and lengthening the diffusion path for electrolytes reaching the substrate.

2.2 Biological Applications of Si_3N_4 Metal Matrix Composites

Due to the difficulty in combining Si_3N_4 with metals, few studies have explored the bioactivity of Si_3N_4 metal matrix composites. One in situ formed exception was an examination of the in vitro bioactivity of Mg/Si_3N_4 compared to Mg/Al_2O_3 composites fabricated by microwave sintering [49]. After fabricating components, the researchers immersed both composites in simulated body fluid (SBF) for 14 days, it was found that the Mg/Si_3N_4 composite generated significantly greater amounts of hydroxyapatite (HAp) on its surfaces than on the Mg/Al_2O_3 composite. Additionally, the Ca/P ratio of the HAp crystals expressed by the Mg/Si_3N_4 composites was essentially equivalent to stoichiometric HAp, whereas the Al_2O_3 composites were deficient in calcium.

The other significant exception includes the use of laser cladding to prepare Si_3N_4 coatings on metal surfaces. Laser cladding uses a high-energy laser beam to partially melt and embed Si_3N_4 particles into the surface of the metallic substrate [68]. The method is not only useful for metals [69], but it also can be applied to other ceramic materials [70], and even polymers [71]. The general methodology for preparing these types of composites is shown in Fig. 4.

The process first involves preparing an aqueous or organic solvent suspension containing a given concentration of Si_3N_4 powder and typically a binder. The suspension is applied to the substrate either as a spray, dip, or paint coating. After curing, a laser beam is then directed at the powder bed using specific energy and

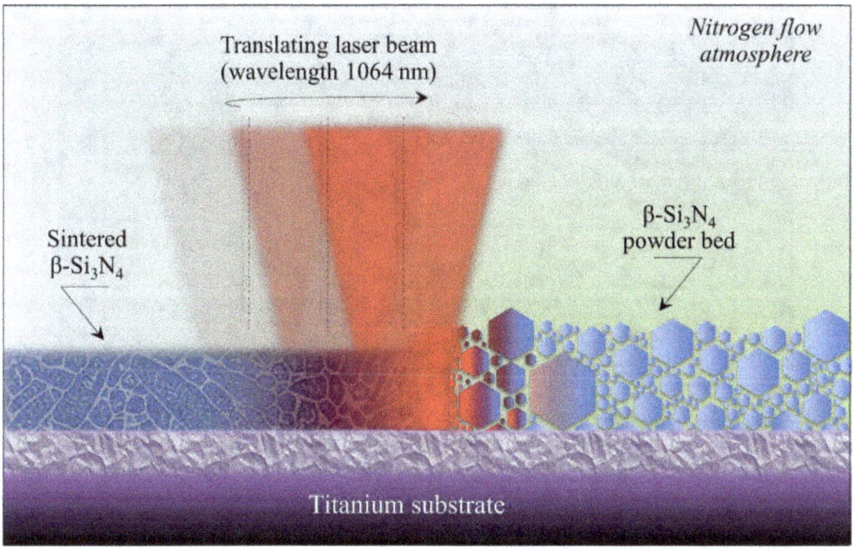

Fig. 4 Schematic diagram of the 3D-additive laser-cladding procedure for producing a dense Si_3N_4 ceramic coating on a titanium substrate. (Used with permission [69])

raster parameters to embed the Si_3N_4 particles into the substrate's surface. During laser cladding, the particulate temperature surpasses the decomposition point of Si_3N_4 (1850 °C) within milliseconds, triggering the dissociation of silicon and nitrogen. While rapid processing with coarse ceramic powder may not fully expel all nitrogen, it leads to a sub-stoichiometric Si_3N_4 coating characterized by a nanocrystalline silicon matrix embedding Si_xN_y crystals. The residual nitrogen content can be effectively modulated by adjusting several parameters including using a finer powder mesh for faster, more uniform melting, tailoring the beam power density, or transitioning the processing atmosphere from air to argon, vacuum, or nitrogen.

Xu et al. were the first to demonstrate the feasibility of this technology [14]. They prepared pure Si_3N_4 and Si_3N_4/tricalcium phosphate (TCP) mixtures in 0.5% polyvinyl alcohol aqueous solutions and spray-coated Ti6Al4V substrates to a thickness of 100 μm. Using three different lasers (i.e., two industrial, and one semiconductor) at different power, frequency, and scan velocities, they systematically explored conditions that provided coherent Si_3N_4 coatings. They found that all three laser types were capable of producing acceptable adherence, but high power and slow raster speeds decomposed the Si_3N_4 and created microcracks in the cladding. Using their best conditions, the cladding effectively doubled the Vickers hardness of the substrate. They also conducted bioactivity testing by immersing selected samples in SBF solutions for between 21 and 28 days. Si_3N_4 claddings significantly increased apatite formation compared to uncoated controls, with Si_3N_4/TCP coatings

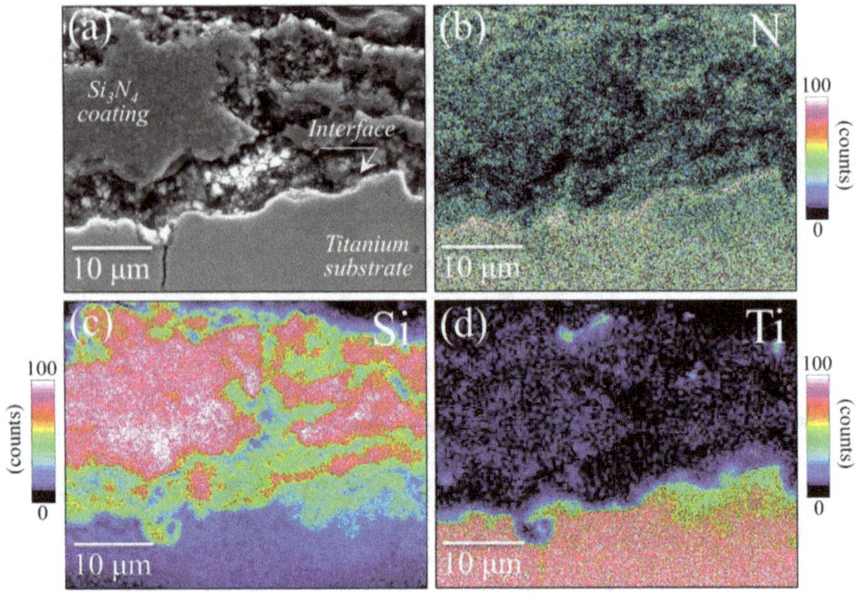

Fig. 5 (**a**) Scanning electron micrograph (SEM) of a cross-sectioned Si_3N_4 cladding on a titanium substrate. The interface exhibits a wavy interlocking morphology between the two materials. (**b–d**) Energy-dispersive X-ray spectroscopy (EDS) maps reveal the distribution of nitrogen (N), silicon (Si), and titanium (Ti) within the cladding and substrate. (Used with permission [69])

demonstrating higher apatite deposition. These findings suggest substantial potential for improved biocompatibility and integration with bone tissue.

A more extensive laser cladding study was conducted by Zanocco et al. [69]. In their approach, after laser cladding Ti6Al4V samples with Si_3N_4, they examined the osteopromotive and bacteriolytic capabilities of the laser-cladded samples in comparison to controls. Samples were prepared according to the methods shown in Fig. 4, but unlike the work of Xu et al. [14], the cladding thickness was only 15 ± 5 µm. Multiple thin spray coatings were sequentially followed by laser embedding steps. A cross-sectional analysis of the cladding is given in Fig. 5.

The dominant feature in the SEM micrograph was the wavy ceramic/metal interlocking interface. In the laser-cladded area, regions rich in silicon were observed, but there was no evidence of diffusional transport of titanium into the laser-cladded region. The low detectability of nitrogen coupled with the rough surface morphology prevented the quantification of diffused nitrogen into the substrate or retained by the coating. Nevertheless, it was evident that the treatment did not result in the formation of titanium nitride (TiN). The claddings were subsequently subjected to a standard scratch test (ASTM C1624) at a load of up to 25 N. High adhesion strength was confirmed by the absence of microcracks or spalling around the scratch in the cladding. This finding exceeds the 20 N scratch resistance requirement stipulated by the standard and demonstrates their suitability for

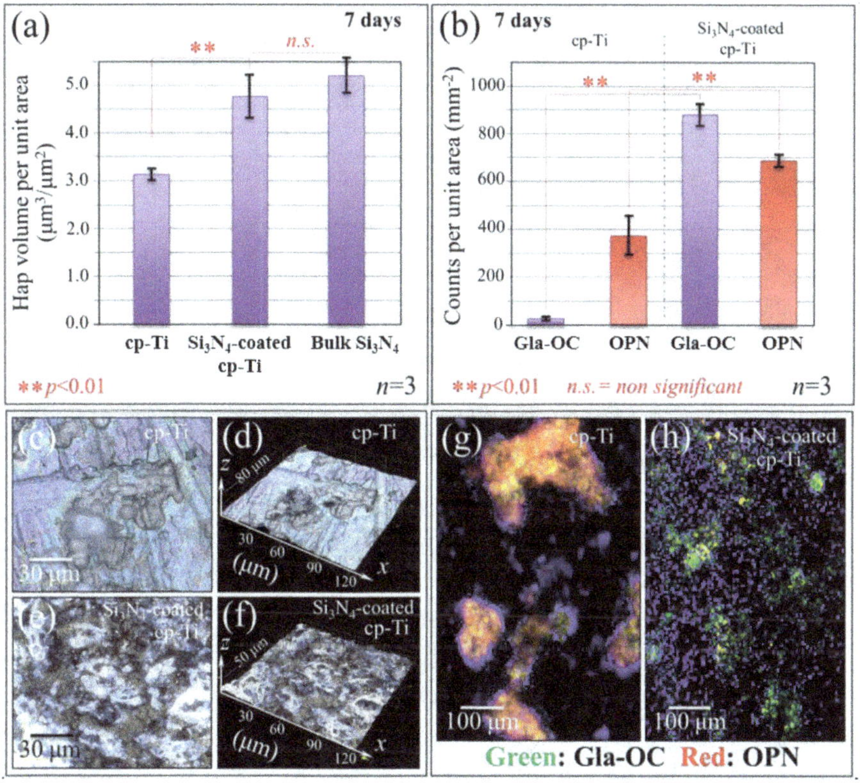

Fig. 6 Graphs of (**a**) bone tissue formation quantified by hydroxyapatite (HAp) deposition per unit area, demonstrating significantly higher levels on Si_3N_4 laser-cladded substrates compared to uncoated controls. (**b**) Compositional analysis of bone tissue revealed elevated levels of Gla-osteocalcin and osteopontin for Si_3N_4 laser-cladded substrates, indicating enhanced mineralization and matrix maturation. Laser microscopy images: (**c**, **d**) Uncoated Ti6Al4V shows sparse bone islands in planar and 3D views, while (**e**, **f**) Si_3N_4 laser-cladded substrates exhibit a near-homogeneous, ~50 μm thick layer of bone apatite. Fluorescence microscopy reveals: (**g**) uncoated Ti6Al4V with limited osteogenic markers, and (**h**) Si_3N_4 laser-cladded Ti6Al4V with abundant Gla-osteocalcin (green) and osteopontin (red) expression, confirming bone formation. (Used with permission [69])

biomedical applications. Next the laser-cladded samples were tested for their ability to upregulate osteogenic activity. Figure 6a quantifies the impact of the Si_3N_4 cladding on bone tissue formation by SaOS-2 cells after 7 days of incubation.

Laser microscopy measurements revealed a significant increase in bone tissue volume per unit area on Si_3N_4 laser-cladded Ti6Al4V substrates compared to both uncoated controls and bulk Si_3N_4. This is visualized in Fig. 6c, d, which show planar and 3D laser microscopy images of the uncoated alloy. Sparse bone islands are evident, indicating limited bioactivity. In contrast, Fig. 6e, f present strikingly different pictures of the Si_3N_4 laser-cladded alloy. Planar and 3D views depict a

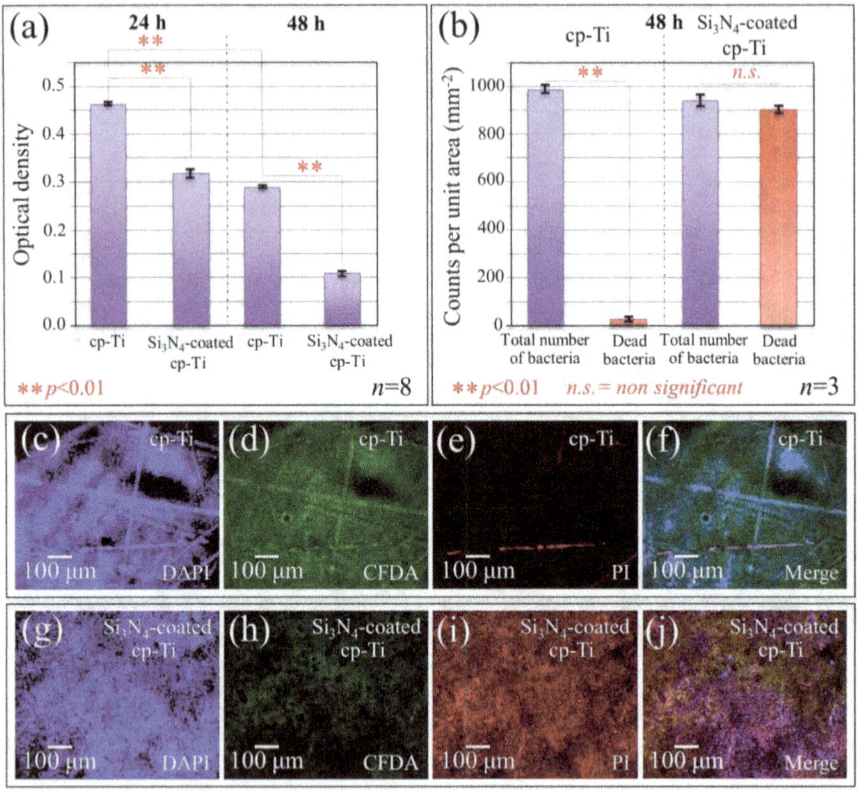

Fig. 7 Graphs of (**a**) Cell viability assessed by MTT assay reveals significantly reduced metabolic activity of *S. epidermidis* on Si_3N_4 laser-cladded substrates compared to uncoated controls at both 24 and 48 h ($p < 0.05$). (**b**) Bacterial viability by propidium iodide (PI) staining demonstrates an increase in dead/live cell ratio on Si_3N_4 laser-cladded substrates at 48 h, indicating extensive bacterial lysis. Fluorescence microscopy analysis (48 h): (**c–f**) Uncoated substrate: 4',6-diamidino-2-phenylindole (DAPI) (blue) stains all nuclei, 6-Carboxyfluorescein diacetate (CFDA) (green) labels viable bacteria, and PI (red) marks dead cells. Sparse green fluorescence and dominant blue fluorescence indicate limited bacterial adhesion and viability. (**g–j**) Si_3N_4 laser-cladded substrate: Overwhelming red fluorescence throughout the image signifies extensive bacterial lysis due to the cladding's antibacterial properties. Minimal green fluorescence highlights the reduced bacterial adhesion compared to the uncoated control. (Used with permission [69])

nearly homogeneous layer of bony apatite covering the surface, reaching a maximum height of approximately 50 µm. This robust bone formation is further corroborated by compositional analysis in Fig. 6b, confirming elevated levels of Gla-osteocalcin and osteopontin, which are key markers of bone mineralization and matrix maturation.

Finally, these laser-cladded samples were then evaluated for their antimicrobial effectiveness. Figure 7a, b assess the metabolic activity and viability of Gram-positive *S. epidermidis* after exposure to uncoated and Si_3N_4 laser-cladded Ti6Al4V substrates at 24 and 48 h.

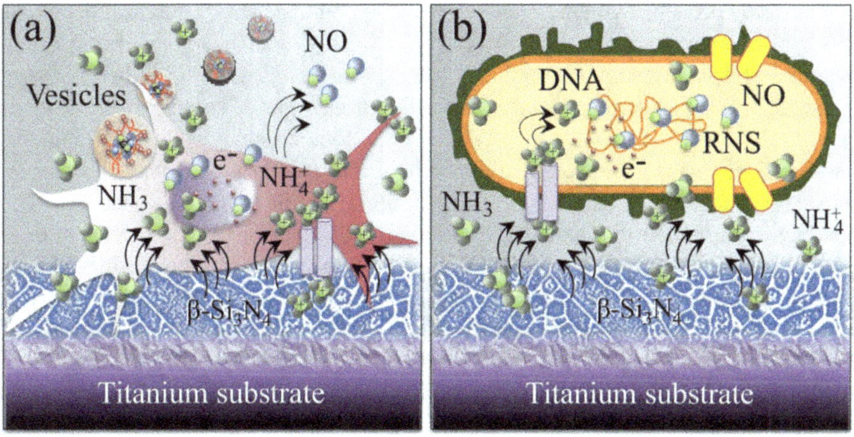

Fig. 8 (**a**) Osteoblast–Si_3N_4 interaction: Eukaryotic osteoblasts utilize ammonium (NH_4^+) ions as a nutrient source at the biomolecular interface with Si_3N_4. These cells possess the enzymatic capability to dissociate NH_3 from NH_4^+, ultimately producing nitric oxide (NO) for signaling osteoblastogenesis. (**b**) Bacteria–Si_3N_4 interaction: In contrast to osteoblasts, ammonia (NH_3) is lethal to prokaryotic cells due to its direct attack on DNA and RNA structures. This effect highlights the biocompatibility and selectivity of Si_3N_4 for eukaryotic cell interactions. (Used with permission [69])

Optical density measurements in Fig. 7a indicate significantly higher bacterial populations on the uncoated substrates compared to the Si_3N_4 laser-cladded samples at both time points. This suggests reduced bacterial colonization and potentially enhanced antimicrobial properties when utilizing Si_3N_4 laser claddings. Figure 7b strikingly reveals the contrasting antibacterial effects of uncoated and Si_3N_4 laser-cladded Ti6Al4V substrates using fluorescence counting methods. At 48 h, the uncoated substrate showed a negligible fraction of dead bacteria, evidenced by faint red fluorescence from propidium iodide (PI), a marker for cell death. In stark contrast, the Si_3N_4 laser-cladded substrate displayed almost complete bacterial lysis, indicated by the overwhelming dominance of red fluorescence, signifying a powerful antibacterial effect. While the Si_3N_4 cladding demonstrably outperformed the uncoated Ti-alloy in terms of both adhesion and lysing of *S. epidermidis*, it is noteworthy that the uncoated control also exhibited a measurable, albeit modest, antibacterial effect. This observed reduction in bacterial adhesion can be attributed to the inherent passive TiO_2 layer that naturally forms on the surface of titanium alloys [72, 73].

The osteopromotive and antibacterial mechanisms of the Si_3N_4 laser-cladded samples are the same as those described earlier in this book (cf., Chaps. 5 and 6). A schematic representation of these mechanisms is shown in Fig. 8a, b.

Briefly, the Si_3N_4 cladding elutes ammonium (NH_4^+) and a small fraction of ammonia (NH_3) at physiologic pH due to surface hydrolysis and homolytic cleavage of Si-N bonds. NH_4^+ ions enter the cytoplasm of cells through specific transporters and at regulated concentrations [74, 75]. Ammonium serves as a crucial nutrient for synthesizing proteins, enzymes, and genetic compounds, thereby supporting cellular differentiation and proliferation [76]. In conjunction with the leaching of orthosilicic acid ($Si(OH)_4$) and related compounds, NH_4^+ promotes osteoblast synthesis of bone tissue [77, 78]. Furthermore, it stimulates the production of collagen type 1, a key structural protein in bone [78, 79]. While highly toxic to many cell types, including bacteria, NH_3 plays a surprising role in supporting bone growth. In healthy eukaryotic cells like human osteoblasts, mitochondria handle unpaired electrons. This sparks a chain reaction where NH_3 is transformed into nitric oxide (NO) through several steps [80–84]. When delivered gradually, NO acts as a powerful signaling molecule, prompting human osteoblasts to mature and build new bone tissue. In contrast to its beneficial role in mammalian cells, highly volatile ammonia (NH_3) freely permeates bacterial cell membranes. Once inside, NH_3 directly targets the stability of DNA and RNA structures, thereby exerting a lethal effect [85]. This slow but steady release of NO helps combat bacterial colonies, even mutant ones, as they lack the necessary structure to assimilate it.

3 Silicon Nitride Bioglass Matrix Composites

The first composites to incorporate Si_3N_4 were glass-based. Si_3N_4 is itself a glass-based composite because of its grain boundary phase that aids densification. A strong argument can be made that the prevalence of secondary compounds like MgO, Al_2O_3, and Y_2O_3 in many commercially available Si_3N_4 formulations qualifies them as ceramic composites [86]. Despite significant efforts to densify Si_3N_4 without additives [87], achieving full density necessitates a matrix glass. However, the amount of the matrix is kept minimal (typically less than 10 wt%) to avoid compromising the material's mechanical properties [88, 89]. Commercial-grade Si_3N_4 ceramics often exhibit intergranular phases, notably SiYAlON oxynitrides formed through interactions with sintering aids like Al_2O_3 and Y_2O_3 [90]. Due to SiYAlON's superior bioactivity relative to stoichiometric Si_3N_4, biomedical-grade Si_3N_4 formulations have been explored by increasing its presence, either integrally or through final SiYAlON glazing [91]. It was this simple understanding that led several researchers to examine higher glass amounts as matrices for Si_3N_4 composites.

Pioneering research by Amaral et al. [92, 93] demonstrated the ability of Si_3N_4-bioglass mixtures to significantly enhance osteoblast production in vitro. They fabricated a composite containing 70 wt% Si_3N_4 and 30 wt% bioglass (45.0% SiO_2, 24.5% Na_2O, 24.5% CaO, and 6.0% P_2O_5) via raw powder mixing followed by pressureless sintering or hot-pressing at 1300–1500 °C for 40 min and 30 MPa. This process yielded a composite with impressive properties: ~97.9% density, 4.4 $MPa{\cdot}m^{1/2}$ fracture toughness, and 383 MPa bending strength. In vitro tests with MG63 and human bone marrow osteoblast-like cells revealed significantly higher proliferation rates compared to controls. While their conclusion that Si_3N_4 was bioinert relative to the amorphous bioglass proved inaccurate in later studies, the combined improvement in mechanical properties and bioactivity compared to conventionally used bioinert alumina was remarkable. Amaral et al. aptly highlighted the composite's potential as "a very promising candidate for high-stress medical applications" [93].

Among bioactive ceramic reinforcements for biomedical applications, 45S5 Bioglass® enjoys significant recognition for its effectiveness. Developed by Prof. Larry Hench, Bioglass® demonstrates exceptional aptitude in promoting osteointegration while simultaneously suppressing bacterial adhesion and proliferation. However, its inherent mechanical limitations restrict its use to nonload-bearing applications. To address this drawback, researchers have investigated combining Bioglass with various polymers, including polyethylene [94], polylactic acid [95], PEEK [96], and PMMA [97] to enhance its mechanical properties and broaden its clinical utility.

Marin and coworkers followed a different route in studying mixtures of Bioglass® and Si_3N_4. In the first of two studies [98], they mixed 5 and 10 mol.% Si_3N_4 and Bioglass® powders, followed by melting, crushing, and comminuting the experimental materials back to glass frits. Following in vitro testing with SaOS-2 human osteosarcoma cells, the osseoinductive and osseoconductive properties of the two compositions were assessed and compared to standard 45S5 Bioglass® and monolithic Si_3N_4. This in vitro biological testing revealed superior bone tissue formation on Si_3N_4-doped materials compared to both standard 45S5 Bioglass® and a titanium alloy. Notably, the bone tissue exhibited lower mineral-to-matrix ratios compared to standard Bioglass®, suggesting a focus on early matrix production. Additionally, the experimental compositions enhanced osteocalcin expression, likely due to environmental alkalinization, and they upregulated overall osteoblastic activity in the SaOS-2 cells. Importantly, this work also demonstrated that by fine-tuning the ratio of Bioglass® to Si_3N_4, it was possible to precisely control both the mineralization rate and osteoconduction of bone tissue surrounding the implants, while maintaining optimal bone volume.

In the second study [99], the researchers employed a novel sandblasting technique to create a controlled, micron-scale texture on dense monolithic biomedical silicon nitride (β-Si_3N_4) disks. The resulting surface featured a matrix of uniform, discrete square trenches reaching a maximum depth of about 250 μm. The trench patterning process is depicted in Fig. 9a, b. After forming trenches by sandblasting (α-SiC, ~40 μm) through stainless steel mesh screens (250 ~ 300 μm), the cavities were filled with an aqueous sol of 45S5 Bioglass® in a pullulan solution (10% in H_2O), and the samples dried at 60 °C for 30 min, (Fig. 9c).

An osteoconductive evaluation was subsequently performed using precultured SaOS-2 osteosarcoma cells seeded onto the disk samples (5×10^5 cells/mL). This allowed for the assessment of cellular activity and bone tissue formation on the textured Si_3N_4 and Bioglass® filled surfaces in comparison to established reference materials. Standard titanium alloy and untextured Si_3N_4 discs of the same dimensions served as controls. All samples were concurrently evaluated for osteoconductivity after an incubation period of 168 h at 37 °C. The results of this assessment are provided in Fig. 10. While monolithic Si_3N_4 is recognized as a

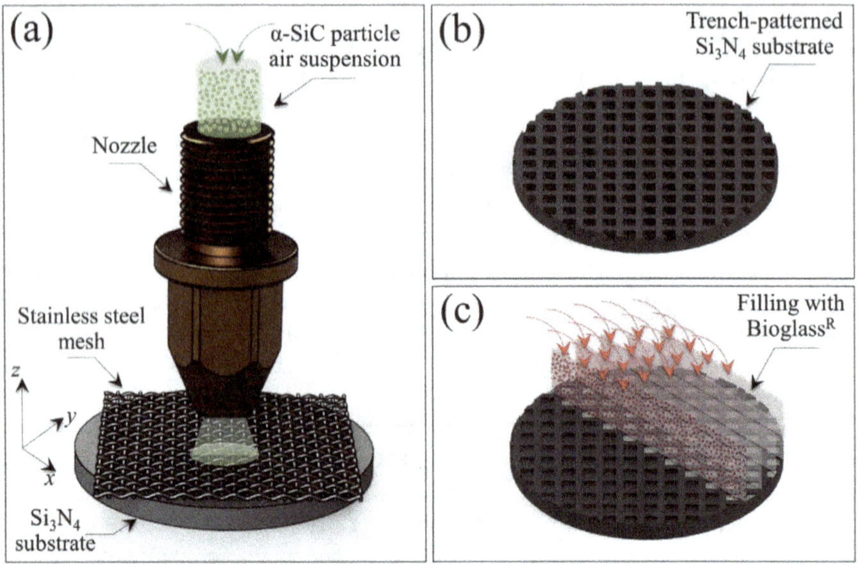

Fig. 9 Schematic diagrams showing: (**a**) the controlled texturing process—the nozzle was positioned perpendicular to the Si_3N_4 surface and traversed in the x-y plane while maintaining a fixed z-distance; (**b**) the textured substrate—following sandblasting, the stainless-steel mesh was removed, revealing the resulting surface pattern of square trenches on the Si_3N_4 substrate; and (**c**) bioglass filling—This final step involved filling the trenches with 45S5 Bioglass® powder. (Used with permission [99])

bioactive ceramic, this study revealed that its osteogenic potential was significantly amplified when combined with Bioglass® (cf., Fig. 10). This synergistic effect suggests that Bioglass®-coated Si_3N_4 offers promising potential for promoting earlier and more robust osseous integration in arthrodesis devices across diverse anatomical locations, including spinal, extremity, cranial, maxillofacial, and endosseous applications. Notably, the two biomaterials possess highly complementary properties. In vivo, the inherent structural stability, reliability, bacteriostatic, and osteopromotive characteristics of Si_3N_4 are anticipated to be markedly enhanced by the noncontiguous Bioglass® coating. The gradual dissolution of Bioglass® and pullulan within the trenches provides critical chemical cues to initiate and accelerate osseous activity, while the exposed Si_3N_4 surfaces act as anchorage points for nascent bone formation.

While the former study focused on surface texturing and Bioglass® filling of Si_3N_4 substrates, other research explored alternative approaches to biomaterial surface functionalization using Si_3N_4 and Bioglass®. As a pertinent example, Pezzotti et al. (2019) investigated the osteogenic potential of zirconia-toughened alumina (ZTA, BIOLOX®*delta*) modified with either laser-patterned Bioglass®-Si_3N_4 powder mixtures or pulse-laser sintered Si_3N_4 coatings [100]. These processes are depicted in Fig. 11a–d. During laser patterning, localized partial melting of the ZTA substrate occurred under the laser beam. This resulted in the ejection of molten

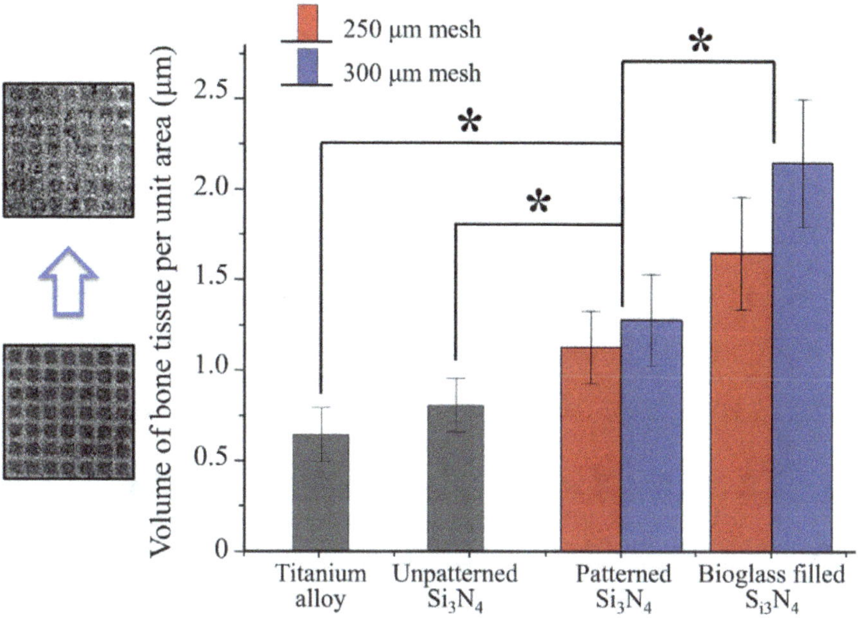

Fig. 10 SaOS-2 bone tissue formation on textured and bioglass-filled Si_3N_4 substrates: This graph illustrates the in vitro bone tissue formation volume per unit area on various Si_3N_4 substrates compared to a titanium alloy control. Each bar represents the average volume measured from ≥3 replicate samples, with error bars denoting standard deviations. An asterisk (*) marks a p-value of ≤0.05. (Used with permission [99])

material, leading to some surface roughening around the well borders. Subsequent filling of the laser-patterned wells involved pressing Bioglass®/Si_3N_4 powder mixtures onto the ZTA surface. Excess material was then manually removed using a sharp blade to ensure precise filling. This process facilitated adhesion of the powder mixtures to the internal walls of the wells. The study employed two distinct Bioglass®/Si_3N_4 compositions, formulated with 5 and 10 wt% Si_3N_4 powder, respectively. The remaining constituent in both compositions was pure 45S5 Bioglass®. The initial powder for laser-sintering of the Si_3N_4 coating onto the ZTA substrate comprised β-Si_3N_4 powder containing 6 wt% Y_2O_3 and 4 wt% Al_2O_3. A homogenous Si_3N_4 coating, exhibiting a thickness of 15 ± 5 μm across the entire ZTA substrate surface, was obtained through repeated laser-sintering operation under a constant nitrogen gas flow to prevent Si_3N_4 degradation. Cultured and seeded SaOS-2 cells (5×10^5/mL) were incubated for 7 days at 37 °C on the test samples to assess their osteogenic potential in comparison to identical ZTA control samples that were neither patterned nor coated.

The key findings from this study were: (1) Osteoblasts cultured directly on pure ZTA underwent apoptosis due to its inherent surface chemistry, hindering osteogenic activity; (2) creating a Ø500 μm and depth cylindrical well grid on the ZTA surface also failed to improve cell proliferation or osteogenesis; (3) filling the

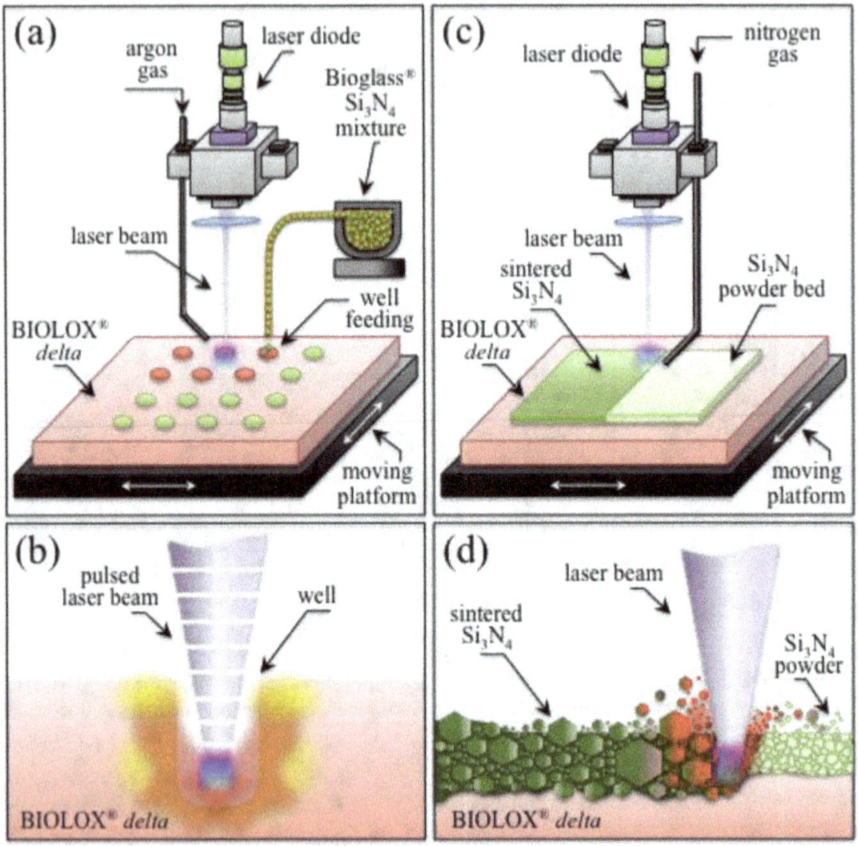

Fig. 11 Schematic diagrams showing two surface functionalization strategies for ZTA: (**a**) laser-patterned Bioglass®-Si_3N_4 wells—this method utilized laser ablation to precisely create wells on the ZTA surface, which were then filled with a biocompatible mixture of Bioglass® and Si_3N_4 powders; (**b**) cross-section view of laser excavation—the diagram provides a closer look at the laser drilling process used to create the wells in (**a**). It showcases the interaction of the laser beam with the ZTA material, revealing the molten zone and the resulting well formation; (**c**) laser-sintered Si_3N_4 coating—this diagram shows the ZTA substrate with a thin Si_3N_4 layer deposited using an automated 3D laser-sintering process. This alternative approach involves applying a uniform Si_3N_4 coating directly onto the ZTA surface using laser sintering. This coating aims to preserve the inherent strength and stability of ZTA while introducing osteopromotive properties of Si_3N_4, and (**d**) cross-section view of laser–Si_3N_4 interaction—this diagram provides a detailed view of the laser-sintering process to deposit the Si_3N_4 coating in (**c**). It demonstrates the laser melting and subsequent solidification of Si_3N_4 powder particles onto the ZTA surface, forming the micro-thin coating. (Used with permission under Creative Common CC BY license [100])

patterned wells with Bioglass® alone increased the amount of mineralized apatite on the ZTA surface, indicating improved mineralization but not cell proliferation; (4) adding 10 wt% Si_3N_4 to the Bioglass® mixture significantly enhanced osteoblast proliferation and resulted in a favorable bone mineral/matrix ratio, and (5) coating

ZTA with a thin layer of sintered Si_3N_4 effectively imparted the osteogenic properties of Si_3N_4 onto the ZTA substrate. This study successfully demonstrated two effective methods for enhancing the osteogenic properties of ZTA: Incorporating Si_3N_4 into Bioglass® mixtures for well-filling and directly coating ZTA with sintered Si_3N_4. These strategies could be promising for improving the integration and performance of ZTA-based implants in various load-bearing orthopedic applications.

Investigators continue to explore various Si_3N_4/glass compositions to enhance mechanical or biological properties [101–106]. As a recent example, Filho et al. examined the mechanical properties and in vitro bioactivity of Si_3N_4 doped with SiO_2, CaO, and MgO additions [107]. They selected these additives presumably on their potential to improve osteogenesis, and they produced four experimental compositions at 20 wt% glass content. After pressureless sintering at 1750 °C for 1 h under N_2, they evaluated comparative mechanical properties and performed biological testing in simulated body fluid (SBF). They found that the compositions with higher SiO_2 contents were more able to form dense composites, while lower SiO_2/MgO mass ratios tended to reduce Young's Modulus and hardness. The samples' bioactivity depended on porosity (i.e., higher was better) and additive composition. High SiO_2 and CaO contents coupled with a low amount of MgO promoted greater apatite formation in SBF and had the best values for hardness (11.7 GPa), fracture toughness (6.8 $MPa \cdot m^{1/2}$), Youngs Modulus (260 GPa), and flexural strength (620 MPa).

4 Silicon Nitride Hydroxyapatite Composites

Similar to Si_3N_4/glass composites, there has also been considerable research on hydroxyapatite (HAp) composites with Si_3N_4 for biomedical applications. This combination aims to leverage the strengths of both materials. It combines Si_3N_4's superior mechanical properties (i.e., strength, fracture toughness, and wear resistance, which make it suitable for load-bearing implants), with HAp that mimics the mineral composition of bone and enhances osseointegration.

One study worth highlighting is the work of Precnerová et al. [108, 109]. Recognizing that Si_3N_4 could be used as a bone scaffold, they elected to prepare porous Si_3N_4 using a polyurethane (PU) sponge replica method. Two Si_3N_4 processes were utilized. In the first, the sponge was impregnated with a 35 vol.% Si_3N_4 sol. After removing excess liquid, samples were dried and fired at 600 and 1250 °C, for 1 h. Sintered reaction-bonded Si_3N_4 (SRBSN) was used for the second process. Silicon and α-Si_3N_4 powders were mixed in a 3:1 ratio, and then dispersed in a 48 wt % sol, followed by impregnation into the sponge. After drying and air firing (600 °C, 1 h), samples were nitrided (1400 °C, 3 h, N_2), then sintered (1750 °C 2 h, N_2). A HAp precursor was prepared by a sol-gel method using triethyl phosphite $(C_2H_5O)_3P$ and calcium nitrate $Ca(NO_3)_2$ dispersed in deionized water. The precursor was infiltrated into the porous Si_3N_4 samples followed by firing at 500 °C in air.

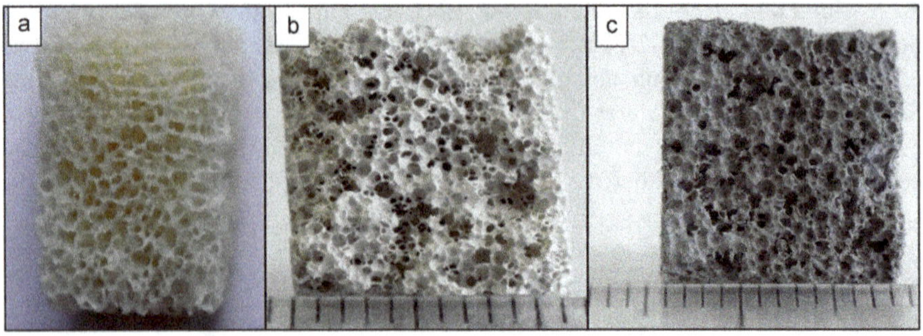

Fig. 12 Macroscopic structure comparison of (**a**) cancellous bone; (**b**) porous air-sintered Si_3N_4 (ASSN); (**c**) porous sintered reaction-bonded Si_3N_4 (SRBSN). (Redrawn and used with permission [109])

Table 1 Comparative properties of trabecular bone and porous HAp-coated Si_3N_4 scaffolds produced using air-sintered Si_3N_4 (ASSN) and sintered reaction bonded Si_3N_4 (SRBSN) [108, 109]

Property	Bone	ASSN	SRBSN
Hardness (GPa)	0.51 ± 0.03	0.46 ± 0.07	6.3 ± 2.3
Young's modulus (GPa)	11.2 ± 0.5	11.4 ± 0.9	92.1 ± 21.3
Compressive strength (MPa)	4.1 ± 1.9	2.9 ± 0.4	7.5 ± 1.5
Porosity (vol.%)	73 ± 17	65	73
Average pore size (μm)	750 ± 250	720 ± 190	750 ± 200

The replica method mimicked native bone as shown in Fig. 12a–c. The physical and mechanical properties of these Si_3N_4/HAp composites were also within the range of cancellous bone, as shown in Table 1.

The average thickness of the HAp coating was 2 ~ 3 μm, but it was discontinuous with some isolated spots about 10 μm in size, particularly towards the center of the samples. To determine their biological response, samples were soaked in SBF for up to 21 days. Following 7 days, the air-sintered Si_3N_4 (ASSN)-HAp composites exhibited complete surface coverage with a newly formed layer of interconnected apatite. In contrast, SRBSN-HAp specimens initially displayed only partial apatite coverage. However, this layer's thickness and continuity increased significantly after 21 days in SBF. Notably, the ASSN-HAp composite demonstrated even greater apatite formation at 21 days, with a thicker and demonstrably denser layer. This suggests superior bioactivity for air-sintered Si_3N_4 due to the incorporation of SiO_2.

Ultimately, both composites displayed the ability to induce bone-like apatite nucleation and growth on their surfaces from SBF. The authors didn't speculate on the effect Si_3N_4 may have had on upregulating apatite formation, but it undoubtedly played a role based on subsequent studies and reviews [110, 111]. The novelty of this study was the unique use of highly porous Si_3N_4 composites as a potential bone scaffold, and its outcomes mimic earlier work by Anderson and Olsen that demonstrated osseous integration of porous Si_3N_4 devices (without HAp) in an ovine model [112].

Several researchers have recognized the potential of a simple thin coating of HAp on Si_3N_4 to form a biomedical composite. For instance, Usinskas et al. used a sol-gel dip-coating technique to apply a 2 μm adherent coating after firing the gel at 650 °C for 2 h [113]. Two additional studies focused on using the biomimetic method [114, 115]. Instead of precoating Si_3N_4 with a HAp precursor, these studies showed that samples exposed to a controlled SBF solution for extended periods naturally formed a thin adherent HAp coating on all exposed Si_3N_4 surfaces.

Other studies have utilized minor fractions of Si_3N_4 to improve the mechanical properties of HAp. Sudan et al. demonstrated that 10 wt% Si_3N_4/90 wt% HAp composition could be combined to produce 3D bone scaffolds using ceramic fused filament fabrication [116]. Zhao used a similar approach of mixing 10 wt% ß-Si_3N_4 with HAp to enhance the mechanical properties of the HAp while preserving its bioactivity [117]. They doped the Si_3N_4 with La^{3+}/Y^{3+} sintering additives that resulted in a remarkable improvement in both hardness (6.44 GPa, a 110% increase) and toughness (1.69 MPa·m$^{1/2}$, a 140% increase) compared to pure HAp. Furthermore, the 10 wt% β-Si_3N_4/HAp composite demonstrated superior protein adsorption and significantly stimulated osteoblast adhesion, proliferation, and osteogenic differentiation. The sintering aid played a dual role, not only facilitating the mechanical property enhancements but also promoting cell-material interactions and osteoblast activity, rendering this composite a promising candidate for bone tissue engineering applications. Undoubtedly, future research will continue to explore the possibilities of combining the bone-forming ability of HAp with the strength and toughness of Si_3N_4.

5 Silicon Nitride Ceramic Matrix Composites

Traditionally, Si_3N_4 ceramic matrix composites have been synthesized by combining various ceramic powders with Si or Si_3N_4 followed by high-temperature hot-pressing or pressureless sintering under N_2 or Ar atmospheres. This preparation method allows for the incorporation of diverse reinforcing phases, tailoring the material's properties for specific applications.

- Alumina: Al_2O_3 is widely used as a refractory material. Adding Si_3N_4 to Al_2O_3 to form a nanocomposite increased its fracture strength, toughness, and hardness at room- and elevated-temperatures [118].
  Titanium carbide: TiC possesses exceptional hardness and strength. Adding 30 vol.% TiC particles to Si_3N_4 increased the composite's toughness to between 7.8 and 9.1 MPa·m$^{1/2}$ [119].
- Titanium nitride: TiN offers high wear resistance, thermal stability, and good electrical conductivity. The addition of between 20 and 40 vol.% TiN dispersoids to hot-pressed Si_3N_4 improved fracture toughness (up to ~9.6 MPa·m$^{1/2}$), flexural strength (up to 1043 MPa) and low electrical resistivity (~9.2×10^{-4} Ω·cm)

[120]. Alternative materials to TiN include metallic Ti and TiO_2. Both can be readily converted to TiN in a nitrogen atmosphere [121].
- Silicon carbide: SiC provides strength, thermal stability, and chemical resistance. Incorporating 20 vol.% 5 μm SiC particles into hot-pressed Si_3N_4 essentially doubled its high-temperature strength [122].
- Boron nitride: BN exhibits both high thermal conductivity and enhanced lubrication. The addition of up to 10 vol.% BN platelets to Si_3N_4 decreased the composite's dry sliding friction coefficient by about 18% [123].

By varying the composition, grain size, and processing parameters, the mechanical, thermal, and chemical properties of the final composite can be precisely tailored for various applications. Si_3N_4-based CMCs excel in applications demanding high hardness, such as metal cutting tools [124], tribological coatings [125], high-temperature components [126], and electronic devices [127]. The incorporation of Si_3N_4, particularly in its elongated hexagonal ß-phase, triggers a synergistic interplay of strengthening and toughening mechanisms including crack deflection, pinning effects, elongated grain pull-out, and mixed fracture modes [128]. Within tribological applications, Si_3N_4's high hardness synergistically pairs with a notably low coefficient of friction, especially under lubricated conditions. This desirable behavior stems from the spontaneous formation of a silica layer on its outermost surface [129]. However, while Si_3N_4 undeniably enhances hardness in composites, alternative materials like carbon nanotubes, graphene, or boron nitride offer superior capabilities in further minimizing friction coefficients [130]. Si_3N_4's high electrical resistivity makes it a valuable material in electronics. Here it primarily serves as an interference barrier, preventing unwanted crosstalk between components. Additionally, Si_3N_4 can significantly enhance the mechanical properties of these composites by acting as a reinforcing constituent. Electronic-based composite matrices incorporating Si_3N_4 often comprise TiN [127], SiC nanofibers [131], $MoSi_2$ [132], or carbon-based materials [133].

Although the foregoing applications of Si_3N_4 CMCs lie outside the biomedical field, their inherent characteristics raise intriguing possibilities for their adaptation to enhance biological functionality within medical devices. There have been some crossovers from industry to medicine. For example, several studies have examined Si_3N_4/TiN composites because of their capability to fabricate complex shapes via electrical discharge machining (EDM). Mazzocchi et al. prepared a dense Si_3N_4/35 vol.% TiN composite and subsequently evaluated its physical, mechanical, and machinability properties [134]. Using uniaxial hot-pressing, they achieved flexural strengths above 800 MPa and hardness values $>$14.7 GPa. More importantly, the composite was easily EDM machined into prototype bone screws as shown in Fig. 13. A separate study by Bucciotti et al. demonstrated a similar outcome [135]. While both Si_3N_4 and TiN are considered highly biocompatible, Monticellia et al. evaluated the composite's wear and corrosion properties [136]. In comparison to biomedical titanium (Ti6Al4V), they showed that the Si_3N_4/TiN composite had equivalent static corrosion resistance in saline and lactic acid solutions. However, under a harsher tribocorrosion scenario, their behaviors diverged. While the

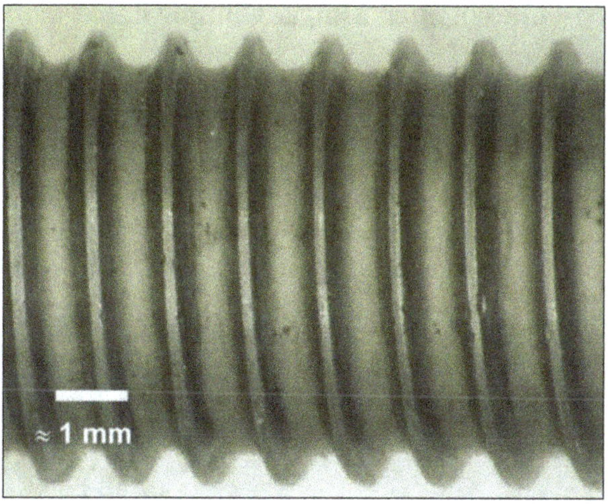

Fig. 13 The intricate geometry of this prototype bone screw demonstrated the potential for machining a 35 vol.% TiN/Si_3N_4 composite by EDM. (Used with permission [134])

composite's corrosion currents experienced a modest 2–three-fold increase, the Ti6Al4V suffered a two-order-of-magnitude surge. Mechanical wear was a significant contributor to overall damage in both materials, accounting for 70–78% in the composite and 95–96% in the alloy. The authors concluded that the Si_3N_4/TiN composite represents a promising biomaterial for hip arthroplasty. Its superior tribocorrosion resistance translates to enhanced durability and holds the potential for long-lasting performance in demanding sliding environments.

The field of medical implant technology has also witnessed a paradigm shift in recent years with the emergence of 3D printing techniques for crafting both monolithic ceramics and CMCs for specific applications [137]. Si_3N_4-based CMCs have great potential in bone tissue engineering due to their excellent mechanical properties, biocompatibility, and osteoconductivity. They can revolutionize osteosynthesis in biomimetic bone scaffolds by providing a high-strength, lightweight, and biocompatible material that can be tailored to match the mechanical properties of natural bone. They can also be used to create porous scaffolds that promote bone ingrowth and regeneration. While still in its nascent stage, Si_3N_4/SiC CMCs appear to be a leading candidate for this transformative application [138, 139]. As research progresses, we can anticipate refined printing techniques, novel CMC compositions with optimized properties, and the integration of bioactive agents for targeted drug delivery and enhanced bone regeneration. The future of osteosynthesis, fueled by 3D-printed Si_3N_4-based CMCs, promises personalized, effective, and minimally invasive bone repair strategies, accompanied by improved patient outcomes and quality of life.

6 Silicon Nitride Polymer Matrix Composites

6.1 Nonbiological Applications of Si_3N_4 Polymer Matrix Composites

Unlike metals and ceramics, polymers offer a distinct advantage for Si_3N_4 composites—lower processing temperatures. This inhibits the oxidation of the Si_3N_4 particles, which is a crucial benefit for retaining their exceptional properties. Ceramic powders are often used as fillers in polymers to improve mechanical properties, thermal stability, wear resistance, or electrical conductivity [140]. Similarly, Si_3N_4 is incorporated as a reinforcement in various polymeric matrices, primarily to enhance strength and impact resistance, thermal and electrical conductivity, and hardness and wear resistance.

As examples of mechanical improvements, Fig. 14a, b depicts the effect of Si_3N_4 particle content on strength and impact resistance. In Fig. 14a, the relative tensile strength of polymethylmethacrylate (PMMA) composites combined with Si_3N_4 powder of differing weight fractions and mesh sizes is provided [141].

Note that approximately a 10% improvement was observed for the smallest weight fraction and mesh size. Coarser Si_3N_4 powders at higher weight fractions engendered a decline in tensile strength. It is important to acknowledge that this observation may not translate to compressively tested specimens, as the underlying fracture-initiating mechanisms diverge between the two loading conditions. Specifically, hard particulates act as crack initiation sites in tension, whereas they primarily augment stiffness and ultimate load under compression [142]. Similarly, in Fig. 14b, very small quantities of nano-sized Si_3N_4 powder added to an epoxy resin were effective in increasing the composite's impact strength. At a loading of ~2 vol.%, the composite's Charpy impact resistance was improved by ~56%.

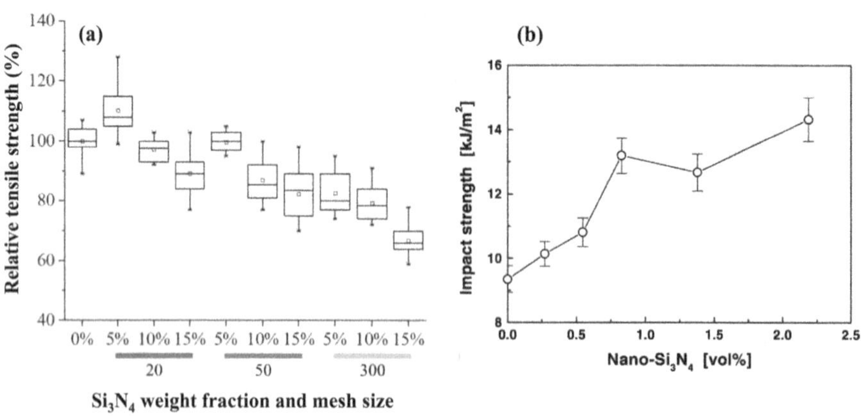

Fig. 14 (**a**) Percentage variation in tensile strength of PMMA/Si_3N_4 composites relative to pristine PMMA (Redrawn and used with permission [141]); and (**b**) unnotched Charpy impact strength of nano-Si_3N_4/epoxy composites as a function of filler content. (Used with permission [35])

Si_3N_4 reinforcements in polyethylene [143], epoxy resin [144], and polystyrene matrices [21] improved thermal conductivity and lowered electrical resistivity with increasing filler content. However, this came at the expense of mechanical properties like ultimate tensile strength and elongation. While initially considered for thermal dissipation and electrical insulation applications in electronic packaging, these composites ultimately lost ground to other existing solutions due to their compromised mechanical performance [145]. In addition to their use in electronic packaging, Si_3N_4 polymer composites have been studied as potential candidates for microwave substrates. When combined with a PTFE matrix, fractions of Si_3N_4 up to 62% have been shown to improve the thermal conductivity of the composite and reduce its dielectric loss. This makes Si_3N_4/PTFE composites a promising material for high-frequency microwave dielectric substrates [146].

The tribological properties of Si_3N_4 polymer matrix composites are dependent on the fraction, size, and morphology of the reinforcement particles, as well as the bonding force between the two phases. This is particularly true for polymeric matrices, where the hardness gap between the two materials can be as high as three orders of magnitude. An increase in wear resistance has been observed for PEEK [29], epoxy resin [35], and nylon matrices [147] when combined with nanometric particulates, as well as for whisker-reinforced polyamide-imide [148]. Si_3N_4 has also been applied as a reinforcement for phenolic and epoxy resin nanowire membranes under bio-tribological conditions [149].

6.2 *Biological Applications of Si_3N_4 Polymer Matrix Composites*

While the above-cited Si_3N_4 polymer matrix composites were mainly investigated and developed for various industrial applications, they have also recently found important uses as biomaterials. Yet, when working with bioactive composite materials, it is important to find a balance between volume- and surface-controlled effects. As the quantity of bioactive particles increases, so does the overall bioactivity, but this often comes at the cost of strength, toughness, and fatigue resistance. Unlike conventional reinforcing agents, where the overall mechanical properties of a composite are predominantly limited by its weakest constituent, even trace amounts of a potent bioactive compound incorporated into a bio-tolerant or bio-inert polymer matrix can significantly influence the composite's overall biological performance. This phenomenon is exemplified by the addition of antibiotics to PMMA matrices, where concentrations as low as 1 wt% effectively inhibit bacterial growth without compromising mechanical properties [150]. However, as the bioactive efficacy of the compound decreases, the required minimum fraction for achieving the desired level of bioactivity increases disproportionately. For instance, while TiO_2 particles dispersed in PLA matrices demonstrated significant *S. aureus* reduction, this effect was only observed at volume fractions of 5% or higher. Furthermore, at a TiO_2

concentration of 10%, the mechanical properties were reported to decline by approximately 30%, dramatically narrowing the window of viable reinforcement concentrations for this biomedical application [151].

The four most common biocompatible structural polymers, PMMA, PEEK, polyethylene, and polypropylene, none of which possess intrinsic bioactive properties, can be functionalized with fractions of Si_3N_4. This functionalization has been shown to increase cellular adhesion, proliferation, and metabolism when tested in vitro [16, 152]. Furthermore, the composite polymers have demonstrated antibacterial [26, 141], antifungal [18], and antiviral capabilities [153]; and contemporary research continues to highlight the biological benefits of combining Si_3N_4 with various polymers [154–161].

Similar to Bioglass® (cf., Sect. 3), Si_3N_4 exhibits two key bioactive properties: promoting osteointegration and thwarting infections from diverse pathogens, including bacteria [162], fungi [18], and viruses [153]. While Bioglass® achieves osteointegration via calcium phosphate precipitation [163] and osteocalcin binding [98], Si_3N_4 stimulates cellular proliferation [164] and metabolism [165]. Notably, the controlled chemical instability of Si_3N_4 leads to the release of NH_3 and other nitrogenous species, which act as potent antimicrobial agents [81] while concurrently elevating local pH [166]. Unlike Bioglass®, where bioactivity compromises mechanical strength, Si_3N_4 retains its excellent mechanical properties and fatigue resistance. This makes bulk Si_3N_4 components attractive for load-bearing applications, while Si_3N_4-based polymer composites offer enhanced strength and durability.

For instance, PEEK is a semi-crystalline aromatic polymer belonging to the polyaryletherketones family. It stands out for its exceptional biocompatibility, meaning it doesn't elicit toxic, mutagenic, or inflammatory responses within the human body [167–170]. This makes it a highly desirable material for medical implants, particularly in spinal surgery. One of PEEK's key advantages lies in its mechanical properties. Its elastic modulus closely resembles that of human cortical bone (~4 GPa for PEEK vs. 5–25.5 GPa for bone) [27]. This similarity minimizes stress shielding, a phenomenon that leads to implant loosening. Furthermore, PEEK's resistance to wear is comparable to that of bone, promoting long-term implant stability and reducing the risk of bone resorption. This combination of biocompatibility and suitable mechanical properties has made PEEK a widely exploited material in spinal fusion cages, intervertebral disc replacements, and other implant applications [171]. However, PEEK has limitations. It is bioinert. Because of its hydrophobic nature, it has poor osteointegration, and it tends to harbor nosocomial pathogens [162, 172–174].

Adding Si_3N_4 powder as a filler to PEEK was examined as a method of retaining its desirable mechanical properties while enhancing its osteopromotive and antimicrobial characteristics. In a simple study, PEEK (Ø25 μm) powder was separately melt-compounded with either 15 vol.% of α-Si_3N_4, ß-Si_3N_4, or ß-SiYAlON powders (~Ø1.0 μm) [16]. Identical geometry test samples were subsequently machined and manually abraded using diamond polishing paper to ensure consistent surface roughness. In vitro antimicrobial and osteopromotive testing was conducted with precultured *S. epidermidis,* (1 × 10^8 CFU/mL, 24 h) and SaOS-2 human

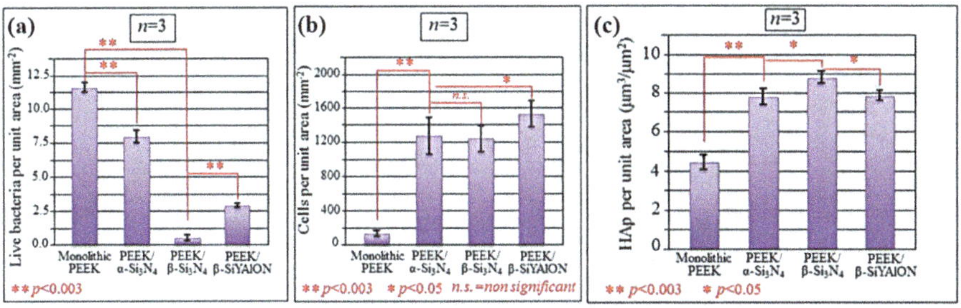

Fig. 15 (**a**) Results of counts for *S. epidermidis* living bacteria on monolithic PEEK and PEEK-matrix composites; (**b**) results of SaOS-2 cell proliferation based on direct cell nuclei counting of fluorescence-stained microscopic images; and (**c**) the amounts of bony apatite on different substrates after 7 days exposure to SaOS-2 cells as assessed by 3D image analysis-assisted laser microscopy. (Redrawn and reproduced with permission [16])

osteosarcoma cells (5×10^5/mL, 7 days) at 37 °C, respectively. Key results of the study are presented in Fig. 15a–c. As anticipated, monolithic PEEK harbored the greatest amount of living *S. epidermidis* pathogens as depicted in Fig. 15a, while the ß-Si_3N_4 and ß-SiYAlON composites showed the lowest amounts of the active pathogen. It was somewhat surprising that the α-Si_3N_4 powder, although effective in reducing *S. epidermidis*, was not as effective as the two ß-phase powders. This may be associated with differences in the kinetic release of NH_3.

In Fig. 15b, the proliferation of SaOS-2 cells indicates that the osteopromotive properties of all three Si_3N_4 composites were far superior to that of monolithic PEEK, with the ß-SiYAlON material eclipsing the α- and ß-Si_3N_4 phases. The reason for this is likely due to the higher glass content in SiYAlON. Silica (SiO_2) and particularly silicic acid ($Si(OH)_4$) are well-known osteopromotive materials [175]. Also, the higher proliferation of osteosarcoma cells on the Si_3N_4 composites led to greater hydroxyapatite formation as shown in Fig. 15c. In this case, all three composites had about twice the HAp formation as monolithic PEEK. This simple in vitro investigation confirmed the initial effectiveness of adding Si_3N_4 particles into PEEK, but it's important to remember that the only way the particles were functionalized was by machining and sanding the composites' surfaces to ensure their exposure. Figure 16 illustrates this relationship, as observed in the in vitro culture of *S. epidermidis* after 24 h of exposure. Notably, the antibacterial activity of the composite increased nonlinearly with the exposed surface area, despite the presence of some data scattering.

Precise control over the volume and dispersion of Si_3N_4 reinforcements may not translate directly to an equivalent exposed surface area. This challenge arises from the tendency of many common polymer matrices to form strong interfacial bonds with the particles, potentially obscuring a significant portion of the Si_3N_4 surface [176]. While hydrophobic polymeric matrices like PEEK, PMMA, and PE exhibit a natural affinity for hydrophilic materials like Si_3N_4, this interaction presents conflicting outcomes. The resulting interface adhesion enhances mechanical properties. However, this benefit comes at the cost of diminished exposed Si_3N_4 surface

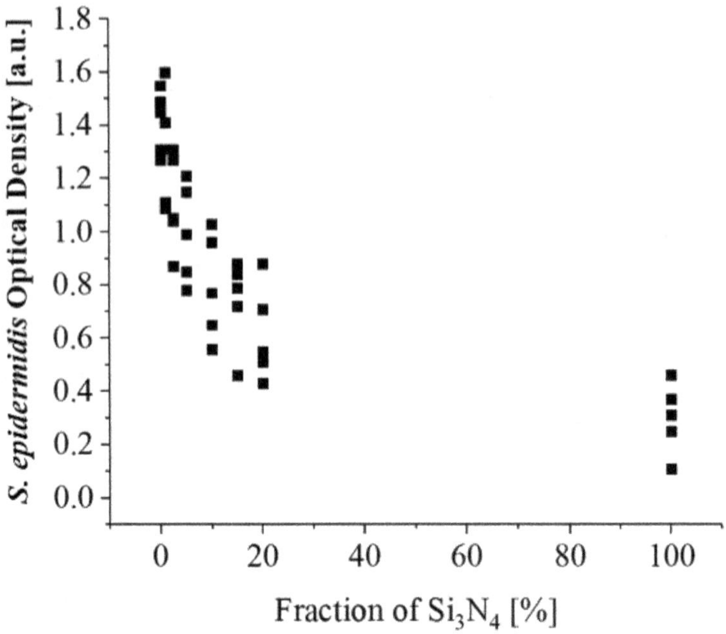

Fig. 16 WST optical density for *S. epidermidis* as a function of the fraction of Si_3N_4 present in the PEEK matrix [141]

area, particularly for smaller particles that are prone to complete encapsulation by the matrix. This poses a challenge for applications where maximizing Si_3N_4's surface properties is crucial. Optimizing Si_3N_4 polymer composites for biomedicine requires navigating a complex interplay of factors. Employing smaller Si_3N_4 particles offers advantages such as enhanced mechanical properties, faster initial nitrogen release, and more potent short-term antibacterial or osteogenic effects. However, finer particles exhibit heightened susceptibility to spontaneous oxidation in humid environments, ultimately diminishing the sustained release of nitrogen. Additionally, while maximizing the exposed surface area of silicon nitride is desirable, a higher degree of exposure may compromise mechanical properties, particularly fatigue resistance.

As an alternative to bulk incorporation, surface modification via Si_3N_4 particle addition can introduce bioactivity while minimizing the mechanical and fatigue properties of the underlying polymer. Several researchers have used different techniques to bind Si_3N_4 particles to the surfaces of polymers.

Laser cladding provides a promising alternative for incorporating bioactivity into polymers while minimizing the loss of their inherent mechanical and fatigue properties. This technique involves directly embedding ceramic particles onto the polymer surface, achieving high coverage and controlled embedment depth. For instance, research by Zanocco et al. demonstrated laser cladding of Si_3N_4 particles onto polyethylene, attaining a surface coverage exceeding 50% with an embedment depth of approximately 50 μm [71]. Compared to traditional fabrication methods involving mixing, laser cladding offers several advantages for achieving optimal bioactivity. The key benefit is the increased fraction of exposed Si_3N_4 surfaces, leading to enhanced interaction with biological tissues. Additionally, the stoichiometry of the nitride remains unaffected by the laser treatment, preserving its biocompatible properties.

The effectiveness of laser cladding is further evidenced by cellular colonization studies. When paired with bulk Si_3N_4 substrates, laser-cladded polyethylene surfaces demonstrated enhanced cell adhesion and proliferation. In vitro testing with SaOS-2 osteosarcoma cell lines also revealed the formation of significantly higher amounts of bone tissue on laser-cladded samples compared to non-functionalized polyethylene [71]. Wu et al. applied a similar approach by using a femtosecond laser on a blended composition of PEEK and Si_3N_4 to induce a micro/nanostructural surface that exposed Si_3N_4 particles, resulting in the upregulation of osteogenic activity [17].

Other surface functionalization techniques have also been investigated. Xu et al. used chemical vapor deposition to deposit a dense amorphous Si_3N_4 coating (~500 μm thick) onto PEEK. The coating had excellent adhesion, and when immersed in SBF, silicon and nitrogen were released, creating a weak alkaline environment. Antibacterial testing demonstrated that the composite was superior to monolithic PEEK. Additionally, the coating promoted adhesion, proliferation, and differentiation of rat bone marrow stromal cells in comparison to pristine PEEK [154].

To enhance the antibacterial and osteogenic properties of poly (etherketoneketone) (PEKK), Wu, et al. employed a concentrated sulfuric acid (H_2SO_4) suspension containing Si_3N_4 microparticles to modify the PEKK surface. Through a sulfonation reaction, they created microporous coatings of PEKK/SN composites on the PEKK substrate. They investigated 5 and 10 wt% Si_3N_4 coatings and found that the higher concentration significantly exhibited antibacterial activity due to the synergistic effects of amino ($-NH_2$) and sulfonic acid ($-SO_3H$) groups, as well as enhanced protein absorption. Furthermore, in vitro studies demonstrated that the 10 wt% coatings stimulated adhesion, proliferation, and osteogenic differentiation of rat bone mesenchymal stem cells [155]. All of these methods hold promise for developing novel medical implants, prosthetics, and other biomedical devices that require both optimal biocompatibility and robust mechanical performance.

A simple method of Si_3N_4 surface functionalization was demonstrated by Boschetto et al. They warm-pressed a mix of particulate sodium chloride (NaCl) salt and β-Si_3N_4 onto the surface of bulk PEEK. Then, the NaCl grains were removed by leaching in water, resulting in a porous PEEK surface embedded with ~15 vol.% β-Si_3N_4 particles [156]. Their process is depicted in Fig. 17a. Then, they proceeded to examine the efficacy of the composite by conducting osteogenic and antibacterial tests using precultured and incubated SaOS-2 cells (5×10^5 cells/mL, 7 days) and Gram-positive *S. epidermidis* bacteria (1×10^8 CFU/mL, 48 h at 37 °C), respectively. SaOS-2 cell counts revealed a low proliferation rate on unmodified PEEK surfaces (data not shown). Introducing a mixture of NaCl and Si_3N_4

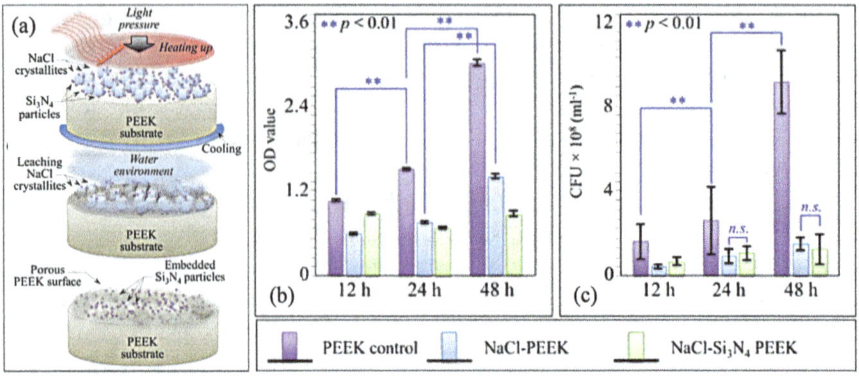

Fig. 17 (**a**) The process involved three steps: (i) light pressing (~10 MPa) for a short duration on a PEEK surface precoated with a mixture of coarse NaCl and fine Si_3N_4 grains; and (ii) subsequent leaching of the NaCl grains using distilled water at room temperature. This resulted in a surface decorated with a coarse open porosity containing a finely dispersed layer of Si_3N_4 particles; (**b**) OD measurement and (**c**) CFU counting on PEEK control sample with smooth surface, NaCl-PEEK sample, and NaCl-Si_3N_4 PEEK sample. (Redrawn and used with permission [156])

dramatically changed the scenario. Cell proliferation on these surfaces skyrocketed to over four times the levels observed with NaCl alone (data not shown). This significant positive impact suggests that Si_3N_4 played a crucial role in promoting cell interaction and proliferation on PEEK substrates. The antibacterial tests are given in Fig. 17b, c. The authors investigated bacterial proliferation on smooth PEEK, NaCl-PEEK, and NaCl-Si_3N_4 PEEK surfaces through optical density (OD) measurements and colony-forming unit (CFU) counting. Following *S. epidermidis* inoculation at time zero, bacterial populations were monitored at 12, 24, and 48 h. Both methods consistently revealed an exponential increase in bacteria on the smooth PEEK surface, indicating optimal proliferation conditions. Conversely, the NaCl-PEEK and NaCl-Si_3N_4-PEEK showed only minimal linear changes with exposure. Note that there was an obvious bactericidal effect associated with NaCl in the absence of Si_3N_4. To clarify this issue, samples were subsequently autoclaved (1 h, 121 °C) to remove residual NaCl and then retested. In this case, the PEEK samples previously exposed to NaCl had greater than 3x count of live bacteria compared to the Si_3N_4 coated sample, confirming the antibacterial potential of this process.

Dai et al. similarly functionalized PEEK by precoating and then melt-bonding nano Si_3N_4 into its surface. Their study demonstrated significant improvements in the composite's surface roughness, hydrophilicity, and protein absorption compared to PEEK. A sustained release of Si ions into simulated body fluid was observed, leading to a slightly alkaline environment due to the controlled dissolution of Si_3N_4. This favorable microenvironment significantly enhanced the adhesion, proliferation, differentiation, and expression of bone-related genes in MC3T3-E1 cells on the composite compared to PEEK. Furthermore, an in vivo assessment by micro-computed tomography and histological evaluation on a rat skull defect model confirmed the composite's ability to promote bone regeneration and osseointegration [177].

In summary, Si_3N_4 polymer composites hold immense promise as revolutionary medical materials due to their unique combination of properties. For Si_3N_4, it offers exceptional mechanical strength and wear resistance, ideal for demanding implants and prosthetics. It also possesses inherent antibacterial properties, reducing the risk of infections associated with implants; and it promotes bone growth and integration with surrounding tissue, crucial for long-term implant success. For the polymers (e.g., PEEK, PMMA, polyethylene, polypropylene), the presence of Si_3N_4 particles stimulates bone growth and integration, overcoming their bioinert limitations. Composites with adjustable elasticity can be chosen to match the surrounding tissue stiffness, minimizing stress shielding and promoting natural bone remodeling. These combined advantages can be used in orthopedics, dental implants, and tissue engineering scaffolds. However, further development is needed to address challenges such as optimized manufacturing techniques, fine-tuning of composite properties, and long-term in vivo studies. Despite these challenges, the potential of Si_3N_4 polymer composites to revolutionize medical care is undeniable. Their unique combination of strength, biocompatibility, and bone-promoting properties makes them a highly promising platform for next-generation medical devices and implants.

7 Conclusion

Traditionally, silicon nitride has found widespread use as a reinforcement in diverse matrices, primarily leveraging its outstanding hardness, toughness, and electrical and thermal properties. Recent discoveries, however, have unveiled its promising biocompatibility, paving the way for the development of various biocompatible composite materials, particularly those utilizing metal, glass, ceramic, or polymer matrices. For each composite, the challenge is to add Si_3N_4 particles as a functional element without compromising the integrity or properties of the host matrix. Achieving a desirable balance between bioactivity (primarily governed by surface characteristics) and mechanical strength (influenced by the bulk properties) poses a significant challenge due to the vast disparity in the elastic modulus and hardness between the matrix and the reinforcement. Fortunately, technologies such as surface functionalization through Si_3N_4 particle embedding offer promising avenues to overcome these limitations and unlock the full potential of Si_3N_4 in the realm of biocompatible material design.

References

1. Gogotsi, Y.G. Particulate Silicon Nitride-Based Composites. *J. Mater. Sci.* **1994**, *29*, 2541–2556, doi:https://doi.org/10.1007/BF00356803.

2. Paramsothy, M.; Chan, J.; Kwok, R.; Gupta, M. Enhanced Mechanical Response of Hybrid Alloy AZ31/AZ91 Based on the Addition of Si_3N_4 Nanoparticles. *Mater. Sci. Eng. A* **2011**, *528*, 6545–6551, doi:https://doi.org/10.1016/j.msea.2011.05.003.
3. Anbuchezhiyan, G.; Mohan, B.; Senthilkumar, N.; Pugazhenthi, R. Synthesis and Characterization of Silicon Nitride Reinforced Al–Mg–Zn Alloy Composites. *Met. Mater. Int.* **2021**, *27*, 3058–3069, doi:https://doi.org/10.1007/s12540-020-00906-3.
4. Sathish, T.; Mohanavel, V.; Karthick, A.; Arunkumar, M.; Ravichandran, M.; Rajkumar, S. Study on Compaction and Machinability of Silicon Nitride (Si_3N_4) Reinforced Copper Alloy Composite through P/M Route. *Int. J. Polym. Sci.* **2021**, *2021*, 7491679, doi:https://doi.org/10.1155/2021/7491679.
5. Razmara, N.; Mohammadzadeh, R. Molecular Dynamics Study of Nnitrogen Diffusion in Nanocrystalline Iron. *J. Mol. Model.* **2016**, *23*, 8, doi:https://doi.org/10.1007/s00894-016-3174-y.
6. De Morton, M.E. Measurement of Nitrogen Diffusion in Chromium by Anelastic Methods. *J. Appl. Phys.* **2004**, *33*, 2768–2770, doi:https://doi.org/10.1063/1.1702545.
7. Marin, E.; Offoiach, R.; Regis, M.; Fusi, S.; Lanzutti, A.; Fedrizzi, L. Diffusive Thermal Treatments Combined with PVD Coatings for Tribological Protection of Titanium Alloys. *Mater. Des.* **2016**, *89*, 314–322, doi:https://doi.org/10.1016/j.matdes.2015.10.011.
8. Balázsi, C.; Kónya, Z.; Wéber, F.; Biró, L.P.; Arató, P. Preparation and Characterization of Carbon Nanotube Reinforced Silicon Nitride Composites. *Mater. Sci. Eng. C* **2003**, *23*, 1133–1137, doi:https://doi.org/10.1016/j.msec.2003.09.085.
9. Ramirez, C.; Figueiredo, F.M.; Miranzo, P.; Poza, P.; Osendi, M.I. Graphene Nanoplatelet/Silicon Nitride Composites with High Electrical Conductivity. *Carbon N. Y.* **2012**, *50*, 3607–3615, doi:https://doi.org/10.1016/j.carbon.2012.03.031.
10. Perevislov, S.N.; Nesmelov, D.D. Properties of SiC and Si_3N_4 Based Composite Ceramic with Nanosize Component. *Glas. Ceram.* **2016**, *73*, 249–252, doi:https://doi.org/10.1007/s10717-016-9867-y.
11. Yuan, B.; Wang, G. Preparation and Properties of Si_3N_4/BN Ceramic Composites. *Procedia Eng.* **2012**, *27*, 1292–1298, doi:https://doi.org/10.1016/j.proeng.2011.12.584.
12. Petzow, G.; Herrmann, M. Silicon Nitride Ceramics. In *High Performance Non-Oxide Ceramics II SE - 2*; Jansen, M., Ed.; Structure and Bonding; Springer Berlin Heidelberg, 2002; Vol. 102, pp. 47–167 ISBN 978-3-540-43132-9.
13. Lukianova, O.A.; Novikov, V.Y.; Parkhomenko, A.A.; Sirota, V. V; Krasilnikov, V. V Microstructure of Spark Plasma-Sintered Silicon Nitride Ceramics. *Nanoscale Res. Lett.* **2017**, *12*, 293, doi:https://doi.org/10.1186/s11671-017-2067-z.
14. Xu, X.; Han, J.; Wang, C.; Huang, A. Laser Cladding of Composite Bioceramic Coatings on Titanium Alloy. *J. Mater. Eng. Perform.* **2016**, *25*, 656–667, doi:https://doi.org/10.1007/s11665-015-1868-4.
15. Zhou, W.; Wang, C.; Ai, T.; Wu, K.; Zhao, F.; Gu, H. A Novel Fiber-Reinforced Polyethylene Composite with Added Silicon Nitride Particles for Enhanced Thermal Conductivity. *Compos. Part A Appl. Sci. Manuf.* **2009**, *40*, 830–836, doi:https://doi.org/10.1016/j.compositesa.2009.04.005.
16. Pezzotti, G.; Marin, E.; Adachi, T.; Lerussi, F.; Rondinella, A.; Boschetto, F.; Zhu, W.; Kitajima, T.; Inada, K.; McEntire, B.J.; *et al.* Incorporating Si_3N_4 into PEEK to Produce Antibacterial, Osteoconductive, and Radiolucent Spinal Implants. *Macromol. Biosci.* **2018**, *18*, 1080033, doi:https://doi.org/10.1002/mabi.201800033.
17. Wu, H.; Liu, T.; Xu, Z.; Qian, J.; Shen, X.; Li, Y.; Pan, Y.; Wang, D.; Zheng, K.; Boccaccini, A.R.; *et al.* Enhanced Bacteriostatic Activity, Osteogenesis and Osseointegration of Silicon Nitride/Polyetherketoneketone Composites with Femtosecond Laser Induced Micro/Nano Structural Surface. *Appl. Mater. Today* **2020**, *18*, 100523, doi:https://doi.org/10.1016/j.apmt.2019.100523.
18. Pezzotti, G.; Asai, T.; Adachi, T.; Ogitani, E.; Yamamoto, T.; Kanamura, N.; Boschetto, F.; Zhu, W.; Zanocco, M.; Marin, E.; *et al.* Antifungal Activity of Polymethyl Methacrylate/Si_3N_4

Composites against *Candida Albicans*. *Acta Biomater.* **2021**, *126*, 259–276, doi:https://doi.org/10.1016/j.actbio.2021.03.023.
19. Xu, G.C.; Wang, J.; Ji, X.L.; Xiong, J.Y.; Li, F. Effect of Nano-Silicon Nitride on the Mechanical and Electric Properties of Polypropylene Nanocomposite. *J. Compos. Mater.* **2007**, *41*, 2213–2223, doi:https://doi.org/10.1177/0021998307074185.
20. Rangari, V.K.; Shaik, M.Y.; Mahfuz, H.; Jeelani, S. Fabrication and Characterization of High Strength Nylon-6/Si_3N_4 Polymer Nanocomposite Fibers. *Mater. Sci. Eng. A* **2009**, *500*, 92–97, doi:https://doi.org/10.1016/j.msea.2008.09.014.
21. He, H.; Fu, R.; Shen, Y.; Han, Y.; Song, X. Preparation and Properties of Si_3N_4/PS Composites Used for Electronic Packaging. *Compos. Sci. Technol.* **2007**, *67*, 2493–2499, doi:https://doi.org/10.1016/j.compscitech.2006.12.014.
22. Tai, Y.; Qian, J.; Miao, J.; Xia, R.; Zhang, Y.; Yang, Z. Preparation and Characterization of Si_3N_4/SBR Nanocomposites with High Performance. *Mater. Des.* **2012**, *34*, 522–527, doi:https://doi.org/10.1016/j.matdes.2011.05.002.
23. Fang, S.-Y.; Yin, S.-H.; Zhang, Y.-C.; Xia, Y.-S.; Yan, J.; Zhang, H.-C. Properties of Surface Modified nano-Si_3N_4/Acrylic Rubber Composites. 合成橡膠工業 **2009**, *09*, 409–411.
24. Zhou, W.; Wang, C.; An, Q.; Ou, H. Thermal Properties of Heat Conductive Silicone Rubber Filled with Hybrid Fillers. *J. Compos. Mater.* **2008**, *42*, 173–187, doi:https://doi.org/10.1177/0021998307086184.
25. Zgalat-Lozynskyy, O.B.; Matviichuk, O.O.; Tolochyn, O.I.; Ievdokymova, O. V; Zgalat-Lozynska, N.O.; Zakiev, V.I. Polymer Materials Reinforced with Silicon Nitride Particles for 3D Printing. *Powder Metall. Met. Ceram.* **2021**, *59*, 515–527, doi:https://doi.org/10.1007/s11106-021-00189-2.
26. Marin, E.; Boschetto, F.; Zanocco, M.; Honma, T.; Zhu, W.; Pezzotti, G. Explorative Study on the Antibacterial Effects of 3D-Printed PMMA/Nitrides Composites. *Mater. Des.* **2021**, *206*, 109788, doi:https://doi.org/10.1016/j.matdes.2021.109788.
27. McEntire, B.J.; Bal, B.S.; Rahaman, M.N.; Chevalier, J.; Pezzotti, G. Ceramics and Ceramic Coatings in Orthopaedics. *J. Eur. Ceram. Soc.* **2015**, *35*, 4327–4369, doi:https://doi.org/10.1016/j.jeurceramsoc.2015.07.034.
28. Lukianova, O.A.; Khmara, A.N.; Perevislov, S.N.; Kolesnikov, D.A.; Krasilnikov, V. V Electrical Resistivity of Silicon Nitride Produced by Various Methods. *Ceram. Int.* **2019**, *45*, 9497–9501, doi:https://doi.org/10.1016/j.ceramint.2018.09.198.
29. Wang, Q.; Xu, J.; Shen, W.; Liu, W. An Investigation of the Friction and Wear Properties of Nanometer Si_3N_4 Filled PEEK. *Wear* **1996**, *196*, 82–86, doi:https://doi.org/10.1016/0043-1648(95)06866-X.
30. Bhagyashekar, M.S.; Rao, R.M.V.G.K. Characterization of Mechanical Behavior of Metallic and Non-Metallic Particulate Filled Epoxy Matrix Composites. *J. Reinf. Plast. Compos.* **2008**, *29*, 30–42, doi:https://doi.org/10.1177/0731684408095034.
31. Verma, V.; Pandey, A.K.; Sharma, C. Fatigue Behavior of Particulate-Reinforced Polymer Composites. In *Advanced Materials and Manufacturing Processes*; Patnaik, A., Banerjee, M. K., Kozeschnik, E., Cavaleiro, A., Davim, J.P., Kukshal, V., Eds.; CRC Press: Boca Raton, FL USA, 2021; p. 17 ISBN 9781003093213.
32. Lin, Q.; Wang, X.; Cai, M.; Yan, H.; Zhao, Z.; Fan, X.; Zhu, M. Enhancement of Si_3N_4@MoS2 Core–Shell Structure on Wear/Corrosion Resistance of Epoxy Resin/Polyacrylate IPN Composite Coating. *Appl. Surf. Sci.* **2021**, *568*, 150938, doi:https://doi.org/10.1016/j.apsusc.2021.150938.
33. Kukshal, V.; Sharma, A.; Kiragi, V.R.; Patnaik, A.; Patnaik, T.K. Erosive Wear Behaviour of Carbon Fiber/Silicon Nitride Polymer Composite for Automotive Application BT - Automotive Tribology. In; Katiyar, J.K., Bhattacharya, S., Patel, V.K., Kumar, V., Eds.; Springer Singapore: Singapore, 2019; pp. 117–129 ISBN 978-981-15-0434-1.
34. Scheller II, W.L.; Wanmuhamad, W. Machining of Green Si_3N_4 Polymer Bonded Ceramic Materials. *Mater. Manuf. Process.* **1996**, *11*, 775–787, doi:https://doi.org/10.1080/10426919608947526.

35. Shi, G.; Zhang, M.Q.; Rong, M.Z.; Wetzel, B.; Friedrich, K. Friction and Wear of Low Nanometer Si_3N_4 Filled Epoxy Composites. *Wear* **2003**, *254*, 784–796, doi:https://doi.org/10.1016/S0043-1648(03)00190-X.
36. Wang, F.; Guo, J.; Li, K.; Sun, J.; Zeng, Y.; Ning, C. High Strength Polymer/Silicon Nitride Composites for Dental Restorations. *Dent. Mater.* **2019**, *35*, 1–10, doi:https://doi.org/10.1016/j.dental.2019.05.022.
37. Yang, Y.; Luo, J.; Song, G.; Liu, Y.; Tang, G. The Experimental Exploration of Nano-Si_3N_4/Paraffin on Thermal Behavior of Phase Change Materials. *Thermochim. Acta* **2014**, *597*, 101–106, doi:https://doi.org/10.1016/j.tca.2014.10.014.
38. Kusunose, T.; Yagi, T.; Firoz, S.H.; Sekino, T. Fabrication of Epoxy/Silicon Nitride Nanowire Composites and Evaluation of their Thermal Conductivity. *J. Mater. Chem. A* **2013**, *1*, 3440–3445, doi:https://doi.org/10.1039/C3TA00686G.
39. Sadej, M.; Gojzewski, H.; Gajewski, P.; Vancso, G.J.; Andrzejewska, E. Photocurable Arylate-Based Composites with Enhanced Thermal Conductivity Containing Boron and Silicon Nitrides. *Express Polym. Lett.* **2018**, *12*, 790–807, doi:https://doi.org/10.3144/expresspolymlett.2018.68.
40. Hou, Z.; Yin, X.; Xu, H.; Wei, H.; Li, M.; Cheng, L.; Zhang, L. Reduced Graphene Oxide/Silicon Nitride Composite for Cooperative Electromagnetic Absorption in Wide Temperature Spectrum with Excellent Thermal Stability. *ACS Appl. Mater. Interfaces* **2019**, *11*, 5364–5372, doi:https://doi.org/10.1021/acsami.8b20023.
41. Hosier, I.L.; Praeger, M.; Vaughan, A.S.; Swingler, S.G. Electrical Properties of Polymer Nano-Composites Based on Oxide and Nitride Fillers. In Proceedings of the 2015 IEEE Electrical Insulation Conference (EIC); 2015; pp. 438–441.
42. Houssat, M.; Villeneuve-Faure, C.; Lahoud Dignat, N.; Locatelli, M.-L.; Cambronne, J.-P. Temperature Influence on PI/Si_3N_4 Nanocomposite Dielectric Properties: A Multiscale Approach. *Polymers (Basel).* 2021, *13*.
43. Zhang, C.; Chen, Y.; Fang, H.; Tian, Z.; Meng, F.; Lin, H. The Effect of Si_3N_4 on the Thermal and Dielectric Properties of Polytetrafluoroethylene/Glass Fiber Composites. *J. Mater. Sci. Mater. Electron.* **2021**, *32*, 21957–21965, doi:https://doi.org/10.1007/s10854-021-06611-0.
44. Balikai, A.; Adarsha, H.; Keshavamurthy, R. Microstructure and Nanoindentation response of Si_3N_4-Reinforced Magnesium-Based Composite Synthesized by Powder Metallurgy Route. *J. Inst. Eng. Ser. D* **2022**, *103*, 235–247, doi:https://doi.org/10.1007/s40033-021-00304-y.
45. Anand, N.; Ramachandran, K.K.; Bijulal, D. Microstructural, Mechanical and Tribological Characterization of Vacuum Stir Cast Mg-4Zn/Si_3N_4 Magnesium Matrix Nanocomposite. *Mater. Today Proc.* **2021**, *46*, 9387–9391, doi:https://doi.org/10.1016/j.matpr.2020.03.053.
46. Paramsothy, M.; Chan, J.; Kwok, R.; Gupta, M. Enhanced Mechanical Response of Magnesium Alloy ZK60A Containing Si_3N_4 Nanoparticles. *Compos. Part A Appl. Sci. Manuf.* **2011**, *42*, 2093–2100, doi:https://doi.org/10.1016/j.compositesa.2011.09.019.
47. Paramsothy, M.; Tan, X.H.; Chan, J.; Kwok, R.; Gupta, M. Si_3N_4 Nanoparticle Addition to Concentrated Magnesium Alloy AZ81: Enhanced Tensile Ductility and Compressive Strength. *ISRN Nanomater.* **2012**, *2012*, 1–8, doi:https://doi.org/10.5402/2012/169850.
48. Pasha, M.B.; Rao, R.N.; Ismail, S.; Gupta, M. Microstructure, Mechanical and Ignition Characteristics of Si_3N_4 Reinforced Magnesium Matrix Nanocomposites. *Appl. Sci.* **2022**, *12*, doi:https://doi.org/10.3390/app12126138.
49. Ghasali, E.; Bordbar-Khiabani, A.; Alizadeh, M.; Mozafari, M.; Niazmand, M.; Kazemzadeh, H.; Ebadzadeh, T. Corrosion Behavior and in-vitro Bioactivity of Porous Mg/Al_2O_3 and Mg/Si_3N_4 Metal Matrix Composites Fabricated Using Microwave Sintering Process. *Mater. Chem. Phys.* **2019**, *225*, 331–339, doi:https://doi.org/10.1016/j.matchemphys.2019.01.007.
50. Mohan, D.; Chinnasamy, B.; Naganathan, S.K.; Nagaraj, N.; Jule, L.T.; Badassa, B.; Ramaswamy, K.; Kathirvel, P.; Murali, G.; Vatin, N.I. Experimental Investigation and Comparative Analysis of Aluminium Hybrid Metal Matrix Composites Reinforced with Silicon Nitride, Eggshell and Magnesium. *Materials (Basel).* **2022**, *15*, doi:https://doi.org/10.3390/ma15176098.

51. Munakata, F.; Takeda, M.; Nemoto, K.; Ookubo, K.; Sato, Y.; Mizukami, Y.; Koga, M.; Abe, S.; Bao, Y.; Kobayashi, R. Multifractal Characteristics of the Self-Assembly Material Texture of β-Si_3N_4/SUS316L Austenitic Stainless Steel Composites. *J. Alloys Compd.* **2021**, *853*, 156570, doi:https://doi.org/10.1016/j.jallcom.2020.156570.
52. Munakata, F.; Ookubo, K.; Takeda, M.; Sato, Y.; Mizukami, Y.; Nemoto, K.; Abe, S.; BAO, Y.; Kobayashi, R. Self-Assembly Process Under a Solid-State Reaction of β-Si_3N_4/Austenitic Stainless-Steel Composites: Stirring Conditions and Material Texture. *J. Compos. Mater.* **2021**, *56*, 455–466, doi:https://doi.org/10.1177/00219983211057337.
53. Guan, D.D.; He, X.B.; Qu, X.H. Fabrication of Si_3N_4 Reinforced 316L Stainless Steel Composites by Powder Injection Moulding. *Adv. Mater. Res.* **2012**, *535–537*, 133–138, doi: https://doi.org/10.4028/www.scientific.net/AMR.535-537.133.
54. Behera, M.P.; Dougherty, T.; Singamneni, S. Selective Laser Melting of Stainless Steel and Silicon Nitride Fibre Metal Matrix Composites. *Proc. Inst. Mech. Eng. Part B J. Eng. Manuf.* **2020**, 1–13, doi:https://doi.org/10.1177/0954405420928684.
55. Bennett, M.J.; Houlton, M.R. The Interaction between Silicon Nitride and Several Iron, Nickel and Molybdenum-Based Alloys. *J. Mater. Sci.* **1979**, *14*, 184–196, doi:https://doi.org/10.1007/BF01028343.
56. Shimoo, T.; Okamura, K. Interaction of Si_3N_4 with Cr. *J. Mater. Sci.* **1994**, *29*, 2231–2237, doi:https://doi.org/10.1007/BF01154703.
57. Rawers, J.C.; Dunning, J.S.; Asai, G.; Reed, R.P. Characterization of Stainless Steels Melted Under High Nitrogen Pressure. *Metall. Trans. A* **1992**, *23*, 2061–2068, doi:https://doi.org/10.1007/BF02647552.
58. Drumond, J.; Girina, O.; da Silva Filho, J.F.; Fonstein, N.; de Oliveira, C.A.S. Effect of Silicon Content on the Microstructure and Mechanical Properties of Dual-Phase Steels. *Metallogr. Microstruct. Anal.* **2012**, *1*, 217–223, doi:https://doi.org/10.1007/s13632-012-0034-8.
59. Singh, H.; Hayat, M.; Huang, S.; Sui, Y.; Cao, P. An In Situ Neutron Diffraction Study of Phase Formation During Ti–Si_3N_4 Powder Sintering. *Adv. Powder Mater.* **2022**, *1*, 100001, doi:https://doi.org/10.1016/j.apmate.2021.09.001.
60. Shimoo, T.; Okamura, K.; Adachi, S. Interaction of Si_3N_4 with Titanium at Elevated Temperatures. *J. Mater. Sci.* **1997**, *32*, 3031–3036, doi:https://doi.org/10.1023/A:1018617810454.
61. QiLiang, H.; Juan, C.; Wei, P.; Jian, C.; Jie, L. In situ Processing of TiN/Si_3N_4 Composites by Ti-Si_3N_4 Solid State Reaction. *Mater. Lett.* **1997**, *31*, 221–225, doi:https://doi.org/10.1016/S0167-577X(96)00277-7.
62. Ahmed, M.A.; Daoush, W.M.; El-Nikhaily, A.E. Fabrication and Characterization of Copper/Silicon Nitride Composites. *Adv. Mater. Res.* **2016**, *5*, 131–140, doi:https://doi.org/10.12989/amr.2016.5.3.131.
63. Eslami, M.; Saghafian, H.; Golestani-fard, F.; Robin, A. Effect of Electrodeposition Conditions on the Properties of Cu–Si_3N_4 Composite Coatings. *Appl. Surf. Sci.* **2014**, *300*, 129–140, doi:https://doi.org/10.1016/j.apsusc.2014.02.021.
64. Eslami, M.; Golestani-fard, F.; Saghafian, H.; Robin, A. Study on Tribological Behavior of Electrodeposited Cu–Si_3N_4 Composite Coatings. *Mater. Des.* **2014**, *58*, 557–569, doi:https://doi.org/10.1016/j.matdes.2014.02.030.
65. Mohankumar, P.C.; Venkatesha, V.T. Surfactants Effect on Zn–Si_3N_4 Coating, Electrochemical Properties, and Their Corrosion Behaviors. *Ind. Eng. Chem. Res.* **2013**, *52*, 12827–12837, doi:https://doi.org/10.1021/ie302525r.
66. Rajiv, E.P.; Iyer, A.; Seshadri, S.K. Tribomechanical Properties of Cobalt-Silicon Nitride Composite Coatings. *Surf. Eng.* **1996**, *12*, 163–166, doi:https://doi.org/10.1179/sur.1996.12.2.163.
67. Xia, Y.; Sasaki, S.; Murakami, T.; Nakano, M.; Shi, L.; Wang, H. Ionic Liquid Lubrication of Electrodeposited Nickel–Si_3N_4 Composite Coatings. *Wear* **2007**, *262*, 765–771, doi:https://doi.org/10.1016/j.wear.2006.06.015.
68. Vilar, R. Laser cladding. *J. Laser Appl.* **1999**, *11*, 64–79, doi:https://doi.org/10.2351/1.521888.

69. Zanocco, M.; Boschetto, F.; Zhu, W.; Marin, E.; McEntire, B.J.; Bal, B.S.; Adachi, T.; Yamamoto, T.; Kanamura, N.; Ohgitani, E.; *et al.* 3D-Additive Deposition of an Antibacterial and Osteogenic Silicon Nitride Coating on Orthopaedic Titanium Substrate. *J. Mech. Behav. Biomed. Mater.* **2020**, *103*, 103557, doi:https://doi.org/10.1016/j.jmbbm.2019.103557.
70. Marin, E.; Zanocco, M.; Boschetto, F.; Santini, M.; Zhu, W.; Adachi, T.; Ohgitani, E.; McEntire, B.J.; Bal, B.S.; Pezzotti, G. Silicon Nitride Laser Cladding: A Feasible Technique to Improve the Biological Response of Zirconia. *Mater. Des.* **2020**, *191*, 108649, doi:https://doi.org/10.1016/j.matdes.2020.108649.
71. Zanocco, M.; Marin, E.; Boschetto, F.; Adachi, T.; Yamamoto, T.; Kanamura, N.; Zhu, W.; McEntire, B.J.; Bal, B.S.; Ashida, R.; *et al.* Surface Functionalization of Polyethylene by Silicon Nitride Laser Cladding. *Appl. Sci.* **2020**, *10*, 2612.
72. Kiwi, J.; Rtimi, S.; Sanjines, R.; Pulgarin, C. TiO2 and TiO2-Doped Films Able to Kill Bacteria by Contact: New Evidence for the Dynamics of Bacterial Inactivation in the Dark and Under Light Irradiation. *Int. J. Photoenergy* **2014**, *2014*, 785037, doi:https://doi.org/10.1155/2014/785037.
73. Vatansever, F.; de Melo, W.C.M.A.; Avci, P.; Vecchio, D.; Sadasivam, M.; Gupta, A.; Chandran, R.; Karimi, M.; Parizotto, N.A.; Yin, R.; *et al.* Antimicrobial Strategies Centered Around Reactive Oxygen Species - Bactericidal Antibiotics, Photodynamic Therapy, and Beyond. *FEMS Microbiol. Rev.* **2013**, *37*, 955–989, doi:https://doi.org/10.1111/1574-6976.12026.
74. Knepper, M.A.; Packer, R.; Good, D.W. Ammonium Transport in the Kidney. *Physiol. Rev.* **1989**, *69*, 179–249.
75. Adeva, M.M.; Souto, G.; Blanco, N.; Donapetry, C. Ammonium Metabolism in Humans. *Metabolism* **2012**, *61*, 1495–1511, doi:https://doi.org/10.1016/j.metabol.2012.07.007.
76. Geisseler, D.; Horwath, W.R.; Joergensen, R.G.; Ludwig, B. Pathways of Nitrogen Utilization by Soil Microorganisms - A Review. *Soil Biol. Biochem.* **2010**, *42*, 2058–2067, doi:https://doi.org/10.1016/j.soilbio.2010.08.021.
77. Olkku, A.; Mahonen, A. Wnt and Steroid Pathways Control Glutamate Signalling by Regulating Glutamine Synthetase Activity in Osteoblastic Cells. *Bone* **2008**, *43*, 483–493, doi:https://doi.org/10.1016/j.bone.2008.04.016.
78. Aminian, A.; Solati-Hashjin, M.; Samadikuchaksaraei, A.; Bakhshi, F.; Gorjipour, F.; Farzadi, A.; Moztarzadeh, F.; Schmücker, M. Synthesis of Silicon-Substituted Hydroxyapatite by a Hydrothermal Method with Two Different Phosphorous Sources. *Ceram. Int.* **2011**, *37*, 1219–1229, doi:https://doi.org/10.1016/j.ceramint.2010.11.044.
79. Reffitt, D.M.; Ogston, N.; Jugdaohsingh, R.; Cheung, H.F.J.; Evans, B.A.J.; Thompson, R.P.H.; Powell, J.J.; Hampson, G.N. Orthosilicic Acid Stimulates Collagen Type 1 Synthesis and Osteoblastic Differentiation in Human Osteoblast-Like Cells *in vitro*. *Bone* **2003**, *32*, 127–135, doi:https://doi.org/10.1016/S8756-3282(02)00950-X.
80. Pezzotti, G. A Spontaneous Solid-State NO Donor to Fight Antibiotic Resistant Bacteria. *Mater. Today Chem.* **2018**, *9*, 80–90, doi:https://doi.org/10.1016/j.mtchem.2018.05.004.
81. Pezzotti, G.; Bock, R.M.; McEntire, B.J.; Adachi, T.; Marin, E.; Boschetto, F.; Zhu, W.; Mazda, O.; Bal, B.S. *In vitro* Antibacterial Activity of Oxide and Non-Oxide Bioceramics for Arthroplastic Devices: I. In situ Time-Lapse Raman Spectroscopy. *Analyst* **2018**, *143*, 3708–3721, doi:https://doi.org/10.1039/C8AN00233A.
82. Pacher, P.; Beckman, J.S.; Liaudet, L. Nitric Oxide and Peroxynitrite in Health and Disease. *Physiol. Rev.* **2007**, *87*, 315–424.
83. Arp, D.J.; Stein, L.Y. Metabolism of Inorganic N Compounds by Ammonia-Oxidizing Bacteria. *Crit. Rev. Biochem. Mol. Biol.* **2003**, *38*, 471–495, doi:https://doi.org/10.1080/10409230390267446.
84. Stein, L.Y.; Yung, Y.L. Production, Isotopic Composition, and Atmospheric Fate of Biologically Produced Nitrous Oxide. *Annu. Rev. Earth Planet. Sci.* **2003**, *31*, 329–356, doi:https://doi.org/10.1146/annurev.earth.31.110502.080901.

85. Abumourad, I.M.K.; Hanna, M.I.; Girgis, S.M. Genotoxicity Assessment of Ammonia in Cultured Oreochromis niloticus Using RAPD Assay. *J. Genet. Eng. Biotechnol.* **2012**, *10*, 209–212, doi:https://doi.org/10.1016/j.jgeb.2012.10.002.
86. Loehman, R.E.; Rowcliffe, D.J. Sintering of Si_3N_4-Y_2O_3-Al_2O_3. *J. Am. Ceram. Soc.* **1980**, *63*, 144–148.
87. Tanaka, I.; Pezzotti, G.; Okamoto, T.; Miyamoto, Y.; Koizumi, M. Hot Isostatic Press Sintering and Properties of Silicon Nitride without Additive. *J. Am. Ceram. Soc.* **1989**, *72*, 1656–1660.
88. Zeng, J.; Yamada, O.; Tanaka, I.; Miyamoto, Y. Hot Isostatic Pressing and High-Temperature Strength of Silicon Nitride-Silica Ceramics. *J. Am. Ceram. Soc.* **1990**, *73*, 1095–1097, doi: https://doi.org/10.1111/j.1151-2916.1990.tb05162.x.
89. Zeng, J.; Tanaka, I.; Miyamoto, Y.; Yamada, O.; Niihara, K. Densification and Microstructural Development of Silicon Nitride-Silica During Hot Isostatic Pressing. *J. Am. Ceram. Soc.* **1992**, *75*, 148–152, doi:https://doi.org/10.1111/j.1151-2916.1992.tb05457.x.
90. Rouxel, T.; Wakai, F.; Izaki, K. Tensile Ductility of Superplastic Al_2O_3–Y_2O_3–Si_3N_4/SiC Composites. *J. Am. Ceram. Soc.* **1992**, *75*, 2363–2372, doi:https://doi.org/10.1111/j.1151-2916.1992.tb05586.x.
91. Bock, R.M.; Marin, E.; Rondinella, A.; Boschetto, F.; Adachi, T.; McEntire, B.J.; Bal, B.S.; Pezzotti, G. Development of a SiYAlON Glaze for Improved Osteoconductivity of Implantable Medical Devices. *J. Biomed. Mater. Res. Part B Appl. Biomater.* **2017**, 1–13, doi:https://doi.org/10.1002/jbm.b.33914.
92. Amaral, M.; Lopes, M.A.; Silva, R.F.; Santos, J.D. Densification Route and Mechanical Properties of Si_3N_4-Bioglass Biocomposites. *Biomaterials* **2002**, *23*, 857–62.
93. Amaral, M.; Costa, M.A.; Lopes, M.A.; Silva, R.F.; Santos, J.D.; Fernandes, M.H. Si_3N_4-Bioglass Composites Stimulate the Proliferation of MG63 Osteoblast-Like Cells and Support the Osteogenic Differentiation of Human Bone Marrow Cells. *Biomaterials* **2002**, *23*, 4897–4906.
94. Wang, M.; Hench, L.L.; Bonfield, W. Bioglass®/High Density Polyethylene Composite for Soft Tissue Applications: Preparation and Evaluation. *J. Biomed. Mater. Res.* **1998**, *42*, 577–586, doi:https://doi.org/10.1002/(SICI)1097-4636(19981215)42:4<577::AID-JBM14>3.0.CO;2-2.
95. Verrier, S.; Blaker, J.J.; Maquet, V.; Hench, L.L.; Boccaccini, A.R. PDLLA/Bioglass® Composites for Soft-Tissue and Hard-Tissue Engineering: An *in vitro* Cell Biology Assessment. *Biomaterials* **2004**, *25*, 3013–3021, doi:https://doi.org/10.1016/j.biomaterials.2003.09.081.
96. Ma, R.; Tang, T. Current Strategies to Improve the Bioactivity of PEEK. *Int. J. Mol. Sci.* **2014**, *15*, 5426–5445, doi:https://doi.org/10.3390/ijms15045426.
97. Ravarian, R.; Wei, H.; Rawal, A.; Hook, J.; Chrzanowski, W.; Dehghani, F. Molecular Interactions in Coupled PMMA-Bioglass Hybrid Networks. *J. Mater. Chem. B* **2013**, *1*, 1835–1845, doi:https://doi.org/10.1039/c2tb00251e.
98. Marin, E.; Adachi, T.; Boschetto, F.; Zanocco, M.; Rondinella, A.; Zhu, W.; Bock, R.; McEntire, B.; Bal, B.S.; Pezzotti, G. Biological Response of Human Osteosarcoma Cells to Si_3N_4-Doped Bioglasses. *Mater. Des.* **2018**, *159*, 79–89, doi:https://doi.org/10.1016/j.matdes.2018.08.020.
99. Marin, E.; Adachi, T.; Zanocco, M.; Boschetto, F.; Rondinella, A.; Zhu, W.; Somekawa, S.; Ashida, R.; Bock, R.M.; McEntire, B.J.; *et al.* Enhanced Bioactivity of Si_3N_4 through Trench-Patterning and Back-Filling with Bioglass®. *Mater. Sci. Eng. C* **2019**, *106*, 110278, doi:https://doi.org/10.1016/j.msec.2019.110278.
100. Pezzotti, G.; Marin, E.; Zanocco, M.; Boschetto, F.; Zhu, W.; McEntire, B.J.; Bal, B.S.; Adachi, T.; Yamamoto, T.; Kanamura, N. Osteogenic Enhancement of Zirconia-Toughened Alumina with Silicon Nitride and Bioglass®. *Ceramics* **2019**, *2*, 554–567, doi:https://doi.org/10.3390/ceramics2040043.

101. Ling, G.; Yang, H. Pressureless Sintering of Silicon Nitride with Magnesia and Yttria. *Mater. Chem. Phys.* **2005**, *90*, 31–34, doi:https://doi.org/10.1016/j.matchemphys.2004.08.041.
102. Frajkorová, F.; Bodišová, K.; Boháč, M.; Bartoníčková, E.; Sedláček, J. Preparation and Characterisation of Porous Composite Biomaterials Based on Silicon Nitride and Bioglass. *Ceram. Int.* **2015**, *41*, 9770–9778, doi:https://doi.org/10.1016/j.ceramint.2015.04.049.
103. Han, W.; Li, Y.; Chen, G.; Yang, Q. Effect of Sintering Additive Composition on Microstructure and Mechanical Properties of Silicon Nitride. *Mater. Sci. Eng. A* **2017**, *700*, 19–24, doi: https://doi.org/10.1016/j.msea.2017.05.100.
104. Guedes-Silva, C.C.; Cecilia, A.; Rodas, D. Microstructure, Mechanical Properties and *in vitro* Biological Behavior of Silicon Nitride Ceramics. *Mater. Res.* **2018**, *21*, e20180266.
105. Guedes-Silva, C.; Rodas, A.; Carvalho, F.; Higa, O.; Ferreira, T. Silicon Nitride with Titania, Calcia and Silica Additives for Orthopaedic Applications. *Process. Appl. Ceram.* **2020**, *14*, 63–70, doi:https://doi.org/10.2298/pac2001063g.
106. Li, Y.; Shi, J.; Man, C.; Han, J. Effect of Sintering Additive Composition on Microstructure, Thermal and Mechanical Property of Multi-Step Spark Plasma Sintered Silicon Nitride. *Mater. Chem. Phys.* **2023**, *307*, 128211, doi:https://doi.org/10.1016/j.matchemphys.2023.128211.
107. Sona Filho, C.R.; Machado de Souza Carvalho, F.; Chaves Guedes-Silva, C. Mechanical Properties and *in vitro* Bioactivity of Silicon Nitride Ceramics with SiO_2, CaO, and MgO Additions. *J. Biomed. Mater. Res. Part B Appl. Biomater.* **2022**, *110*, 507–516, doi:https://doi.org/10.1002/jbm.b.34930.
108. Bodišová, K.; Kašiarová, M.; Domanická, M.; Hnatko, M.; Lenčéš, Z.; Nováková, Z.V.; Vojtaššák, J.; Gromošová, S.; Šajgalík, P. Porous Silicon Nitride Ceramics Designed for Bone Substitute Applications. *Ceram. Int.* **2013**, *39*, 8355–8362, doi:https://doi.org/10.1016/j.ceramint.2013.04.015.
109. Precnerová, M.; Bodišová, K.; Frajkorová, F.; Galusková, D.; Varchulová Nováková, Z.; Vojtaššák, J.; Lenčéš, Z.; Šajgalík, P. *In vitro* Bioactivity of Silicon Nitride–Hydroxyapatite Composites. *Ceram. Int.* **2015**, *41*, 8100–8108, doi:https://doi.org/10.1016/j.ceramint.2015.03.011.
110. Pezzotti, G. Silicon Nitride as a Biomaterial. *J. Ceram. Soc. Japan* **2023**, *131*, 398–428, doi: https://doi.org/10.2109/jcersj2.23057.
111. Heimann, R.B. Silicon Nitride, A Close to Ideal Ceramic Material for Medical Application. *Ceramics* **2021**, *4*, 208–223.
112. Anderson, M.C.; Olsen, R. Bone Ingrowth into Porous Silicon Nitride. *J. Biomed. Mater. Res.* **2010**, *92A*, 1598–1605, doi:https://doi.org/10.1002/jbm.a.32498.
113. Usinskas, P.; Stankeviciute, Z.; Niaura, G.; Maminskas, J.; Juodzbalys, G.; Kareiva, A. Sol–Gel Processing of Calcium Hydroxyapatite Thin Films on Silicon Nitride (Si_3N_4) Substrate. *J. Sol-Gel Sci. Technol.* **2017**, *83*, 268–274, doi:https://doi.org/10.1007/s10971-017-4431-y.
114. Guedes e Silva, C.C.; da Silva Rigo, E.C.; Marchi, J.; de Almeida Bressiani, A.H.; Bressiani, J.C. Hydroxyapatite Coating on Silicon Nitride Surfaces using the Biomimetic Method. *Mater. Res.* **2008**, *11*, 47–50.
115. Kushan Akin, Ş.R. Hydroxyapatite Deposition on Silicon Nitride and SiAlON Ceramic Substrates via Biomimetic Method. *Cankaya Univ. J. Sci. Eng.* **2023**, *20*, 146–151.
116. Sudan, K.; Singh, P.; Gökçe, A.; Balla, V.K.; Kate, K.H. Processing of Hydroxyapatite and its Composites using Ceramic Fused Filament Fabrication (CF3). *Ceram. Int.* **2020**, *46*, 23922–23931, doi:https://doi.org/10.1016/j.ceramint.2020.06.168.
117. Zhao, H.; Xing, H.; Lai, Q.; Song, T.; Tang, X.; Zhu, K.; Deng, Y.; Zhao, Y.; Liu, W.; Xue, R. Mechanical Properties, Microstructure, and Bioactivity of β-Si_3N_4/HA Composite Ceramics for Bone Reconstruction. *Ceram. Int.* **2021**, *47*, 34225–34234, doi:https://doi.org/10.1016/j.ceramint.2021.08.332.
118. Nakahira, A.; Niihara, K.; Ohkijima, J.; Hirai, T. The Al_2O_3/Si_3N_4 Nano-Composites. *J. Japan Soc. Powder Powder Metall.* **1989**, *36*, 239–242.

119. Huang, J.-L.; Chiu, H.-L.; Lee, M.-T. Microstructure and Chemical Reaction in a Si_3N_4–TiC Composite. *J. Am. Ceram. Soc.* **1994**, *77*, 705–710, doi:https://doi.org/10.1111/j.1151-2916.1994.tb05353.x.
120. Bellosi, A.; Guicciardi, S.; Tampieri, A. Development and Characterization of Electroconductive Si_3N_4-TiN Composites. *J. Eur. Ceram. Soc.* **1992**, *9*, 83–93, doi:https://doi.org/10.1016/0955-2219(92)90049-J.
121. Hojo, J.; Kugimoto, H.; Uehara, M.; Enomoto, N. Fabrication of Si_3 N_4-TiN Nanocomposites via Various Routes. *Key Eng. Mater.* **2006**, *317–318*, 191–194, doi:https://doi.org/10.4028/www.scientific.net/kem.317-318.191.
122. Lange, F.F. Effect of Microstructure on Strength of Si_3N_4-SiC Composite System. *J. Am. Ceram. Soc.* **1973**, *56*, 445–450, doi:https://doi.org/10.1111/j.1151-2916.1973.tb12520.x.
123. Carrapichano, J.M.; Gomes, J.R.; Silva, R.F. Tribological Behaviour of Si_3N_4–BN Ceramic Materials for Dry Sliding Applications. *Wear* **2002**, *253*, 1070–1076, doi:https://doi.org/10.1016/S0043-1648(02)00219-3.
124. Olszyna, A.; Smolik, J. Nanocrystalline Diamond-Like Carbon Coatings Produced on the Si_3N_4–TiC Composites Intended for the Edges of Cutting Tools. *Thin Solid Films* **2004**, *459*, 224–227, doi:https://doi.org/10.1016/j.tsf.2003.12.088.
125. Veprek, S.; Reiprich, S.; Shizhi, L. Superhard Nanocrystalline Composite Materials: The TiN/Si_3N_4 System. *Appl. Phys. Lett.* **1995**, *66*, 2640–2642, doi:https://doi.org/10.1063/1.113110.
126. Klemm, H.; Herrmann, M.; Schubert, C. Silicon Nitride Composites Materials with an Improved High Temperature Oxidation Resistance. In *Proceedings of the 21st Annual Conference on Composites, Advanced Ceramics, Materials, and Structures—A: Ceramic Engineering and Science Proceedings*; Ceramic Engineering and Science Proceedings; 1997; pp. 615–623 ISBN 9780470294437.
127. Gao, L.; Li, J.; Kusunose, T.; Niihara, K. Preparation and Properties of TiN–Si_3N_4 Composites. *J. Eur. Ceram. Soc.* **2004**, *24*, 381–386, doi:https://doi.org/10.1016/S0955-2219(03)00218-8.
128. Bai, X.; Huang, C.; Wang, J.; Zou, B.; Liu, H. Fabrication and Characterization of Si_3N_4 Reinforced Al_2O_3-Based Ceramic Tool Materials. *Ceram. Int.* **2015**, *41*, 12798–12804, doi:https://doi.org/10.1016/j.ceramint.2015.06.115.
129. Li, L.; Ding, M.; Lin, B.; Zhang, B.; Zhang, Y.; Sui, T. Influence of Silica Nanoparticles on Running-in Performance of Aqueous Lubricated Si_3N_4 Ceramics. *Tribol. Int.* **2021**, *159*, 106968, doi:https://doi.org/10.1016/j.triboint.2021.106968.
130. Chen, W.; Shi, H.; Xin, H.; He, N.R.; Yang, W.; Gao, H. Friction and Wear Properties of Si_3N_4-hBN Ceramic Composites Using Different Synthetic Lubricants. *Ceram. Int.* **2018**, *44*, 16799–16808, doi:https://doi.org/10.1016/j.ceramint.2018.06.113.
131. Zhou, W.; Long, L.; Bu, G.; Li, Y. Mechanical and Microwave-Absorption Properties of Si_3N_4 Ceramic with SiCNFs Fillers. *Adv. Eng. Mater.* **2019**, *21*, 1800665, doi:https://doi.org/10.1002/adem.201800665.
132. Manukyan, K.V.; Kharatyan, S.L.; Blugan, G.; Kocher, P.; Kuebler, J. MoSi2–Si_3N_4 Composites: Influence of Starting Materials and Fabrication Route on Electrical and Mechanical Properties. *J. Eur. Ceram. Soc.* **2009**, *29*, 2053–2060, doi:https://doi.org/10.1016/j.jeurceramsoc.2008.12.015.
133. Saleem, A.; Iqbal, R.; Hussain, A.; Javed, M.S.; Ashfaq, M.Z.; Imran, M.; Hussain, M.M.; Akbar, A.R.; Jun, S.; Majeed, M.K. Recent Advances and Perspectives in Carbon-Based Fillers Reinforced Si_3N_4 Composite for High Power Electronic Devices. *Ceram. Int.* **2022**, *48*, 13401–13419, doi:https://doi.org/10.1016/j.ceramint.2022.02.050.
134. Mazzocchi, M.; Bellosi, A. On the Possibility of Silicon Nitride as a Ceramic for Structural Orthopaedic Implants. Part I: Processing, Microstructure, Mechanical Properties, Cytotoxicity. *J. Mater. Sci. Mater. Med.* **2008**, *19*, 2881–2887, doi:https://doi.org/10.1007/s10856-008-3417-2.

135. Bucciotti, F.; Mazzocchi, M.; Bellosi, A. Perspectives of the Si_3N_4-TiN Ceramic Composite as a Biomaterial and Manufacturing of Complex-Shaped Implantable Devices by Electrical Discharge Machining (EDM). *J. Appl. Biomater. Biomech.* **2010**, *8*, 28–32.
136. Monticelli, C.; Zucchi, F.; Tampieri, A. Triboelectrochemical Behaviour of a Si_3N_4–TiN Ceramic Composite and a Titanium Alloy Commonly Used in Biomedical Applications. *Wear* **2009**, *266*, 327–336, doi:https://doi.org/10.1016/j.wear.2008.07.005.
137. Aguirre, T.G.; Cramer, C.L.; Mitchell, D.J. Review of Additive Manufacturing and Densification Techniques for the Net- and Near Net-Shaping of Geometrically Complex Silicon Nitride Components. *J. Eur. Ceram. Soc.* **2022**, *42*, 735–743, doi:https://doi.org/10.1016/j.jeurceramsoc.2021.11.001.
138. Luo, M.; Hou, G.-Y.; Yang, J.-F.; Fang, J.-Z.; Gao, J.-Q.; Zhao, L.; Li, X. Manufacture of fibrous β-Si_3N_4-Reinforced Biomorphic SiC Matrix Composites for Bioceramic Scaffold Applications. *Mater. Sci. Eng. C* **2009**, *29*, 1422–1427, doi:https://doi.org/10.1016/j.msec.2008.11.012.
139. Xiao, S.; Mei, H.; Han, D.; Yuan, W.; Cheng, L. Porous (SiCw-Si_3N_4w)/(Si_3N_4-SiC) Composite with Enhanced Mechanical Performance Fabricated by 3D Printing. *Ceram. Int.* **2018**, *44*, 14122–14127, doi:https://doi.org/10.1016/j.ceramint.2018.05.011.
140. Mark, J.E. Ceramic-Reinforced Polymers and Polymer-Modified Ceramics. *Polym. Eng. Sci.* **1996**, *36*, 2905–2920, doi:https://doi.org/10.1002/pen.10692.
141. Marin, E.; Boschetto, F.; Honma, T.; Yang, Q.; Ajunwa, O.M.; Marsili, E.; Zhu, W.; Pezzotti, G. Particulate Reinforced Bone Cements: The Potential Antibacteric Role of Nitrides. *Materialia* **2023**, *28*, 101748, doi:https://doi.org/10.1016/j.mtla.2023.101748.
142. Hayward, M.R.; Johnston, J.H.; Dougherty, T.; De Silva, K. Interfacial Adhesion: Improving the Mechanical Properties of Silicon Nitride Fibre - Epoxy Polymer Composites. *Compos. Interfaces* **2019**, *26*, 263–273, doi:https://doi.org/10.1080/09276440.2018.1499328.
143. An, Q.; Qi, S.; Zhou, W. Thermal, Electrical, and Mechanical Properties of Si_3N_4 Filled LLDPE Composite. *Polym. Compos.* **2009**, *30*, 866–871, doi:https://doi.org/10.1002/pc.20688.
144. Shi, Z.; Fu, R.; Agathopoulos, S.; Gu, X.; Zhao, W. Thermal Conductivity and Fire Resistance of Epoxy Molding Compounds Filled with Si_3N_4 and $Al(OH)_3$. *Mater. Des.* **2012**, *34*, 820–824, doi:https://doi.org/10.1016/j.matdes.2011.07.012.
145. Wan, Y.-J.; Li, G.; Yao, Y.-M.; Zeng, X.-L.; Zhu, P.-L.; Sun, R. Recent Advances in Polymer-Based Electronic Packaging Materials. *Compos. Commun.* **2020**, *19*, 154–167, doi:https://doi.org/10.1016/j.coco.2020.03.011.
146. Yuan, Y.; Li, Z.; Cao, L.; Tang, B.; Zhang, S. Modification of Si_3N_4 Ceramic Powders and Fabrication of Si_3N_4/PTFE Composite Substrate with High Thermal Conductivity. *Ceram. Int.* **2019**, *45*, 16569–16576, doi:https://doi.org/10.1016/j.ceramint.2019.05.194.
147. Shiva Kumar, K.; Chennakesava Reddy, A. Mechanical and tribological behavior of particulate filled silicon nitride reinforced nylon-6 polymer composites. *Int. J. Eng. Adv. Technol.* **2019**, *8*, 3951–3955, doi:https://doi.org/10.35940/ijeat.F9061.088619.
148. Jia, W.; Shangguan, J.; Lang, F.; Li, D.; Yang, H. Study on the Tribological Properties of Si_3N_4/Polyamide-Imide Composites. *High Perform. Polym.* **2022**, *34*, 616–625, doi:https://doi.org/10.1177/09540083221082599.
149. Liu, Y.; Zhang, L.; Nie, H.; Sheng, H.; Li, H. Balanced Mechanical and Biotribological Properties of Polymer Composites Reinforced by a 3D Interlocked Si_3N_4 Nanowire Membrane. *ACS Appl. Mater. Interfaces* **2022**, *14*, 56203–56212, doi:https://doi.org/10.1021/acsami.2c19535.
150. Weisman, D.L.; Olmstead, M.L.; Kowalski, J.J. *In vitro* Evaluation of Antibiotic Elution from Polymethylmethacrylate (PMMA) and Mechanical Assessment of Antibiotic-PMMA Composites. *Vet. Surg.* **2000**, *29*, 245–251, doi:https://doi.org/10.1053/jvet.2000.4389.
151. Kaseem, M.; Hamad, K.; Rehman, Z.U. Review of Recent Advances in Polylactic Acid/TiO2 Composites. *Materials (Basel).* **2019**, *12*, doi:https://doi.org/10.3390/ma12223659.

152. Marin, E.; Boschetto, F.; Zanocco, M.; Adachi, T.; Toyama, N.; Zhu, W.; McEntire, B.J.; Bock, R.M.; Bal, B.S.; Pezzotti, G. KUSA-A1 Mesenchymal Stem Cells Response to PEEK-Si_3N_4 Composites. *Mater. Today Chem.* **2020**, *17*, 100316.
153. Pezzotti, G.; Boschetto, F.; Ogitani, E.; Fujita, Y.; Zhu, W.; Marin, E.; McEntire, B.J.; Bal, B. S.; Mazda, O. Silicon Nitride: A Potent Solid-State Bioceramic Inactivator of ssRNA Viruses. *Sci. Rep.* **2021**, *11*, 2977, doi:https://doi.org/10.1038/s41598-021-82608-3.
154. Xu, Z.; Wu, H.; Wang, F.; Kaewmanee, R.; Pan, Y.; Wang, D.; Qu, P.; Wang, Z.; Hu, G.; Zhao, J.; et al. A Hierarchical Nanostructural Coating of Amorphous Silicon Nitride on Polyetheretherketone with Antibacterial Activity and Promoting Responses of rBMSCs for Orthopedic Applications. *J. Mater. Chem. B* **2019**, *7*, 6035–6047, doi:https://doi.org/10.1039/c9tb01565e.
155. Wu, H.; Yang, L.; Qian, J.; Wang, D.; Pan, Y.; Wang, X.; Nabanita, S.; Tang, T.; Zhao, J.; Wei, J. Microporous Coatings of PEKK/SN Composites Integration with PEKK Exhibiting Antibacterial and Osteogenic Activity, and Promotion of Osseointegration for Bone Substitutes. *ACS Biomater. Sci. Eng.* **2019**, acsbiomaterials.8b01508, doi:https://doi.org/10.1021/acsbiomaterials.8b01508.
156. Boschetto, F.; Marin, E.; Ohgitani, E.; Adachi, T.; Zanocco, M.; Horiguchi, S.; Zhu, W.; McEntire, B.J.; Mazda, O.; Bal, B.S.; *et al.* Surface Functionalization of PEEK with Silicon Nitride. *Biomed. Mater.* **2020**, *16*, 015015, doi:https://doi.org/10.2139/ssrn.3447526.
157. Hu, G.; Zhu, Y.; Xu, F.; Ye, J.; Guan, J.; Jiang, Y.; Di, M.; Li, Z.; Guan, H.; Yao, X. Comparison of Surface Properties, Cell Behaviors, Bone Regeneration and Osseointegration Between Nano tantalum/PEEK Composite and Nano Silicon Nitride/PEEK Composite. *J. Biomater. Sci. Polym. Ed.* **2022**, *33*, 35–56.
158. Zhao, Q.; Gao, S. Poly (Butylene Succinate)/Silicon Nitride Nanocomposite with Optimized Physicochemical Properties, Biocompatibility, Degradability, and Osteogenesis for Cranial Bone Repair. *J. Funct. Biomater.* 2022, *13*.
159. Zhang, H.; Wei, Q.; Ji, R.; Xie, E.; Sun, A.; Xiao, B.; Huang, C.; Ma, S.; Wei, J.; Yang, X.; *et al.* Biodegradable Composites of Poly(propylene carbonate) Mixed with Silicon Nitride for Osteogenic Activity of Adipose-Derived Stem Cells and Repair of Bone Defects. *J. Mater. Chem. B* **2023**, *11*, 6922–6933, doi:https://doi.org/10.1039/D3TB00017F.
160. Mohammed, A.A.; Habeeb, M.A. Modification and Development of the Structural, Optical and Antibacterial Characteristics of PMMA/Si_3N_4/TaC Nanostructures. *Silicon* **2023**, 1–12, doi:https://doi.org/10.1007/s12633-023-02426-2.
161. Vidakis, N.; Petousis, M.; Michailidis, N.; Papadakis, V.; Mountakis, N.; Argyros, A.; Dimitriou, E.; Charou, C.; Moutsopoulou, A. Polylactic Acid/Silicon Nitride Biodegradable and Biomedical Nanocomposites with Optimized Rheological and Thermomechanical Response for Material Extrusion Additive Manufacturing. *Biomed. Eng. Adv.* **2023**, *6*, 100103, doi:https://doi.org/10.1016/j.bea.2023.100103.
162. Bock, R.M.; Jones, E.N.; Ray, D.A.; Bal, B.S.; Pezzotti, G.; McEntire, B.J. Bacteriostatic Behavior of Surface-Modulated Silicon Nitride in Comparison to Polyetheretherketone and Titanium. *J. Biomed. Mater. Res. Part A* **2017**, *105*, 1521–1534, doi:https://doi.org/10.1002/jbm.a.35987.
163. Andersson, H.; Kangasniemi, I. Calcium Phosphate Formation at the Surface of Bioactive Glass *in vitro*. *J. Biomed. Mater. Res.* **1991**, *25*, 1019–1030, doi:https://doi.org/10.1002/jbm.820250808.
164. Pezzotti, G.; McEntire, B.J.; Bock, R.; Boffelli, M.; Zhu, W.; Vitale, E.; Puppulin, L.; Adachi, T.; Yamamoto, T.; Kanamura, N.; *et al.* Silicon Nitride: A Synthetic Mineral for Vertebrate Biology. *Sci. Rep.* **2016**, *6*, 31717, doi:https://doi.org/10.1038/srep11384.
165. Pezzotti, G.; Bock, R.M.; Adachi, T.; Rondinella, A.; Boschetto, F.; Zhu, W.; Marin, E.; McEntire, B.; Bal, B.S.; Mazda, O. Silicon Nitride Surface Chemistry: A Potent Regulator of Mesenchymal Progenitor Cell Activity in Bone Formation. *Appl. Mater. Today* **2017**, *9*, 82–95, doi:https://doi.org/10.1016/j.apmt.2017.05.005.

166. Pezzotti, G.; Puppulin, L.; La Rosa, A.; Boffelli, M.; Zhu, W.; McEntire, B.J.; Marunaka, Y. Effect of pH and Monovalent Cations on the Raman Spectrum of Water: Basics Revisited and Application to Measure Concentration Gradients at Water/Solid Interface in Si_3N_4 Biomaterials. *J. Phys. Chem. C* **2015**, *463*, 120–136.
167. Wenz, L.M.; Merritt, K.; Brown, S.A.; Moet, A.; Steffee, A.D. *In vitro* Biocompatibility of Polyetheretherketone and Polysulfone Composites. *J. Biomed. Mater. Res.* **1990**, *24*, 207–215, doi:https://doi.org/10.1002/jbm.820240207.
168. Rivard, C.H.; Rhalmi, S.; Coillard, C. *In vivo* Biocompatibility Testing of PEEK Polymer for a Spinal Implant System: A Study in Rabbits. *J. Biomed. Mater. Res.* **2002**, *62*, 488–498, doi: https://doi.org/10.1002/jbm.10159.
169. Nieminen, T.; Kallela, I.; Wuolijoki, E.; Kainulainen, H.; Hiidenheimo, I.; Rantala, I. Amorphous and Crystalline Polyetheretherketone: Mechanical Properties and Tissue Reactions During a 3-Year Follow-Up. *J. Biomed. Mater. Res. Part A* **2008**, *84A*, 377–383, doi: https://doi.org/10.1002/jbm.a.31310.
170. Katzer, A.; Marquardt, H.; Westendorf, J.; Wening, J. V.; von Foerster, G. Polyetheretherketone—Cytotoxicity and Mutagenicity *in vitro*. *Biomaterials* **2002**, *23*, 1749–1759, doi:https://doi.org/10.1016/S0142-9612(01)00300-3.
171. Kurtz, S.M.; Devine, J.N. PEEK Biomaterials in Trauma, Orthopedic, and Spinal Implants. *Biomaterials* **2007**, *28*, 4845–4869.
172. Khonsari, R.H.; Berthier, P.; Rouillon, T.; Perrin, J.-P.; Corre, P. Severe Infectious Complications after PEEK-Derived Implant Placement: Report of Three Cases. *J. Oral Maxillofac. Surgery, Med. Pathol.* **2014**, *26*, 477–482, doi:https://doi.org/10.1016/j.ajoms.2013.04.006.
173. Webster, T.J.; Patel, A.A.; Rahaman, M.N.; Bal, B.S. Anti-Infective and Osteointegration Properties of Silicon Nitride, Poly (Ether Ether Ketone), and Titanium Implants. *Acta Biomater.* **2012**, *8*, 4447–4454.
174. Gorth, D.J.; Puckett, S.; Ercan, B.; Webster, T.J.; Rahaman, M.; Bal, B.S. Decreased Bacteria Activity on Si_3N_4 Surfaces Compared with PEEK or Titanium. *Int. J. Nanomedicine* **2012**, *7*, 4829–4840, doi:https://doi.org/10.2147/IJN.S35190.
175. Marin, E.; Rondinella, A.; Boschetto, F.; Zanocco, M.; McEntire, B.J.; Bal, B.S.; Pezzotti, G. Understanding Silicon Nitride's Biological Properties: From Inert to Bioactive Ceramic. *Key Eng. Mater.* **2018**, *782*, 289–296, doi:https://doi.org/10.4028/www.scientific.net/KEM.782.289.
176. Fu, S.-Y.; Feng, X.-Q.; Lauke, B.; Mai, Y.-W. Effects of Particle Size, Particle/Matrix Interface Adhesion and Particle Loading on Mechanical Properties of Particulate–Polymer Composites. *Compos. Part B Eng.* **2008**, *39*, 933–961, doi:https://doi.org/10.1016/j.compositesb.2008.01.002.
177. Dai, Y.; Guo, H.; Chu, L.; He, Z.; Wang, M.; Zhang, S.; Shang, X. Promoting Osteoblasts Responses *in vitro* and Improving Osteointegration *in vivo* through Bioactive Coating of Nanosilicon Nitride on Polyetheretherketone. *J. Orthop. Transl.* **2020**, *24*, 198–208, doi: https://doi.org/10.1016/j.jot.2019.10.011.

Silicon Nitride Coatings and Biologic Applications

Charlotte Skjöldebrand, Håkan Engqvist, Cecilia Persson, and Bryan J. McEntire

1 Introduction

Beyond its potential as a bulk material, silicon nitride (Si_3N_4) is also a promising coating material for biomedical implants [1]. This approach leverages the substrate material's inherent mechanical and structural integrity while providing a surface tailored to interact with the biological environment. Silicon nitride's exceptional wear resistance, osteogenic properties, and inherent antimicrobial activity make it a compelling material for medical implants, particularly in orthopedics [2]. However, successfully harnessing these benefits hinges on selecting the appropriate deposition technique. A diverse array of techniques exists for depositing coatings or films onto surfaces. Among these are physical vapor deposition (PVD), methods like sputtering and evaporation, chemical vapor deposition (CVD), electrophoretic deposition, sol-gel processing, and thermal or plasma spraying [3]. Each technique offers distinct advantages and limitations, and the optimal choice hinges heavily on the desired application. For instance, cutting tool coatings aimed at wear reduction often utilize PVD sputtering [4], while ceramics like hydroxyapatite for promoting bone growth are commonly deposited via thermal spraying techniques [5]. This chapter

C. Skjöldebrand
Department of Medical Devices, Swedish Medical Products Agency, Uppsala, Sweden

H. Engqvist
Department of Materials Science and Engineering, Applied Materials Science, Ångströmlaboratoriet, Uppsala, Sweden

C. Persson
Department of Materials Science and Engineering, Biomedical Engineering, Ångströmlaboratoriet, Uppsala, Sweden

B. J. McEntire (✉)
SINTX Technologies, Inc., Salt Lake City, UT, USA
e-mail: bmcentire@sintx.com

B. S. Bal et al. (eds.), *Silicon Nitride Bioceramics*,
https://doi.org/10.1007/978-3-031-67047-3_8

delves into the three most prevalent methods for depositing silicon nitride coatings: sputtering, chemical vapor deposition (CVD), and laser sintering, exploring their advantages and limitations. By understanding the strengths and weaknesses of these techniques, researchers and engineers can make informed decisions about selecting the optimal deposition method for their specific silicon nitride coating applications.

2 Coating Techniques

Optimizing the functionality of Si_xN_y thin films necessitates a thorough understanding of the selected deposition technique, the coating's inherent stoichiometry, and the chemical and mechanical properties governing substrate adhesion. Si_xN_y coatings exhibit a range of stoichiometries, each influencing its properties. For instance, silicon-rich (i.e., Si-rich Si_xN_y) films offer enhanced electrical conductivity [6], while nitrogen-rich (i.e., N-rich Si_xN_y) coatings have superior wear resistance [7]. The optimal deposition technique often relies on achieving the desired stoichiometry balance to maximize performance for the specific application. The selection of sputtering as the deposition method, with optimized target compositions and nitrogen flow ratios, can fine-tune film stoichiometry, while plasma-enhanced CVD provides precise control over precursor gas mixtures to achieve specific nitrogen-to-silicon ratios [3]. The deposition method of choice also needs to consider the limitations of the underlying substrate. Thermal spraying, while offering good conformity for complex geometries, is not suitable for temperature-sensitive substrates. Sputtering's line-of-sight deposition can struggle with coating uniformity on complex components. CVD with its excellent conformality may be preferred for intricate geometries. This section delves into the key factors influencing technique selection and explores the prominent choices for depositing Si_xN_y coatings.

2.1 Physical Vapor Deposition (PVD)

PVD, especially sputtering, is the most commonly used deposition technique for medical implants. PVD includes several techniques where solid materials are deposited onto the substrate through evaporation or ejection of atoms and particles. With a history stretching over a century [4], PVD and sputtering, in particular, have established themselves as reliable and well-understood technologies. In 1941, the innovation of plasma-assisted PVD further propelled its capabilities, as evidenced by Burkhard's patent [8]. This advancement paved the way for precise control over film composition and thickness, crucial for tailoring Si_xN_y coatings to specific implant needs. Sputter deposition is carried out in a vacuum chamber at a low base pressure. A sputtering gas (e.g., argon) is introduced to achieve a low-pressure inert atmosphere. A target made of Si_3N_4 is bombarded with energetic ions generated in the plasma. These ions dislodge atoms from the target, which then travel across the

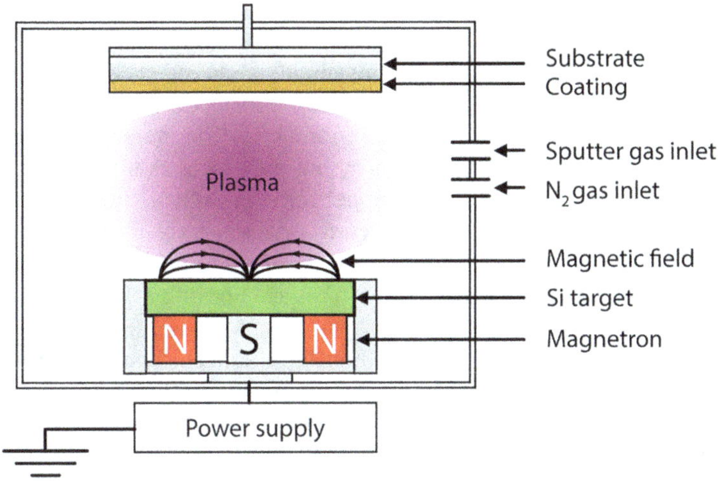

Fig. 1 A schematic illustration of the sputter deposition process. The sputter gas is ionized to generate a plasma that causes material from the target to deposit onto the substrate surface

vacuum chamber and condense onto the implant surface or chamber walls, forming a thin, adherent coating (cf., Fig. 1) [9]. The utilization of a high-intensity magnetic field significantly increases the deposition rate. By magnetically confining electrons in the vicinity of the target surface, collisions are amplified, leading to the phenomenon known as magnetron sputtering (MS). This technique enables highly efficient material deposition. There are several variations in MS including direct current (DC), pulsed DC, and radio frequency (RF) [10].

Achieving a very high-density coating is paramount for its effectiveness as a metal ion barrier. High-power pulsed magnetron sputtering (HiPIMS) addresses this challenge by utilizing pulsed discharges with exceptionally high-power densities. These short bursts facilitate the creation of dense, well-adhered films [11]. However, there are two additional sputtering variations for Si_xN_y films that are worth mentioning: (1) Reactive Sputtering: This process utilizes a silicon target bombarded by sputtering ions (often argon). Simultaneously, a reactive gas, nitrogen (N_2), is introduced into the chamber. The sputtered silicon atoms collide with the N_2, forming silicon nitride molecules that subsequently deposit on the substrate, forming the desired coating. This method offers precise control over film stoichiometry and thickness, making it a popular choice for microelectronics and protective coatings. (2) Compound Target Sputtering: While less common, a pre-formed silicon nitride compound target can be used for sputtering. This eliminates the need for a separate reactive gas source, simplifying the process, but controlling the film stoichiometry can be more challenging with this approach. The optimal technique for Si_xN_y deposition depends on the specific application requirements. Reactive sputtering offers greater control and versatility, making it suitable for demanding applications in microelectronics and other fields. Compound target sputtering might be preferred for simpler applications where stoichiometry is less critical.

In general, sputtering offers several advantages for medical implant coatings [4], including (1) Precise Control Over Film Thickness and Composition: This allows for fine-tuning of wear resistance, biocompatibility, and other crucial properties; (2) Excellent Adhesion: The strong bond between the coating and the implant substrate ensures long-term stability and performance; (3) Line-of-Sight Deposition: This provides control over the coated area, but only surfaces that are in the direct line-of-sight are coated; and (4) Low-Temperature Deposition: The ability to deposit coatings at low temperatures with sputtering makes it ideal for applications with thermally fragile substrates. However, sputtering also has limitations, such as potential line-of-sight shadowing on intricate surfaces and higher material costs compared to some other deposition techniques. Despite these limitations, sputtering remains the dominant choice for Si_xN_y coatings in medical implants due to its proven track record, versatility, and precise control over film properties. Ongoing advancements in target materials, plasma sources, and process optimization continue to push the boundaries of this mature technology, ensuring its relevance for future generations of high-performance medical implants.

2.2 *Chemical Vapor Deposition (CVD)*

Chemical vapor deposition is an alternative to sputtering. In CVD, the precursors are supplied as gases that react on or near a heated substrate. However, the temperatures required for CVD can be too high for many intended substrates. Conventional CVD often necessitates temperatures exceeding the tolerance of materials like polyetheretherketone (PEEK) or even cobalt-chromium molybdenum alloys (CoCrMo). To address this limitation, plasma-enhanced CVD (PECVD) emerged as a valuable alternative. In PECVD, a plasma of the reacting gases is employed, significantly lowering the required temperature without compromising film quality. This technique allows for the deposition of Si_xN_y films onto heat-sensitive materials like PEEK and CoCrMo, expanding the versatility of CVD applications [12]. The most common precursors for Si_xN_y deposition are silane (SiH_4) for the silicon source and ammonia (NH_3) for the nitrogen source. Film formation involves a five-step process: (1) precursor transport of SiH_4 and NH_3 to the substrate surface; (2) species adsorption onto the surface; (3) surface diffusion of the adsorbed species; (4) chemical reaction between the adsorbed species; and (5) desorption of the reaction products. A higher deposition temperature generally leads to increased diffusion lengths, resulting in improved step coverage, meaning the film conforms better to the underlying topography. However, for substrates with limited thermal tolerance, PECVD can be a valuable alternative. It enhances the reactivity of the precursors, thereby promoting reactions at lower temperatures without compromising film quality. This advantage makes PECVD suitable for depositing Si_xN_y onto heat-sensitive materials [13, 14]. Another variant of CVD is low-pressure chemical vapor deposition (LPCVD). It utilizes a much lower pressure environment, typically 0.01–1.0 Pa, whereas CVD operates near atmospheric pressure (~101 kPa). The reduction in gas pressure allows for a more controlled deposition process and purer films of uniform thickness.

A key advantage of CVD over PVD lies in its conformality, meaning its ability to deposit uniform coatings on complex geometries. Unlike PVD, which exhibits a line-of-sight dependence, CVD can effectively coat intricate surfaces. This characteristic arises from the gas-phase nature of the precursors in CVD, allowing them to penetrate deep into recesses and conform to the substrate's topography. Furthermore, CVD often excels in producing adherent and high deposition rate coatings. These combined attributes make CVD a versatile technique for diverse applications across various fields [12].

2.3 *Additive Laser Sintering*

As presented and discussed in the prior chapter, laser sintering is an alternative to conventional deposition techniques like CVD and PVD [15]. Laser sintering offers the versatility to fabricate intricate three-dimensional structures from a powdered precursor and deposit coatings directly onto a substrate. In the referenced case, a carefully formulated powder blend containing Si_3N_4, Y_2O_3, and Al_2O_3 was utilized. The laser selectively melted individual particles within this mixture, fusing them to create the desired coating layer. Crucially, the entire process takes place within a controlled nitrogen gas atmosphere. This protective shroud effectively minimizes the risk of oxidation, a detrimental phenomenon that can significantly alter the material properties and functionality of the resulting sintered Si_3N_4 layer. By carefully balancing the laser power, scan speed, and powder composition, this technique allows for precise control over the final morphology, thickness, and composition of the deposition. However, the fundamentals of laser sintering deviate significantly from those of both PVD and CVD. In the context of this book compilation, a clear distinction is drawn between laser-sintered coatings and those deposited via PVD and CVD methods. This distinction hinges on the fundamental differences in their material structure and properties. Laser-sintered coatings, typically exceeding 10 μm in thickness and built from pre-reacted, pre-coated Si_3N_4 powder, exhibit characteristics more akin to composites. Conversely, PVD and CVD techniques produce thin, adherent coatings ($\leq$ 5 μm) directly on the substrate through in situ chemical reactions or physical atomistic deposition. Given their distinct characteristics, laser-sintered coatings were addressed in the previous chapter under the umbrella of composites, while PVD and CVD coatings are discussed within this focused chapter as thin film technologies.

3 Properties of the Coating

3.1 *Wear Properties*

Wear resistance is a critical property for coatings intended for articulating surfaces. One study demonstrated that most amorphous Si_xN_y coatings exhibited wear rates (1.3×10^{-7} mm^3/Nm to 4.8×10^{-7} mm^3/Nm) comparable to bulk Si_3N_4 (1×10^{-7} mm^3/Nm) and significantly lower than CoCrMo (60×10^{-7} mm^3/Nm) under hard-on-hard contact conditions simulated by a revolving ball-on-disk setup with a Si_3N_4 ball counterface [16]. These wear differences are shown in Fig. 2a–c.

This finding corroborates prior reports of similar wear rates for Si_xN_y coatings (2.7×10^{-7} mm^3/Nm to 3.2×10^{-7} mm^3/Nm) [17]. However, a subsequent study using the same deposition technique observed higher wear rates for Si_xN_y coatings, which were attributed to variations in testing conditions (i.e., contact pressure, linear or circular movements, and wear distance) and potentially differing coating densities (non-verified) [18]. The latter study proposed that increased substrate rotation during deposition might have contributed to the reduced density.

Most artificial joints employ a "hard-on-soft" configuration, where the polymer component (such as ultra-high-molecular-weight polyethylene, UHMWPE) generates the majority of wear debris. Therefore, assessing the performance of Si_xN_y coatings and their counter-surfaces paired with UHMWPE is crucial. Multidirectional pin-on-disc tests against Si_xN_y coatings revealed UHMWPE pin wear rates below 0.37 mm^3/million cycles (MC) [19]. However, some coatings exhibited wear patterns indicative of a surface transformation into silicon oxide. This phenomenon, attributed to oxygen incorporation in low-density, columnar-structured coatings upon air exposure, led to significantly higher UHMWPE wear on those surfaces [19]. These findings highlight the importance of dense coatings with minimal impurity uptake for optimal wear performance. Efforts to improve wear performance have involved incorporating additional elements such as carbon, niobium, and chromium [16, 19]. While some promising results emerged, certain coatings containing carbon or niobium exhibited unexpected wear and even failure.

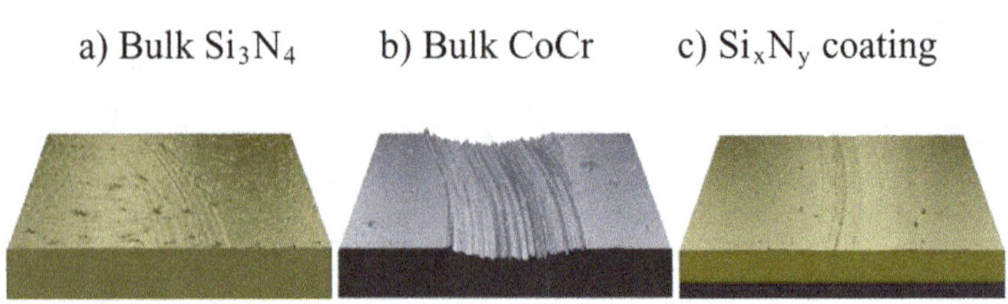

Fig. 2 Wear tracks after 10,000 revolutions in a ball-on-disc setup of (**a**) bulk Si_3N_4, (**b**) CoCrMo, and (**c**) $Si_{ix}N_y$-coated CoCrMo. The wear volume is similar for the bulk Si_3N_4 and coated sample, while the uncoated sample shows comparably severe signs of wear. (Used with permission from [16])

Subsequent investigations suggest that oxygen incorporation may be the culprit, potentially promoting higher reactivity and compromising the coating's integrity [19].

Sputter deposition faces unique challenges when applied to three-dimensional (3D) substrates. Non-flat surfaces necessitate substrate rotation in multiple directions to achieve uniform coating thickness. This can inadvertently affect the coating's density, potentially leading to the formation of silicon oxide and increased wear rates on the counter surface [20]. To mitigate these issues, two strategies have been proposed: (1) Increasing the Sputter Power Applied to the Silicon Target: Higher power can enhance the energy of sputtered particles, allowing them to penetrate deeper into shadowed areas and achieve better film thickness uniformity on complex geometries; and (2) Enrich the Sputter Gas with Nitrogen: Nitrogen plays a crucial role in the formation of the Si_xN_y film. Increasing its concentration can promote denser, more robust coatings with improved wear resistance. By optimizing these parameters, researchers can unlock the full potential of sputter coating complex 3D components while maintaining desired coating properties and minimizing wear.

As wear debris is believed to be the primary contributor to reduced lifespan in joint implants, understanding the wear particle characteristics of Si_xN_y coatings is crucial. Despite exhibiting exceptionally low wear rates, both bulk ceramic and coated joint implants generate wear debris. While the concentration of these particles is often minimal, their isolation and analysis remain crucial for understanding their potential biological impact. Yet, traditional isolation methods have often proven to be inadequate due to the low debris volume. To address this challenge, Lal et al. developed a novel method that successfully isolated wear particles from joint simulator tests, even with wear rates as low as 0.01 mm^3 per million cycles (MC) [21]. This innovative approach leverages enzymatic digestion and custom sodium polytungstate gradients. Its key advantages include (1) an ability to overcome the limitations of traditional techniques, enabling researchers to analyze wear debris even when its concentration is minimal; (2) it successfully isolates both Si_xN_y and CoCrMo particles, demonstrating its versatility for various implant materials; (3) by using enzymes to break down surrounding tissues and lubricants, the method efficiently retrieves wear particles without compromising their integrity; and (4) the process allows for precise separation of wear particles based on their density, facilitating their isolation and subsequent analysis.

3.2 Dissolution and Corrosion of the Coating and Substrate

Recent studies have shed light on the size, morphology, and potential biocompatibility of Si_xN_y wear debris. Hard-on-hard contact testing with an alumina ball as the counter surface revealed that Si_xN_y wear particles range in size from 0.15 to 1.96 μm, with a median diameter of 0.59 μm. Closer examination revealed that these particles were composed of agglomerates, ranging from 0.01 to 0.5 μm, with a median diameter of 0.04 μm [22]. However, a significant advantage of Si_xN_y coatings is

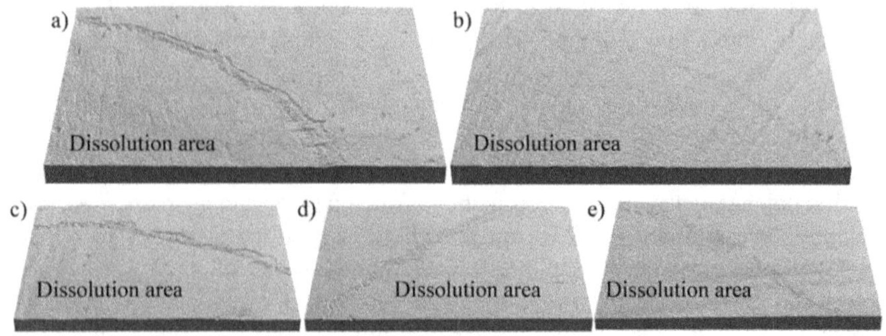

Fig. 3 Images obtained by vertical scanning interferometry of samples after 60 days of exposure to simulated body fluid. The references, (**a**) a standard Si_xN_y coating with an N/Si ratio of 0.8 and (**b**) CoCrMo, were compared with Si_xN_y samples of different N/Si ratios (**c**) 0.3, (**d**) 0.8, and (**e**) 1.1. The sample with the highest N/Si displayed lower signs of dissolution. (Used with permission from [23])

its solubility in aqueous solutions. This raises the hope that generated wear particles might dissolve, minimizing their negative impact on joint wear and the immune system. In vitro studies using model silicon nitride particles of relevant sizes (agglomerates ranging from 0.05 to 0.51 μm and individual particles ranging from 0.02 to 0.08 μm) have shown promising results, demonstrating their dissolution in simulated body fluid [22]. These findings highlight the potential of Si_xN_y coatings for joint implants. Their wear particles are relatively small, and their solubility in aqueous solutions suggests a potentially reduced inflammatory response compared to other implant materials.

Dissolution is not only an important aspect of the particles but also the coating itself. Si_xN_y coatings offer potential benefits for joint implants due to their wear resistance and biocompatibility. Yet, a crucial factor to consider is the dissolution behavior of both the wear particles and the coating itself. Striking a balance between desired particle dissolution for biocompatibility and minimal coating dissolution for wear protection is critical. Dissolution of the coating directly impacts its ability to function as a barrier against wear and protect the underlying implant material. Studies have shown that the composition of the coating plays a significant role in its dissolution rate. For example, amorphous coatings with a high N/Si ratio (1.1) exhibit a lower dissolution rate compared to those with a lower ratio (0.3) [23]. This difference is attributed to the increased presence of stable Si-N bonds, which possess higher binding energies compared to Si-Si bonds [23]. This effect is demonstrated in Fig. 3a–e.

Modifying the composition of the Si_xN_y coating also represents a potential avenue for mitigating ion release. Notably, incorporating elements like carbon and, more effectively, chromium, has been demonstrated to reduce ion release. This phenomenon can be attributed to chromium's propensity to form a passivating oxide layer that hinders further dissolution [24]. Notably, mitigating Si_xN_y coating dissolution may be important depending on the composition of the coating. Its dissolution can rapidly release ammonia, a compound known to be detrimental to surrounding tissues in high concentrations [25]. While indirect cytotoxicity tests with fibroblasts suggest a favorable biological response from Si_3N_4, it is important to recognize that these results represent relative effects, not absolute measures [26, 27]. This distinction emphasizes the need for further investigation to evaluate the material's safety across different cell types and exposure scenarios.

Si_xN_y coatings demonstrate promise in mitigating metal ion release from underlying substrates. Inductively coupled plasma analysis confirmed this potential, using cobalt ion release as a marker for the effectiveness of Si_xN_y coatings. Compared to uncoated CoCrMo, coated samples exhibited a two-order-of-magnitude reduction in cobalt ions [23]. Moreover, silicon nitride coatings incorporating niobium and chromium displayed similar efficacy in another study where coated and reference CoCrMo samples were exposed to cell media for 14 days. All coated samples in this study demonstrated reduced cobalt release [24]. This ability to impede metal ion release could be particularly beneficial for patients with metal hypersensitivity. Therefore, optimizing the coating composition to achieve a balance between biocompatibility, wear protection, and dissolution kinetics is crucial for the long-term success of Si_xN_y-coated implants.

3.3 *Biocompatibility*

Despite promising wear and dissolution properties, thorough biocompatibility evaluation is vital before the clinical implementation of Si_xN_y-based coatings. Indirect in vitro studies using cell media exposed to Si_xN_y coatings have been conducted. Mouse fibroblast cells (L929) were used as per ISO 10993 [28]. Although the standard only recommends an extraction time of 24 h, in this study, the total extraction time was 14 days [24]. The extended period was used to ensure sufficient ion release that could be reasonably detected. The results showed cell viability for samples from Si_xN_y coatings with the incorporation of niobium and chromium that were comparable to extracts from CoCrMo. There were no clear trends when comparing coatings with different compositions.

3.4 *Adhesion*

Adhesion, a perennial concern for most coatings, poses a unique challenge for Si_xN_y films. Delamination could exacerbate wear and ion release, ultimately limiting implant lifespan. The common method for evaluating adhesion is a scratch test. It utilizes a diamond stylus (Ø200 μm, 120° apex angle) pressed and laterally displaced across the sample surface under a controlled load (typically 0–100 N). Under optical examination, the observed failure modes dictate the coating's adhesive strength.

Chipping or flaking, especially at the edges, represents the most frequent failure [29]. Studies revealed critical loads for Si_xN_y coatings ranging from 0 to 88 N, which were generally lower than those observed for titanium nitride coatings on CoCrMo (45–70 N) [30]. Notably, coatings deposited with higher target power exhibited lower critical loads compared to those deposited with lower power. This may be linked to increased residual stresses arising from higher nitrogen levels or denser coatings [19].

Rockwell indentation, employing a diamond tip (Ø200 μm, 120° apex angle) pressed non-laterally onto the surface with a 100 or 150 kgf load, offers another adhesion evaluation technique. Subsequent optical analysis assigns a grade from 0 to 3 (ISO 0: no cracking/delamination, ISO 1: cracking without delamination, ISO 2: partial delamination with/without cracking, ISO 3: complete delamination) [31]. Studies using Rockwell indentations to assess the adhesion of Si_xN_y films directly deposited onto CoCrMo reported significant delamination (ISO 3). However, employing a chromium nitride interlayer demonstrably mitigated this issue [18].

3.5 Osteogenic and Antipathogenic Properties

In its solid form, Si_3N_4 is known to accelerate bone repair [32–35]. The osteogenic mechanism in mesenchymal and osteosarcoma cells has been explained by the presence of silicic acid ($Si(OH)_4$) and reactive nitrogen species (NH_3, NH_4^+, and NO) which are endocytotically incorporated into newly formed hydroxyapatite by osteoblastic action [30–33]. Similar results have been observed when Si_3N_4 has been applied as a thick coating [15]. As an example, one study investigated the osteogenic potential of a Si_3N_4 coating on zirconia-toughened alumina (ZTA) using 3D additive laser sintering. Even though the resulting coating exhibited partial decomposition of the Si_3N_4 powder, leading to a mixture of Si_3N_4 and free silicon clusters, the analysis revealed that when the coated ZTA sample was exposed to SaOS-2 osteoblasts for 1 week, bone tissue formation was observed exclusively on the Si_3N_4 coated surface, as shown in Fig. 4a–d. This finding supports the hypothesis that regardless of whether it is solid Si_3N_4 or a thick Si_3N_4 coating, its presence enhances osteogenic properties [36].

According to various other studies, solid Si_3N_4 also demonstrates antibacterial and antiviral properties [37–45]. As explained in prior chapters of this compilation, surface dissolution of Si_3N_4 leads to the splitting of water, formation of silanols and free electrons, and the pH-dependent release of NH_3 and NH_4^+, all of which play roles in the material's antipathogenic characteristics. These properties highlight the potential benefits of silicon nitride for implants, as they could offer improved bone regeneration and reduced risk of post-surgical infections.

The obvious question is as follows: Do these osteogenic and antipathogenic properties carry over to thin PVD and CVD Si_xN_y coatings? A 2015 research paper sought to answer this question [46]. This PECVD study investigated the

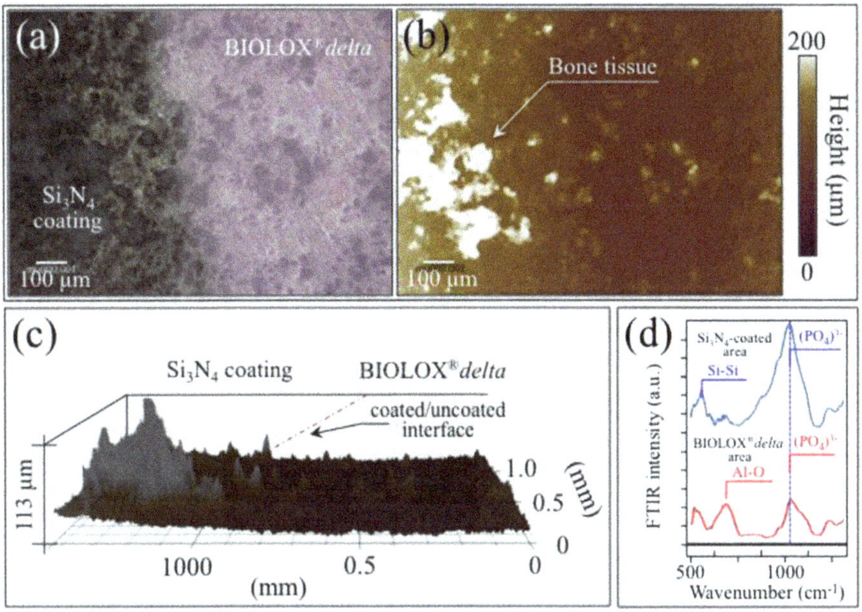

Fig. 4 (**a**) Optical micrograph, (**b**) laser micrograph, and (**c**) three-dimensional laser microscopy views of the same bone tissue island; and (**d**) average FTIR spectra collected near the interface between Si_3N_4-coated and ZTA uncoated zones, highlighting the $(PO_4)^{3-}$ band of mineral apatite. (Used with permission under Creative Commons License [36])

osteogenic potential of silicon oxynitride ($Si_x(ON)_y$) coatings compared to silicon oxide (SiO_x) and silicon nitride (Si_xN_y) thin films. Each film was deposited on silicon wafers at a thickness of approximately 100 nm. Human periosteal cells were then cultured on these samples for up to 4 weeks. All surfaces induced expression of osteoblast markers, including bone morphogenetic protein 2 (BMP2) and osterix (OSX), essential for osteoblast differentiation and bone formation. Within 6 h of exposure, this in vitro study demonstrated significant surface transformation, leading to the formation of hydroxycarbonate apatite. The surface chemistry of silicon oxynitride ($Si_x(ON)_y$) overlays further induced osteogenic gene expression in human periosteal cells, facilitating rapid "bone-mimetic" biomineralization within 4 weeks. Enhanced osteogenesis and biomineralization were attributed to the incorporation of nitrogen within the $Si(ON)_y$ surface. Notably, $Si_x(ON)_y$ coatings elicited the highest expression levels of these markers, suggesting their enhanced potential for promoting osteogenesis.

Other studies soon followed. In 2019, Xu and colleagues published research that examined the antimicrobial and osteogenic response of 500-nm-thick Si_xN_y coatings deposited by inductively coupled plasma-enhanced chemical vapor deposition (ICPECVD) onto PEEK substrates. Pre-cultured *Staphylococcus aureus* (*S. aureus*) and *Escherichia coli* (*E. coli*) (both at 1×10^6 CFU/mL) were incubated on Si_xN_y-coated PEEK substrates for up to 24 h at 37 °C and compared to uncoated

PEEK samples (controls). Using standard CFU counting methods, the results showed between 90 and 95% reduction in both pathogens compared to the controls ($p < 0.05$). To evaluate the coating's osteopromotive properties, pre-cultured rat bone marrow stromal cells (rBMSCs) (2×10^4 cells/well) were seeded onto both test and control substrates and incubated from 1 to 14 days at homeostatic conditions. Afterward, the cultured cells were fixed, dyed, and examined by fluorescent microscopy for adhesion, proliferation, and alkaline phosphate (ALP) activity. At each timepoint, and for each measurement, the Si_xN_y-coated samples showed better adhesion, greater proliferation, and a higher amount of ALP activity than the controls ($p < 0.05$). The authors concluded that Si_xN_y-coated PEEK "exhibited not only excellent antibacterial activity, but also cytocompatibility"; they suggested that the material could be a novel implant for orthopedic applications [47].

Later, Marin et al. investigated the impact of varying stoichiometric and non-stoichiometric Si/N ratios of PVD-deposited thin Si_xN_y films on glass substrates [48]. Their study hypothesized that silicon-rich coatings might promote osteogenesis, while nitrogen-rich coatings would exhibit enhanced antibacterial efficacy. All coatings were deposited at a thickness of approximately 2 μm. No significant differences in surface morphology were observed across all samples, with no detectable droplets or pinholes visible at any magnification levels. Bacterial testing utilized *E. coli* and *Staphylococcus epidermidis* (*S. epidermidis*) as the in vitro pathogens and standard CFU counting methods at time points of 24 and 48 h. PEEK along with monolithic ß-Si_3N_4 and uncoated glass microscope slides were utilized as negative and positive controls, respectively. The viability and proliferation of KUSA-A1 bone marrow stromal was assessed using standard colorimetric assays after 10 days of incubation, and ALP activity was determined at 48 h using standard staining methods. Results of the antipathogenic tests are provided in Fig. 5a, b.

At both time points, an approximate threefold reduction in bacterial counts (99.9%) was observed for each test material in comparison with the PEEK control. Both *E. coli* and *S. epidermidis a*dhesion to all the PVD-coated samples was minimal, exhibiting values comparable to the positive glass controls. This suggests near-complete bacterial detachment, likely attributable to both the presence of the nitride film and the nanometric surface roughness of the coatings, which discourages bacterial attachment and biofilm formation [49]. For *E. coli*, only the stoichiometric Si_3N_4 coating displayed statistically significant bacterial adhesion at the 24-h timepoint, (cf., Fig. 5a), indicating some potential differences in surface properties compared to the other coated samples. For *S. epidermidis*, the monolithic polycrystalline ß-Si_3N_4 sample exhibited minimal bacterial growth, similar to the low roughness glass control and PVD Si_3N_4 films. After 24 h, the stoichiometric PVD Si_3N_4 film's CFU concentration was modestly higher than the glass control and other Si_xN_y-coated samples. However, at 48 h, the bacterial amount decreased significantly (cf., Fig. 5b). Overall, there were no clear correlations that suggested higher concentrations of nitrogen in these coatings improved antipathogenicity. Figure 6a, b presents results for hydroxyapatite growth and ALP concentrations on Si-rich and

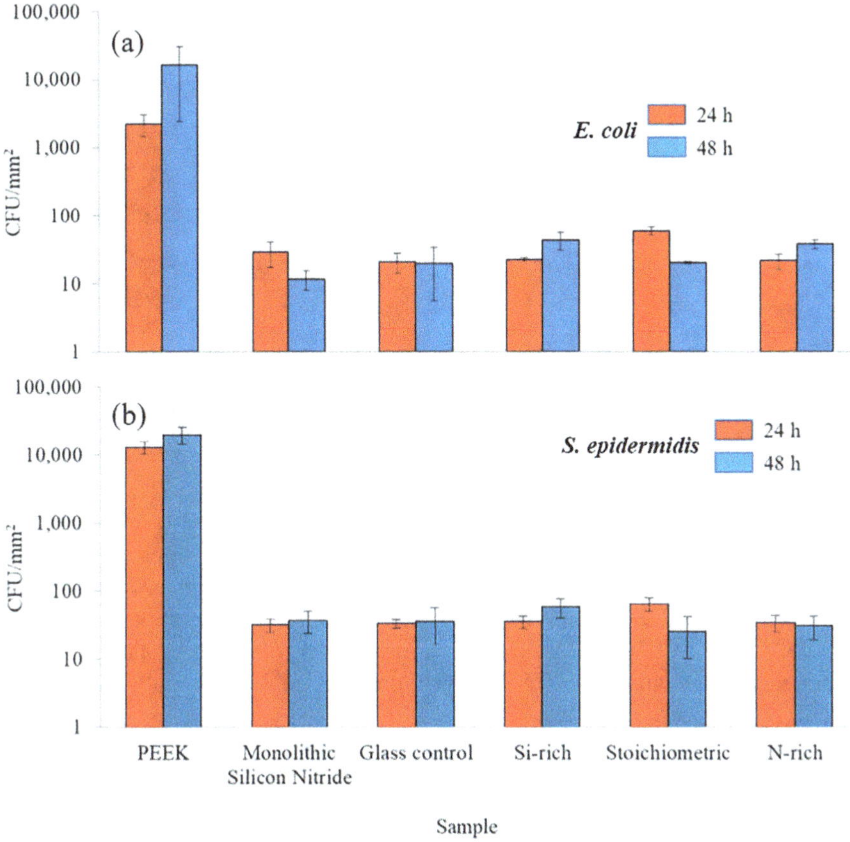

Fig. 5 CFU/mm^2 as measured after 24 and 48 h of in vitro exposure of PEEK, monolithic Si_3N_4, glass, and PVD-deposited Si_xN_y-coated glass samples to (**a**) *E. coli* and (**b**) *S. epidermidis*. The error bars on the chart depict the calculated standard error for each sample group. (Used with permission from [48])

N-rich Si_xN_y coatings in comparison with monolithic Si_3N_4 samples (positive control) and glass substrates (negative control).

The silicon-rich surface exhibited the largest amount of specific bone tissue volume, reaching an average amount five times higher than the glass reference. The other two PVD coatings showed statistically lower amounts of particular bone tissue formed, about 3.5 times greater than the glass controls. Similar trends were observed for the ALP tests. Again, the silicon-enriched conditions showed the highest values, while stoichiometric and nitrogen-enriched coatings showed no statistically significant differences. ALP reflects the cellular activity on the surface and represents a marker for osteoblast differentiation and mineralization. Higher ALP values for the silicon-rich PVD coating reflect enhanced stimulation of cellular differentiation [50, 51]. Note that the results presented in Fig. 6a, b suggest that N-rich films may have an osteopromotive effect. Overall, this study suggests that

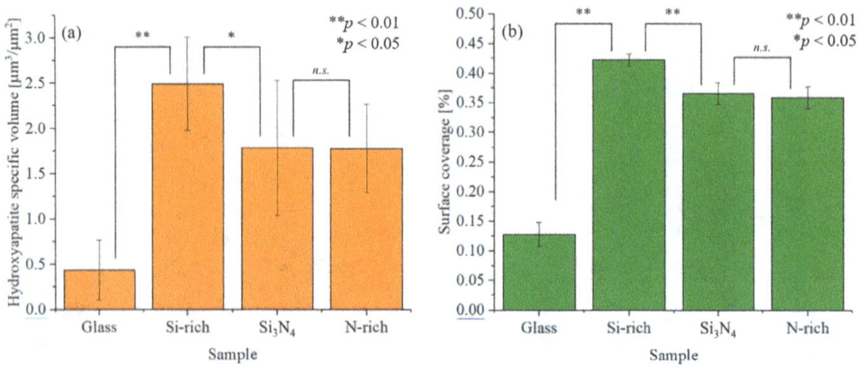

Fig. 6 (**a**) Hydroxyapatite specific volume as calculated using laser microscopy on the surface of the different samples, after in vitro testing with KUSA-A1 cells; (**b**) ALP area coverage testing results. (Used with permission from [48])

tailoring the surface properties of Si_xN_y films can enhance their bioactivity. However, both Si-rich and N-rich compositions may offer distinct advantages and drawbacks, necessitating application-specific optimization. In arthroplasty, where bioinert materials are frequently employed, Si-rich films may enhance osteointegration, while N-enriched stoichiometries might potentially reduce the risk of perioperative infections.

4 Applications

Significant progress has been made in the development of hard ceramic coatings for biomedical applications, particularly for SixNy thin films; recent reviews have discussed both the opportunities and challenges associated with these technologies [30, 52]. Si_xN_y coatings hold promise across various biomedical applications, from orthopedics to neurology, and their versatility continues to be of interest for future medical advancements.

4.1 Artificial Joints

Due to their ability to reduce wear, promote bone growth, and inhibit pathogens, Si_xN_y-based coatings hold significant promise for applications in the replacement and treatment of injured or diseased joints. They demonstrate significantly reduced wear rates compared to conventional CoCrMo alloys [16–20], making them highly promising for articulation joints. Research in this area has primarily focused on coatings deposited using reactive sputtering, often employing the high-power impulse magnetron sputtering (HiPIMS) technique. This method enables the deposition of amorphous and dense films, critical for achieving optimal wear resistance, adhesion, and minimal ion release from the underlying substrate. However, other deposition processes also hold considerable promise, including CVD methods, although significant optimization may still be required [53]. Si_xN_y coatings also demonstrate tunable dissolution properties through adjustments in composition [22–24]. This versatility positions them as highly desirable candidates for many joint replacement applications. At low concentrations, Si and N ions are biocompatible, and excessive amounts are excreted from the body through normal pathways [54–56]. This unique property can mitigate the negative effects of the body's immunological response to wear particles. Further investigations utilizing inflammatory cells would be valuable to elucidate the immunological response triggered by wear particles and their contribution to osteolysis. Examining these interactions could provide additional insights into biocompatibility beyond traditional cytotoxicity assays. It is also crucial to remember that osteolysis also involves complex interplay with the immune system, acting as a foreign body response to the implant itself. Moreover, deposition parameters exert a significant influence on the coating's composition, residual stresses, and density, fundamentally impacting its functional properties like wear resistance and adhesion. Consequently, comprehensive optimization of these parameters through tailored deposition techniques is imperative to ensure the coating's suitability for clinical applications. Nevertheless, significant advancements are expected in the development of functional Si_xN_y coatings for joint prostheses over the next decade.

4.2 *Bone Scaffolds and Plugs*

Although autograft, allograft, and xenograft remain the accepted standards for bone scaffolds, plugs, and flaps due to their inherent biocompatibility and osteoinductivity, abiotic materials are gaining traction because of their intrinsic advantages. Calcium phosphates, synthetic hydroxyapatite, bioglasses, specific polymers (e.g., polylactic acid, polyglycolic acid, and PEEK), and titanium alloys are emerging as acceptable alternatives to natural bone sources [57, 58]. Abiotic materials eliminate the risk of disease transmission associated with allograft and xenograft. However, their lower biological effectiveness and fabrication difficulties have historically limited their application.

The advent of 3D printing technology has revolutionized this field by enabling the fabrication of customized abiotic implants with complex geometries tailored to individual patient needs. Furthermore, surface functionalizing these 3D-printed designs with Si_xN_y coatings (via PVD or CVD methods) holds promise in upregulating bone growth and inhibiting bacterial biofilm formation. These advancements pave the way for the development of next-generation implants that have the potential to not only alleviate the pain associated with autograft harvesting but also eliminate the need for allografts and xenografts, offering a safer and potentially more effective solution for bone repair and regeneration [59–62].

4.3 Arthrodesis Devices

Chronic joint pain in the back, neck, wrists, and feet is a prevalent issue for both congenital and age-related causes. For severe cases, surgical interventions like spinal arthrodesis aim to fuse adjacent bones for pain relief and stability. This procedure typically involves disk removal and implantation of a spacer between the spinal segments to promote bone growth and fusion. Among the commonly used biomaterials for these spacers are PEEK, titanium, and monolithic Si_3N_4 [63]. While PEEK boasts an elastic modulus close to native bone, its bioinert nature hinders osseointegration and bacterial resistance. On the other hand, titanium, despite superior bioactivity compared to PEEK, also falls short in promoting bone-implant integration and bacterial inhibition. The functionalization of these materials with SixNy thin films deposited by PVD or CVD techniques presents a potential avenue for performance improvement, although further research is necessary for verification. 3D printing is also expected to play an important role for these devices [64]. Not only will it produce implants with complex bone-mimicking geometries, but it will also provide for improved fusion through rapid osseous ingrowth. These modifications aim to enhance integration and minimize perioperative infections, leading to potentially improved implant performance and clinical outcomes.

4.4 Dental Implants

Modern dental implants are predominately made of titanium. They typically have threaded cylindrical or conical shapes. The inclusion of threads on implant surfaces not only increases their area for direct bone-implant integration but also demonstrably enhances long-term implant stability. Furthermore, the introduction of macroporosity to implant surfaces has shown promise in promoting osseointegration. Multiple techniques have been employed to achieve enhanced surface porosity to mimic the structure of adjacent bone. Surface porosity enlarges bone-implant contact, theoretically enhancing both bone ingrowth (growth within pores) and bone ongrowth (adhesion to the surface). This increased surface area is accomplished through grit blasting using silica, hydroxyapatite powder, alumina, or titanium particles. A subsequent acid etch is performed to increase porosity and remove blast residue [65]. However, long-term clinical studies are still required to definitively validate the potential of highly porous implants, particularly in situations with compromised bone. The application of PVD or CVD Si_xN_y coatings to these highly porous surfaces potentially offers a meaningful upgrade to these implants. Although research is ongoing, the proposed advantages include improved osseointegration, enhanced bioactivity, and microbial resistance [66, 67].

4.5 Hematologic Coatings

Researchers have found that thin Si_xN_y coatings may have antithrombogenic properties. Adherence of blood clots (thrombi) to the surfaces of biomedical devices in contact with blood represents a significant unmet clinical challenge, as it can lead to device malfunction and potentially life-threatening complications. In one study, Wan et al. deposited Si-N-O films (~200 nm) on Si wafers using a modified PVD technique that combines plasma immersion ion implantation and cathodic arc deposition (PIII&D). They evaluated the blood compatibility of the coating in comparison with low-temperature isotropic pyrolytic carbon (LTIC) [68]. LTIC is a commonly used material for artificial heart valves. Both materials were assessed for blood platelet adhesion after a 2-h incubation period. Platelet aggregation, pseudopodium development, and spreading, as visualized by scanning electron microscopy (SEM), are well-established markers of material thrombogenicity, occurring before the actual formation of a thrombus [69]. While the Si–N–O films showed decreased platelet aggregation, all samples exhibited pseudopodia formation, with LTIC displaying the most extensive presence. The authors concluded that the observed platelet behavior and fibrinogen conformational changes suggest enhanced blood compatibility of the Si-N-O films compared to LTIC. This indicates their potential suitability for application in blood-contacting medical devices such as biological micro-electromechanical devices (bioMEMS), biosensors, and artificial implants. Coincidentally, in another study published several years later, Shi et al. showed essentially identical outcomes when comparing sputter-coated Si-N-O to Si-C-N thin films [70]. Their adhesion tests showed significant differences in the behavior of platelets on the Si_xN_y films deposited under different N_2 and CF_4 gas flow conditions. Similar to the earlier work, they concluded that Si–N–O coatings are great candidates for developing anti-thrombogenic surfaces in blood-contacting devices.

4.6 Neurophotonic Probes

Si_xN_y thin films have been used in neurophotonic probes for monitoring neural activity. These probes can be implanted in the brain to study neural circuits and facilitate communication between neurons. The probe consists of a thin, implantable shank made from materials compatible with neural tissue. The PECVD- or LPCVDD-deposited Si_xN_y film is a key component. It acts as a waveguide which is integrated onto the probe to deliver light to specific brain regions. The neurophotonic probe leverages optogenetics, a technique where neurons are genetically modified to express light-sensitive proteins (e.g., channelrhodopsin-2). When light is delivered through the Si_xN_y waveguide, it activates these proteins, allowing precise control over neural activity. By stimulating or silencing specific neurons, researchers can study neural circuits, behavior, and disease mechanisms.

Neurophotonic probes enable mapping of neural circuits by selectively activating or inhibiting specific neurons. This helps unravel brain connectivity and function. By stimulating specific brain regions, behavioral responses can be observed. Neurophotonic probes are used in disease models (e.g., epilepsy, Parkinson's) to investigate abnormal neural activity patterns and test potential therapies. In closed-loop systems, the probe detects neural activity (e.g., seizures) and responds by delivering light to modulate it. This approach holds promise for personalized treatments. Neurophotonic probes can also serve as imaging tools. By incorporating fluorophores, they allow simultaneous stimulation and imaging of neural activity. The advantage of a Si_xN_y waveguide in a neuronal probe is its high biocompatibility and its visible light operating range (430–645 nm), which covers the excitation spectra of opsins and fluorophores commonly used in neuroscience. Additionally, the probes can be miniaturized and suitable for implantation in animal or human brains. Neurophotonic probes incorporating Si_xN_y waveguides provide powerful tools for studying neural function, disease mechanisms, and potential therapeutic interventions. Their ability to precisely manipulate neural activity using light opens new avenues in neuroscience research and clinical applications [71–73].

4.7 Intraocular Pressure Sensors

Glaucoma, a leading cause of irreversible blindness, is characterized by increased intraocular pressure; continuous monitoring is crucial for managing glaucoma patients. Implantable intraocular pressure sensors provide real-time data, allowing physicians to tailor treatment plans; these sensors aid researchers in studying eye conditions and understanding pressure-related effects on ocular health. The functional part of the sensor consists of a thin deformable PECVD or LPCVD Si_xN_y thin film membrane (< Ø800 μm) that resides at the bottom of a small cavity in the sensor. The membrane acts as a partial mirror, reflecting part of the incident light, while allowing the remaining light to be reflected off of a glass substrate at the bottom of the cavity. As the membrane deforms based on ocular pressure, it generates interference fringes between reflected and transmitted light. Using these interference patterns, this micro-electromechanical device (MEMS) can precisely determine ocular pressure. Si_xN_y thin films are preferred because of their outstanding mechanical strength, biocompatibility, and optical transparency [74]. However, it was recently discovered that these devices exhibited small but noticeable drifts in calibration due to minute but continuous dissolution of the Si_xN_y coating. This resulted in a change in the Si_xN_y film thickness which altered the fringe pattern of the device. This drift was recognized and arrested by sputtering a thin (2 nm) titanium dioxide (TiO_2) coating onto the Si_xN_y film in advance of device assembly [75]. The TiO_2 barrier prevented direct contact of the ocular fluid with the Si_xN_y membrane.

5 Summary

Various deposition techniques (e.g., sputter coating, PECVD, LPCVD, and laser sintering) have been employed to create both stoichiometric and non-stoichiometric Si_xN_y-coated devices for various biomedical applications. Each technique offers distinct advantages and challenges, necessitating careful selection based on the intended use. These coatings share the inherent benefits of monolithic solid Si_3N_4, including high hardness, improved wear resistance, and bacteriostatic and osteogenic properties. Notably, the solubility of Si_xN_y thin films may offer a unique advantage: Wear particles can dissolve, potentially mitigating the adverse immune response observed with conventional metallic or oxide-ceramic implants. By leveraging their unique combination of biocompatibility, mechanical strength, bacteriostatic, osteogenic, tunable dissolution, and reduced wear properties, Si_xN_y thin films hold immense promise for various medical interventions. Their potential spans a diverse and growing list of applications, including artificial joints and bone scaffolds, arthrodesis devices, biocompatible and osseointegrated dental implants, functional hematologic coatings, high-precision neurophotonic probes, and next-generation intraocular pressure sensors.

References

1. Heimann, R.B. Silicon Nitride, A Close to Ideal Ceramic Material for Medical Application. *Ceramics* **2021**, *4*, 208–223.
2. Du, X.; Lee, S.S.; Blugan, G.; Ferguson, S.J. Silicon Nitride as a Biomedical Material: An Overview. *Int. J. Mol. Sci.* **2022**, *23*, 6551, https://doi.org/10.3390/ijms23126551.
3. Li, P.H.; Chu, P.K. Thin Film Deposition Technologies and Processing of Biomaterials. In *Thin Film Coatings for Biomaterials and Biomedical Applications*; Griesser, H.J.B.T.-T.F.C. for B. and B.A., Ed.; Woodhead Publishing, 2016; pp. 3–28 ISBN 978-1-78242-453-6.
4. Baptista, A.; Silva, F.; Porteiro, J.; Míguez, J.; Pinto, G. Sputtering Physical Vapour Deposition (PVD) Coatings: A Critical Review on Process Improvement and Market Trend Demands. *Coatings* 2018, *8*.
5. Sun, L.; Berndt, C.C.; Gross, K.A.; Kucuk, A. Material Fundamentals and Clinical Performance of Plasma-Sprayed Hydroxyapatite Coatings: A Review. *J. Biomed. Mater. Res. Appl. Biomater.* **2001**, *58*, 570–592, doi:https://doi.org/10.1002/jbm.xxxx.
6. Vila, M.; Prieto, C.; Ramírez, R. Electrical Behavior of Silicon Nitride Sputtered Thin Films. *Thin Solid Films* **2004**, *459*, 195–199, doi:https://doi.org/10.1016/j.tsf.2003.12.082.
7. Zhou, H.; Persson, C.; Xia, W.; Engqvist, H. Effects of N/Si Ratio on Mechanical Properties of Amorphous Silicon Nitride Coating. *Mater. Res. Express* **2023**, *10*, 115403, doi:https://doi.org/10.1088/2053-1591/ad0eac.
8. Berghaus, B.; Burkhardt, W. Method of Coating Metallic Articles. *US Pat. 2,251,510* **1941**, 1–3.
9. Mattox, D.M. *Handbook of Physical Vapor Deposition (PVD) Processing*; Mattox, D.M., Ed.; Second Edi.; Elsevier: Burlington, MA, 2010; ISBN 978-0-8155-2037-5.
10. Kelly, P.J.; Arnell, R.D. Magnetron Sputtering: A Review of Recent Developments and Applications. *Vacuum* **2000**, *56*, 159–172, doi:https://doi.org/10.1016/S0042-207X(99)00189-X.

11. Sarakinos, K.; Alami, J.; Konstantinidis, S. High Power Pulsed Magnetron Sputtering: A Review on Scientific and Engineering State of the Art. *Surf. Coatings Technol.* **2010**, *204*, 1661–1684, doi:https://doi.org/10.1016/j.surfcoat.2009.11.013.
12. Choy, K.L. Chemical Vapour Deposition of Coatings. *Prog. Mater. Sci.* **2003**, *48*, 57–170, doi: https://doi.org/10.1016/S0079-6425(01)00009-3.
13. Kaloyeros, A.E.; Jové, F.A.; Goff, J.; Arkles, B. Review—Silicon Nitride and Silicon Nitride-Rich Thin Film Technologies: Trends in Deposition Techniques and Related Applications. *ECS J. Solid State Sci. Technol.* **2017**, *6*, P691–P714, doi:https://doi.org/10.1149/2.0011710jss.
14. Kaloyeros, A.E.; Pan, Y.; Goff, J.; Arkles, B. Review—Silicon Nitride and Silicon Nitride-Rich Thin Film Technologies: State-of-the-Art Processing Technologies, Properties, and Applications. *ECS J. Solid State Sci. Technol.* **2020**, *9*, 063006, doi:https://doi.org/10.1149/2162-8777/aba447.
15. Zanocco, M.; Boschetto, F.; Zhu, W.; Marin, E.; McEntire, B.J.; Bal, B.S.; Adachi, T.; Yamamoto, T.; Kanamura, N.; Ohgitani, E.; *et al.* 3D-Additive Deposition of an Antibacterial and Osteogenic Silicon Nitride Coating on Orthopaedic Titanium Substrate. *J. Mech. Behav. Biomed. Mater.* **2020**, *103*, 103557, doi:https://doi.org/10.1016/j.jmbbm.2019.103557.
16. Pettersson, M.; Tkachenko, S.; Schmidt, S.; Berlind, T.; Jacobson, S.; Hultman, L.; Engqvist, H.; Persson, C. Mechanical and Tribological Behavior of Silicon Nitride and Silicon Carbon Nitride Coatings for Total Joint Replacements. *J. Mech. Behav. Biomed. Mater.* **2013**, *25*, 41–7, doi:https://doi.org/10.1016/j.jmbbm.2013.05.002.
17. Olofsson, J.; Pettersson, M.; Teuscher, N.; Heilmann, A.; Larsson, K.; Grandfield, K.; Persson, C.; Jacobson, S.; Engqvist, H. Fabrication and Evaluation of SixNy Coatings for Total Joint Replacements. *J. Mater. Sci. Mater. Med.* **2012**, *23*, 1879–89, doi:https://doi.org/10.1007/s10856-012-4625-3.
18. Filho, L.; Schmidt, S.; Leifer, K.; Engqvist, H.; Högberg, H.; Persson, C. Towards Functional Silicon Nitride Coatings for Joint Replacements. *Coatings* **2019**, *9*, 73, doi:https://doi.org/10.3390/coatings9020073.
19. Filho, L.C.; Schmidt, S.; Goyenola, C.; Skjöldebrand, C.; Engqvist, H.; Högberg, H.; Toble, M.; Persson, C. The Effect of N, C, Cr, and Nb Content on Silicon Nitride Coatings for Joint Applications. *Materials (Basel).* **2020**, *13*, doi:https://doi.org/10.3390/MA13081896.
20. Filho, L.C.; Schmidt, S.; Lopez, A.; Cogrel, M.; Leifer, K.; Engqvist, H.; Högberg, H.; Persson, C. The Effect of Coating Density on Functional Properties of SiNx Coated Implants. *Materials (Basel).* **2019**, *12*, 3370, doi:https://doi.org/10.3390/ma12203370.
21. Lal, S.; Hall, R.M.; Tipper, J.L. A Novel Method for Isolation and Recovery of Ceramic Nanoparticles and Metal Wear Debris from Serum Lubricants at Ultra-Low Wear Rates. *Acta Biomater.* **2016**, *42*, 420–428, doi:https://doi.org/10.1016/j.actbio.2016.07.004.
22. Pettersson, M.; Skjoldebrand, C.; Filho, L.; Engqvist, H.; Persson, C. Morphology and Dissolution Rate of Wear Debris from Silicon Nitride Coatings. *ACS Biomater. Sci. Eng.* **2016**, 1–28.
23. Pettersson, M.; Bryant, M.; Schmidt, S.; Engqvist, H.; Hall, R.M.; Neville, A.; Persson, C. Dissolution Behaviour of Silicon Nitride Coatings for Joint Replacements. *Mater. Sci. Eng. C* **2016**, *62*, 497–505, doi:https://doi.org/10.1016/j.msec.2016.01.049.
24. Skjöldebrand, C.; Echeverri, E.; Hulsart-Billström, G.; Persson, C. Tailoring the Dissolution Rate and *in vitro* Cell Response of Silicon Nitride Coatings through Combinatorial Sputtering with Chromium and Niobium. *Biomater. Sci.* **2022**, *10*, 3757–3769, doi:https://doi.org/10.1039/d1bm01978c.
25. Kundra, A.; Jain, A.; Banga, A.; Bajaj, G.; Kar, P. Evaluation of Plasma Ammonia Levels in Patients with Acute Liver Failure and Chronic Liver Disease and its Correlation with the Severity of Hepatic Encephalopathy and Clinical Features of Raised Intracranial Tension. *Clin. Biochem.* **2005**, *38*, 696–699, doi:https://doi.org/10.1016/j.clinbiochem.2005.04.013.
26. Mazzocchi, M.; Bellosi, A. On the Possibility of Silicon Nitride as a Ceramic for Structural Orthopaedic Implants. Part I: Processing, Microstructure, Mechanical Properties, Cytotoxicity. *J. Mater. Sci. Mater. Med.* **2008**, *19*, 2881–2887, doi:https://doi.org/10.1007/s10856-008-3417-2.

27. Hnatko, M.; Hi, M.; Labudová, M.; Galusková, D.; Sedlá, J.; Len, Z.; Pavol, Š. Bioactive Silicon Nitride by Surface Thermal Treatment. *J. Eur. Ceram. Soc.* **2019**, doi:https://doi.org/10.1016/j.jeurceramsoc.2019.12.053.
28. ISO 10993 Biological Evaluation of Medical Devices. *Int. Stand. Organ.* **2009**.
29. ISO 20502 Fine Ceramics (Advanced Ceramics, Advanced Technical Ceramics) – Determination of Adhesion of Ceramic Coatings by Scratch Testing. *Int. Stand. Organ.* **2015**.
30. Skjöldebrand, C.; Tipper, J.L.; Hatto, P.; Bryant, M.; Hall, R.M.; Persson, C. Current Status and Future Potential of Wear-Resistant Coatings and Articulating Surfaces for Hip and Knee Implants. *Mater. Today Bio* **2022**, *15*, 100270, doi:https://doi.org/10.1016/j.mtbio.2022.100270.
31. ISO 26443: Fine Ceramics (Advanced Ceramics, Advanced Technical Ceramics) – Rockwell Indentation Test for Evaluation of Adhesion of Ceramic Coatings. *Int. Stand. Organ.* **2008**.
32. Pezzotti, G.; McEntire, B.J.; Bock, R.; Boffelli, M.; Zhu, W.; Vitale, E.; Puppulin, L.; Adachi, T.; Yamamoto, T.; Kanamura, N.; *et al.* Silicon Nitride: A Synthetic Mineral for Vertebrate Biology. *Sci. Rep.* **2016**, *6*, 31717, doi:https://doi.org/10.1038/srep11384.
33. Pezzotti, G.; Bock, R.M.; Adachi, T.; Rondinella, A.; Boschetto, F.; Zhu, W.; Marin, E.; McEntire, B.; Bal, B.S.; Mazda, O. Silicon Nitride Surface Chemistry: A Potent Regulator of Mesenchymal Progenitor Cell Activity in Bone Formation. *Appl. Mater. Today* **2017**, *9*, 82–95, doi:https://doi.org/10.1016/j.apmt.2017.05.005.
34. Pezzotti, G.; Marin, E.; Adachi, T.; Rondinella, A.; Boschetto, F.; Zhu, W.-L.; Sugano, N.; Bock, R.M.; McEntire, B.J.; Bal, B.S. Bioactive Silicon Nitride: A New Therapeutic Material for Osteoarthropathy. *Sci. Rep.* **2017**, *7*, 44848.
35. Pezzotti, G.; Oba, N.; Zhu, W.; Marin, E.; Rondinella, A.; Boschetto, F.; McEntire, B.J.; Yamamoto, K.; Bal, B.S. Human Osteoblasts Grow Transitional Si/N Apatite in Quickly Osteointegrated Si_3N_4 Cervical Insert. *Acta Biomater.* **2017**, *64*, 411–420.
36. Pezzotti, G.; Marin, E.; Zanocco, M.; Boschetto, F.; Zhu, W.; McEntire, B.J.; Bal, B.S.; Adachi, T.; Yamamoto, T.; Kanamura, N. Osteogenic Enhancement of Zirconia-Toughened Alumina with Silicon Nitride and Bioglass®. *Ceramics* **2019**, *2*, 554–567, doi:https://doi.org/10.3390/ceramics2040043.
37. Pezzotti, G.; Bock, R.M.; McEntire, B.J.; Adachi, T.; Marin, E.; Boschetto, F.; Zhu, W.; Mazda, O.; Bal, B.S. *In vitro* Antibacterial Activity of Oxide and Non-Oxide Bioceramics for Arthroplastic Devices: I. In situ Time-Lapse Raman Spectroscopy. *Analyst* **2018**, *143*, 3708–3721, doi:https://doi.org/10.1039/C8AN00233A.
38. Boschetto, F.; Toyama, N.; Horiguchi, S.; Bock, R.M.; McEntire, B.J.; Adachi, T.; Marin, E.; Zhu, W.; Mazda, O.; Bal, B.S.; *et al. In vitro* Antibacterial Activity of Oxide and Non-Oxide Bioceramics for Arthroplastic Devices: II. In situ Time-Lapse Fourier Transform Infrared Spectroscopy. *Analyst* **2018**, *143*, 2128–2140, doi:https://doi.org/10.1039/C8AN00234G.
39. Pezzotti, G. Silicon Nitride: A Bioceramic with a Gift. *ACS Appl. Mater. Interfaces* **2019**, *11*, 26619–26636, doi:https://doi.org/10.1021/acsami.9b07997.
40. Pezzotti, G.; Boschetto, F.; Ohgitani, E.; Fujita, Y.; Shin-Ya, M.; Adachi, T.; Yamamoto, T.; Kanamura, N.; Marin, E.; Zhu, W.; *et al.* Mechanisms of Instantaneous Inactivation of SARS-CoV-2 by Silicon Nitride Bioceramic. *Mater. Today Bio* **2021**, 100144, doi:https://doi.org/10.1016/j.mtbio.2021.100144.
41. Pezzotti, G.; Boschetto, F.; Ogitani, E.; Fujita, Y.; Zhu, W.; Marin, E.; McEntire, B.J.; Bal, B.S.; Mazda, O. Silicon Nitride: A Potent Solid-State Bioceramic Inactivator of ssRNA Viruses. *Sci. Rep.* **2021**, *11*, 2977, doi:https://doi.org/10.1038/s41598-021-82608-3.
42. Katsaros, I.; Zhou, Y.; Welch, K.; Xia, W.; Persson, C. Bioactive Silicon Nitride Implant Surfaces with Maintained Antibacterial Properties. *J. Funct. Biomater.* **2022**, *13*, 129.
43. Katsaros, I.; Ling, J.; Donzel-Gargand, O.; Akusjärvi, G.; Lundkvist, Å.; Persson, C.; Xia, W.; Engqvist, H. Antiviral Properties of Oxidized Silicon Nitride Against SARS-CoV-2, Uppsala University: Uppsala, Sweden, 2023.

44. Katsaros, I.; Carlsson, A.; Persson, C.; Akusjärvi, G.; Xia, W.; Engqvist, H. Contact Inactivation of Human Adenovirus Type 5 by Silicon Nitride Ceramics, Uppsala University: Uppsala, Sweden, 2023.
45. Pezzotti, G. Silicon Nitride as a Biomaterial. *J. Ceram. Soc. Japan* **2023**, *131*, 398–428, doi: https://doi.org/10.2109/jcersj2.23057.
46. Ilyas, A.; Lavrik, N. V; Kim, H.K.W.; Aswath, P.B.; Varanasi, V.G. Enhanced Interfacial Adhesion and Osteogenesis for Rapid "Bone-like" Biomineralization by PECVD-Based Silicon Oxynitride Overlays. *ACS Appl. Mater. Interfaces* **2015**, *7*, 15368–15379, doi:https://doi.org/10.1021/acsami.5b03319.
47. Xu, Z.; Wu, H.; Wang, F.; Kaewmanee, R.; Pan, Y.; Wang, D.; Qu, P.; Wang, Z.; Hu, G.; Zhao, J.; *et al.* A Hierarchical Nanostructural Coating of Amorphous Silicon Nitride on Polyetheretherketone with Antibacterial Activity and Promoting Responses of rBMSCs for Orthopedic Applications. *J. Mater. Chem. B* **2019**, *7*, 6035–6047, doi:https://doi.org/10.1039/c9tb01565e.
48. Marin, E.; Boschetto, F.; Zanocco, M.; Zhu, W.; Adachi, T.; Kanamura, N.; Yamamoto, T.; McEntire, B.J.; Jones, E.N.; Powell, C.; *et al.* Biological Responses to Silicon and Nitrogen-Rich PVD Silicon Nitride Coatings. *Mater. Today Chem.* **2021**, *19*, 100404, doi:https://doi.org/10.1016/j.mtchem.2020.100404.
49. An, Y.H.; Friedman, R.J. Concise Review of Mechanisms of Bacterial Adhesion to Biomaterial Surfaces. *J. Biomed. Mater. Res.* **1998**, *43*, 338–348, doi:https://doi.org/10.1002/(SICI)1097-4636(199823)43:3<338::AID-JBM16>3.0.CO;2-B.
50. Ilmer, M.; Karow, M.; Geissler, C.; Jochum, M.; Neth, P. Human Osteoblast—Derived Factors Induce Early Osteogenic Markers in Human Mesenchymal Stem Cells. *Tissue Eng.* **2009**, *15*, 2397–2409.
51. Belbachir, K.; Noreen, R.; Gouspillou, G.; Petibois, C. Collagen Types Analysis and Differentiation by FTIR Spectroscopy. *Anal. Bioanal. Chem.* **2009**, *395*, 829–837, doi:https://doi.org/10.1007/s00216-009-3019-y.
52. Heimann, R.B. Silicon Nitride Ceramics: Structure, Synthesis, Properties, and Biomedical Applications. *Materials (Basel).* **2023**, *16*, 1–16, doi:https://doi.org/10.3390/ma16145142.
53. Zhou, H.; Persson, C.; Xia, W.; Engqvist, H. SiNx Coating Deposition on CoCr by Plasma-Enhanced Chemical Vapor Deposition. *Biomed. Mater. Devices* **2023**, 1–8, doi:https://doi.org/10.1007/s44174-023-00083-y.
54. Koeppen, B.M. Renal Regulation of Acid-Base Balance. *APS Refresh. Course Rep. - Am. Physiol. Soceity* **1998**, *20*, S132–S141, doi:https://doi.org/10.14842/jpnjnephrol1959.22.772.
55. Vladescu, A.; Titorencu, I.; Dekhtyar, Y.; Jinga, V.; Pruna, V.; Balaceanu, M.; Dinu, M.; Pana, I.; Vendina, V.; Braic, M. *In vitro* Biocompatibility of Si Alloyed Multi-Principal Element Carbide Coatings. *PLoS One* **2016**, *11*, 1–18, doi:https://doi.org/10.1371/journal.pone.0161151.
56. Shekaari, A.; Jafari, M. Biocompatibility of 2D Silicon Nitride: Interaction at the Nano-Biointerface. *Mater. Res. Express* **2021**, *8*, doi:https://doi.org/10.1088/2053-1591/ac24cc.
57. Oryan, A.; Alidadi, S.; Moshiri, A.; Maffulli, N. Bone Regenerative Medicine: Classic Options, Novel Strategies, and Future Directions. *J. Orthop. Surg. Res.* **2014**, *9*, 1–27, doi:https://doi.org/10.1186/1749-799X-9-18.
58. Haugen, H.J.; Lyngstadaas, S.P.; Rossi, F.; Perale, G. Bone Grafts: Which is the Ideal Biomaterial? *J. Clin. Periodontol.* **2019**, *46*, 92–102, doi:https://doi.org/10.1111/jcpe.13058.
59. Nyberg, E.L.; Farris, A.L.; Hung, B.P.; Dias, M.; Garcia, J.R.; Dorafshar, A.H.; Grayson, W.L. 3D-Printing Technologies for Craniofacial Rehabilitation, Reconstruction, and Regeneration. *Ann. Biomed. Eng.* **2017**, *45*, 45–57, doi:https://doi.org/10.1007/s10439-016-1668-5.
60. Maroulakos, M.; Kamperos, G.; Tayebi, L.; Halazonetis, D.; Ren, Y. Applications of 3D Printing on Craniofacial Bone Repair: A Systematic Review. *J. Dent.* **2019**, *80*, 1–14, doi: https://doi.org/10.1016/j.jdent.2018.11.004.

61. Singh, S.; Prakash, C.; Ramakrishna, S. 3D Printing of Polyether-Ether-Ketone for Biomedical Applications. *Eur. Polym. J.* **2019**, *114*, 234–248, doi:https://doi.org/10.1016/j.eurpolymj.2019.02.035.
62. Jariwala, S.H.; Lewis, G.S.; Bushman, Z.J.; Adair, J.H.; Donahue, H.J. 3D Printing of Personalized Artificial Bone Scaffolds. *3D Print. Addit. Manuf.* **2015**, *2*, 56–64, doi:https://doi.org/10.1089/3dp.2015.0001.
63. Warburton, A.; Girdler, S.J.; Mikhail, C.M.; Ahn, A.; Cho, S.K. Biomaterials in Spinal Implants: A Review. *Neurospine* **2020**, *17*, 101–110, doi:https://doi.org/10.14245/ns.1938296.148.
64. Wilcox, B.; Mobbs, R.J.; Wu, A.-M.; Phan, K. Systematic Review of 3D Printing in Spinal Surgery: The Current State of Play. *J. Spine Surg.* **2017**, *3*, 433–443, doi:https://doi.org/10.21037/jss.2017.09.01.
65. Alghamdi, H.S.; Jansen, J.A. The Development and Future of Dental Implants. *Dent. Mater. J.* **2020**, *39*, 167–172, doi:https://doi.org/10.4012/dmj.2019-140.
66. Raza, S.M.; Khurshid, Z.; Zafar, M.S.; Najeeb, S.; Ul Yaqin, S.A. 15 - Silicon Nitride (SiN): An Emerging Material for Dental Implant Applications. In *Woodhead Publishing Series in Biomaterials*; Zafar, M.S., Khurshid, Z., Khan, A.S., Najeeb, S., Sefat, F.B.T.-D.I., Eds.; Woodhead Publishing, 2020; pp. 287–299 ISBN 978-0-12-819586-4.
67. Mohammadi, H.; Beddu, S.; Petrů, M.; Sepantafar, M.; Ebadi, M.; Yap, B.K.; Bang, L.T.; Yong, T.C.; Ramesh, S.; Rahimian Koloor, S.S.; *et al.* Advances in Silicon Nitride Ceramic Biomaterials for Dental Applications—A Review. *J. Mater. Res. Technol.* **2024**, *28*, 2778–2791, doi:https://doi.org/10.1016/j.jmrt.2023.12.186.
68. Wan, G.J.; Huang, N.; Kwok, S.C.H.; Shao, Z.Y.; Zhao, A.S.; Yang, P.; Chu, P.K. Si-N-O Films Synthesized by Plasma Immersion Ion Implantation and Deposition (PIII&D) for Blood-Contacting Biomedical Applications. *IEEE Trans. Plasma Sci.* **2006**, *34*, 1160–1165.
69. Gorbet, M.B.; Sefton, M. V Biomaterial-Associated Thrombosis: Roles of Coagulation Factors, Complement, Platelets and Leukocytes. *Biomaterials* **2004**, *25*, 5681–5703, doi:https://doi.org/10.1016/j.biomaterials.2004.01.023.
70. Shi, Z.; Wang, Y.; Du, C.; Huang, N.; Wang, L.; Ning, C. Silicon Nitride Films for the Protective Functional Coating: Blood Compatibility and Biomechanical Property Study. *J. Mech. Behav. Biomed. Mater.* **2012**, *16*, 9–20, doi:https://doi.org/10.1016/j.jmbbm.2012.06.020.
71. Froeter, P.; Huang, Y.; Cangellaris, O. V.; Huang, W.; Dent, E.W.; Gillette, M.U.; Williams, J. C.; Li, X. Toward Intelligent Synthetic Neural circuits: Directing and Accelerating Neuron Cell Growth by Self-Rolled-Up Silicon Nitride Microtube Array. *ACS Nano* **2014**, *8*, 11108–11117, doi:https://doi.org/10.1021/nn504876y.
72. Rudmann, L.; Alt, M.T.; Ashouri Vajari, D.; Stieglitz, T. Integrated Optoelectronic Microprobes. *Curr. Opin. Neurobiol.* **2018**, *50*, 72–82, doi:https://doi.org/10.1016/j.conb.2018.01.010.
73. Sacher, W.D.; Luo, X.; Yang, Y.; Chen, F.-D.; Lordello, T.; Mak, J.C.C.; Liu, X.; Hu, T.; Xue, T.; Guo-Qiang Lo, P.; *et al.* Visible-Light Silicon Nitride Waveguide Devices and Implantable Neurophotonic Probes on Thinned 200 mm Silicon Wafers. *Opt. Express* **2019**, *27*, 37400–37418, doi:https://doi.org/10.1364/OE.27.037400.
74. Phan, A.; Truong, P.; Trumpp, J.; Talke, F.E. Design of an Optical Pressure Measurement System for Intraocular Pressure Monitoring. *IEEE Sens. J.* **2018**, *18*, 61–68, doi:https://doi.org/10.1109/JSEN.2017.2767539.
75. Schade, C.; Phan, A.; Joslin, K.; Talke, F.E. Material Loss of Silicon Nitride Thin Films in a Simulated Ocular Environment. *Microsyst. Technol.* **2021**, *27*, 2263–2268, doi:https://doi.org/10.1007/s00542-020-05096-x.

Silicon Nitride for Spinal Arthrodesis

Micah W. Smith and Jasmine M. Higgins

1 Introduction

Nell Tays, a 78-year-old woman, was experiencing debilitating lower back pain that continued to worsen and eventually extended to her legs. Her pain profoundly impaired her mobility and quality of life, causing her to decline invitations to the activities she loved. She was unable to attend her granddaughters' performances and sports events due to the severe pain she was experiencing. Nell's pain became so unbearable that it impaired her ability to buy groceries, keep up with housework, and engage in any physical activity. After a thorough investigation of her spine, the doctor pinpointed the origin of the pain to a nerve that was being irritated at the L5 vertebral level, located just above the buttocks. The severity of the pain dictated immediate surgery to decompress the nerve root and fuse the vertebrae at that level. When Nell awoke from her surgery, she experienced immediate relief. Six months later, Nell attended Grandparents' Day at her granddaughter's school with her mobility fully restored. Her quality of life improved significantly, and she was able to return to the activities she loved.[1]

As illustrated by Ms. Tays' story, low back pain can severely impact many aspects of life. An individual's back pain may be so severe that they may be unable to work, finding it difficult to pay bills and purchase groceries in addition to being unable to make the trip to obtain such goods. Lower back pain can have a significant monetary impact on individuals, but it can also limit their ability to participate in fulfilling activities like sports or spending time with family. The cumulative effect of these factors can significantly diminish patients' quality of life, highlighting the

[1] This illustrative clinical case reflects an actual patient's testimonial submitted to Fort Wayne Orthopedics.

M. W. Smith (✉) · J. M. Higgins
Orthopedics Northeast SpineONE, Fort Wayne, IN, USA
e-mail: msmith@orthone.com

B. S. Bal et al. (eds.), *Silicon Nitride Bioceramics*,
https://doi.org/10.1007/978-3-031-67047-3_9

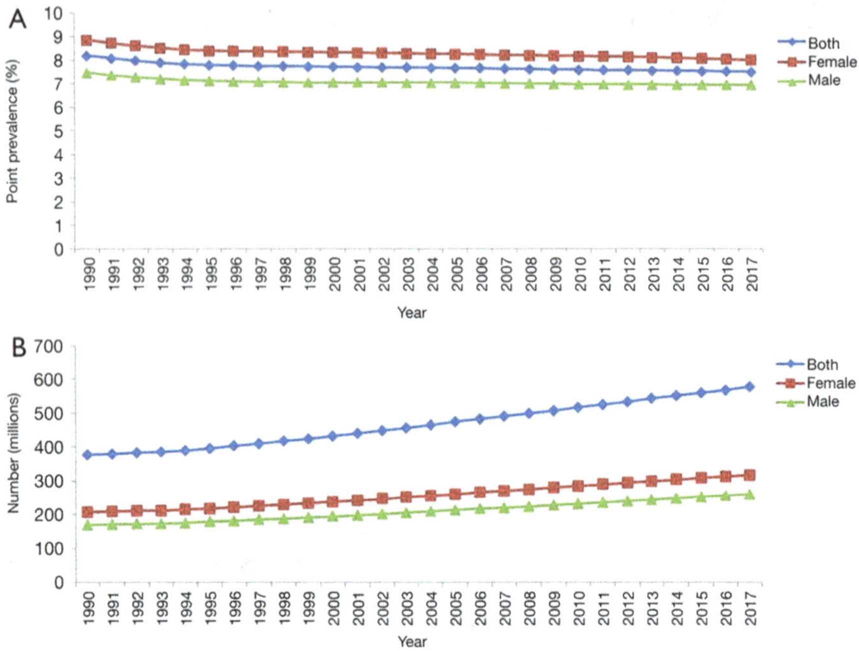

Fig. 1 The prevalence trend of lower back pain. (**a**) The age-standardized point prevalence of low back pain from 1990 to 2017; and (**b**) The estimated prevalent number of people with low back pain from 1990 to 2017. (Used with permission under CC BY-NC-ND 4.0 DEED International License [1])

critical need for novel techniques and biomaterials that demonstrably improve pain relief and functional outcomes.

Worldwide, over 500 million people are currently experiencing debilitating low back pain which is the most common musculoskeletal problem as well as the leading global cause of years living with disability. Back pain severely limits activity and results in absenteeism from work, making it a considerable medical and economic burden. Shown in Fig. 1 are the age-standardized point prevalence data related to lower back pain from 1990 to 2017 and the estimated prevalence of people with low back pain from 1990 to 2017, both by gender and number [1]. Figure 2 illustrates the global burden of years lived with this disability. As depicted, the years lived with lower back pain increased from 42.5 million in 1990 to 64.9 million in 2017. This burden was consistently higher for women compared to men at both time points. The data suggest a higher prevalence among individuals aged 30–59 years, coinciding with their prime working years when high productivity is often expected and needed. In the United States, 65 million Americans have reported back pain as a health issue and low back pain is the number one cause of absenteeism in the US workforce. The financial and medical burden on the individual translates to a national burden since back pain is one of the most expensive conditions in the US, costing an estimated $34 billion yearly in direct healthcare costs [2]. Research into surgical and

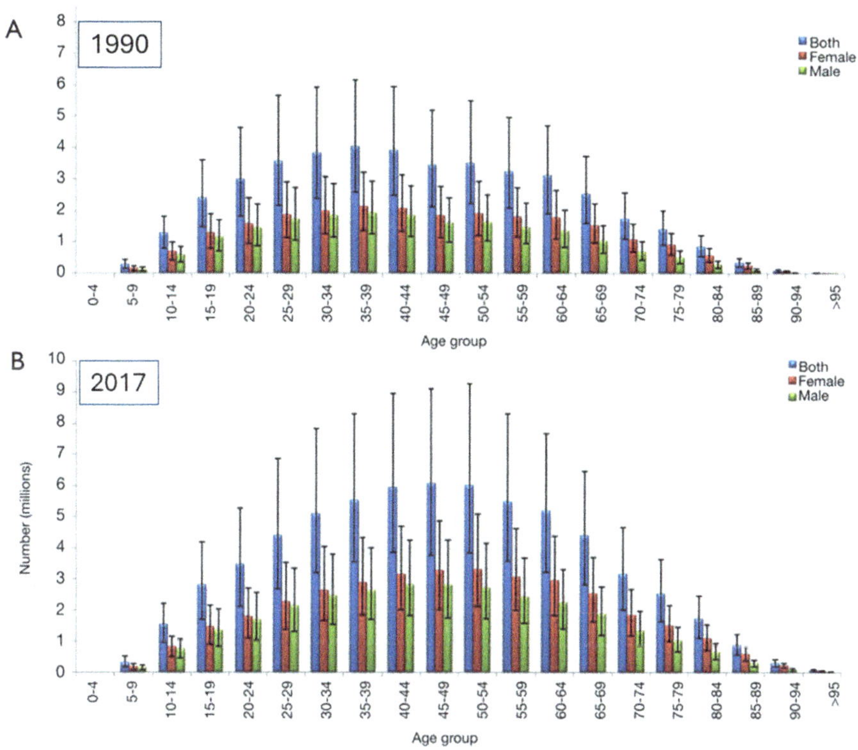

Fig. 2 The age-specific number of years lived with disability by age and gender (with uncertainty intervals) in (**a**) 1990 and (**b**) 2017. (Modified and used with permission under CC BY-NC-ND 4.0 DEED International License [1])

nonsurgical means of addressing back pain can address significant individual and societal burdens. Currently, lumbar spinal arthrodesis, or bone fusion, is the main surgical treatment to alleviate lower back pain caused by degenerative spine disorders, including spondylolisthesis, where one vertebra slips on the adjacent one, resulting in pain and instability [3].

The Spine Patient Outcomes Research Trial (SPORT), a hallmark randomized controlled trial demonstrated significant improvement in pain and function from surgical intervention when compared to nonoperative management for the three most prevalent conditions (i.e., intervertebral disc herniation, spinal stenosis, and degenerative spondylolisthesis) for periods of 6 weeks to 4 years [4, 5]. Optimal outcomes for spine fusions rely on achieving a solid arthrodesis (i.e., bone healing between one vertebral body and the adjacent segment). Recent technological advancements have demonstrably facilitated the achievement of this goal, and ongoing research holds promise for even more favorable clinical outcomes. Since the early 1900s, doctors and scientists have recognized the need for improvements in spinal implants. These implants, also called spacers, occupy the space between adjacent vertebral bodies

Table 1 Timeline of the introduction of spinal interbody materials

Timeline	Introduction of spinal interbody implant material
1930	Autograft
1950	Tri-corticate autograft
1960	Stainless steel
1980	Titanium and alloys
1990	Carbon-fiber-PEEK, silicon nitride
2000	Nitinol, tantalum, cobalt–chromium–molybdenum

when the interposed diseased disc is surgically removed. Implant materials are chosen to optimize the chances of achieving a solid bony union between the vertebral bodies.

In the 1930s, the first surgical interventions used bone autograft for spinal fusion. By 1960, doctors began utilizing stainless steel (SS), and by the 1980s, titanium alloys were adopted as biomaterials for spinal fusion implants [6]. At the turn of the century, carbon fiber polyetheretherketone (PEEK), and to a lesser extent silicon nitride (Si_3N_4) began to be incorporated into spinal implants.

Most recently, additive manufacturing and materials such as nitinol, tantalum, and cobalt–chromium–molybdenum have entered the ring as contenders for the ideal biomaterial for spinal fusion implants [7]. These materials are still being investigated with limited clinical implementation. Table 1 represents the timeline of these material science advancements.

Today, the most popular abiotic materials used in spinal spacers are titanium alloys and PEEK [8], but these materials have their limitations. Finding the best biomaterial for spinal implants is important to the future success of spine surgery. This is where Si_3N_4 may have advantages over today's implants. Traditionally, it has been used in high-temperature combustion engines, rocket thrusters, and gas turbines [9, 10]. However, Si_3N_4's high fracture toughness, low wear, and strength make it an appealing compound for biomedical purposes [11]. Its novel use as a spinal implant began in the mid-1980s. At that time, the first research team to use Si_3N_4 in spinal implants found significant pain reduction and spinal fusion across reaction-bonded Si_3N_4 intervertebral spacers [12]. In addition to its mechanical properties, Si_3N_4 has been shown to promote osseointegration which may contribute to faster fusion times and lower implant subsidence rates [13, 14]. Its surface texture and chemistry are easily modifiable to inhibit microbe growth [15, 16]. Alongside these favorable mechanical and chemical properties, it has improved radiolucency making it easy to capture and analyze with modern imaging modalities, such as X-rays, CT, and MRI [17]. Silicon nitride, at first glance, is comparable to other traditional spinal implant materials. This chapter will cover the essential properties of intervertebral spacers, analyze the currently employed biomaterials in this application, and explore the specific chemical and mechanical characteristics of Si_3N_4. Subsequently, the chapter critically evaluates the clinical trial outcomes associated with Si_3N_4 use in intervertebral spacers and the broader suitability of Si_3N_4 as a biomaterial for various orthopedic and dental implants.

2 Required and Desirable Properties of Intervertebral Spacers

To assess the properties of intervertebral spacers, it is instructive to revisit the history and development of spinal implants [18–20]. The first successful spinal fusion was performed by the American surgeon Berthold Hadra in 1891. Segmental fusion was achieved by wrapping wires around the C6-C7 cervical vertebrae for stabilization. Eighteen years later, a German surgeon, Fritz Lange, straightened the spine of a scoliosis patient with steel wires and silver fasteners. Unfortunately, this method led to increased rates of infection and irritation from the sharp wire, highlighting the need for materials that could reduce these complications. This led to the use of celluloid bars, steel, and silk wiring. In the early 1910s, two New York-based surgeons, Russell Hibbs and Fred Albee, proposed grafting bone to enhance spinal fusion in patients with Pott's disease (i.e., tuberculosis of the spine). Without the fusion of healthy spine segments, diseased vertebrates in Pott's disease can deteriorate and collapse, leading to paraplegia. To avoid this devastating lower body paralysis, Hibbs and Albee developed separate techniques for incorporating autograft bone fragments into the disc space. Later, Hibbs used the patient's spinous process and Albee used the patient's tibia bone, over a fusion bed in the interspinous space (cf., Fig. 3). At the time, Albee recognized that bone grafts "derived from the same individual were most trustworthy," without understanding the biological basis for this otherwise accurate observation [19]. Patients who had undergone the Hibbs method began to experience loss of spinal curvature correction prompting a need for stronger, more resilient implants than autograft. Then, in 1955, Dr. Edward Harrington introduced the Harrington rod, a spinal implant crafted from Vitallium, a cobalt-chromium alloy. This device utilized a series of ratchet mechanisms to achieve spinal realignment and maintain kyphosis correction, (cf., Fig. 3) [20]. In the 1970s, Dr. Eduardo Luque of Mexico City further developed spinal instrumentation by employing a system of flexible rods and wires. These were threaded through each vertebral level to achieve enhanced correction and stability compared to the Harrington rod. Notably, this method did not require a postoperative corrective brace like the previous technique. It did, however, cause a greater risk of neurological injury due to implant proximity to the spinal canal. Ten years after the development of the "Luque rods," the1980s saw an alternative that led to the development of titanium plates with specially designed screw slots [18].

Concurrently, the 1980s also saw an increased popularity of intervertebral spacers commonly referred to as intervertebral cages. These cages provide anterior column support to the spine as part of the arthrodesis construct and have a central donut-like hole to retain bone graft and biological augments that facilitate bone fusion. The increased use of cages encouraged another physician, Dr. Harms, to develop an innovative spinal cage construct that combined bone graft and titanium. The placement of the cage in the disc space and its inclusion of bone graft provided structural integrity of the anterior column of the spine. This specialized cage also provided a biologically active surface in the bone graft to help achieve solid fusion [20].

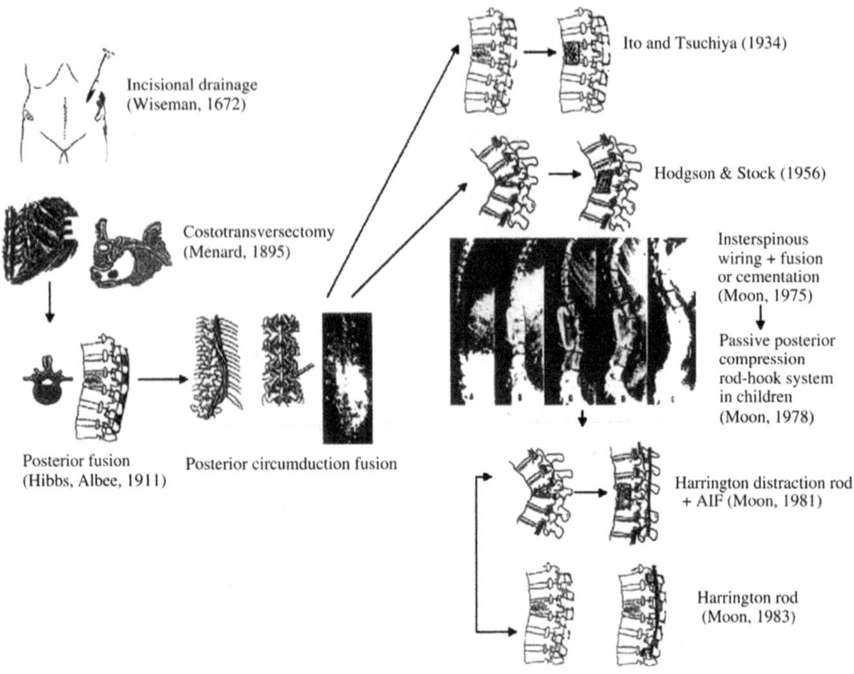

Fig. 3 Progression of surgical techniques and spinal implant materials. (Used with permission [21])

By the end of the twentieth century, intervertebral plastic cages became widely accepted. PEEK, a high-performance thermopolymer was recognized as an inert, rigid material with a similar elastic modulus to cancellous bone [20]. A material's elastic modulus is a ratio that describes its degree of stiffness. Due to their superior stiffness, metal alloys can bear higher loads without deformation. However, this alters the biomechanical stress distribution in the surrounding bone, potentially leading to periprosthetic bone resorption through a phenomenon known as "stress shielding" [8]. PEEK's stiffness similarity to cancellous bone is desirable because it reduces the level of stress shielding seen in more rigid materials. PEEK remains a widely used material for spinal implants due to its favorable properties. However, its inherent hydrophobicity has been identified as a potential barrier to osseointegration, necessitating ongoing research for biomaterials with enhanced biocompatibility [20].

Throughout the development of spinal implants—intervertebral spacers, cages, rods, screws, plates, and wires (cf., Fig. 4)—surgeons gained a better understanding of the required and desirable properties of spinal fusion implants. Dr. Lange (1909) recognized the need for nonirritating, anti-infective properties of spinal implants. Dr. Albee and Dr. Hibbs (1910) noticed a decreased risk of infection by using autografts. However, their technique emphasized the need for stronger, more flexible implants that could maintain curvature over time and resist corrosion over longer lifespans [18]. To consider the most appropriate material and structure of

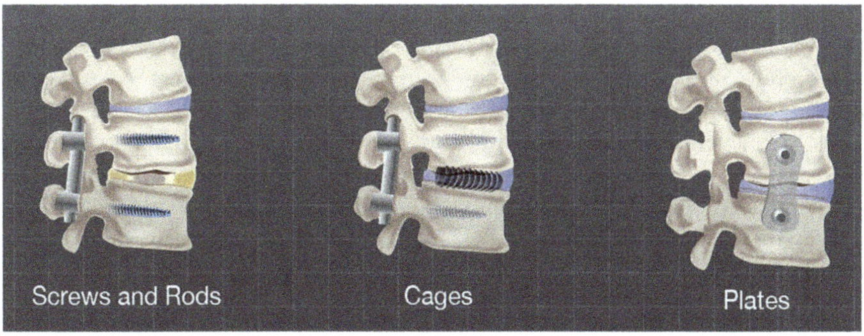

Fig. 4 Different types of spinal implants. (Courtesy BrainLab.org©, https://www.brainlab.org/get-educated/spine/explore-spine-treatment/what-is-a-spinal-implant/; Accessed May 25, 2024; used with permission)

intervertebral spacers in the modern age, potential devices must possess all the aforementioned qualities as well as high compressive/flexural strength, high capacity for tissue integration, and increased radiolucency. The best materials have all these properties, with reasonable costs, simplicity of manufacture, and ease of implantation [22].

2.1 *Mechanical Strength*

Mechanically strong spacers can withstand the strain and stress of axial compression in the spine. The compressive force on a healthy spine ranges from 200 N in the supine position to 1400 N in the relaxed standing spine. This translates to roughly 20 kg (45 lbs.) and 143 kg (315 lbs.), respectively. When considering static or dynamic weightlifting, this compressive force experienced by the spine increases substantially. Ideally, intervertebral spacers should provide the strength to support much larger axial loads compared to healthy bone. The necessary strength to resist fracture against large forces is the most important mechanical quality of a spacer. Weak materials and implants that cannot withstand stress will not progress to the next step in development.[2]

2.2 *Flexural Strength*

While a material needs to be mechanically strong, it should also be highly compressive with increased flexural strength [8]. Increased stiffness may exacerbate stress on

[2] See: https://musculoskeletalkey.com/biomechanics-of-the-spine/. Accessed, May 25, 2024.

the bone leading to higher rates of implant subsidence (i.e., implant migration into adjacent bone). Stiffness is measured by elastic modulus, which expresses the level of deformity of the spacer under a given pressure. Subsidence refers to the level of penetration of the interbody cage into the adjacent vertebrae. Cages were developed to provide anterior spinal column support, and when they fail, severe subsidence may occur leading to recurrent stenosis (i.e., narrowed nerve root foramina) that results in nerve root compression and pain. The recurrence of stenosis may prompt the return of low back pain and loss of sagittal plane alignment necessitating reoperation [22].

2.3 Resistance to Corrosion

Ideal spinal cage biomaterials require both high flexural strength to withstand bending forces and excellent corrosion resistance to ensure long-term implant integrity. Corrosion of the spinal implant may compromise the integrity of the biomaterial reducing its mechanical and flexural strength. Corrosion may also encourage local tissue reaction as particles of the biomaterial may be distributed into the vertebral space, and beyond. Notably, corrosion is primarily seen in metal implants [8].

2.4 Osseointegrative Properties

A high probability of tissue integration or osseointegration with the spacer is important at the interface between the bone and the implant to limit implant migration, (i.e., from bone nonunion), and tissue reaction. Osseointegration involves several complex steps, but most importantly, the recruitment of mesenchymal stem cells for tissue regeneration and the formation of fibrin clots that serve to fill voids between the bone and the implant. Further recruitment of osteoblasts and maturation of osteocytes results in solid arthrodesis [23].

2.5 Anti-infective Properties

Anti-infective properties are particularly important to avoid post-operative complications related to bacterial colonization of the implant. Implants that become colonized with bacteria, usually *Staphylococcus*, may lead to serious outcomes like osteomyelitis, a deep bone infection that is very difficult to eradicate. When bone is infected and becomes swollen, it can lead to ischemia and tissue necrosis. Bone infections manifest as severe pain, fever, loss of appetite, and lethargy. Major complications like these increase the need to create materials that can inherently resist bacterial adhesion and growth [24].

2.6 *Radiolucency*

Radiographic imaging of the spine is done routinely after surgery. Therefore, spinal implants must be amenable to radiographic visualization. The use of postoperative imaging to monitor progress and osseointegration is an integral part of clinical care in spine surgery. Radiolucent materials are those that cannot be seen or impart a hazy shadow on a given radiographic imaging modality. Ideally, an implant should be partially radiolucent, such that its outline is visible on radiographs, while the underlying bone anatomy is not blocked by the implant. Metal spine implants distort CT and MRI images, resulting in image scatter and difficulty in image interpretation. Therefore, implants that are not ferromagnetic and easily seen on both CT and MRI scans are highly desirable for spinal interbody spacers [8].

3 Materials Used as Intervertebral Spacers

Since the introduction of spinal implants in the late nineteenth and early twentieth centuries, many varied materials have been evaluated and used. Stainless steel, titanium alloys, cobalt chrome, tantalum, and PEEK have withstood the test of time and continue to be among the most popular materials used in orthopedic and spine implants. These materials are widely used due to their inherent strength and biocompatibility, plus proven clinical performance. They can be used to make rods, screws, plates, and cages [8]. Screws hold the vertebrae together and attach to plates and rods allowing for pedicle fixation. Rods add stability to the implant and are typically used in spinal fusion and correctional scoliosis surgeries. Plates offer overall stability while cages stabilize and restore the space between vertebral bodies. Cages are usually porous allowing bone ingrowth, thereby contributing to segmental stability once fusion has occurred [25]. Comparative views of a typical PEEK and titanium cage are provided in Fig. 5. Titanium and titanium alloys continue to exhibit excellent biocompatibility as well as overall corrosion resistance and strength. Their high flexural and compressive strengths make titanium materials stronger than cortical bone, and fully capable of withstanding the weight and stresses experienced by the human spine. However, their almost five-fold increase in elastic modulus may lead to a risk of bone resorption due to stress shielding (cf. Fig. 6). Other downfalls of titanium and its alloys are the lack of consistent bone-bonding ability and high-relative radiodensity. Since titanium is a metal, it introduces artifacts in imaging modalities like MRI and CT. Additionally, the radiopacity of titanium makes it difficult to image subtle bone changes [8]. To address the above concerns, the biomaterial industry has developed porous titanium. Introducing pores into the

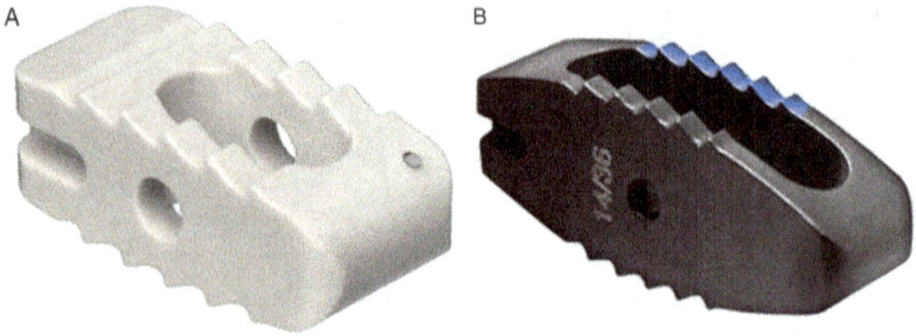

Fig. 5 Intervertebral spinal cages: (**a**) PEEK and (**b**) titanium. (Redrawn and used with permission [25])

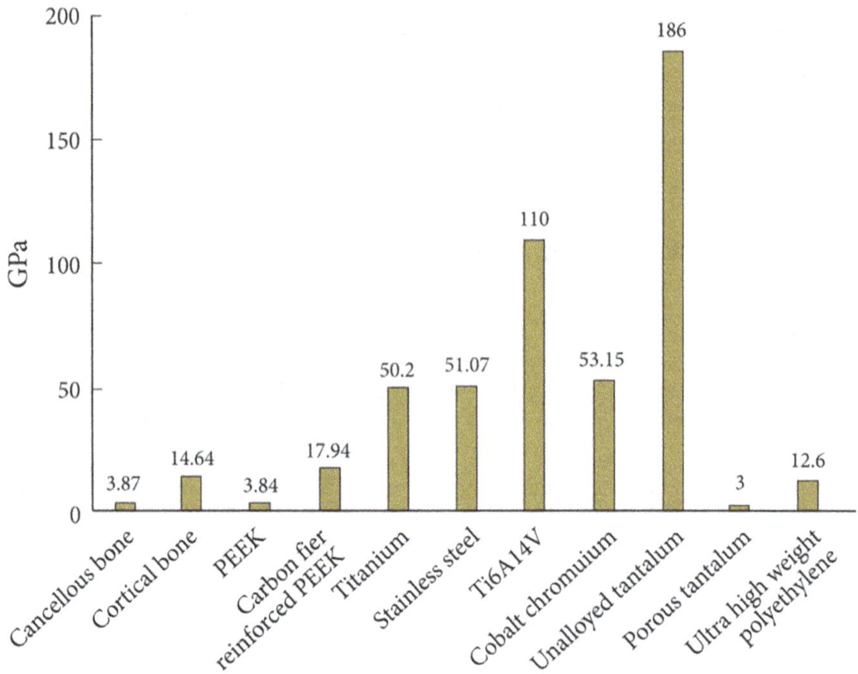

Fig. 6 The elastic moduli of common biomaterials compared to cancellous and cortical bone. (Used with permission under Creative Commons Attribution Noncommercial License [8])

titanium implant reduces its modulus of elasticity bringing it closer to cortical bone, thereby reducing concerns related to stress shielding [26]. Additionally, in vivo studies have shown increased cell growth response compared to nonporous titanium implants [27].

In contrast to titanium, PEEK has an elastic modulus similar to cancellous bone (cf., Fig. 6) [8]. Additionally, PEEK is essentially radiotransparent. Embedded metal markers allow for the identification of the implant on X-rays without obscuring underlying bony details [28]. The main drawback of PEEK is its hydrophobic nature, making it difficult for the implant to heal to native bone. This deficiency increases the likelihood of nonunion and cage migration. Compared to titanium, PEEK has lower compression strength but correspondingly lower cage subsidence rates [8]. Yet, like titanium, a porous version of PEEK has been developed. Porous PEEK has been shown to facilitate cell adhesion, proliferation, and osseointegration in both in vitro and in vivo studies [29]. A critical consideration for porous PEEK is its resistance to high-strain loading. Solid PEEK possesses a relatively low elastic modulus (cf., Fig. 6), and its porous counterpart exhibits further reduced stiffness. Studies employing supraphysiological compressive loads have demonstrated some degree of pore deformation in PEEK implants. However, a sizable portion of the implants retained their original structure. Interestingly, porous PEEK has also been shown to possess shear strength exceeding that of trabecular bone, suggesting potentially improved fatigue life compared to native bone [29]. However, PEEK screws have a higher prevalence of failures and earlier reoperation rates likely due to their relatively low shear strength compared to titanium [8].

To obtain the strength of titanium and improve the osseointegrative properties of PEEK, a hybrid PEEK titanium composite has been developed for lumbar fusion cages (cf., Fig. 7). Animal studies show significant reductions in ranges of motion and increased stiffness due to ingrowth into the porous composite [30]. Clinical studies have shown that these Ti/PEEK composite implants are safe and effective. However, radiographic fusion rates remain comparable to standard PEEK [31].

While titanium and PEEK are the most commonly used biomaterials, tantalum is another widely used metal as a cage material. Porous tantalum showed better osseointegration and osteogenesis than porous titanium in both in vitro and in vivo studies. Like porous PEEK, porous tantalum exhibits a modulus of elasticity similar to that of cancellous bone (cf., Fig. 6). Compared to autograft, porous tantalum implants showed similar rates of fusion and lower revision rates [30].

Cobalt chromium is known to be a strong, corrosion-resistant material, and has been used to make rods for the correction of scoliosis due to its increased rigidity. Yet, it does not have improved imaging clarity and has greater distortion on MRI [8]. However, cobalt chromium has improved antibacterial properties compared to titanium [31].

One of the more desirable properties of a biomaterial is its relative radiolucency on plain X-rays, a quite common and routinely used imaging modality. X-ray imaging offers a quick and accurate evaluation of the position of the implant, its fusion status, and its degree of subsidence. Given in Fig. 8 is a radiograph of a hand with overlying implant materials to allow plain X-ray comparisons of different biomaterials [28]. From this image, it is apparent that the PEEK/ß-Si_3N_4 composite was comparable to the monolithic Si_3N_4 imaging. However, note that the composite had significantly improved visibility compared to PEEK on its own, which shows up as a faint outline only. PEEK/ß-SiYAlON (another ceramic) was as visible as the titanium samples. The ideal radiolucent image allows for visualization of the implant as well as a view of the underlying bone.

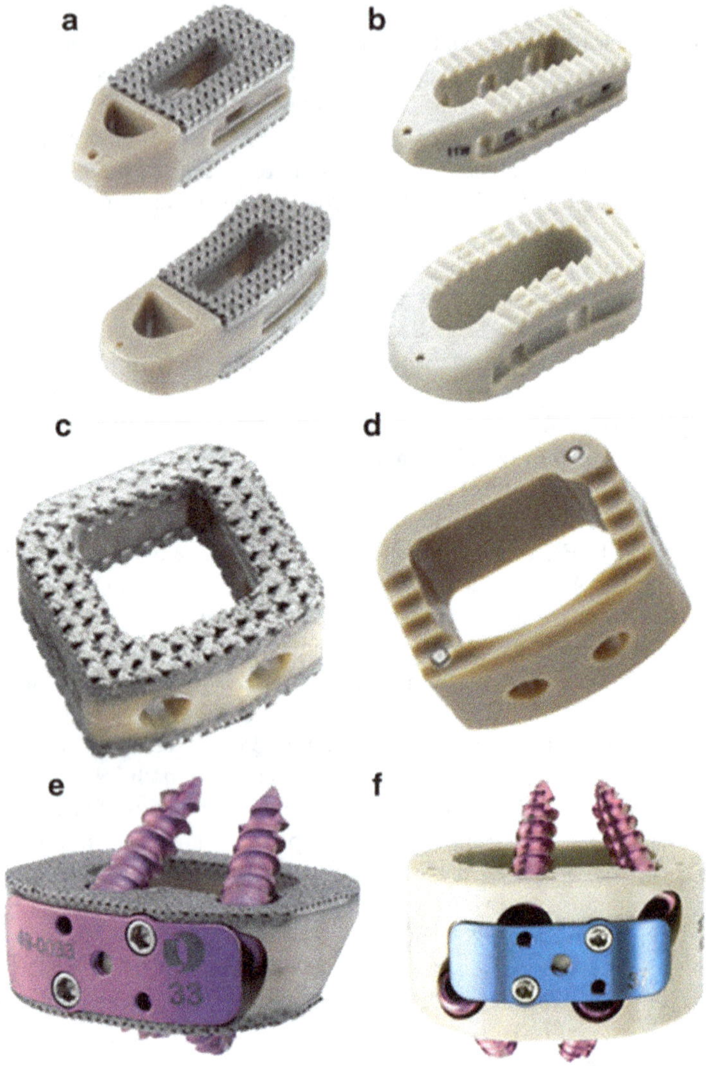

Fig. 7 PTC devices that use Ti6Al4V endplates and PEEK equivalents. (**a**) Lumbar FORZA®PTC Spacer System. (**b**) Lumbar FORZA®PEEK Spacer System. (**c**) Cervical CONSTRUX®Mini PTC Spacer System. (**d**) Cervical CONSTRUX®Mini PEEK Spacer System. (**e**) Lumbar Stand-Alone PILLAR® SA PTC Spacer System. (**f**) Lumbar Stand-Alone PILLAR® SA PEEK Spacer System. (Used with permission [32])

Metals such as titanium, tantalum, and chromium cobalt are easily visible on imaging but are also very radio-opaque making it difficult to visualize the underlying bone. They are also known to introduce artifacts on other imaging modalities such as MRI and CT. PEEK implants are essentially radiotransparent, and not easily defined by any imaging modality. They typically include embedded metal radiographic

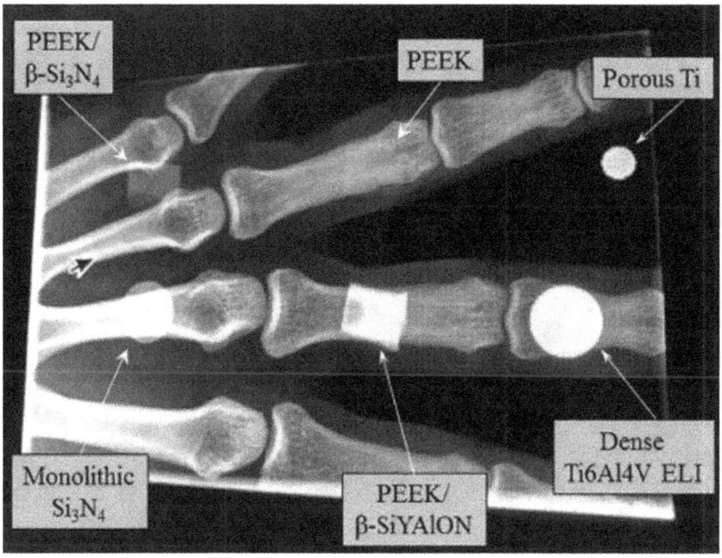

Fig. 8 X-ray radiograph compares the translucency performance of three composites with monolithic PEEK and other biomaterials employed as orthopedic implants. (Used with permission [28])

Table 2 Implants and biomaterials used in spinal arthrodesis

Implant type	Implementation procedure	Standard material
Cage	Anterior/posterior interbody fusion—restores height of the collapsed disc	Titanium PEEK Ceramic Acrylic
Screws	Pedicle screw fusion—holds vertebrae together to attach to plates and rods	Titanium doped with hydroxyapatite, calcium phosphate, extracellular matrix, tantalum
Rods	Spinal fusion—adds stability to a spinal implant	Titanium Chromium cobalt PEEK Stainless steel
Plates	Spinal stabilization	Titanium

markers to indicate their anatomical positions. Each biomaterial has defining characteristics that make it suitable for continued clinical use. Comparisons of the different biomaterials and types of available implants are shown in Tables 2 and 3.

Titanium and PEEK, in particular, have seen advancements both biologically and mechanically [8], which makes them desirable and popular materials for use in spinal implants. However, there are drawbacks to both materials even when they have been mechanically and biologically altered to enhance performance. Hybrid

Table 3 Biomaterials used in spinal arthrodesis

Materials	Advantages	Disadvantages	Applications
Stainless steel	Strong, stiff, easily doped, inexpensive	Corrosion, poor biocompatibility, high artifacts on imaging	Scoliosis correction Use in rods, screws
Titanium	Lightweight, strong, flexible, biocompatible, easily doped	Relatively expensive, some artifacts on imaging	Screws, rods, plates, cages
PEEK	Lightweight, flexible, inexpensive, biocompatible, easily doped, low artifacts on imaging, radiotransparent	Low Young's modulus, some grafting issues	Rods, cages, disc replacement
Chromium cobalt	Strong, flexible, biocompatible	Expensive, high artifacts on imaging	Adolescent scoliosis correction
Ceramics	Inexpensive, biocompatible, wear-resistant, easily doped, low artifacts on imaging, radiolucent	Brittle, grafting issues	Cages
Tantalum	High frictional characteristics, low elastic modulus	Expensive, not stiff enough for some corrections	Not frequently used

versions of these implants strive for the best of both materials but are still hindered by individual material's shortcomings as observed in Ti/PEEK composites [33].

4 Silicon Nitride and Silicon Nitride Composites

Silicon nitride is a synthetic compound that has been historically used in internal combustion engines and gas turbines. This robust material is characterized by high fracture toughness, high strength, high-temperature stability, high wear and corrosion resistance, and is inert to most chemicals making it suitable to withstand extreme temperatures and prolonged friction [34]. These features limit the generation of wear debris in environments prone to constant motion, such as human joints. In turn, this minimizes the biological response to foreign particles. Si_3N_4's microstructure is composed of elongated matrix grains that are interlocked and oriented randomly [34]. At room temperature, it has three crystallographic structures, trigonal (alpha), hexagonal (beta), or cubic (gamma) phases (cf., Chap. 2). The gamma structure is the hardest, followed by the alpha and beta structures. Each hexagonal modification is arranged in a stacking sequence with alpha modifications containing twice as many atoms per unit cell as the beta polymorph, yet the beta form is easier to transform into a porous mass and is the desired phase for spinal implants (cf., Chaps. 1 and 2) [35].

There are multiple methods for Si_3N_4 to be manufactured into a biomaterial. Three prominent techniques include reaction-bonded silicon nitride (RBSN), dense sintered reaction-bonded silicon nitride (SRBSN), and dense sintered silicon nitride (SSN), (cf., Chap. 3). Notably, reaction-bonded silicon nitride can be used as the porous inner body of intervertebral spacers. Using these processing methods, Si_3N_4 can be formulated in both porous and hard-bearing forms. The porous form is typically used for biomedical applications where it can support direct appositional bone ingrowth for biological fixation to the skeleton [12]. The dense form is more suitable for prosthetic hip and knee joints where low wear rates and low frictional coefficients are desired [17, 36], (cf., Chaps. 10 and 11). Following densification, a texturing technique can be used to reinforce the mechanical and thermal performance of the final biomaterial used in spacers. The surface texture and chemistry of Si_3N_4 are modifiable allowing for physiochemical modulation to optimize osseointegration and microbial resistance. Its chemical makeup is thought to encourage host cell adhesion and osteoconductive potential (cf., Chaps. 4, 5, and 6). The hydrophilic properties of Si_3N_4 aid it in achieving interbody fusion because extracellular fluid containing bone growth mediating proteins is attracted to the ceramic's surface, thus encouraging osteoblast activation and osteoclast suppression. Silicon nitride's bacterial resistance is also due to its unique surface chemistry, surface pH, and electrical charge. Its basic surface pH has shown increased lysis of bacteria due to the local release of various nitrogen moieties (cf., Chaps. 4, 5, and 6) [11]. In the following subsections, additional details are provided on Si_3N_4's desirable features.

4.1 Resilience

Silicon nitride's intrinsic properties have made it advantageous as an engineering material. It has a low density compared to steel, low friction, and corrosion resistance, and can perform under extreme environmental conditions. Its resilient properties have made it suitable for high-performance aircraft and space vehicles where it can withstand extreme temperatures [35]. While none of the conditions are common for biomedical applications, it has demonstrated its resilience by withstanding a surgical autoclave at 121 °C for 100 h. This test mimicked accelerated aging and demonstrated no change in the material's flexural strength or phase composition [37]. Its proven performance under these conditions shows its resilience and ability to withstand severe human stress and strain.

Silicon nitride is subject to two modes of wear: mechanical and tribochemical. The mechanical mode is observed at high load and low speed and stop-start motions, whereas the tribochemical wear occurs during the dissolution of Si_3N_4 in aqueous lubricating environments. Under these conditions, a layer of hydrated silicon oxide forms between articulating surfaces at low loads and speeds and continuous motion (cf., Chaps. 10 and 11). In practice, surface degradation is dependent on friction, local heating, micro-separation, microfracture, chemical dissolution, and biofilm formation at the contact interface [38]; and studies regarding bioceramics have concluded that wear and tear is predominately dependent on surface fracture toughness [39].

4.2 Surface Fracture Toughness and Flexural Strength

Silicon nitride's surface fracture toughness is dependent on the size and morphology of its microstructure. High fracture toughness is observed with reasonably coarse grains. Improvements in Si_3N_4 processing have allowed for the production of elongated grains as well as compositional modifications to grain boundaries. Ideal strength, fracture toughness, and flexural strength can be above 1 GPa and over 10 $MPa{\cdot}m^{1/2}$, respectively. Today's manufacturing methods have resulted in dense Si_3N_4 materials with strength and toughness that are at least two times greater than modern alumina (Al_2O_3) orthopedic implants [40], with comparable properties to alumina-matrix composites (AMC) (cf., Chaps. 10 and 11). Silicon nitride, in a 3D fabricated state, with a porous architecture, has also been shown to have reasonably high flexural strength and toughness [41]. Silicon nitride can also be combined with other metals and polymers to form composites (cf., Chap. 7). For example, recent studies have shown that polymer-infiltrated Si_3N_4 composites possess high flexural strengths and moderate hardness. Such composites could be potential candidates for spinal or dental implants and other high-load medical devices [42].

4.3 Anti-infective Properties and Host Cell Response

One of the more impressive properties of Si_3N_4 is its ability to inhibit bacterial growth. Silicon nitride experimental murine tibial implants exposed to MRSA in vitro showed little to no bacterial adherence. This in vitro study looked at both as-fired and machined implants. The study demonstrated that native as-fired Si_3N_4 implants inhibit direct adhesion despite increased surface area and topography. Notably, machining the surface to create a flatter topography showed bacterial adhesion similar to that of stainless steel (cf. Fig. 9) [43].

While native, as-fired Si_3N_4 was effective in preventing bacterial adhesion, it demonstrated a highly reactive host response. In contrast, the machined Si_3N_4 did not display significant reactivity with the host. In another in vivo study, no live bacteria were present on Si_3N_4 samples implanted in the calvaria of Wistar rats exposed to *Staphylococcus epidermidis* (*S. epidermidis*) which is known to synthesize a biofilm layer that elicits a host immune response [44]. As previously mentioned, the hydrophilic properties of Si_3N_4's surface can influence serum protein adsorption. Adsorbed proteins can suppress *S. epidermidis* adhesion and hinder its ability to produce a biofilm [24]. The net surface charge and the presence of amine groups lead to bacterial membrane disruption and lysis [45]. While Si_3N_4 is effective in inhibiting bacterial growth, it is also biocompatible and noncytotoxic when compared to other implant materials like metals, and it does not induce a host immune response to its wear particles [40].

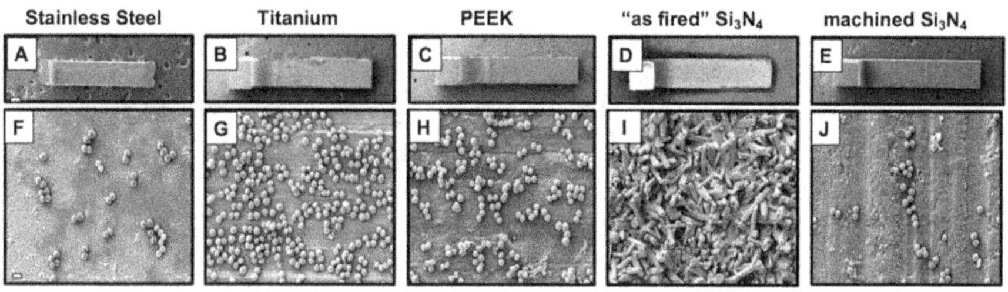

Fig. 9 Native as-fired Si_3N_4 implant surfaces are highly resistant to MRSA adherence in vitro. SS, Ti, PEEK, as-fired Si_3N_4, and machined Si_3N_4 implants ($n = 4$) were submerged in an overnight culture (O.D. 50.7 at 630 nm) of USA300LAC, air-dried and processed for SEM imaging. Representative low magnification images of the L-shaped implants are shown at x30 ([**a**–**e**]; scale bar in **a** = 200 μm), and higher magnification images of regions containing adherent MRSA are shown at x5000 ([**f**–**j**]; scale bar in **f** = 1 μm). No bacteria were directly adhering to native as-fired Si_3N_4 implant surfaces (**i**). Used with permission [43]

4.4 *Osseointegration*

The osseointegrative property of Si_3N_4 is probably its most impressive feature as a biomedical implant (cf., Chap. 5). In one study, Si_3N_4 samples had high densities of extracellular matrix (ECM) and mineral formation after 30 and 60 days compared to titanium and PEEK. Additionally, they were covered in collagen bundles and amorphous calcium phosphate. The presence of ECM and collagen formation on Si_3N_4 can be attributed to its surface bioactivity, hydrophilic nature, and higher surface energy. PEEK surfaces did not support collagen or mineral formation due to their hydrophobic surface and surface topography leading to low protein adsorption and cell adhesion. Increased formation of collagen signifies osteoblastic induction of collagen. An integrin mediator binds the collagenous matrix to the osteogenic cell membrane inducing protein kinases. Activation of protein kinases results in upregulation of bone-specific transcription factors for bone formation and mineralization [46].

4.5 *Radiolucency*

Silicon nitride is partially radiolucent allowing for visualization of the implant, adjacent bone, and the implant-bone interface. Additionally, it is nonmagnetic, enabling the use of MRI of soft tissues adjacent to the implant. Comparatively, metals and metal alloys are radio-opaque and introduce MRI artifacts [40]. A cervical implant and its radiographic image are provided in Fig. 10.

In summary, Si_3N_4 may be a formidable competitor and replacement for existing materials used to make various biomedical implants in orthopedic surgery and dentistry. Future applications will likely include substitutes for bone grafts, bone

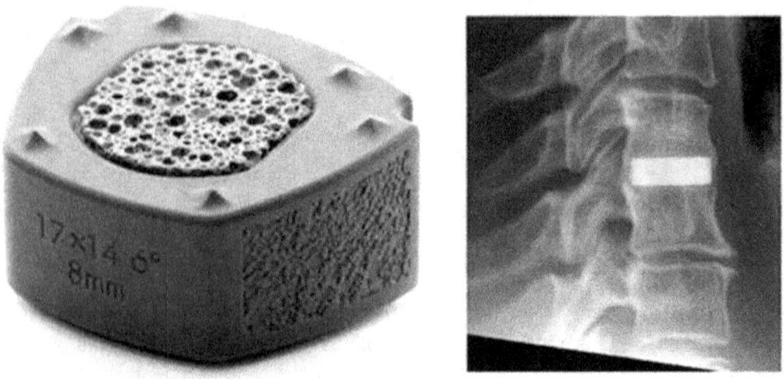

Fig. 10 Silicon nitride cervical interbody fusion device (Valeo CC^{s}C®, left) with the radiographic image of the implanted device (right). (Used with permission under Creative Commons Attribution 4.0 International License [11, 47])

scaffolds, knee and hip endoprosthesis implants, and porous intervertebral spacers. Its use in dental implants is still being researched, but the material holds considerable promise in dentistry [34]. Its resilience, flexural strength, surface fracture toughness, anti-infective and osseointegrative properties, and imaging advantages present a highly desirable combination of features that are ideally suited for implants similar to those used in spinal arthrodesis.

5 Clinical Results for Silicon Nitride Spinal Implants

Surgical intervention for symptomatic lumbar spinal stenosis is effective in reducing leg pain and disability compared to conservative nonoperative interventions [48]. Several small randomized controlled trials and the large SPORT trial mentioned earlier have concluded that there is a significant persistent early advantage for those undergoing surgery vs nonoperative treatment, at least up to 6 years of follow-up. Beyond 6 years, the advantages of surgery were no longer statistically significant based on clinical outcomes [4]. One large, randomized control trial compared interspinous spacers to continued nonoperative therapy and found that at 2 years, 48% of the spacer group and only 5% of the nonoperative group showed improved symptom severity, improved physical function, and satisfaction with treatment. However, the spacer group reported more side effects. A comparison of different surgical types shows lower rates of complications but higher rates of reoperation in intervertebral spacers compared to decompressive surgery [49]. Consequently, using the best material for intervertebral spacers could improve outcomes and help reduce reoperation rates.

Most intervertebral spacers used today are made of either titanium alloys, PEEK, or allograft. Silicon nitride is a relatively new implant material and is yet to be widely adopted. Ironically, it was the first abiotic material to be used as a spinal spacer in a

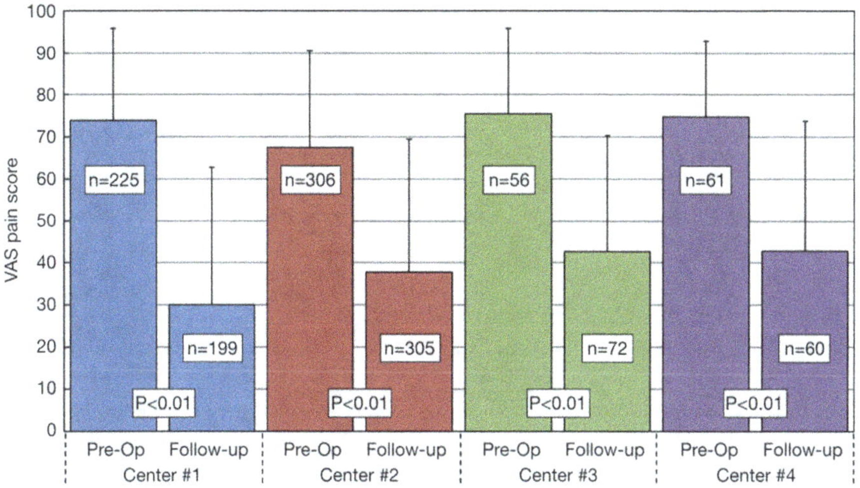

Fig. 11 Pre-op and last follow-up cervical VAS scores for Si_3N_4 intervertebral spacers from the four participating surgical centers. (Used with permission under Creative Commons Attribution-NonCommercial-NoDerivs License (CC BY-NC-ND 4.0) [51])

small (30-patient) Australian clinical trial initiated in 1986. At 5- 10- 15-, and even 30-year follow-ups, it demonstrated significant pain reduction and solid interbody fusion [12, 50].

Notably, in 2020, two large retrospective multicenter studies (1310 patients) evaluated Si_3N_4 for both lumbar and cervical arthrodesis. Within these studies, the researchers conducted a systematic review and meta-analysis of clinical outcomes for Si_3N_4 compared to other biomaterials. The Si_3N_4 spacers were statistically effective in reducing pain scores (visual analog scale, VAS) from pre-op to the last follow-up for both cervical and lumbar devices, (cf. Fig. 11 for cervical spacers) [51]; and a Forest plot comparing lumbar pain scores is provided in Fig. 12.

They found no differences in pain scores between the Si_3N_4 implants and the other spacers [52]. Not only did these studies show that Si_3N_4 was at least comparable to other biomaterials in its effectiveness in reducing VAS pain scores, but they also showed that Si_3N_4 was non-inferior in terms of expected complications and reoperations [52]. This was an important finding since new biomaterials often show unexpected complications and adverse events.

In a contemporary clinical trial, patients undergoing transforaminal lumbar interbody fusion (TLIF) for degenerative lumbar disc disease demonstrated similar clinical outcomes and recovery rates regardless of whether PEEK or Si_3N_4 cages were implanted. No statistically significant differences in clinical measures were observed during the 24-month follow-up period (cf., Fig. 13a–c) [53]. The primary hypothesis of this study was to establish non-inferiority of the Si_3N_4 cages based on the Roland Morris Disability Questionnaire (RMDQ) scores. Although there was a general trend of higher RMDQ scores for the Si_3N_4 implants, this was not statistically different from the PEEK implants. Nevertheless, the non-inferiority hypothesis

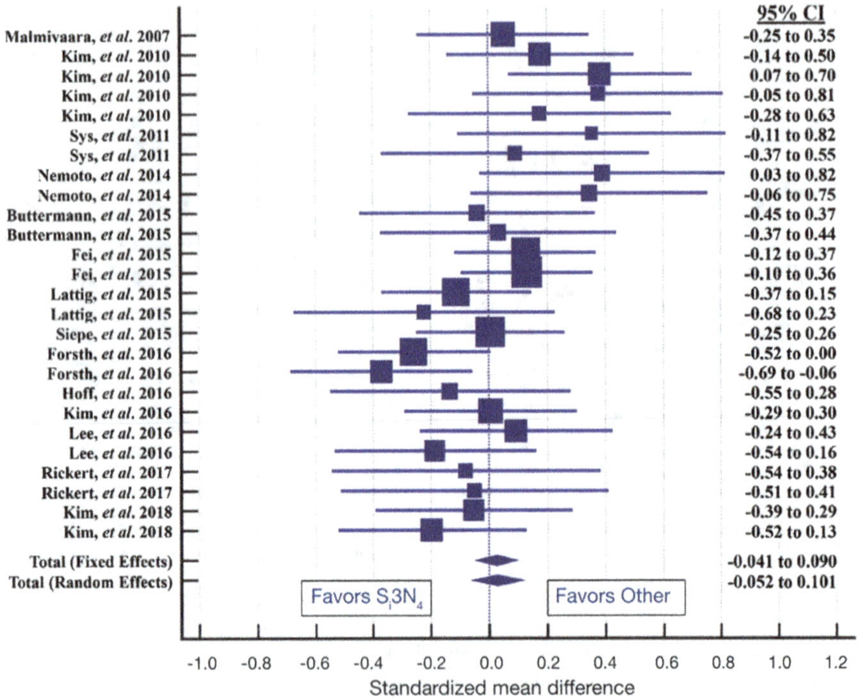

Fig. 12 Forest plot comparing changes in VAS pain scores for meta-data and four surgical centers comparing Si_3N_4 lumbar spinal spacers to other commonly used arthrodesis methods and materials. VAS visual analog scale. (Used with permission under Creative Commons Attribution-NonCommercial-NoDerivs License (CC BY-NC-ND 4.0) [52])

could not be definitively confirmed using the initial protocol design due to lower-than-anticipated patient retention and higher-than-anticipated RMDQ standard deviations at the 12-month mark. These variances reduced the study's statistical power from a planned 90% to just 50%. A post hoc analysis and review of relevant literature led to a revision of the non-inferiority margin based on the loss of patients to follow-up and higher-than-expected RMDQ standard deviations. Under the revised criterion, the non-inferiority of Si_3N_4 cages was affirmed. Both PEEK and Si_3N_4 cages appeared to offer comparable clinical efficacy for lumbar spinal fusion.

A separate non-inferiority study investigated porous Si_3N_4 and PEEK cages for anterior cervical discectomy and fusion (ACDF) [47]. The key results from the study are shown in Fig. 14a, b. The researchers concluded that there were no significant differences in clinical outcomes at 24 months between patients with symptomatic disc herniation or spondylosis treated with anterior cervical discectomy and interbody fusion using either Si_3N_4 or PEEK implants. While some intermediate outcome measures favored PEEK, these differences did not reach the level of clinical significance. Fusion rates for both implant materials improved over time, with no statistically significant difference observed in terms of overall fusion or subsidence.

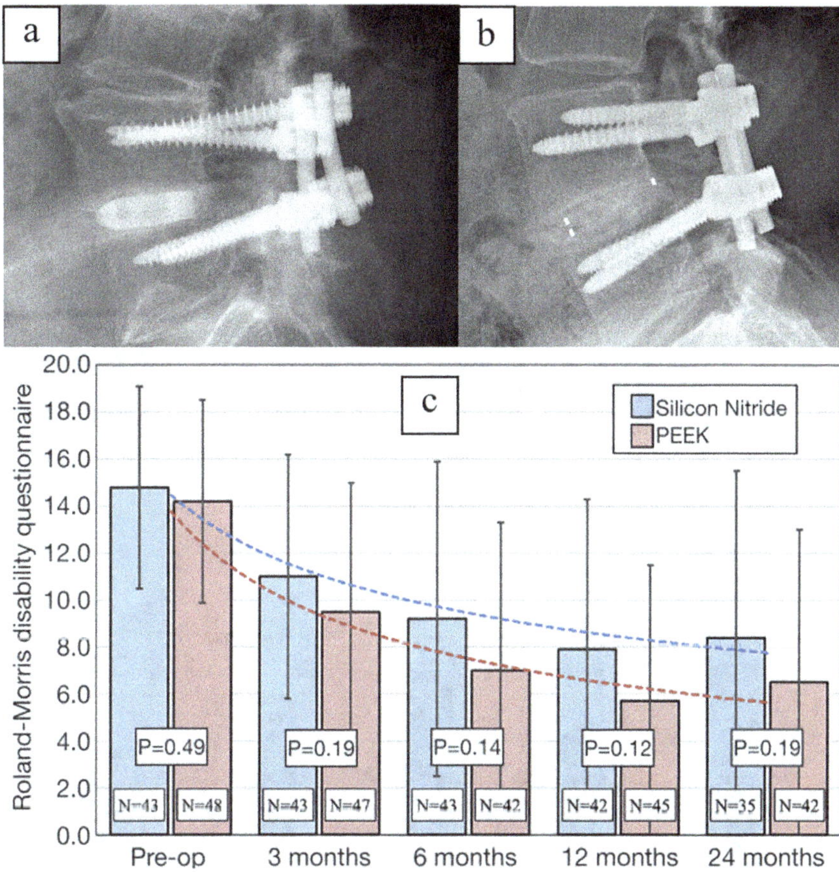

Fig. 13 Comparable lateral X-rays of L4–L5 fusion at 24 months for (**a**) Si_3N_4 cage and (**b**) PEEK cage. Note that solid fusion was achieved with both cage types as indicated by bone bridging between the endplates; and (**c**) RMDQ scores for patient follow-up periods. RMDQ, Roland-Morris Disability Questionnaire. There were no significant differences in RMDQ scores at any follow-up period. (Redrawn and used with permission under Creative Commons Attribution 4.0 International License [53])

These findings supported the primary hypothesis that both Si_3N_4 and PEEK interbody implants achieve similar functional improvements as measured by the Neck Disability Index (NDI) at 2 years post-surgery [47].

While PEEK and titanium alloys are the dominant biomaterials for spinal cages, other spacers, like those made from allograft, continue to be used clinically and are therefore worth comparing. In 2018, a single-center retrospective clinical evaluation compared the efficacy of traditional allograft spacers and Si_3N_4 cages in 92 patients undergoing ACDF [54]. Radiographic outcomes were assessed at 3, 6, 12, and 24 months post-surgery, focusing on fusion rates, osseous integration, and subsidence. The key results of this study are presented in Figs. 15 and 16.

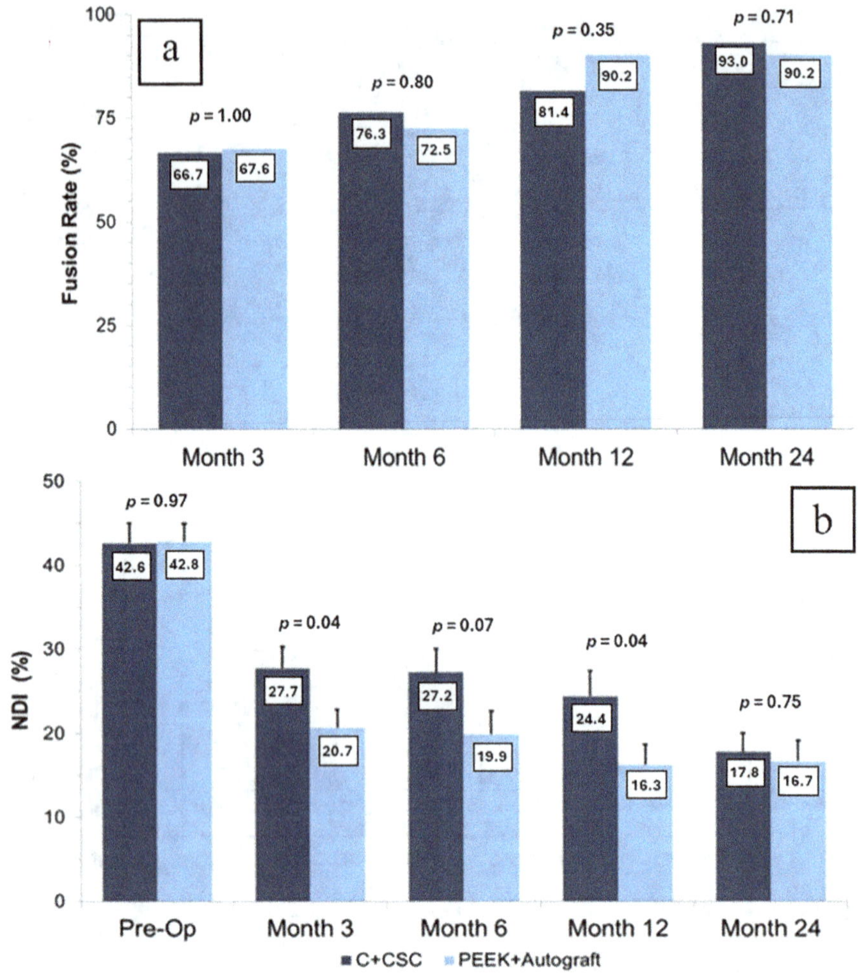

Fig. 14 (**a**) Comparative fusion rates vs follow-up for Si_3N_4 (C + CSC) and PEEK+Autograft cervical spacers. Fusion rate was defined as rotation ≤4° and ≤1.25 mm translation on flexion-extension films on dynamic X-ray. In both groups, the fusion rate increased over time and there was no significant difference between both groups and (**b**) Neck and Disability Index scores during the 24-month follow-up period. At 3 and 12 months there were small significant differences in favor of PEEK, although not clinically relevant. (Redrawn and used with permission under CC BY-NC-ND 4.0 license [47])

Both materials achieved acceptable levels of fusion and osseous integration at ≥12 months. However, Si_3N_4 cages demonstrated significantly faster rates of fusion and osseous integration at earlier time points (3, 6, and 12 months) as measured by both interspinous distance (ISD) and Cobb angle methods. Notably, the time for full osseointegration was significantly shorter for Si_3N_4 compared to allograft by almost 1 year. Additionally, the Si_3N_4 group exhibited lower subsidence at all follow-up

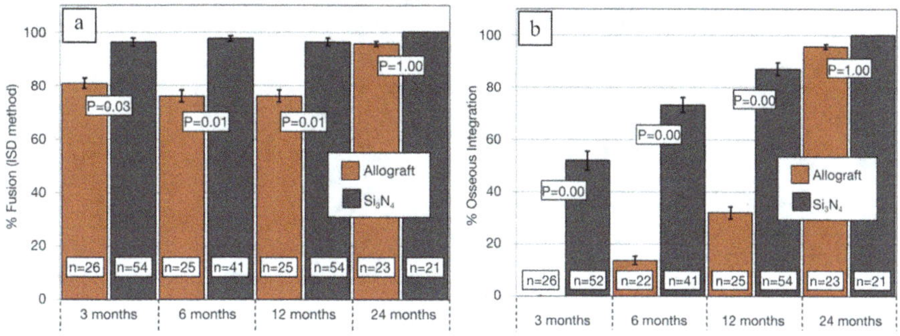

Fig. 15 (**a**) Time series of interspinous distance (ISD) based fusion rates for allograft spacers and Si_3N_4 cages, and (**b**) Time series comparison of osseous integration between allograft spacers and Si_3N_4 cages. P-values indicate statistically significant differences between groups at each time point. n-values represent the number of surgical levels analyzed. Error bars depict binomial standard deviations. (Redrawn and used with permission under CC BY-NC-ND 4.0 license [54])

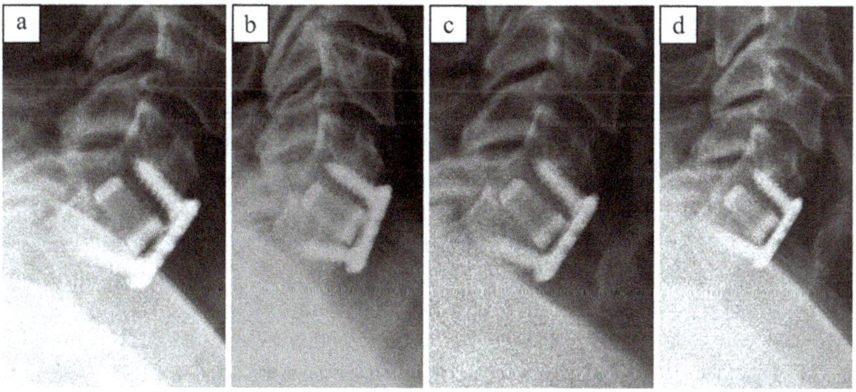

Fig. 16 Progressive osseous integration of a Si_3N_4 cage with the posterior vertebral elements. Panel (**a**) depicts the cage at 2 weeks post-surgery. Panels (**b**–**d**) demonstrate progressive fusion at 6 weeks, 3 months, and 6 months, respectively. Panel (**d**) shows complete bridging of posterior bone by 6 months. (Redrawn and used with permission under CC BY-NC-ND 4.0 license [54])

intervals. The average subsidence was 0.51 mm for Si_3N_4 and 2.71 mm for allograft cohorts at 1-year follow-up. Figure 16 is particularly instructive because it shows complete posterior bone bridging of the Si_3N_4 implant at 6 months, which demonstrates excellent osseous integration. This study suggests that Si_3N_4 cages are at least as effective as traditional allograft spacers in promoting fusion and integration in ACDF patients, with a potential advantage in achieving these outcomes at earlier time points.

Silicon nitride's enhanced fusion and antibacterial properties make it an ideal biomaterial for use in treating lumbar pyogenic discitis. One case series showed positive outcomes for patients with Si_3N_4 implants to treat disc degeneration from bacterial discitis (cf., Fig. 17) [55]. Silicon nitride's surface shows increased lysis of

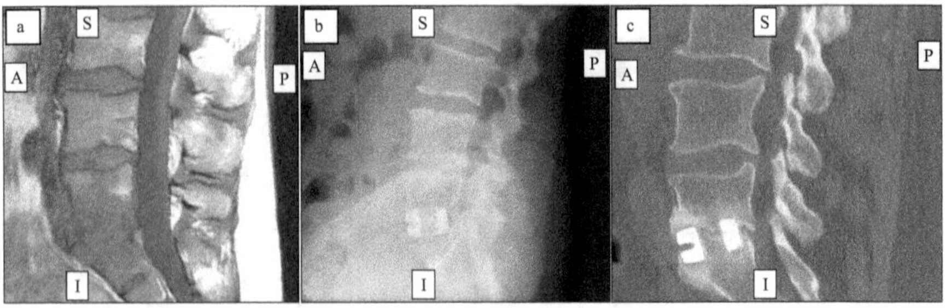

Fig. 17 In this case series, (**a**) A patient's pre-op noncontrasted MRI shows marked disc degeneration and adjacent vertebral body inflammation at L5-S1, consistent with discitis, (**b**) at 7 months post-op, a lateral X-ray image showed stabilization of the fused segments obtained by the Si_3N_4 interbody spacer without instrumentation. Note the radiographic visibility of the implant on this view, and (**c**) at 9 months post-op, a contrasted CT image showed successful fusion without migration of the Si_3N_4 interbody spacer. (Reconfigured and used with permission [55])

gram-negative phylum due to the conversion of ammonia in the local environment. Ammonia disrupts bacterial membranes leading to bacterial death [15]. Additionally, the surface texture prevented the attachment of gram-positive bacteria and biofilm formation [43].

As shown in Fig. 17, Si_3N_4 offers a distinct advantage in postoperative magnetic resonance imaging (MRI) due to its inherent biocompatibility and nonmagnetic properties. This allows for clear visualization of adjacent soft tissues without artifacts, which can be crucial for evaluating postoperative healing, identifying potential complications like infections or recurrent disc herniation, and monitoring implant stability. Conversely, metallic cages and those composed of PEEK present limitations due to their varying degrees of radiolucency, radiopacity, and susceptibility to MRI-induced signal distortion. Radiolucent implants may not be readily visualized on MRI scans, while radiopaque implants can obscure surrounding soft tissues. PEEK implants, while not typically ferromagnetic, can still introduce signal voids or geometric distortions that can hinder image quality [11].

Clinical data, including case studies and noninferiority trials, suggest that Si_3N_4 performs at least as well as PEEK in achieving spinal fusion. This is evidenced by comparable rates of fusion, subsidence, and visual analog scale pain scores. Additionally, Si_3N_4 appears to offer potential advantages, including faster bone fusion, enhanced osseointegration, and accelerated bone formation. These findings warrant further investigation in larger, prospective clinical studies. Beyond these potential benefits, Si_3N_4 possesses superior imaging MRI compatibility due to its nonmagnetic properties. This allows for clear visualization of adjacent soft tissues, which is crucial for postoperative evaluation. Furthermore, Si_3N_4 exhibits inherent antibacterial properties, suggesting a potential advantage in reducing the risk of implant-associated infections. Hopefully, future research and additional clinical data will definitively establish these previously observed benefits.

6 Conclusions

Silicon nitride has emerged as a highly promising biomaterial for diverse orthopedic and dental applications. Its potential extends beyond spinal implants to foot and ankle fusion procedures, as well as oral and craniomaxillofacial reconstructions. This versatility stems from a unique combination of desirable properties: exceptional mechanical strength, favorable chemical stability, and radiolucency. These attributes make Si_3N_4 particularly suitable for scenarios requiring long-term implant durability, resistance to implant infection, and predictable bone healing. Si_3N_4's elastic modulus and flexural strength render it capable of withstanding the substantial loads borne by the lumbar spine. Conversely, its low friction coefficient minimizes wear in prosthetic hip and knee bearings, potentially extending their longevity. Notably, the inherent antibacterial properties of Si_3N_4 hold significant promise for reducing the risk of implant-associated infections, a major complication in orthopedic surgery. This advantage is further augmented by superior osseointegration, increased extracellular matrix and mineral densities, and abundant collagen bundles. These favorable biological properties translate to enhanced clinical outcomes, with studies demonstrating superior osseointegration compared to traditional allografts. Beyond its material properties, Si_3N_4 offers distinct advantages in postoperative monitoring. Its radiolucency facilitates clear radiographic visualization, with minimal to no image distortion. This allows for unobstructed evaluation of the underlying bone and surrounding soft tissues, crucial for assessing post-surgical progress and identifying potential complications. While ongoing research explores novel biomaterials for implants, Si_3N_4 stands out for its potential to usher in a new and improved generation of orthopedic devices. These advancements can revolutionize patient care by alleviating pain, significantly reducing the burden of infections, and accelerating recovery through faster bone healing.

References

1. Wu, A.; March, L.; Zheng, X.; Huang, J.; Wang, X.; Zhao, J.; Blyth, F.M.; Smith, E.; Buchbinder, R.; Hoy, D. Global Low Back Pain Prevalence and Years Lived with Disability from 1990 to 2017: Estimates from the Global Burden of Disease Study 2017. *Ann. Transl. Med.* **2020**, *8*, 299–299, doi:https://doi.org/10.21037/atm.2020.02.175.
2. Gaskin, D.J.; Richard, P. The Economic Costs of Pain in the United States. *J. Pain* **2012**, *13*, 715–724, doi:https://doi.org/10.1016/j.jpain.2012.03.009.
3. Lykissas, M.G.; Aichmair, A. Currrent Concepts on Spinal Arthrodesis in Degenerative Disorders of the Lumbar Spine. *World J. Clin. Cases* **2013**, *1*, 4–12, doi:https://doi.org/10.12998/wjcc.v1.i1.4.
4. Weinstein, J.N.; Tosteson, T.D.; Lurie, J.D.; Tosteson, A.N.A.; Hanscom, B.; Skinner, J.S.; Abdu, W.A.; Hilibrand, A.S.; Boden, S.D.; Deyo, R.A. Surgical vs Nonoperative Treatment for Lumbar Disk Herniation: the Spine Patient Outcomes Research Trial (SPORT). A Randomized Trial. *JAMA* **2006**, *296*, 2441–2450, doi:https://doi.org/10.1001/jama.296.20.2441.
5. Tosteson, A.N.A.; Lurie, J.D.; Tosteson, T.D.; Skinner, J.S.; Herkowitz, H.; Albert, T.; Boden, S.D.; Bridwell, K.; Longley, M.; Andersson, G.B.; *et al.* Surgical Treatment of Spinal Stenosis

with and without Degenerative Spondylolisthesis: Cost-Effectiveness after 2 Years. *Ann. Intern. Med.* **2008**, *149*, 845–853, doi:https://doi.org/10.7326/0003-4819-149-12-200812160-00003.
6. Hallab, N.J.; Jacobs, J.J. 2.5.4 - Orthopedic Applications. In *Biomaterials Science*; Wagner, W. R., Sakiyama-Elbert, S.E., Zhang, G., Yaszemski, M.J.B.T.-B.S. (Fourth E., Eds.; Academic Press, 2020; pp. 1079–1118 ISBN 978-0-12-816137-1.
7. Meena, V.K.; Kumar, P.; Kalra, P.; Sinha, R.K. Additive Manufacturing for Metallic Spinal Implants: A Systematic Review. *Ann. 3D Print. Med.* **2021**, *3*, 100021, doi:https://doi.org/10.1016/j.stlm.2021.100021.
8. Warburton, A.; Girdler, S.J.; Mikhail, C.M.; Ahn, A.; Cho, S.K. Biomaterials in Spinal Implants: A Review. *Neurospine* **2020**, *17*, 101–110, doi:https://doi.org/10.14245/ns.1938296.148.
9. McEntire, B.J.; Hengst, R.R.; Collins, W.T.; Taglialavore, A.P.; Yeckley, R.L. Ceramic Component Processing Development for Advanced Gas Turbine Engines. *J. Eng. Gas Turbines Power* **1993**, *115*, 1–8.
10. Richerson, D.W. Historical Review of Addressing the Challenges of Use of Ceramic Components in Gas Turbine Engines 2006, 241–254.
11. Heimann, R.B. Silicon Nitride, A Close to Ideal Ceramic Material for Medical Application. *Ceramics* **2021**, *4*, 208–223.
12. Mobbs, R.J.; Rao, P.J.; Phan, K.; Hardcastle, P.; Choy, W.J.; McCartney, E.R.; Druitt, R.K.; Mouatt, C.A.L.; Sorrell, C.C. Anterior Lumbar Interbody Fusion Using Reaction Bonded Silicon Nitride Implants: Long Term Case Series of the First Synthetic ALIF Spacer Implanted in Humans. *World Neurosurg.* **2018**, *120*, 256–264, doi:https://doi.org/10.1016/j.wneu.2018.08.237.
13. Pezzotti, G.; Oba, N.; Zhu, W.; Marin, E.; Rondinella, A.; Boschetto, F.; McEntire, B.J.; Yamamoto, K.; Bal, B.S. Human Osteoblasts Grow Transitional Si/N Apatite in Quickly Osteointegrated Si_3N_4 Cervical Insert. *Acta Biomater.* **2017**, *64*, 411–420.
14. Ament, J.D.; Vokshoor, A.; Yee, R.; Johnson, J.P. A Systematic Review and Meta-Analysis of Silicon Nitride and Biomaterial Modulus as it Relates to Subsidence Risk in Spinal Fusion Surgery. *North Am. Spine Soc. J.* **2022**, *12*, 100168, doi:https://doi.org/10.1016/j.xnsj.2022.100168.
15. Bock, R.M.; Jones, E.N.; Ray, D.A.; Bal, B.S.; Pezzotti, G.; McEntire, B.J. Bacteriostatic Behavior of Surface-Modulated Silicon Nitride in Comparison to Polyetheretherketone and Titanium. *J. Biomed. Mater. Res. Part A* **2017**, *105*, 1521–1534, doi:https://doi.org/10.1002/jbm.a.35987.
16. Katsaros, I.; Zhou, Y.; Welch, K.; Xia, W.; Persson, C. Bioactive Silicon Nitride Implant Surfaces with Maintained Antibacterial Properties. *J. Funct. Biomater.* **2022**, *13*, 129.
17. Ghalme, S.; Falak, Y. Review on Evolution of Silicon Nitride in the Field of Orthopedics. *Curr. Trends Biomed. Eng. Biosci.* **2018**, *11*, 1–3, doi:https://doi.org/10.19080/CTBEB.2018.11.555824.
18. Tarpada, S.P.; Morris, M.T.; Burton, D.A. Spinal Fusion Surgery: A Historical Perspective. *J. Orthop.* **2017**, *14*, 134–136, doi:https://doi.org/10.1016/j.jor.2016.10.029.
19. Hampel, G.A.; Yilmaz, E.; Massrey, C.; Clifton, W.; Iwanaga, J.; Loukas, M.; Tubbs, R.S. History of Bone Grafts in Spine Surgery. *Cureus* **2022**, *14*, 1–10, doi:https://doi.org/10.7759/cureus.24655.
20. Virk, S.; Qureshi, S.; Sandhu, H. History of Spinal Fusion: Where We Came from and Where We are Going. *HSS Journal®* **2020**, *16*, 137–142, doi:https://doi.org/10.1007/s11420-020-09747-7.
21. Moon, M.S. Development in the Management of Tuberculosis of the Spine. *Curr. Orthop.* **2006**, *20*, 132–140, doi:https://doi.org/10.1016/j.cuor.2005.12.002.
22. Chatham, L.S.; Patel, V. V; Yakacki, C.M.; Dana Carpenter, R. Interbody Spacer Material Properties and Design Conformity for Reducing Subsidence During Lumbar Interbody Fusion. *J. Biomech. Eng.* **2017**, *139*, doi:https://doi.org/10.1115/1.4036312.

23. Gittens, R.A.; Olivares-Navarrete, R.; Schwartz, Z.; Boyan, B.D. Implant Osseointegration and the Role of Microroughness and Nanostructures: Lessons for Spine Implants. *Acta Biomater.* **2014**, *10*, 3363–3371, doi:https://doi.org/10.1016/j.actbio.2014.03.037.
24. Ribeiro, M.; Monteiro, F.J.; Ferraz, M.P. Infection of Orthopedic Implants with Emphasis on Bacterial Adhesion Process and Techniques Used in Studying Bacterial-Material Interactions. *Biomatter* **2012**, *2*, 176–194, doi:https://doi.org/10.4161/biom.22905.
25. Jain, S.; Eltorai, A.E.M.; Ruttiman, R.; Daniels, A.H. Advances in Spinal Interbody Cages. *Orthop. Surg.* **2016**, *8*, 278–284, doi:https://doi.org/10.1111/os.12264.
26. Ferraris, S.; Spriano, S. Porous Titanium by Additive Manufacturing: A Focus on Surfaces for Bone Integration. *Metals (Basel).* **2021**, *11*, 1–12, doi:https://doi.org/10.3390/met11091343.
27. Torres-Sanchez, C.; Al Mushref, F.R.A.; Norrito, M.; Yendall, K.; Liu, Y.; Conway, P.P. The Effect of Pore Size and Porosity on Mechanical Properties and Biological Response of Porous Titanium Scaffolds. *Mater. Sci. Eng. C* **2017**, *77*, 219–228, doi:https://doi.org/10.1016/j.msec.2017.03.249.
28. Pezzotti, G.; Marin, E.; Adachi, T.; Lerussi, F.; Rondinella, A.; Boschetto, F.; Zhu, W.; Kitajima, T.; Inada, K.; McEntire, B.J.; *et al.* Incorporating Si_3N_4 into PEEK to Produce Antibacterial, Osteoconductive, and Radiolucent Spinal Implants. *Macromol. Biosci.* **2018**, *18*, 1080033, doi:https://doi.org/10.1002/mabi.201800033.
29. Torstrick, F.B.; Safranski, D.L.; Burkus, J.K.; Chappuis, J.L.; Lee, C.S.D.; Guldberg, R.E.; Gall, K.; Smith, K.E. Getting PEEK to Stick to Bone: The Development of Porous PEEK for Interbody Fusion Devices. *Tech. Orthop.* **2017**, *32*, 158–166.
30. Han, Q.; Wang, C.; Chen, H.; Zhao, X.; Wang, J. Porous Tantalum and Titanium in Orthopedics: A Review. *ACS Biomater. Sci. Eng.* **2019**, *5*, 5798–5824, doi:https://doi.org/10.1021/acsbiomaterials.9b00493.
31. Watanabe, K.; Fukuzaki, S.; Sugino, A.; Benson, N.; Metcalf, N.; Nakamura, M.; Matsumoto, M. Cobalt–Chromium Alloy Has Superior Antibacterial Effect Than Titanium Alloy: *In vitro* and *in vivo* Studies. *Spine (Phila. Pa. 1976).* **2021**, *46*.
32. Waldorff, E.I.; Fang, S.; Zhang, N.; Visai, L.; Imbriani, M.; Magalini, E.; Preve, E.; Robotti, P.; Raines, A.L.; Goldberg, E.; *et al.* PEEK Titanium Composite (PTC) for Spinal Implants. In *Orthopedic Biomaterials: Advances and Applications*; Li, B., Webster, T., Eds.; Springer International Publishing: Cham, 2017; pp. 427–465 ISBN 978-3-319-73664-8.
33. Ma, H.; Suonan, A.; Zhou, J.; Yuan, Q.; Liu, L.; Zhao, X.; Lou, X.; Yang, C.; Li, D.; Zhang, Y. PEEK (Polyether-Ether-Ketone) and its Composite Materials in Orthopedic Implantation. *Arab. J. Chem.* **2021**, *14*, 102977, doi:https://doi.org/10.1016/j.arabjc.2020.102977.
34. Heimann, R.B. Silicon Nitride Ceramics: Structure, Synthesis, Properties, and Biomedical Applications. *Materials (Basel).* **2023**, *16*, 1–16, doi:https://doi.org/10.3390/ma16145142.
35. Petzow, G.; Herrmann, M. Silicon Nitride Ceramics. In *High Performance Non-Oxide Ceramics II SE - 2*; Jansen, M., Ed.; Structure and Bonding; Springer Berlin Heidelberg, 2002; Vol. 102, pp. 47–167 ISBN 978-3-540-43132-9.
36. Bal, B.S.; Rahaman, M.N. Orthopedic Applications of Silicon Nitride Ceramics. *Acta Biomater.* **2012**, *8*, 2889–2898, doi:https://doi.org/10.1016/j.actbio.2012.04.031.
37. Bal, B.S.; Khandkar, A.; Lakshminarayanan, R.; Clarke, I.; Hofmann, A.A.; Rahaman, M.N. Testing of Silicon Nitride Ceramic Bearings for Total Hip Arthroplasty. *J. Biomed. Mater. Res. Part B Appl. Biomater.* **2008**, *87*, 447–454, doi:https://doi.org/10.1002/jbm.b.31123.
38. McEntire, B.J.; Lakshminarayanan, R.; Ray, D.A.; Clarke, I.C.; Puppulin, L.; Pezzotti, G.; Bal, B.S.; Lakshminarayanan, A.; Bock, R. Silicon Nitride Bearings for Total Joint Arthroplasty. *Lubricants* **2016**, *4*, 1–24, doi:https://doi.org/10.3390/lubricants4040035.
39. McEntire, B.J.; Enomoto, Y.; Zhu, W.; Boffelli, M.; Marin, E.; Pezzotti, G. Surface Toughness of Silicon Nitride Bioceramics: II, Comparison with Commercial Oxide Materials. *J. Mech. Behav. Biomed. Mater.* **2016**, *54*, 346–359, doi:https://doi.org/10.1016/j.jmbbm.2015.08.044.
40. Du, X.; Lee, S.S.; Blugan, G.; Ferguson, S.J. Silicon Nitride as a Biomedical Material: An Overview. *Int. J. Mol. Sci.* **2022**, *23*, 6551, doi:https://doi.org/10.3390/ijms23126551.

41. Zeng, X.; Sipaut, C.S.; Ismail, N.M.; Liu, Y.; Farm, Y. yan; He, J. Mechanical Properties and Biological Activity of 3D Printed Silicon Nitride Materials. *Ceram. Int.* **2024**, *50*, 16704–16713, doi:https://doi.org/10.1016/j.ceramint.2024.02.041.
42. Wang, F.; Guo, J.; Li, K.; Sun, J.; Zeng, Y.; Ning, C. High Strength Polymer/Silicon Nitride Composites for Dental Restorations. *Dent. Mater.* **2019**, *35*, 1–10, doi:https://doi.org/10.1016/j.dental.2019.05.022.
43. Ishikawa, M.; Bentley, K.L. de M.; McEntire, B.J.; Bal, B.S.; Schwarz, E.M.; Xie, C.; Avenue, E.; McEntire, B.J.; Schwarz, E.M.; Xie, C. Surface Topography of Silicon Nitride Affects Antimicrobial and Osseointegrative Properties of Tibial Implants in a Murine Model. *J. Biomed. Mater. Res. A* **2017**, *105*, 3413–3421, doi:https://doi.org/10.1002/jbm.a.36189.
44. Webster, T.J.; Patel, A.A.; Rahaman, M.N.; Bal, B.S. Anti-Infective and Osteointegration Properties of Silicon Nitride, Poly (Ether Ether Ketone), and Titanium Implants. *Acta Biomater.* **2012**, *8*, 4447–4454.
45. Gorth, D.J.; Puckett, S.; Ercan, B.; Webster, T.J.; Rahaman, M.; Bal, B.S. Decreased Bacteria Activity on Si_3N_4 Surfaces Compared with PEEK or Titanium. *Int. J. Nanomedicine* **2012**, *7*, 4829–4840, doi:https://doi.org/10.2147/IJN.S35190.
46. Ahuja, N.; Awad, K.R.; Brotto, M.; Aswath, P.B.; Varanasi, V. A Comparative Study on Silicon Nitride, Titanium and Polyether Ether Ketone on Mouse Pre-osteoblast Cells. *Med. Devices Sensors* **2021**, *4*, 1–12, doi:https://doi.org/10.1002/mds3.10139.
47. Arts, M.P.; Wolfs, J.F.C.; Corbin, T.P. Porous Silicon Nitride Spacers versus PEEK Cages for Anterior Cervical Discectomy and Fusion: Clinical and Radiological Results of a Single-Blinded Randomized Controlled Trial. *Eur. Spine J.* **2017**, *26*, 2372–2379, doi:https://doi.org/10.1007/s00586-017-5079-6.
48. Ibrahim, T.; Tleyjeh, I.M.; Gabbar, O. Surgical versus Non-Surgical Treatment of Chronic Low Back Pain: A Meta-Analysis of Randomised Trials. *Int. Orthop.* **2008**, *32*, 107–113, doi:https://doi.org/10.1007/s00264-006-0269-6.
49. Zucherman, J.F.; Hsu, K.Y.; Hartjen, C.A.; Mehalic, T.F.; Implicito, D.A.; Martin, M.J.; Johnson, D.R.I.I.; Skidmore, G.A.; Vessa, P.P.; Dwyer, J.W.; *et al.* A Multicenter, Prospective, Randomized Trial Evaluating the X STOP Interspinous Process Decompression System for the Treatment of Neurogenic Intermittent Claudication: Two-Year Follow-Up Results. *Spine (Phila. Pa. 1976).* **2005**, *30*.
50. Sorrell, C.C.; Hardcastle, P.H.; Druitt, R.K.; Howlett, C.R.; McCartney, E.R. Results of 15-Year Clinical Study of Reaction Bonded Silicon Nitride Intervertebral Spacers. *Proc. 7th World Biomater. Conf.* **2004**, 1872.
51. Calvert, G.C.; Huffmon III, G. V; Rambo Jr, W.M.; Smith, M.W.; McEntire, B.J.; Bal, B.S. Clinical Outcomes for Anterior Cervical Discectomy and Fusion with Silicon Nitride Spine Cages: A Multicenter Study. *J. Spine Surg.* **2019**, *5*, 504–519, doi:https://doi.org/10.21037/jss.2019.11.17.
52. Calvert, G.C.; Huffmon, G.V.B.I.; Rambo, W.M.J.; Smith, M.W.; McEntire, B.J.; Bal, B.S. Clinical Outcomes for Lumbar Fusion using Silicon Nitride versus other Biomaterials. *J. Spine Surg.* **2020**, *6*, 33–48, doi:https://doi.org/10.21037/jss.2019.12.11.
53. McEntire, B.J.; Maislin, G.; Bal, B.S. Two-Year Results of a Double-Blind Multicenter Randomized Controlled Non-Inferiority Trial of Polyetheretherketone (PEEK) versus Silicon Nitride Spinal Fusion Cages in Patients with Symptomatic Degenerative Lumbar Disc Disorders. *J. Spine Surg.* **2020**, *6*, 523–540, doi:https://doi.org/10.21037/jss-20-588.
54. Smith, M.W.; Romano, D.R.; McEntire, B.J.; Bal, B.S. A Single Center Retrospective Clinical Evaluation of Anterior Cervical Discectomy and Fusion Comparing Allograft Spacers to Silicon Nitride Cages. *J. Spine Surg.* **2018**, *4*, 349–360, doi:https://doi.org/10.21037/jss.2018.06.02.
55. Rambo, W.M. Treatment of Lumbar Discitis using Silicon Nitride Spinal Spacers: A Case Series and Literature Review. *Int. J. Surg. Case Rep.* **2018**, *43*, 61–68, doi:https://doi.org/10.1016/j.ijscr.2018.02.009.

Silicon Nitride for Total Hip Arthroplasty

Kengo Yamamoto, Yasuhito Takahashi, Toshiyuki Tateiwa, Makiko Yorifuji, Bryan J. McEntire, and Giuseppe Pezzotti

1 Introduction

Total hip arthroplasty (THA) is one of the most successful and reliable surgical treatments for patients suffering from hip joint arthritis. During THA, the worn-out or damaged hip joint is excised entirely, and replaced with metallic, polymeric, and/or ceramic prosthesis, which generally restores pain-free movement and active lifestyles. In cementless THA, a typical hip prosthesis is composed of an acetabular liner housed within a porous metallic shell which is coupled to a spherical femoral head attached to a rodlike, porous metallic femoral stem (cf., Fig. 1). Metal porosity is designed to promote direct bone ingrowth with long-term implant stability, although polymethylmethacrylate (PMMA) can also be used to cement the prosthesis within acetabular and/or femoral bone.

The long-term success of THA is multifactorial, including patient-specific variables, the operation itself in terms of accurate implant positioning, and the wear characteristics of the bearing surfaces. The prosthetic design and physical-chemical characteristics can significantly influence clinical outcomes. Multiple materials are

K. Yamamoto · T. Tateiwa · M. Yorifuji
Department of Orthopedic Surgery, Tokyo Medical University, Tokyo, Japan

Y. Takahashi
Department of Orthopedic Surgery, Tokyo Medical University, Tokyo, Japan

Department of Bone and Joint Biomaterial Research, Tokyo Medical University, Tokyo, Japan

B. J. McEntire
SINTX Technologies Inc., Salt Lake City, UT, USA

G. Pezzotti (✉)
Ceramic Physics Laboratory, Kyoto Institute of Technology, Sakyo-ku, Matsugasaki, Kyoto, Japan
e-mail: pezzotti@kit.ac.jp

B. S. Bal et al. (eds.), *Silicon Nitride Bioceramics*,
https://doi.org/10.1007/978-3-031-67047-3_10

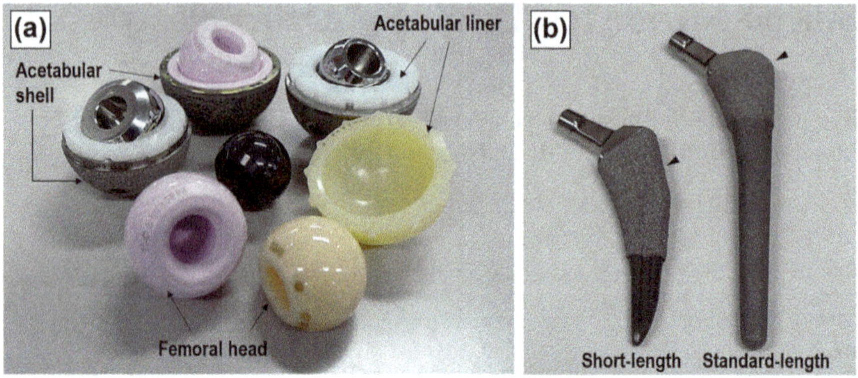

Fig. 1 Components of a total hip prosthesis: (**a**) metallic, polymeric, and ceramic bearing surfaces; and (**b**) metallic femoral stems (arrowheads indicate porous coating areas). (Courtesy Tokyo Medical University)

available for arthroplasty surgeons, but the most commonly accepted bearing components in contemporary THA are cobalt chromium (CoCr) alloys for the prosthetic ball, ultra-high molecular weight polyethylene (UHMWPE, mostly as a crosslinked formulation) for the prosthetic socket. Alumina-based ceramics and composites (e.g., Al_2O_3, ZTA) can be used as alternative bearing surfaces, replacing the traditional articulation of a metal ball against UHMWPE. Different combinations of femoral head and acetabular surface materials can be used based on the following bearing concepts: *Hard-on-soft* (metal-on-polyethylene, and ceramic-on-polyethylene), and *Hard-on-hard* (metal-on-metal, ceramic-on-ceramic, and ceramic-on-metal). The choice of these bearing couples depends on several factors such as patient age, activity and complications, and cost-effectiveness. In particular, young and active patients who participate in a variety of sports still present a challenge for implant materials in terms of their stability and durability. For the young patient, the prosthetic hip construct is required to last for multiple decades of in vivo service [1].

The *hard-on-soft* bearing concept developed by British orthopedic surgeon, Sir John Charnley in 1961, has been the gold standard in THA because it has a relatively safe and predictable clinical outcome. Wear of the UHMWPE acetabular socket insert is a major obstacle limiting implant function and longevity. Reducing wear has been an important motivation for choosing *hard-on-hard* bearings; but, in general, fewer of these bearing couples are implanted. Ceramic-on-ceramic articulation potentially has the most promising tribological properties with high wear- and scratch-resistance as well as low friction under physiological loading and lubrication conditions. However, catastrophic implant fracture, joint noise (squeaking), and increased cost may be encountered with this bearing couple [2–4]. Metal-on-metal articulations also have excellent mechanical and tribological properties with the longest history of clinical use since the 1930s, but various immune-mediated adverse reactions attributed to the release of metal ions/particles and corrosion products have

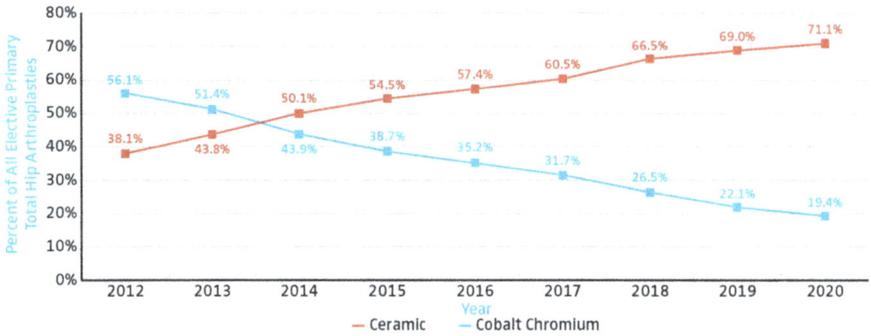

Fig. 2 Composition of femoral heads for all elective primary total hip arthroplasty procedures (excluding dual mobility) by year, 2012–2020 ($n = 574,182$). It depicts the significant shift from the use of cobalt-chromium heads to ceramic heads. (Courtesy AAOS [7])

essentially eliminated the use of metal-on-metal hip bearings. Since the late 2000s, worldwide recalls of some large-diameter metal-on-metal hip implants have been issued due to the unequivocal evidence of high failure rates and adverse tissue reactions [5]. Also, a new mode of failure was subsequently identified at modular interfaces of metal-on-polyethylene bearings, known as mechanically assisted crevice corrosion (MACC). Since then, experts have been calling for metal implants, especially metal-on-metal articulations, to be banned for clinical use [5, 6]. Consequently, in the last decade, surgeons have shifted to using insulating ceramic heads coupled with UHMWPE acetabular components to avoid these corrosion-related failures. At present, femoral heads made of oxide ceramics are used in over 70% of all primary THAs (excluding dual mobility bearings) in the United States (cf., Fig. 2).

In this chapter, material related complications are reviewed to provide an overview of major limitations and challenges regarding contemporary THA articulation. The chapter then introduces silicon nitride (Si_3N_4) as an alternative ceramic femoral head that has the potential to improve prosthetic joint outcomes and longevity.

2 Limitations and Challenges of Contemporary Hard-on-Soft THA Articulation

2.1 *Wear, Oxidation, and Fatigue Failure of Crosslinked UHMWPE*

The routine movement of a *hard* femoral head on a *soft* acetabular liner produces submicron wear particles of UHMWPE. These particles are phagocytized, activating macrophages with the subsequent release of chemokines, cytokines, and other inflammatory mediators (cf., Fig. 3). Their continuous release promotes an

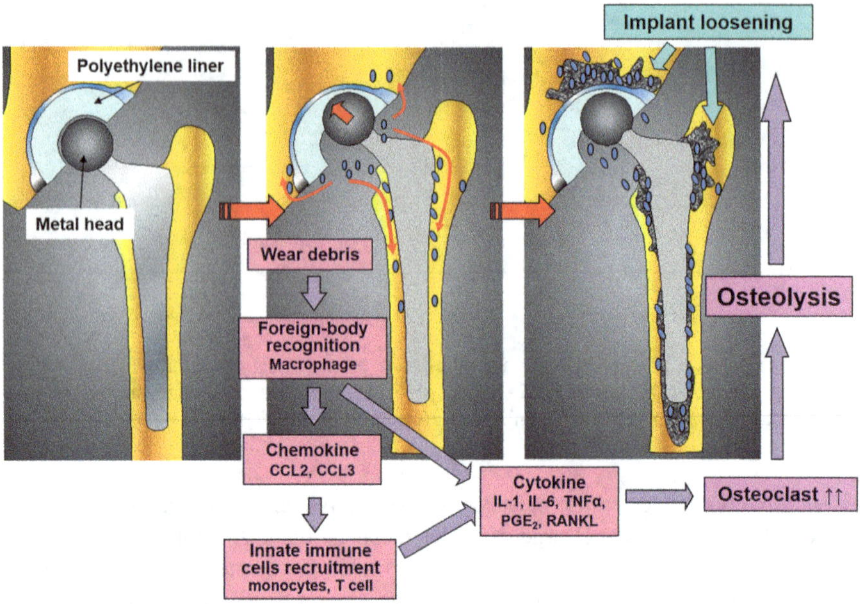

Fig. 3 Schematic diagrams of the biological response to UHMWPE wear particles, causing periprosthetic osteolysis and aseptic implant loosening. (Courtesy Tokyo Medical University)

inflammatory microenvironment in which osteoclasts are stimulated, resulting in bone resorption (osteolysis) surrounding the acetabular and/or femoral components [8–12].

The development of periprosthetic osteolysis leads to aseptic loosening (i.e., loss of implant fixation without a mechanical cause or infection). Osteolysis and aseptic loosening have been the prime reasons for revisions, (i.e., repeat surgeries) [13]. Although wear particles of metal and ceramics may invoke a similar cell-mediated biological response, these materials are less susceptible to abrasion than UHMWPE [8, 10, 12]. Detailed cytological examination of the lytic tissues demonstrates that 70% to 90% of the debris is predominantly spherical UHMWPE particles of less than 1 μm (>90%) with a mean size of 0.5 μm [9, 10, 12]. In clinical practice, an average wear rate of 0.1 mm per year is commonly considered to be a threshold for UHMWPE acetabular components. The frequency of osteolysis dramatically increases above this limit [14]. The biological response to wear debris is influenced not only by its size, shape, concentration, and composition but also by host characteristics, such as genetic variations that can modulate immune system response [10, 12]. Wear of UHMWPE acetabular liners occurs due to the break-up of fibrils formed by local plastic strain and molecular orientation related to the reciprocating motion of the joint [15–17]. Under multidirectional sliding, the oriented microstructure perpendicular to the molecular chain axis is mechanically weak.

The presence of a textured surface layer is responsible for producing wear debris. With that scientific discovery, radiation crosslinking of UHMWPE with irradiation energies greater than sterilization doses (i.e., >40 kGy) was introduced to increase abrasion resistance. This reduced adhesive wear by constraining chain movement followed by molecular texturing [18, 19]. However, ionizing radiation for crosslinking leaves a detectable concentration of free radicals in the final product, which increases the risk of oxidation during shelf storage and in vivo service [20, 21]. Post-irradiation heating performed around the melting temperature of UHMWPE (i.e., remelting or annealing), resulted in a complete or partial reduction of residual free radicals [22].

In 1998, radiation crosslinked, and thermally treated UHMWPE was clinically introduced as the first-generation highly crosslinked polyethylene (HXLPE). Today, it is currently the most implanted material worldwide in THA. According to a systematic review of short to mid-term radiographic outcomes performed by Kurtz et al. [23], these first-generation HXLPE liners decreased the mean wear rate by 69% compared to noncrosslinked (conventional) UHMWPE liners (i.e., 0.042 mm/year based on 28 studies, $n = 1503$ hips, vs. 0.137 mm/year based on 18 studies $n = 695$ hips). In this same paper, they reported an 87% lower risk of osteolysis using HXLPEs. The 2021 Australian Orthopaedic Association National Joint Replacement Registry (AOANJRR) Annual Report showed that HXLPE has a lower rate of revision for primary THA compared to conventional UHMWPE after 9 months [24]. Their cumulative revision percentages were 7.8% and 18.2% at 20 years, respectively.

Today, some drawbacks have been recognized in the first-generation HXLPEs. Indeed, the dramatic improvement of their wear resistance comes at the expense of other important load-bearing mechanical properties. High-dose irradiation has been shown to degrade ductility, elastic modulus, toughness, crack propagation resistance, and ultimate tensile strength [22, 25, 26]. Post irradiation remelting further decreases the tensile strength and fatigue resistance, resulting in catastrophic clinical fractures of remelted HXLPEs. The majority of fractured cases were used in small acetabular shells (≤54 mm) combined with large diameter heads (≥36 mm), (i.e., thin-walled HXLPE acetabular liners) [27]. Moreover, there was an increasing trend of subsurface oxidation with the in vivo implantation period in both annealed and remelted HXLPEs despite a low presence or complete absence of residual free radicals at the end of the manufacturing process [28, 29]. Cyclic loading under high-stress conditions and absorption of synovial lipids (e.g., squalene) have been proposed as possible mechanisms for producing new free radicals in vivo which can then propagate oxidation [30–32]. The degradation reactions induce chain scission and the formation of oxidation products, further decreasing mechanical strength and toughness [22, 33, 34]. These findings subsequently motivated the development of a next-generation HXLPE with antioxidant stabilization to further increase in vivo oxidation resistance and fatigue loading without compromising wear resistance. In the early 2000s, vitamin E (dl-α-tocopherol) was incorporated as an antioxidant agent either into the UHMWPE resin powder or into the consolidated and crosslinked material (cf., Fig. 4a). Vitamin E serves as a scavenger and contributes to neutralizing free radicals within the host polymer by donating a hydrogen atom from the OH group on its ring structure (cf., Fig. 4b). This mechanism worked well

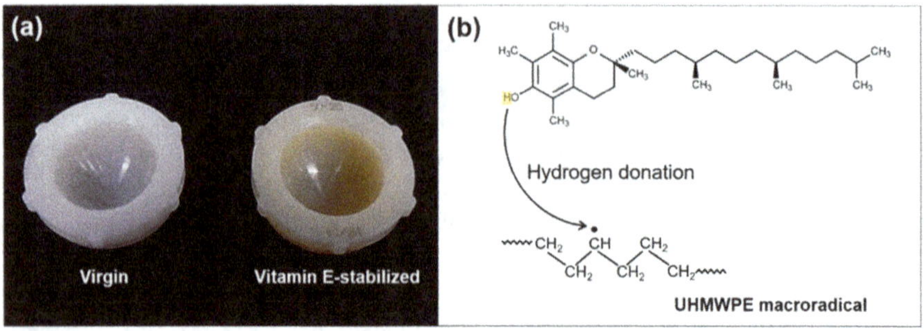

Fig. 4 (**a**) Photographs of virgin (left) and vitamin E-stabilized HXLPE (right). The right one has a light reddish-orange color due to the existence of blended vitamin E (3,000 ppm); (**b**) the stabilization mechanism of vitamin E is through a hydrogen donation from its phenolic OH group to a free radical in UHMWPE. (Modified and used with permission [37])

under absorbed lipid as well as cyclic loading conditions [35]. Vitamin E also acts as a plasticizing agent and partially compensates for the loss of mechanical properties related to crosslinking [36].

However, one early concern was that the presence of vitamin E at ≥3000 ppm caused a significant loss of crosslinking efficiency even with increasing irradiation doses. To mitigate or solve this problem some researchers then proposed increasing irradiation temperature (warm irradiation) [38], post-irradiation annealing [39], or postirradiation vitamin E diffusion and homogenization [40]. So far, laboratory studies suggest the following potential advantages of vitamin E: (1) a dramatic reduction in oxidative degradation [41–43]; (2) an improvement of fatigue resistance [36, 44]; (3) a reduction in the biological activity of wear debris (reduced inflammatory and osteolytic responses) [45–48]; and (4) a reduction in bacterial adhesion and subsequent biofilm formation [48–51]. These advantages are dependent on the vitamin E concentration, but excessive addition may lead to deterioration of the elastic modulus, ultimate tensile strength, and possibly the tribological performance of HXLPE by acting as a softening agent [37, 52, 53]. Takahashi et al. [37] demonstrated that the addition of 3000 ppm vitamin E increased chain mobility and the degree of surface texturing in HXLPE after severe uniaxial compression, potentially elevating cross-shear stresses in hip joint kinematics. The average content of vitamin E ranges from 1000 to 5000 ppm in commercial HXLPE acetabular components, but its optimal concentration has to be balanced against crosslink density. The merit of using this antioxidant remains clinically unproven. Long-term follow-up studies are underway, but its use is also currently associated with increased implant cost. In 2020, vitamin E-stabilized HXLPE was only used in 7.74% of all primary THA cases performed in the USA [7]. In the current era of cost containment, a lack of in vivo evidence is likely a factor limiting its widespread usage. Further investigation of its clinical outcomes in comparison to nonantioxidant HXLPEs is needed.

2.2 Metal Corrosion and Adverse Local Tissue Reactions

Metal corrosion and fretting damage of hip prostheses have raised particular concerns during the last two decades (cf., Fig. 5). Corrosive attack can weaken implant mechanical strength, leading to in vivo fatigue, fracture, and dissociation of the intraoperatively assembled components [54–56].

It should be noted that the release of ionized metals, fretting debris, and corrosive products into the human body may cause several adverse local tissue reactions (ALTR), including osteolysis, necrosis of bone, muscle, and tendon, capsular thickening, cystic lesions, excessive fluid collections, aseptic lymphocyte-dominated vasculitis-associated lesions (ALVAL), and soft tissue masses (pseudotumors) [57–59]. These reactions may be associated with elevated concentrations of intra-articular metal ions and chemokines [60–62], and clinically manifested as subsequent pain. Increased metal ion levels have also been noted in serum, whole blood, cerebrospinal fluid, urine, and solid organs, but it is not yet clearly understood whether the reactions are dose-dependent or an innate hypersensitivity response. These problems have been evident since modularity was introduced to femoral stems in THA. This allowed customized implant selection and fitting for each patient (i.e., a head–neck junction—trunnion, neck–stem junction—dual modular implants, or metaphyseal–diaphyseal junction) [63]. For example, the use of modular stems allows for an intraoperative adjustment of leg length and hip offset via the head–neck junction and femoral anteversion via the neck–stem junction. Moreover, there is no need to exchange a well-fixed femoral stem in revision THA. Modular junctions in total hip components are therefore here to stay.

The modular interfaces consist of cobalt chromium molybdenum (CoCr alloy) and/or titanium-based alloys (Ti6Al4V). CoCr alloys are commonly used for the femoral head and femoral stem. They have excellent mechanical properties and wear resistance. Titanium-based alloys are used mostly for femoral stems due to their light

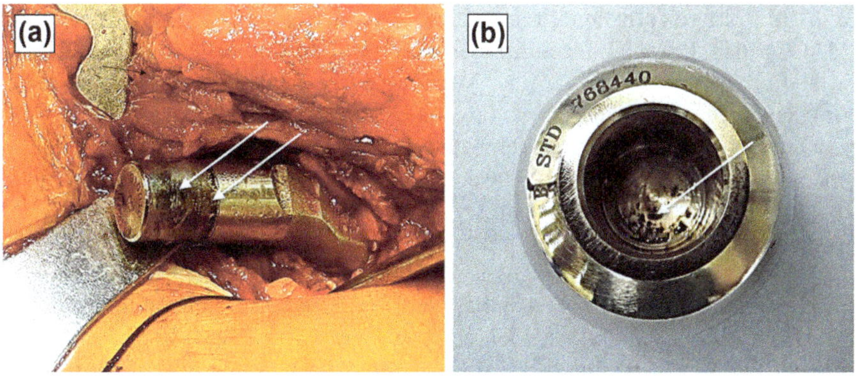

Fig. 5 Intraoperative photographs showing metal corrosion with black metal debris (white arrows) around the femoral neck (trunnion) (**a**) and femoral head bore (**b**). (Courtesy Tokyo Medical University)

weight, flexibility, good fatigue strength, biocompatibility, and osseointegration ability. These alloy surfaces are normally covered by a thin protective oxide layer (i.e., passivation) that prevents redox reactions, thus providing excellent resistance to electrochemical corrosion. If this passivation layer is damaged in an oxygen-rich environment, the implant surfaces immediately reform the protective layer. However, in a hypoxic environment, such as inside modular junctions, and/or under frictional conditions, the surface cannot re-passivate, and corrosion processes may initiate via exposure of the unoxidized surfaces to bodily fluids. All modular junctions have micro-gaps (i.e., crevices) between the two metal passivated surfaces. These are potential sites for crevice corrosion. It is generally triggered by an oxygen imbalance between the outside and the inside of the crevice [60, 62]. The lack of oxygen in the crevice solution leads to anodic and cathodic processes inside and outside the crevice, respectively. These processes create current even if the two interfaces are made of the same alloy. The basic underlying reaction is the loss of electrons from metal atoms in the crevice solution (metal dissolution as expressed by: $M \leftrightarrow M^{n+} + ne^-$). The electron generated at the anodic site (inside the crevice) can migrate to the cathodic site (outside the crevice) and react with oxygen ($3O_2 + 6H_2O + 12e^- \rightarrow 12OH^-$). As a result, oxygen concentration decreases on the metal surfaces outside the crevice. Conversely, the concentration of positive metal ions (M^{n+}) generated inside the crevice becomes excessive, which is then balanced by an influx of negatively charged chloride ions (Cl^-) from the outside. The concentrations of metal, chloride, and hydrogen ions eventually become high in the oxygen-poor environment, causing a decrease in pH within the crevice solution. This results in disruption of the passive oxide layer (de-passivation) and subsequent corrosive attack inside the crevice.

Besides fluid chemistry, the de-passivation of the implant's surface can be facilitated by mechanical stresses or micro–/macro-motions generated at the taper interface during daily activities. Due to taper wear (i.e., fretting), the mechanically de-passivated substrate spontaneously re-oxidizes, leading to additional oxygen depletion inside the crevice and acceleration of crevice and pitting corrosion [65]. This form of corrosion is referred to as mechanically assisted crevice corrosion (MACC) [64]. It is now recognized as a cause of symptomatic implant failure in metal-on-polyethylene THA. Also, MACC may be facilitated by other modes of corrosion such as galvanic effects (i.e., an electrical potential difference in contact between two dissimilar metals or between alloys with different composition or processing) [65, 66]. MACC has been studied primarily on head–neck junctions in total hip replacements. In contemporary modular metal-on-polyethylene THA, the prevalence of MACC has been reported to be in the range of 1.1% to 4.7% for specific components [56, 57, 67, 68]. Multiple factors can affect the overall prevalence of MACC, including femoral heads $\geq$ Ø36 mm, femoral head offsets $>$1.5 mm, taper geometry, neck length, head–neck material couple, material properties of the base alloy, the force of head impaction at the time of surgery, varus position, lateral, offset, patient's body mass index, and time in vivo [69–77]. In vitro [78] and retrieval [79] studies have demonstrated that MACC may be mitigated by using a ceramic femoral head (an electrical insulator), although MACC is not eliminated. Consequently, the use of ceramic femoral heads has become more popular to minimize the risk of symptomatic MACC in THA (cf., Fig. 2). This trend will likely continue until the emergence of new modular devices based on a fuller understanding of the failure mechanisms.

2.3 *Phase Transformation and Fracture of Oxide Ceramics*

One significant weakness of ceramic bearings lies in their inherently low tensile strength and toughness, leading to a concern for brittle fracture during their long in vivo service (cf., Fig. 6a).

The fracture of a ceramic component is a serious complication requiring immediate revision surgery, which may often be complicated. It is impossible to remove all the minute ceramic fragments during revision THA. Some fragments may be embedded in the articulating surface, potentially causing third-body wear in the revised joint [80]. Ceramic head fracture is also problematic in terms of its damage to the UHMWPE and stem neck components (cf., Fig. 6b, c). Overcoming brittleness has been the main objective for developing better orthopedic ceramics for load-bearing applications. The current medical-grade alumina (Al_2O_3) used in THA has been significantly improved over the period 1970s–1990s owing to modifications of

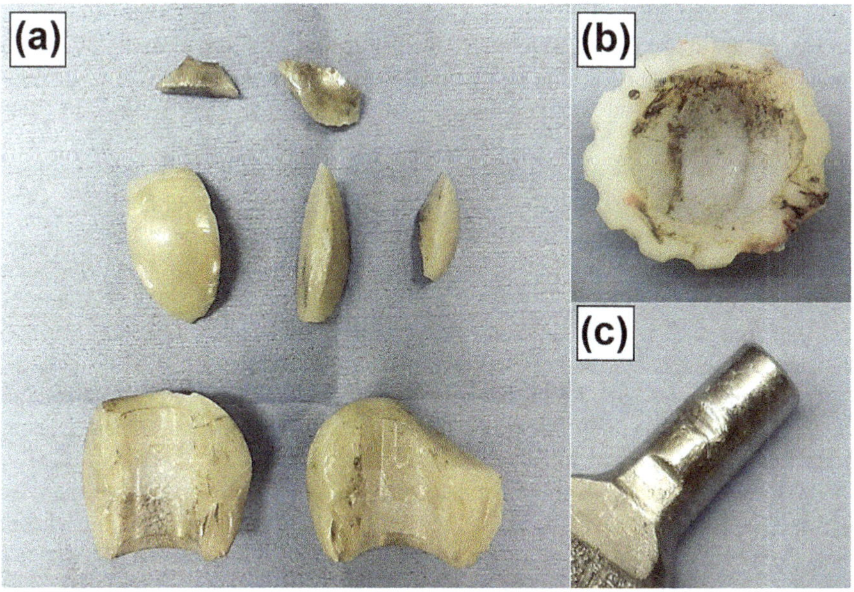

Fig. 6 Intraoperative photographs showing fragmentation of the ceramic femoral head (**a**) and severe damage to the UHMWPE acetabular liner (**b**) and trunnion of the femoral stem (**c**). (Courtesy Tokyo Medical University)

sintering processes, resulting in increased density, hardness, and strength, with a decrease in average grain size and a narrower grain size distribution [81]. Pure alumina is still implanted today because it is less expensive than the latest composite ceramics.

In the 1980s, partially stabilized zirconia (ZrO_2) with the addition of yttrium oxide (Y_2O_3) was widely used as femoral heads coupled with UHMWPE acetabular liners [82–84]. Mechanistically, the metastable tetragonal ZrO_2 transforms into its stable monoclinic crystalline structure with 3% ~ 5% volume expansion and ~7% shear strain. This generates strong compressive forces that oppose crack propagation (i.e., phase transformation toughening) [85]. Zirconia thus has a higher fracture toughness and mechanical strength than alumina. However, it was discovered that this phase transformation may also spontaneously initiate in a biological (aqueous) environment, even in the absence of external stress and cracks. This hydrothermally induced transformation is referred to as low-temperature degradation (LTD) or simply aging [86]. The occurrence of LTD at the bearing surface deteriorates the tribological performance via surface roughening, while the fracture risk within the head increases due to internal stress accumulation. In 2001, European and American regulatory agencies recalled millions of zirconia heads distributed under the trade name of Prozyr® due to their unprecedentedly high fracture rates due to LTD [84, 87]. After this recall, monolithic zirconia heads fell out of favor, and accordingly, two alternatives to zirconia became available after the turn of the century: (1) oxidized zirconium (OxZr); and (2) alumina-matrix composites (AMC) or zirconia-toughened alumina (ZTA) and alumina-toughened zirconia (ATZ).

OxZr is a zirconium-niobium alloy (Zr-2.5Nb) covered with a thin ceramicized layer (~5 μm thick) [88]. LTD is not an issue for OxZr because over 95% of the oxidized surface consists of the stable monoclinic phase. Also, head fractures do not occur because of the ductile metallic substrate. According to the 2021 AOANJRR Annual Report, OxZr heads coupled with HXLPE provided the best long-term results, showing the lowest cumulative percent revision at 17 years among all bearing combinations in primary THA [24]. Nevertheless, several authors pointed out that severe abrasion and scratching of the ceramicized outer surface may occur under certain clinical conditions such as recurrent hip dislocation and subsequent attempts at reduction [89–92]. This occurs because the ceramicized surface layer is too thin and the metallic substrate too soft [93]. When this superficial layer is damaged or penetrated, the exposed soft metallic core and the UHMWPE counterpart are at significant risk for accelerated wear [89–94]. Tribe et al. [92] showed an extreme case of a severely damaged OxZr head and HXLPE counterpart, causing metallosis and local tissue inflammation (seemingly, nontoxic).

ZTA or AMC is currently the most widely used composite ceramic in THA, typically consisting of an alumina matrix phase (80–85 vol.%) with a homogeneous dispersion of a zirconia secondary phase (15–20 vol.%). The presence of tetragonal zirconia dispersoids in ZTA contributes to transformation toughening, resulting in increased mechanical strength and fracture toughness over monolithic alumina [95]. In addition, the tetragonal zirconia grains in the ZTA composite have higher metastability against hydrothermal conditions than those in monolithic zirconia, resulting in improved surface topologic stability [96, 97]. However, head fractures, although rare, continue to be a concern even in ZTA. It occurs in about 1 in 1000 primary THAs [98]. Most of the reported ZTA head fractures are associated with postoperative traumatic events [99]. However, there are reports of atraumatic fractures of large ZTA heads (Ø32 and Ø40 mm) following ceramic-on-polyethylene THA [100, 101] and also an intraoperative fracture of a Ø40 mm ZTA head during impaction onto a well-fixed femoral stem [102]. The tetragonal-monoclinic phase transformation is also a plausible factor affecting ZTA fracture [103, 104]. Several studies suggest that aliovalent cations from body fluids (e.g., Ca^{2+}, Mg^{2+}) [105] and transition metal ion contamination (e.g., Co, Cr, Fe, or Ti) [106–109] could be additional contributors to grain boundary embrittlement and increased destabilization kinetics in the composite architecture.

3 Lifetime of Hard-on-Soft THA Prostheses and the Potential Use of Si_3N_4

THA has been given the accolade of "operation of the century" [110], and this surgical intervention continues to benefit over a million patients annually on a worldwide basis [7]. However, as the prior subsections demonstrate, considerable challenges remain in extending the lifetime of these prostheses to the lifespans of the most demanding, typically younger patients. This is where silicon nitride (Si_3N_4), a non-oxide bioceramic, may have an important future role as a femoral head. As discussed in the remainder of this chapter and elsewhere within this book, Si_3N_4 has promising advantages including excellent mechanical properties (i.e., high strength and toughness obtained without phase transformation—for improved fracture resistance and use of larger head sizes), a chemistry that is inherently osteoconductive and bacteriostatic (for improved biocompatibility and reduced incidence of infection), and an inherently hard, smooth, corrosion resistant, and antioxidative surface (for elimination of MACC and reduction of long-term articular wear against HXLPE). Unlike Al_2O_3 and ZTA which are considered to be "inert" bioceramics, Si_3N_4 is bioactive [111]. Recent independent reviews of Si_3N_4 as a bioceramic have highlighted these and other biological advantages [112, 113].

Although Si_3N_4 is currently clinically used as a biomaterial in spinal arthrodesis, it has yet to be cleared by regulatory authorities for THA applications. Traditionally, in vitro wear testing in hip simulators was the gating requirement for regulatory approval of new femoral heads. However, over the past two decades, testing hurdles have dramatically increased. Devices, such as Al_2O_3 on polyethylene or Al_2O_3 on Al_2O_3, were originally wear-tested to only five million cycles (Mc). These materials were cleared for use provided their wear rates were substantially equivalent to comparable predicates (e.g., metal on polyethylene or metal on metal). But as these new couples became commonplace in the market, clinical issues such as head fracture, bearing noise, metallosis, crevice corrosion of the modular components, and third-body wear became more important. Consequently, additional requirements were added to the in vitro test criteria. This included up to 10-Mc of standard hip simulator testing and an additional 3-Mc of third-body wear. Now, regulatory agencies also require clinical data to substantiate equivalence, including up to a 2-year prospective study of the proposed THA couple against appropriate predicates. Si_3N_4 now falls within these new regulatory demands, and the financial burden associated with conducting these extended tests is generally prohibitive for all but the largest orthopedic companies. Furthermore, corporations with the financial capability to perform these tests typically base their decisions on return on investment (ROI). Their ROI is dependent upon market demand for the new bearing couple. In turn, any adoption of the new device is contingent upon surgeon acceptance and cost-effectiveness. These are difficult barriers that entrench the status quo over newer technologies. Bozic et al. developed a financial analysis that suggests that new bearing technologies *should be* considered for patients based on age [114]. They showed a sliding scale of cost-effectiveness. Provided the new device has at least a 19% increase in 20-year survivorship, a 50-year-old patient would definitively benefit over the status quo. A 65-year-old patient would still marginally benefit, but no benefit would be realized for patients >75 years of age. This suggests that Si_3N_4-bearing couples should be targeted for younger and more active patients. However, the key obstacle to overcome is to demonstrate improvement in 20-year survivorship. This is a classic conundrum. While clinical results are required to substantiate cost-effectiveness, data acquisition first requires a prospective clinical trial, followed by regulatory clearance, and then multi-decade patient monitoring. Unfortunately, this is an untenable timeline which further entrenches the status quo. It is only resolvable if preclinical data coupled with limited clinical trials are acceptable surrogates for long-term clinical results. It is hoped that the information presented in the following sections of this chapter and elsewhere within this book will potentially serve as surrogates to open an effectual door for the use of this unique bioceramic in total joint arthroplasty.

4 Potential Application of Silicon Nitride in Hard-on-Soft THA

As a non-oxide ceramic, Si_3N_4 was developed for heat-engine and various industrial applications during the 1980–90s [115–119]. However, research was concurrently initiated into its clinical benefits; and remarkably, a human spinal arthrodesis study (without significant regulatory oversight) was initiated in Australia in 1986 [120]. A 30-year review of the surviving patients from that study showed excellent clinical and radiographic outcomes [121]. However, it wasn't until 2008 that Si_3N_4 was commercially developed for a range of orthopedic applications (cf., Fig. 7). As described in the following sections, this bioceramic could be a very promising candidate for use as femoral heads in THA bearings primarily due to its favorable

Fig. 7 Silicon nitride orthopedic implants. (Courtesy of SINTX Technologies Corporation, Salt Lake City, USA)

combination of high mechanical stability and tribological characteristics [122–124]. While many of its positive biological properties are covered elsewhere within this book, the remaining sections of this chapter will focus solely on aspects that pertain to THA.

4.1 Wear Performance of Silicon Nitride Bearing Couples

As described in Sect. 2.1, the most important factor in the longevity of THA joints is the wear performance of the mating femoral and acetabular components. Over the past two decades, extensive preclinical wear testing of various silicon nitride couples has been performed. The first evaluation was conducted by Bal et al. in 2008–2009 [122, 125]. It was performed using *hard-on-hard* (Si_3N_4–Si_3N_4) couples. At that time, the use of cobalt-chromium (CoCr) femoral heads and acetabular liners was commonplace in the hip arthroplasty market. Subsequently, four additional hip simulator studies were conducted and reported—three by McEntire et al. [126] and one by Yorifuji et al. [127]. Of these, one test examined Si_3N_4–Si_3N_4 bearings, and the remaining utilized *hard-on-soft* couples (Si_3N_4–UHMWPE, Si_3N_4–XLPE, or Si_3N_4–HXLPE).

4.1.1 Si_3N_4–Si_3N_4 Hip Simulator Wear Studies

The purpose of the first Si_3N_4–Si_3N_4 wear study by Bal et al. [126, 129] was to compare an all-Si_3N_4 couple and a combined couple using CoCr as the femoral head and Si_3N_4 as the acetabular liner to wear rates for CoCr–CoCr and Al_2O_3–Al_2O_3 couples. Testing was conducted using Ø28 mm bearings in an inverted hip simulator (Shore Western Manufacturing, Monrovia, CA, SWM) according to established protocols [125]. Because this was an initial study, wear testing was only performed

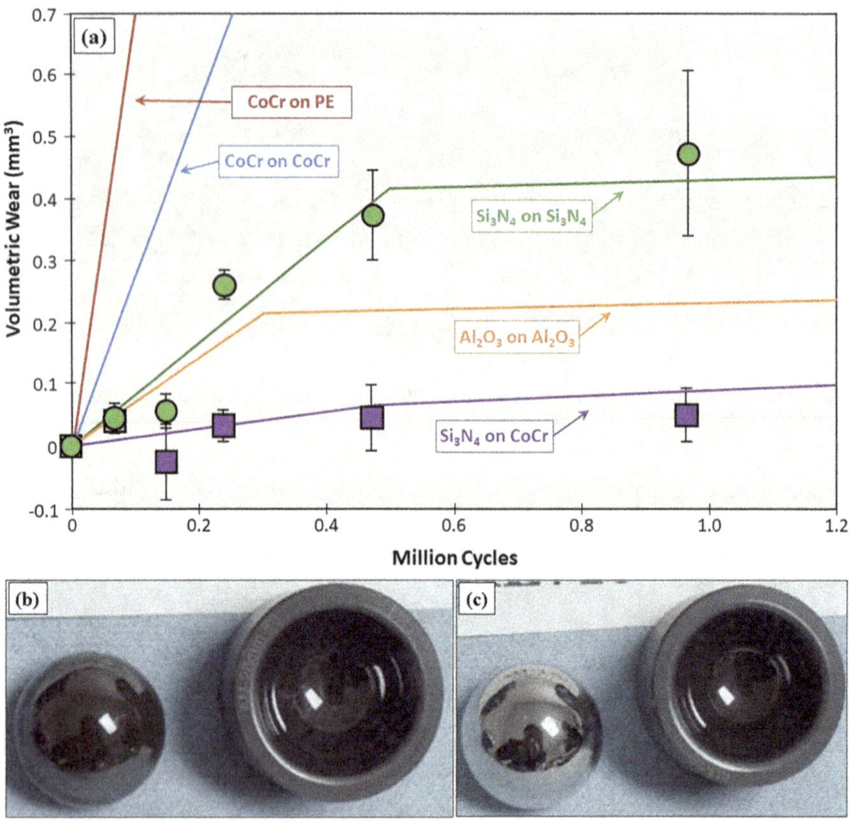

Fig. 8 THA wear study for bearings containing Si_3N_4: (**a**) Si_3N_4–Si_3N_4 and Si_3N_4–CoCr couples compared to Al_2O_3–Al_2O_3 and CoCr–CoCr under identical testing conditions; and optical photographs of post-tested couples, (**b**) Si_3N_4–Si_3N_4, and (**c**) CoCr–Si_3N_4. (Modified and used with permission [122])

to 1-Mc (cf., Fig. 8a–c). The results showed that the wear rates for Si_3N_4–Si_3N_4 and CoCr–Si_3N_4 bearings were essentially equivalent to commercially available Al_2O_3–Al_2O_3 bearings, and significantly lower than either CoCr–CoCr or CoCr–UHMWPE couples. While this test confirmed that Si_3N_4 had potential as a bearing surface, it was surprising to note that the CoCr–Si_3N_4 pairs had both lower wear and lower variability than Si_3N_4–Si_3N_4 or Al_2O_3–Al_2O_3 bearings. Based on comparative hardness values, the Al_2O_3–Al_2O_3 should have had the lowest wear rate. A possible explanation for this observation was postulated in a subsequent Si_3N_4–Si_3N_4 study [126].

The next hard-on-hard test also employed a SWM hip simulator [126]. It examined the wear performance of both Ø28 mm and Ø40 mm matched Si_3N_4 femoral and acetabular pairs with Ø28 mm Al_2O_3–Al_2O_3 couples as controls. The test was initiated and monitored but then suspended at 3-Mc due to essentially no wear for

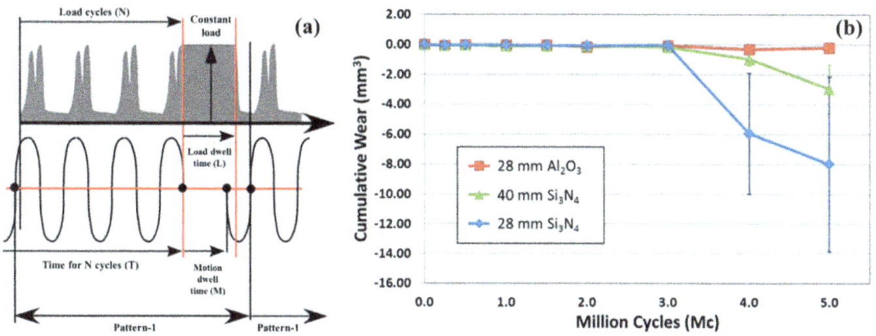

Fig. 9 (**a**) Nonstandard hip simulator study which included continuous motion and stop-dwell-start segments; and (**b**) wear results for Ø28 mm Al_2O_3 on Al_2O_3, and both Ø28 mm and Ø40 mm Si_3N_4 on Si_3N_4 bearing couples. Note that under the continuous motion segment, the wear performance of all three bearing types was equivalent. The introduction of the stop-dwell-start sequence after three million cycles resulted in significant wear, particularly for the Si_3N_4 on Si_3N_4 couples. However, the Al_2O_3 on Al_2O_3 couple was also negatively affected. (Modified and used with permission under Creative Commons CC-BY license [126])

either the Si_3N_4 or Al_2O_3 bearing types. Instead of continuing the study using the same conditions, it was elected to introduce a *stop-dwell-start* sequence within the standard Paul cycle [128] (cf., Fig. 9a). It was postulated that applying a static load during the stop-dwell would displace serum lubricant from the contact zone between the head-liner pairs, and the subsequent restart would require the bearing to overcome low lubricated static friction.

It was assumed that this sequence would create a worst-case for wear. This was observed, but the outcomes were surprisingly different for the two bearing materials. No differences in wear were noted for the Si_3N_4 and Al_2O_3 couples until this change in the protocol at 3-Mc. Then, with the introduction of the *stop-dwell-start* sequence, slightly increased wear was seen for the Ø28 mm Al_2O_3 controls, but a marked increase in wear was observed for both the Ø28 mm and Ø40 mm Si_3N_4 pairs (cf., Fig. 9b). It is recognized that the wear of Si_3N_4 in an aqueous environment is due to a hygroscopic tribochemical reaction that releases silicic acid and ammonia [129, 130]:

$$Si_3N_4 + 12H_2O \rightarrow 3Si(OH)_4 + 4NH_3 \quad (1)$$

Under continuous bearing motion, silicic acid provides a viscous low-friction lubricating gel within the tribolayer [131, 132]. However, the *stop-dwell* sequence returns the bearing to boundary lubrication conditions [133, 134]. With the reduction of lubricating film under this static load, contacting asperities between the surfaces increase. This leads to plowing and microfracture of the mating components during *restart*. This wear pattern was observed on the Si_3N_4 bearings, and its appearance is consistent with findings from other researchers [133, 135–137]. After *restarting*, the tribochemical film reforms and lowers dynamic friction until the next stop in the cycle. This *stop-dwell-start* sequence resulted in repetitive *stick-slip* behavior and noise for the remainder of the test. It should be noted that the Al_2O_3 controls also exhibited this phenomenon but to a much smaller extent than the Si_3N_4 couples.

These two Si_3N_4–Si_3N_4 hip simulator studies suggest that Si_3N_4 may be an inappropriate bearing material for *hard-on-hard* applications. Using finite element analyses, other investigators have recognized the importance of coupling dissimilar materials when constructing effective high-performance bearings [138–140]. Materials with differing chemical affinities, elastic moduli, and hardness increase contact area and lower stress between mating components, thereby reducing overall wear. Furthermore, preferred hip implants today are composed of *hard-on-soft* couples, predominately either metal or ceramic on UHMWPE, XLPE, or HXLPE. The following section discusses the wear testing of these types of couples against Si_3N_4.

4.1.2 Si_3N_4–UHMWPE, XLPE, and HXLPE Hip Simulator Studies

Of the remaining three hip simulator studies, one was performed using standard UHMWPE acetabular liners. The second study used a combination of UHMWPE and XLPE liners, and the third study exclusively utilized vitamin E-doped-HXLPE liners. The reason three different polyethylene materials were used is because the studies were conducted during a time (i.e., 2005 to 2020) when significant strides were being made in improving polyethylene for joint arthroplasty.

The first Si_3N_4–UHMWPE study (circa 2005) utilized a 12-station SWM hip simulator. The test set-up included Ø28 mm Si_3N_4 and CoCr femoral heads paired with GUR 1050 UHMWPE liners (gamma sterilized 25–40 kGy). Identical liners were used as static-load-soak controls, and testing was performed to 5-Mc. Volumetric wear rates for the liners were 17.6 ± 0.7 mm^3/Mc and 21.3 ± 4.1 mm^3/Mc for CoCr and Si_3N_4 femoral heads, respectively (Table 1, [126]). Even though the CoCr pairs showed overall lower wear, this result was not statistically significant ($p = 0.20$).

The second combined Si_3N_4–UHMWPE and Si_3N_4–XLPE study (circa 2011) employed an 18-station hip simulator (Advanced Mechanical Technology, Inc., Watertown, MA, USA, AMTI) using Ø28 mm and Ø40 mm Si_3N_4 and Ø28 mm CoCr femoral heads paired with sets of conventional GUR 1050 UHMWPE liners subjected to different sterilization methods (ethylene oxide, ETO; or gamma irradiation 25–40 kGy), and GUR 1050 XLPE liners that had been previously crosslinked

Table 1 Polyethylene wear for the SWM Si_3N_4–UHMWPE five-million cycle wear study

Femoral head material	Testing machine	Bearing size	UHMWPE and sterilization method	Wear rate (mm^3/Mc)	Range (mm^3/Mc)
CoCr	SWM	28	Conventional 25–40 KGy	17.6 ± 0.7	16.7–18.1
Si_3N_4	SWM	28	Conventional 25–40 KGy	21.3 ± 4.1	16.7–24.3

Table 2 Polyethylene WEAR for the AMTI Si_3N_4–UHMWPE and Si_3N_4–XLPE five-million cycle wear study

Femoral head material	Testing machine	Bearing size	UHMWPE sterilization method	Wear rate (mm^3/Mc)	Range (mm^3/Mc)
Si_3N_4	AMTI	28	Conventional ETO	66.0 ± 5.8	60.3–71.9
CoCr	AMTI	28	Conventional ETO	53.1 ± 4.8	47.5–56
Si_3N_4	AMTI	28	XLPE—78.5 KGy ETO	5.0 ± 1.3	3.5–6.0
Si_3N_4	AMTI	40	XLPE—78.5 KGy ETO	9.1 ± 1.6	8.1–10.9

using 78.5 kGy irradiation and subsequently annealed in nitrogen at 85–90 °C for 24 h. These XLPE liners were also ETO sterilized before wear testing. As expected, the wear results for these components varied based on their size, crosslinking irradiation, and sterilization (Table 2).

However, direct comparative data were only acquired for the Ø28 mm conventional ETO sterilized liners. They showed wear with values of 66.0 ± 5.8 mm^3/Mc and 53.1 ± 4.8 mm^3/Mc for the Si_3N_4 and CoCr femoral heads, respectively (cf., Table 2, $p = 0.04$). Even though this difference was statistically significant, it was not considered clinically relevant. Nowadays, ETO-sterilized conventional UHMWPE is rarely used for THA due to its high wear rate. Note that the crosslinked XLPE liners showed approximately a ten-fold reduction in wear compared to the conventional liners (cf., Table 2). Collectively, the UHMWPE wear results from these two Si_3N_4–UHMWPE tests were within the ranges of other published hip simulator studies [126]. However, the data also showed lower wear for the Ø28 mm CoCr pairs in comparison to the Ø28 mm Si_3N_4 couples in both studies (cf. Tables 1 and 2). This consistency is likely due to the preferential adhesion of UHMWPE debris and protein precipitation. Protein adsorption onto bearing surfaces can be thermally or mechanically activated by high shear, pressure, or frictional heating [141, 142]. Bowsher et al. [143] and Tsai et al. [144] separately demonstrated that the local interfacial temperature between the head and the liner can vary depending on the thermal conductivity of the head material. It is believed that this difference in thermal conductivity (i.e., CoCr ~9 W/m°K versus Si_3N_4 ~ 30 W/m°K) likely led to higher interfacial temperatures for the CoCr–UHMWPE pairs which resulted in greater protein precipitation. The deposited proteins became a protective layer on the head and therefore lowered overall wear.

The third Si_3N_4–HXLPE study (circa 2020) examined the wear behavior of advanced vitamin E-diffused HXLPE acetabular liners coupled to Ø28 mm oxide (ZTA, BIOLOX®*delta*) and non-oxide (Si_3N_4, *MC*²®) ceramic femoral heads [127]. The test was conducted by applying the kinematics recommended by ISO

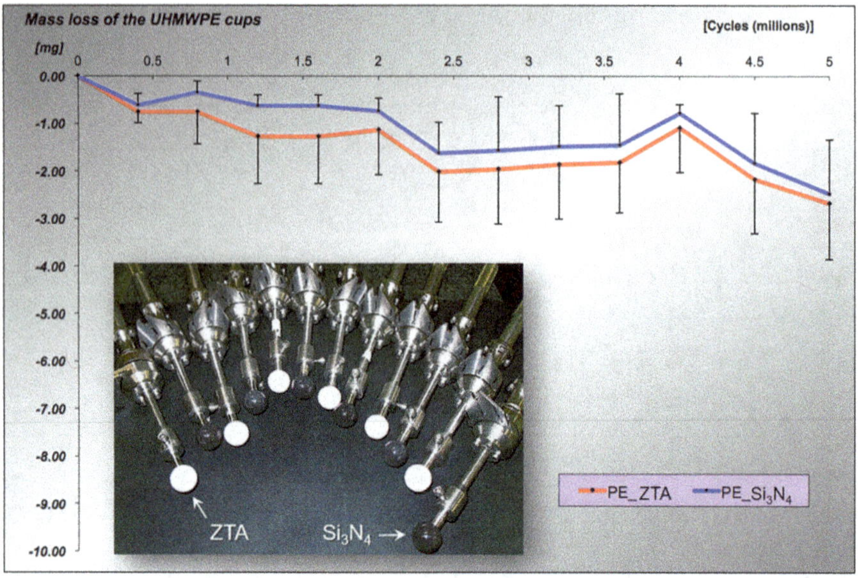

Fig. 10 Total mass loss of vitamin E-diffused HXLPE acetabular liners coupled to ZTA and Si_3N_4 ceramic femoral heads after 5-Mc of simulated gait. No significance was noted between the ZTA and Si_3N_4 groups. (Reconfigured and used with permission under Creative Common CC BY license [127])

14242-1 for 5-Mc using a 12-station hip simulator (IORSynthe, Bologna, Italy). Vitamin E-diffused HXLPE liners in contact with Si_3N_4 heads produced slightly lower wear than identical liners with ZTA heads (cf., Fig. 10).

The mean wear rates (corrected for fluid absorption) of the liners coupled with ZTA and Si_3N_4 heads were 0.53 ± 0.24 and 0.49 ± 0.23 mg/Mc after 5-Mc, respectively ($p = 0.82$). Raman spectroscopy detected a slight increase (3.3%) in the monoclinic zirconia phase in ZTA heads after the 5-Mc wear test, which did not significantly affect their surface roughness. The extent of surface roughening after testing was greater for Si_3N_4 versus ZTA heads although this difference was not statistically significant. This is likely due to a higher Vickers hardness of the ZTA compared to Si_3N_4 heads (i.e., 19.1 GPa vs. 13–16 GPa [145]). ZTA's higher hardness was partially achieved by the addition of chromium oxide into the Al_2O_3 phase.

It is noteworthy that the smoother ZTA heads exhibited a higher mean wear rate on the HXLPE than the Si_3N_4 heads. As previously discussed, wear damage is expected to decrease with lower thermal conductivity femoral heads due to reduced frictional heating at the bearing interface [147]. In turn, lower interface temperatures lead to less precipitation of serum proteins, and it's the proteins that protect the articulation surface from abrasive wear [150, 151]. This type of thermal artifact is more likely to occur in ZTA with its lower thermal conductivity than Si_3N_4 (~17 W/m°K vs. ~30 W/m°K, respectively [146]).

In other words, as compared to ZTA, Si_3N_4 with its higher thermal conductivity may present a more severe wear condition due to less protein precipitates. However, it remarkably exhibited less wear in the 5-Mc simulation test than ZTA. According to a study by Zhou et al. [147], the steady-state frictional coefficient of Si_3N_4 against itself in water was much smaller than that of alumina against alumina in water (0.001 vs. 0.08). Xu and Kato [148] reported that the wear mode of Si_3N_4 sliding in water changes from mechanically dominated to tribochemical-dominated wear as sliding distance increases. They showed that a silica layer is formed on the surface which can significantly reduce friction during wear testing (cf., Eq. 1). It is entirely likely that the formation of a "soft" silica layer on the Si_3N_4 heads reduced friction, leading to the observed lower wear of the Vitamin-E HXLPE liners. However, such a superficial layer in Si_3N_4 heads might be more susceptible to microdamage (i.e., surface roughening) as compared to the harder surface of ZTA.

4.1.3 Other Wear Studies Using Si_3N_4

Although Si_3N_4 has yet to be cleared for hip arthroplasty purposes, worldwide interest in Si_3N_4 for THA continues, with an average of over 120 publications annually during the past 4 years.[1] Of note, several researchers in India and China have completed recent studies on the biocompatibility and tribology of Si_3N_4 components for use in THA applications. For instance, Shankar et al. published an experimental and finite element study on the tribological behavior of *hard-on-hard* Si_3N_4–Al_2O_3 bearings using various lubricant conditions [149]. They showed that this unique couple produced very low wear and was mechanically a good choice for patients who engage in physically demanding activities. Another study, by Kong, et al., investigated the in vitro and in vivo biocompatibility and suitability of Si_3N_4 for hip implants [150]. After completing extensive biocompatibility testing per ISO-10993 and finding their Si_3N_4 to be wholly acceptable, they proceeded to surgically implant Si_3N_4-UHMWPE THA devices into minipigs within a small randomized controlled trial. The purpose of the trial was to compare arthroplasty to arthrotomy over 3 months. Although wear wasn't monitored, they showed that the implants in the test group were functioning well, and the animal's ambulation was comparable to the control group. CT scans during the study showed the implants to be well-fixated and in good position. Upon completion of the study, they found no pathological differences between the test and control groups. They concluded that due to its combination of favorable properties, Si_3N_4 is an excellent choice for THA; and they plan to conduct extensive wear and other preclinical tests to qualify Si_3N_4 as an alternative to oxide ceramics for THA in China.

[1] Google search conducted on September 27, 2022, for the 4-year period beginning in 2018, using the keywords "silicon nitride" and "arthroplasty."

4.2 *Impact of Si_3N_4 on Polyethylene Oxidation*

Polyethylene remains the "weak link" within *hard-on-soft* THA prostheses [151]. Clinical results with ~10-year follow-up demonstrate that first-generation HXLPE dramatically reduces wear-related complications over conventional polyethylene, but wear and osteolysis are still the leading reasons for revision surgeries [152]. Clinical data (i.e., $\leq$5 years) is currently insufficient to conclude that vitamin E-infused HXLPE is an improvement. See Sect. 2.1. However, clinicians remain hopeful, and longer-term studies are underway [153].

Until recently, ceramic femoral heads were not considered to be factors affecting the oxidation of HXLPE liners. But now, their chemical interaction with liners is an emerging topic in THA bearings. This goes beyond simple mechanical behavior or wear. One of the poorly understood interactions is the oxygen affinity of ceramic femoral heads. It was recently demonstrated that oxide (Al_2O_3, ZrO_2, and ZTA) and non-oxide (Si_3N_4) ceramic heads release or attract nascent oxygen species, respectively [154]. Five independent tests showed that oxygen release from oxide ceramics results in the oxidation of HXLPE. Conversely, it appears that Si_3N_4 may play a protective role in minimizing HXLPE oxidation. These five tests include: (1) static hydrothermal testing of femoral heads; (2) static hydrothermal testing of ceramic–HXLPE couples; (3) dynamic frictional testing of ceramic–HXLPE couples; (4) assessment of standard hip-simulator couples; and (5) retrieval analysis.

4.2.1 Static Hydrothermal Testing of Femoral Heads

Using a combination of X-ray photoelectron (XPS) and cathodoluminescence (CL) spectroscopy, surface chemical changes in oxide and non-oxide femoral heads were assessed before and after simple static hydrothermal treatments at 121 °C for up to 96 h. After 24 h, XPS was performed to determine changes in surface chemical bonds for Al_2O_3, ZTA, and Si_3N_4 heads [154]. The Al2p and Si2p spectra along with their deconvolution into Voigtian sub-bands were, respectively, utilized to confirm atomic bonding. Results for the oxide ceramics showed that these materials react with water to form hydroxylated bonds (O-Al-O-H) and oxygen vacancies (O-Al-V_O). Conversely, for Si_3N_4, an opposite trend was observed. It was found that silica bonds (O-Si-O and N-Si-O) increased at the expense of silicon-nitride (N-Si-N) bonds. Then, CL spectroscopy quantitatively confirmed the loss of oxygen from the oxide ceramics while Si_3N_4 gained oxygen. In other words, under hydrothermal conditions, oxygen flows away from Al_2O_3 and ZTA, whereas it flows towards Si_3N_4.

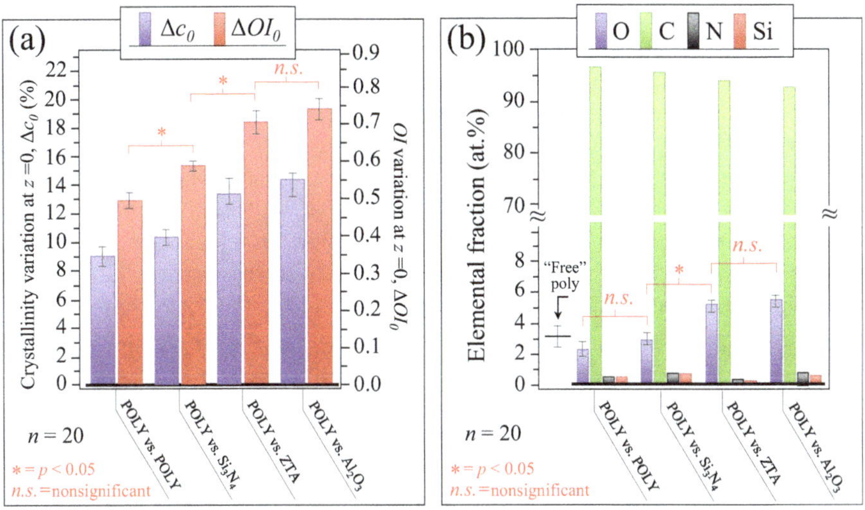

Fig. 11 (**a**) Variations of crystallinity and oxidation indices as detected by vibrational spectroscopy for the HXLPE liners statically coupled to oxide and non-oxide ceramic heads for 24 h in an autoclave; and (**b**) XPS analyses of the same liners in (a). (Used with permission [154])

4.2.2 Static Hydrothermal Testing of Ceramic–HXLPE Couples

Following the previous test on only femoral heads, a unique follow-on study was conducted using a range of oxide and silicon nitride femoral heads coupled with HXLPE liners [154, 155]. A direct correlation was made between oxygen emissions from the ceramics' surfaces and oxidation of the HXLPE surface using a simple static hydrothermal test (24 h; 121 °C; low-fixed load, cf., Fig. 11) [154].

Crystallinity, Δc_O, and oxidation indices, ΔOI_0, were determined at the surfaces of the sequentially irradiated and annealed HXLPE liners and compared with their pristine values for four different couples: (1) HXLPE versus HXLPE, (2) HXLPE versus Si_3N_4, (3) HXLPE versus ZTA, and (4) HXLPE versus Al_2O_3. The HXLPE vs. HXLPE couple was used as a control. It had the same geometrical configuration as the other couples. The null hypothesis was that all of the tested ceramics (if completely bioinert) would induce the same variations in Δc_O and ΔOI_0 as the control couples. The results for Δc_O and ΔOI_0 diverged from the null hypothesis (cf., Fig. 11a, b). The main features of these plots can be summarized as follows: (1) In couples containing the oxide ceramics, a significant increase in surface crystallization (~55%) and oxidation (~45%) of the HXLPE was observed. The results were statistically valid when compared to the controls, while the difference between the two oxide-containing couples was not significant; (2) the liners coupled to the Si_3N_4 heads experienced ~30% lower increase in their oxidation indices than liners coupled to the oxide ceramics; and they were only ~14% higher than the control couples; (3) the XPS data at the liner surfaces were consistent with vibrational data. They showed the highest amount of oxygen at the surfaces of liners coupled to ceramic oxide heads (i.e., about twice the amount that was detected in the all-polyethylene couples), with no statistically significant difference between liners coupled to Al_2O_3 or ZTA; (4) the oxygen content detected by XPS at the surfaces of the liners coupled to Si_3N_4 was only slightly higher (with no statistical relevance) than values detected at the surface of the control couples; and (5) XPS found traces of N and Si on the surface of all tested liners. This was presumably due to annealing and polishing of the HXLPE components, respectively, during their manufacture. Assuming that the environmental loading on all of the samples was geometrically and thermodynamically identical, it follows that the observed increase in polyethylene oxidation for the oxide ceramic couples (compared to the controls) was due to oxygen released from the ceramics' surfaces.

4.2.3 Dynamic Frictional Testing of Ceramic–HXLPE Couples

This test sequentially evaluated Ø28 mm ZTA and Si_3N_4 femoral heads coupled to HXLPE liners in a single station in-plane hip simulator—a simple rocker style device (+20° to −20°) operating at 1 Hz with a 0.25 s pause at both extremes under a constant load of 1700 N. Testing was conducted in a lubricant consisting of deionized water, NaCl, $Na_2HPO_4 \cdot 7H_2O$, and bovine proteins at ambient temperature to 5×10^5 cycles [154]. Before, during, and after the test, the components were examined for surface chemical changes using XPS. Similar to the prior static tests, the ZTA surfaces drifted into nonstoichiometry with the creation of oxygen vacancies. They showed a ~28% decrease in O–Al–O bonds and a commensurate increase in O–Al–V_o vacancies. Conversely, a similar magnitude in off-stoichiometry was observed for the Si_3N_4 heads, but in the opposite direction. A decrease of ~27% in Si–N–Si–N bonds was accompanied by a nearly identical increase in Si–N–Si–O bonds. It was found that the HXLPE liners increased in crystallinity and oxidation when paired with either Si_3N_4 or ZTA (cf., Fig. 12); but polymeric oxidation and crystallization were greater for liners paired with ZTA in both nonwear (NWZ) and main-wear (MWZ) zones. Note that degradation appeared to reach a plateau for the Si_3N_4 coupled liners in the MWZ, but it was always lower than the ZTA pairs, especially in the NWZ (cf., Fig. 12, $\Delta OI \sim 1.2$ vs. 0.4 after 5×10^5 cycles, respectively). Within this test, ZTA *released* oxygen to the tribological environment whereas Si_3N_4 *scavenged* it. In turn, this correlated to reduced degradation for the HXLPE liners coupled to the Si_3N_4 heads.

4.2.4 Assessment of Standard Hip Simulator Couples

Crystallinity and oxidation assessments were also made of the vitamin-E-infused HXLPE liners used in the 5 Mc test reported in Sect. 4.1.2 (cf., Fig. 10). The purpose of this examination was to determine if there were differing polymeric changes for this advanced HXLPE when paired with either ZTA or Si_3N_4 femoral heads. Although both couples showed good wear performance, it was discovered that the

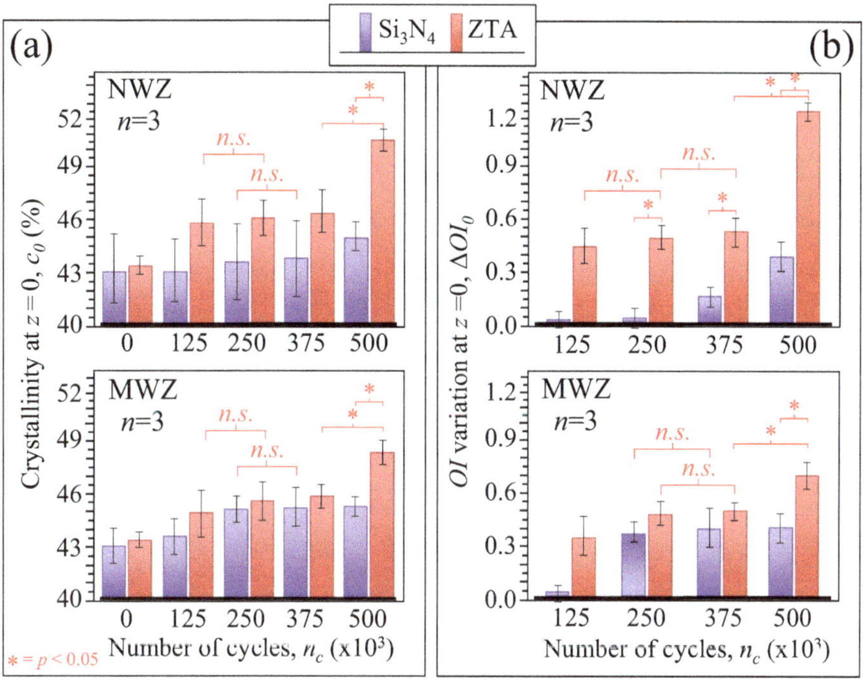

Fig. 12 Crystallinity (**a**) and oxidation (**b**) at the surface of HXLPE liners coupled to Si_3N_4 and ZTA as a function of the number of reciprocating cycles for nonwear (NWZ) and main wear (MWZ) zones. (Used with permission [154])

ZTA-coupled liners had significantly greater changes in both crystallization and oxidation than the Si_3N_4 pairs (cf., Fig. 13). Note that the degradation of this advanced HXLPE due to its articulation against ZTA was pronounced both at the surface and up to a depth of 200 μm, both for the NWZ and MWZ. Conversely, no crystallization and minimal oxidation were observed for the Si_3N_4 coupled liners, regardless of depth. These results are consistent with data presented earlier in Sects. 4.2.2 and 4.2.3.

4.2.5 Retrieval Analysis

Lastly, ex vivo surface chemical analyses were performed on two retrieved oxide femoral heads—Al_2O_3 and ZTA [154]. The selected heads were typical of a relatively large number of oxide femoral head retrievals. Using similar protocols as described in the prior sections, XPS was performed to assess changes in their surface chemical bond structure (cf., Table 3). The Al_2O_3 head was in vivo for 26.3 y, articulating against a first-generation UHMWPE liner, whereas the ZTA head was obtained after only 20 months of in vivo service against a second-generation HXLPE liner. For the Al_2O_3 head, note the reduction in O–Al–O bonds with the

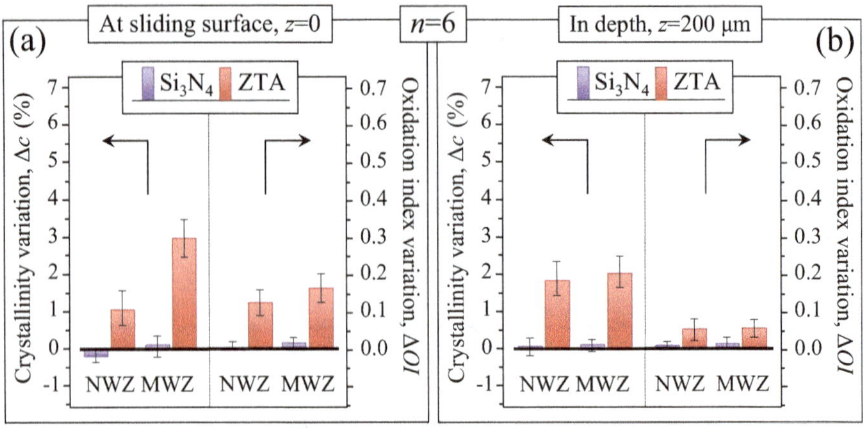

Fig. 13 HXLPE crystallinity and oxidation at the surfaces (**a**) and at 200 μm depth (**b**) of vitamin E-doped HXLPE liners coupled to Si_3N_4 and ZTA femoral heads after a 5 Mc cycles of standard hip simulator testing. (Used with permission from Ref. [154])

Table 3 XPS results of the two oxide femoral head retrievals

Al2p	Monolithic Al_2O_3		Percentage (%)		
Band	Assignment	Energy (eV)	Pristine	NWZ	MWZ
1	O-Al-O	74.2	66.3	48.8	49.3
2	O-Al-V_O	74.4	13.8	33.9	34.9
3	Al-O-H	75.2	19.9	15.2	15.8
Al2p	ZTA		Percentage (%)		
Band	Assignment	Energy (eV)	Pristine	NWZ	MWZ
1	O-Al-O	74.2	83.0	42.4	40.0
2	O-Al-V_O	74.4	7.9	24.2	24.7
3	Al-O-H	75.1	9.1	33.4	35.3

corresponding increase in O–Al–V_o vacancies and the decrease in Al–O–H bonds. This trend was also remarkably apparent but more pronounced for the ZTA head. Vacancy formation for the Al_2O_3 and ZTA heads amounted to increases of 153% and 213%, respectively. The off-stoichiometry changes for these two heads were more than what was previously observed in the in vitro static and articulation tests. This is likely the result of being in an acidic hydrothermal environment under higher frictional forces, which can be typical of diseased joints. Nevertheless, these results are wholly consistent with the prior independent tests given in Sects. 4.2.1 and 4.2.3.

4.2.6 Summary of HXLPE Oxidation Studies

The foregoing five tests sequentially build a strong case for the impact Si_3N_4 femoral heads may have on polyethylene longevity. The in vitro data provide limited but credible evidence of the clinical relevance of the effect of oxygen *pollution* from oxide ceramic heads and further suggest that Si_3N_4 could act as an oxygen *scavenger* within the hypoxic environment of an artificial hip. The results also demonstrate that oxygen released from oxide femoral heads flows across the tribolayer and degrades the polyethylene counter surface. This trend was unequivocally confirmed by an examination of ex vivo retrievals. Conversely, for Si_3N_4 couples, the flow of oxygen is in the reverse direction. Consequently, the use of Si_3N_4 ceramics as femoral heads may represent a viable path for delaying HXLPE oxidation, thereby elongating the lifetime of *hard-on-soft* articulations.

4.3 *Effect of Si_3N_4 Femoral Heads on MACC*

An in vitro study hypothesized that femoral heads made from Si_3N_4 may be better than CoCr in minimizing mechanically assisted crevice corrosion (MACC) [156]. The authors conducted a series of tests on identically sized heads (Ø32 mm +8 mm offset) coupled to Ti6Al4V trunnions (12/14, 5°40′ taper angle). The tests included displacement seating, cyclic fretting corrosion, disassembly, and post-test imaging analyses [157]. The results demonstrated that there was no significant difference in static seating displacement between the two groups (i.e., Si_3N_4: 115 ± 29 μm versus CoCr: 123 ± 40 μm, $p = 0.61$), suggesting both femoral head materials engaged the trunnions equally. However, for cyclic fretting, there was a marked difference in the onset load for MACC. Corrosion began at a load of 1740 ± 360 N for the CoCr heads, whereas it occurred at 2200 ± 188 N for Si_3N_4 ($p = 0.002$). Furthermore, the CoCr couples exhibited substantially higher average corrosion currents at all applied loads above onset (cf., Fig. 14a, b), $p < 0.05$).

A one-million cycle corrosion test was performed at a constant load of 3200 N and 15 Hz. Fretting currents were monitored throughout the test, and the results consistently showed lower overall corrosion for the Si_3N_4 versus the CoCr couples. The last hour of the test and the overall results highlight these differences (cf., Fig. 14c, d, $p < 0.05$). The force to disassemble the head–trunnion pairs was then measured and found to be 2600 ± 230 N and 1450 ± 180 N for Si_3N_4 versus CoCr couples, respectively ($p < 0.001$), suggesting that the Si_3N_4 pairs were more tightly embedded in their respective trunnions. Lastly, post-test imaging analyses confirmed that there were smaller interfacial contact areas between the Si_3N_4 heads and their trunnions versus the CoCr pairs, presumably due to the higher elastic modulus and hardness of the ceramic. Furthermore, there was also less visual evidence of corrosion for the Si_3N_4 pairs. These tests demonstrate that the in vitro corrosion

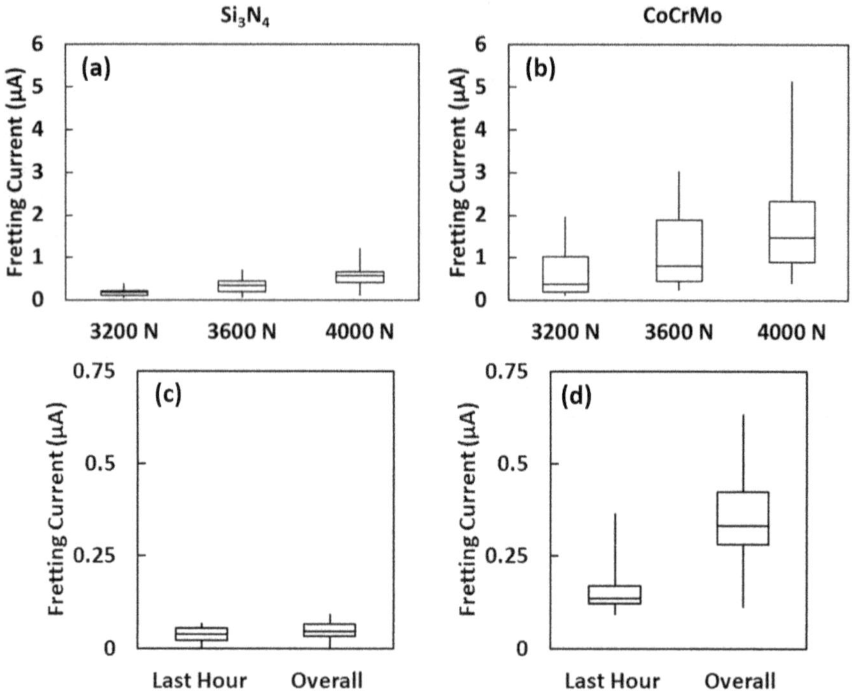

Fig. 14 (**a**) and (**b**) Fretting currents at three different loads for Si_3N_4 and CoCr couples, respectively ($p < 0.05$ for each group); and (**c**) and (**d**) Fretting currents for the last hour and overall, in a one million cycle test for Si_3N_4 and CoCr couples, respectively. (Rearranged and used with permission [156])

performance of Si_3N_4 is superior to CoCr, and this should translate to an increase in in vivo longevity.

However, a pertinent question has yet to be answered: Are Si_3N_4–trunnion pairs superior to ZTA pairs? A direct comparison is currently lacking. Despite its dominant market position, there have been remarkably few publications on MACC for ZTA heads. An early in vitro examination of CoCr and ZrO_2 heads mated to CoCr stems showed less fretting with ZrO_2 [78], and more recent retrieval studies of Al_2O_3 or ZTA (without differentiating between the two) and CoCr heads also demonstrated lower corrosion for the ceramics [79, 158], but only two recent studies specifically compared ZTA to CoCr. Aldinger, et al. conducted a 10-Mc fatigue test comparing ZTA, CoCr, and a ceramicized metal head (DHOxZr) coupled to Ti6Al4V hip stems [159]. They found fretting was greater for the ZTA couple as compared to CoCr, although it was not statistically significant. Their best result was for the DHOxZr couples, which were better than both the ZTA and CoCr pairs ($p < 0.05$). However, even though their comparative test used the same taper geometry, the head sizes of the three pairs were different (i.e., ZTA Ø32 mm; CoCr Ø28 mm; and DHOxZr Ø40 mm). This may have played a confounding role in the study's outcome.

The second paper, by Kyomoto, et al., was a hip-simulator study comparing ZTA and CoCr heads coupled to Ti6Al4V stems for 5-Mc [160]. They monitored galvanic current, pH, and metal ion release during the test, and subsequently performed post-test visual assessments of the pairs. The results showed no significant differences in current or pH over the test's duration, but there were increases in both Co and Cr ions within the lubricant for the CoCr pairs. In particular, the Co concentration exponentially increased by six-fold (i.e., ~100 to ~600 ng/mL). There were no significant increases or differences in Al or Zr ion concentrations. Post-test visual and SEM evaluations of the trunnions showed pronounced corrosion of the CoCr pairs whereas little or no damage was present on the ZTA couples.

In summary, these studies generally conclude that ceramics are preferred for minimizing MACC (see Sect. 2.2), but it remains to be seen which ceramic is better. It can be concluded that Si_3N_4 is likely as good as or better than ZTA or Al_2O_3, but additional testing remains to be completed.

5 Summary and Conclusion

Total hip arthroplasty is the most advanced and most practiced prosthetic surgery known to mankind; and it will continue to provide relief to patients suffering from congenital, traumatic, or age-related maladies. Over the past 50 years, surgical techniques and materials have evolved to address critical needs including improved mobility, faster recovery, and longer device lifetimes. However, while marked advancements have been made in operative methods and patient ambulation, the predicted survivability of THA devices has stymied at $\lesssim$20 y. The use of Si_3N_4 as a femoral head in *hard-on-soft* bearing couples may restore an upward trajectory in prosthesis lifetimes to make them equivalent to patients' lifespans, particularly for those who are younger and more active. This chapter has pointed out how this novel ceramic may address that goal:

1. The wear performance of Si_3N_4 femoral heads in *hard-on-soft* articulations for a range of traditional and advanced polyethylene liners was at least equivalent to the best available metallic or ceramic alternatives. In fact, against the most advanced vitamin E-stabilized HXLPE, Si_3N_4 was slightly better or at least equivalent to that of the current gold standard ZTA heads after a 5-Mc test.
2. Si_3N_4 is capable of *scavenging* active oxygen species from the tribolayer, thus protecting the HXLPE sliding counterpart from degradation. Conversely, oxide ceramics (i.e., Al_2O_3 and ZTA) *release* oxygen into the tribolayer thereby contributing to polyethylene damage and shortened life.
3. Si_3N_4 substantially reduces fretting crevice corrosion when compared to CoCr and it is likely at least as good as other currently marketed ceramic heads.
4. Considering Si_3N_4's high fracture strength, and a fracture toughness that is superior to any other bioceramic so far evaluated [145], the in vitro data presented

in this chapter represent positive news for the future development of long-lasting hip components.

Si_3N_4 *is* a very promising candidate for femoral head components in *hard-on-soft* THA, and it *may be* the ideal biomaterial for this application [112].

References

1. Healy, W.L.; Sharma, S.; Schwartz, B.; Iorio, R. Athletic Activity after Total Joint Arthroplasty. *J. Bone Jt. Surg.* **2008**, *90*, 2245–2252, https://doi.org/10.2106/JBJS.H.00274.
2. Gallo, J.; Goodman, S.B.; Lostak, J.; Janout, M. Advantages and Disadvantages of Ceramic on Ceramic Total Hip Arthroplasty: A Review. *Biomed. Pap.* **2012**, *156*, 204–212, https://doi.org/10.5507/bp.2012.063.
3. Mai, K.; Verioti, C.; Ezzet, K. a; Copp, S.N.; Walker, R.H.; Colwell, C.W. Incidence of "Squeaking" after Ceramic-on-Ceramic Total Hip Arthroplasty. *Clin. Orthop. Relat. Res.* **2010**, *468*, 413–7, https://doi.org/10.1007/s11999-009-1083-4.
4. Carnes, K.J.; Odum, S.M.; Troyer, J.L.; Fehring, T.K. Cost Analysis of Ceramic Heads in Primary Total Hip Arthroplasty. *J Bone Jt. Surg Am* **2016**, *98*, 1794–1800.
5. Smith, A.J.; Dieppe, P.; Vernon, K.; Porter, M.; Blom, A.W. Failure Rates of Stemmed Metal-on-Metal Hip Replacements: Analysis of Data from the National Joint Registry of England and Wales. *Lancet* **2012**, *379*, 1199–1204, https://doi.org/10.1016/S0140-6736(12)60353-5.
6. Cohen, D. How Safe are Metal-on-Metal Hip Implants? *BMJ* **2012**, *344*, e1410, https://doi.org/10.1136/bmj.e1410.
7. American Academy of Orthopedic Surgeons, A.J.R.R. *The Eighth Annual Report of the AJRR on Hip and Knee Arthroplasty*; Rosemont, IL USA, 2021;
8. Archibeck, M.J.; Jacobs, J.J.; Roebuck, K.A.; Glant, T.T. The Basic Science of Periprosthetic Osteolysis. *Instr. Course Lect.* **2013**, *62*, 1478–1489.
9. Jacobs, J.J.; Roebuck, K.A.; Archibeck, M.; Hallab, N.J.; Glant, T.T. Osteolysis: Basic Science. *Clin. Orthop. Relat. Res.* **2001**, 71–77.
10. Jacobs, J.J.; Campbell, P.A.; T Konttinen, Y. How has the Biologic Reaction to Wear Particles Changed with Newer Bearing Surfaces? *J. Am. Acad. Orthop. Surg.* **2008**, *16 Suppl 1*, S49–S55, https://doi.org/10.5435/00124635-200800001-00011.
11. Gallo, J.; Goodman, S.B.S.; Konttinen, Y.T.Y.; Raska, M. Particle Disease: Biologic Mechanisms of Periprosthetic Osteolysis in Total Hip Arthroplasty. *Innate Immun.* **2013**, *19*, 213–224, https://doi.org/10.1177/1753425912451779.Particle.
12. Bitar, D.; Parvizi, J. Biological Response to Prosthetic Debris. *World J. Orthop.* **2015**, *6*, 172–189, https://doi.org/10.5312/wjo.v6.i2.172.
13. Delaunay, C.; Hamadouche, M.; Girard, J.; Duhamel, A. What are the Causes for Failures of Primary Hip Arthroplasties in France? *Clin. Orthop. Relat. Res.* **2013**, *471*, 3863–3869, https://doi.org/10.1007/s11999-013-2935-5.
14. Dumbleton, J.H.; Manley, M.T.; Edidin, A.A. A Literature Review of the Association between Wear Rate and Osteolysis in Total Hip Arthroplasty. *J. Arthroplasty* **2002**, *17*, 649–661, https://doi.org/10.1054/arth.2002.33664.
15. Wang, A.; Sun, D.C.; Yau, S.-S.; Edwards, B.; Sokol, M.; Essner, A.; Polineni, V.K.; Stark, C.; Dumbleton, J.H. Orientation Softening in the Deformation and Wear of Ultra-High Molecular Weight Polyethylene. *Wear* **1997**, *203–204*, 230–241.
16. Wang, A.; Essner, A.; Polineni, V.K.; Stark, C.; Dumbleton, J.H. Lubrication and Wear of Ultra-High Molecular Weight Polyethylene in Total Joint Replacements. *Tribol. Int.* **1998**, *31*, 17–33, https://doi.org/10.1016/S0301-679X(98)00005-X.

17. Edidin, A.A.; Pruitt, L.; Jewett, C.W.; Crane, D.J.; Roberts, D.; Kurtz, S.M. Plasticity-Induced Damage Layer is a Precursor to Wear in Radiation-Cross-Linked UHMWPE Acetabular Components for Total Hip Replacement. *J. Arthroplasty* **1999**, *14*, 616–627.
18. McKellop, H.; Shen, F.W.; Lu, B.; Campbell, P.; Salovey, R. Development of an Extremely Wear-Resistant Ultra High Molecular Weight Polyethylene for Total Hip Replacements. *J. Orthop. Res.* **1999**, *17*, 157–167, https://doi.org/10.1002/jor.1100170203.
19. Muratoglu, O.K.; Bragdon, C.R.; O'Connor, D.O.; Jasty, M.; Harris, W.H.; Rizwan, G.; McGarry, F. Unified Wear Model for Highly Crosslinked Ultra-High Molecular Weight Polyethylenes (UHMWPE). *Biomaterials* **1999**, *20*, 1463–1470, https://doi.org/10.1016/S0142-9612(99)00039-3.
20. Sutula, L.C.; Collier, J.P.; Saum, K.A.; Currier, B.H.; Currier, J.H.; Sanford, W.M.; Mayor, M. B.; Wooding, R.E.; Sperling, D.K.; Williams, I.R.; et al. The Otto Aufranc Award: Impact of Gamma Sterilization on Clinical Performance of Polyethylene in the Hip. *Clin. Orthop. Relat. Res.* **1995**, *319*, 28–40.
21. Bargmann, L.S.; Bargmann, B.C.; Collier, J.P.; Currier, B.H.; Mayor, M.B. Current Sterilization and Packaging Methods for Polyethylene. *Clin. Orthop. Relat. Res.* **1999**, *369*, 49–58, https://doi.org/10.1097/00003086-199912000-00006.
22. Collier, J.P.; Currier, B.H.; Kennedy, F.E.; Currier, J.H.; Timmins, G.S.; Jackson, S.K.; Brewer, R.L. Comparison of Cross-Linked Polyethylene Materials for Orthopaedic Applications. *Clin. Orthop. Relat. Res.* **2003**, 289–304, https://doi.org/10.1097/01.blo.0000073343.50837.03.
23. Kurtz, S.M.; Gawel, H.A.; Patel, J.D. History and Systematic Review of Wear and Osteolysis outcomes for First-Generation Highly Crosslinked Polyethylene. *Clin. Orthop. Relat. Res.* **2011**, *469*, 2262–77, https://doi.org/10.1007/s11999-011-1872-4.
24. Australian Orthopaedic Association National Joint Replacement Registry (AO-ANJRR) *Hip, Knee & Shoulder Arthroplasty: 2021 Annual Report*; Adelaide, Australia, 2021;
25. Pruitt, L.A. Deformation, Yielding, Fracture and Fatigue Behavior of Conventional and Highly Cross-Linked Ultra High Molecular Weight Polyethylene. *Biomaterials* **2005**, *26*, 905–915, https://doi.org/10.1016/j.biomaterials.2004.03.022.
26. Ries, M.D.; Pruitt, L. Effect of Cross-Linking on the Microstructure and Mechanical Properties of Ultra High Molecular Weight Polyethylene. *Clin. Orthop. Relat. Res.* **2005**, *440*, 149–156, https://doi.org/10.1097/01.blo.0000185310.59202.e5.
27. Ast, M.P.; John, T.K.; Labbisiere, A.; Robador, N.; Valle, A.G. Della Fractures of a Single Design of Highly Cross-Linked Polyethylene Acetabular Liners: An Analysis of Voluntary Reports to the United States Food and Drug Administration. *J. Arthroplasty* **2014**, *29*, 1231–1235, https://doi.org/10.1016/j.arth.2013.12.022.
28. Currier, B.H.; Van Citters, D.W.; Currier, J.H.; Collier, J.P. In vivo Oxidation in Remelted Highly Cross-Linked Retrievals. *J. Bone Jt. Surg.* **2010**, *92*, 2409–2418.
29. MacDonald, D.; Sakona, A.; Ianuzzi, A.; Rimnac, C.M.; Kurtz, S.M. Do First-Generation Highly Crosslinked Polyethylenes Oxidize in vivo? *Clin. Orthop. Relat. Res.* **2011**, *469*, 2278–2285, https://doi.org/10.1007/s11999-010-1728-3.
30. Muratoglu, O.K.; Wannomae, K.K.; Rowell, S.L.; Micheli, B.R.; Malchau, H. Ex Vivo Stability Loss of Irradiated and Melted Ultra-High Molecular Weight Polyethylene. *J. Bone Jt. Surg.* **2010**, *92*, 2809–2816.
31. Oral, E.; Ghali, B.W.; Neils, A.; Muratoglu, O.K. A New Mechanism of Oxidation in Ultrahigh Molecular Weight Polyethylene Caused by Squalene Absorption. *J. Biomed. Mater. Res. Part B Appl. Biomater.* **2012**, *100B*, 742–751, https://doi.org/10.1002/jbm.b.32507.
32. Medel, F.; Kurtz, S.; MacDonald, D.; Pascual, F.; Puértolas, J.A. Does Cyclic Stress Play a Role in Highly Crosslinked Polyethylene Oxidation? *Clin. Orthop. Relat. Res.* **2015**, *473*, 1022–1029, https://doi.org/10.1007/s11999-015-4153-9.

33. Berry, D.J.; Currier, B.H.; Mayor, M.B.; Collier, J.P. Gamma-Irradiation Sterilization in an Inert Environment A Partial Solution. *Clin. Orthop. Relat. Res.* **2012**, *470*, 1805–1813, https://doi.org/10.1007/s11999-011-2150-1.
34. Oral, E.; Neils, A.L.; Doshi, B.N.; Fu, J.; Muratoglu, O.K. Effects of Simulated Oxidation on the in vitro Wear and Mechanical Properties of Irradiated and Melted Highly Crosslinked UHMWPE. *J. Biomed. Mater. Res. Part B Appl. Biomater.* **2016**, *104*, 316–322, https://doi.org/10.1002/jbm.b.33368.
35. Popoola, O.O.; Mimnaugh, K.; Orozco-Villasenor, D.A.; Peiserich, M. Vitamin E-Grafted Crosslinked Ultrahigh-Molecular Weight Polyethylene (VE-HXPE) Inhibits Fluid Absorption and has Long Term Oxidative Stability. *Biotribology* **2015**, *4*, 12–17, https://doi.org/10.1016/j.biotri.2015.09.006.
36. Takahashi, Y.; Tateiwa, T.; Pezzotti, G.; Shishido, T.; Masaoka, T.; Yamamoto, K. Improved Resistance to Neck-Liner Impingement in Second-Generation Highly Crosslinked Polyethylene—The Role of Vitamin E and Crosslinks. *J. Arthroplasty* **2016**, *31*, 2926–2932, https://doi.org/10.1016/j.arth.2016.05.049.
37. Takahashi, Y.; Yamamoto, K.; Pezzotti, G. Effects of Vitamin E Blending on Plastic Deformation Mechanisms of Highly Crosslinked Ultrahigh Molecular Weight Polyethylene (HXL-UHMWPE) in Total Hip Arthroplasty. *Acta Biomater.* **2015**, *15*, 227–236, https://doi.org/10.1016/j.actbio.2014.12.022.
38. Oral, E.; Neils, A.L.; Rowell, S.L.; Lozynsky, A.J.; Muratoglu, O.K. Increasing Irradiation Temperature Maximizes Vitamin E Grafting and Wear Resistance of Ultrahigh Molecular Weight Polyethylene. *J. Biomed. Mater. Res. - Part B Appl. Biomater.* **2013**, *101B*, 436–440, https://doi.org/10.1002/jbm.b.32807.
39. Uetsuki, K.; Sugimoto, T.; Turner, A.C.; Hyon, S.; Tomita, N. Simultaneous Improvement in the Mechanical Strength and Oxidative Resistance of Vitamin E Blended UHMWPE. *Trans. 57th Orthop. Res. Soc. San Fr.* **2012**, 1073.
40. Oral, E.; Muratoglu, O.K. Vitamin E Diffused, Highly Crosslinked UHMWPE: A Review. *Int. Orthop.* **2011**, *35*, 215–223, https://doi.org/10.1007/s00264-010-1161-y.
41. Wolf, C.; Maninger, J.; Lederer, K.; Frühwirth-Smounig, H.; Gamse, T.; Marr, R. Stabilisation of Crosslinked Ultra-High Molecular Weight Polyethylene (UHMW-PE)-Acetabular Components with α-Tocopherol. *J. Mater. Sci. Mater. Med.* **2006**, *17*, 1323–1331, https://doi.org/10.1007/s10856-006-0607-7.
42. Wolf, C.; Macho, C.; Lederer, K. Accelerated Ageing Experiments with Crosslinked and Conventional Ultra-High Molecular Weight Polyethylene (UHMW-PE) Stabilised with α-Tocopherol for Total Joint Arthroplasty. *J. Mater. Sci. Mater. Med.* **2006**, *17*, 1333–1340, https://doi.org/10.1007/s10856-006-0608-6.
43. Bracco, P.; Brunella, V.; Zanetti, M.; Luda, M.P.; Costa, L. Stabilisation of Ultra-High Molecular Weight Polyethylene with Vitamin E. *Polym. Degrad. Stab.* **2007**, *92*, 2155–2162, https://doi.org/10.1016/j.polymdegradstab.2007.02.023.
44. Tomita, N.; Kitakura, T.; Onmori, N.; Ikada, Y.; Aoyama, E. Prevention of Fatigue Cracks in Ultrahigh Molecular Weight Polyethylene Joint Components by the Addition of Vitamin E. *J. Biomed. Mater. Res.* **1999**, *48*, 474–478, https://doi.org/10.1002/(SICI)1097-4636(1999)48:4<474::AID-JBM11>3.0.CO;2-T.
45. Renò, F.; Cannas, M. UHMWPE and Vitamin E Bioactivity: An Emerging Perspective. *Biomaterials* **2006**, *27*, 3039–3043, https://doi.org/10.1016/j.biomaterials.2006.01.016.
46. Bladen, C.L.; Teramura, S.; Russell, S.L.; Fujiwara, K.; Fisher, J.; Ingham, E.; Tomita, N.; Tipper, J.L. Analysis of Wear, Wear Particles, and Reduced Inflammatory Potential of Vitamin E Ultrahigh-Molecular-Weight Polyethylene for Use in Total Joint Replacement. *J. Biomed. Mater. Res. - Part B Appl. Biomater.* **2013**, *101B*, 458–466, https://doi.org/10.1002/jbm.b.32904.
47. Bichara, D.A.; Malchau, E.; Sillesen, N.H.; Cakmak, S.; Nielsen, G.P.; Muratoglu, O.K. Vitamin E-Diffused Highly Cross-Linked UHMWPE Particles Induce Less Osteolysis

Compared to Highly Cross-Linked Virgin UHMWPE Particles in vivo. *J. Arthroplasty* **2014**, *29*, 232–237.
48. Chen, W.; Bichara, D.A.; Suhardi, J.; Sheng, P.; Muratoglu, O.K. Effects of Vitamin E-Diffused Highly Cross-Linked UHMWPE Particles on Inflammation, Apoptosis and Immune Response Against S. aureus. *Biomaterials* **2017**, *143*, 46–56.
49. Gómez-Barrena, E.; Esteban, J.; Molina-Manso, D.; Adames, H.; Martínez-Morlanes, M.J.; Terriza, A.; Yubero, F.; Puértolas, J.A. Bacterial Adherence on UHMWPE with Vitamin E: An in vitro Study. *J. Mater. Sci. Mater. Med.* **2011**, *22*, 1701–1706, https://doi.org/10.1007/s10856-011-4340-5.
50. Banche, G.; Bracco, P.; Bistolfi, A.; Allizond, V.; Boffano, M.; Costa, L.; Cimino, A.; Cuffini, A.M.; Del Prever, E.M.B. Vitamin E Blended UHMWPE May Have the Potential to Reduce Bacterial Adhesive Ability. *J. Orthop. Res.* **2011**, *29*, 1662–1667, https://doi.org/10.1002/jor.21432.
51. Banche, G.; Bracco, P.; Allizond, V.; Bistolfi, A.; Boffano, M.; Cimino, A.; Brach del Prever, E.M.; Cuffini, A.M. Do Crosslinking and Vitamin E Stabilization Influence Microbial Adhesions on UHMWPE-based Biomaterials? *Clin. Orthop. Relat. Res.* **2015**, *473*, 974–986, https://doi.org/10.1007/s11999-014-4024-9.
52. Wolf, C.; Krivec, T.; Blassnig, J.; Lederer, K.; Schneider, W. Examination of the Suitability of α-Tocopherol as a Stabilizer for Ultra-High Molecular Weight Polyethylene Used for Articulating Surfaces in Joint Endoprostheses. *J. Mater. Sci. Mater. Med.* **2002**, *13*, 185–189, https://doi.org/10.1023/A:1013834113967.
53. Turner, A.; Okubo, Y.; Teramura, S.; Niwa, Y.; Ibaraki, K.; Kawasaki, T.; Hamada, D.; Uetsuki, K.; Tomita, N. The Antioxidant and Non-Antioxidant Contributions of Vitamin E in Vitamin E Blended Ultra-High Molecular Weight Polyethylene for Total Knee Replacement. *J. Mech. Behav. Biomed. Mater.* **2014**, *31*, 21–30, https://doi.org/10.1016/j.jmbbm.2012.12.006.
54. Gilbert, J.L.; Buckley, C.A.; Jacobs, J.J.; Bertin, K.C.; Zernich, M.R. Intergranular Corrosion-Fatigue Failure of Cobalt-Alloy Femoral Stems. *J. Bone Jt. Surg.* **1994**, *76*, 110–115.
55. Banerjee, S.; Cherian, J.J.; Bono, J. V.; Kurtz, S.M.; Geesink, R.; Meneghini, R.M.; Delanois, R.E.; Mont, M.A. Gross Trunnion Failure after Primary Total Hip Arthroplasty. *J. Arthroplasty* **2015**, *30*, 641–648, https://doi.org/10.1016/j.arth.2014.11.023.
56. Urish, K.L.; Hamlin, B.R.; Plakseychuk, A.Y.; Levison, T.J.; Higgs, G.; Kurtz, S.; DiGioia, A.M. Trunion Failure of the Recalled LFIT Cobalt Chromium Alloy Femoral Head. *J. Arthroplasty* **2017**, *32*, 2857–2863, https://doi.org/10.1016/j.arth.2017.03.075.
57. Cooper, H.J.H.; Della Valle, C.J.; Berger, R.A.; Tetreault, M.; Paprosky, W.G.; Sporer, S.M.; Jacobs, J.J. Corrosion at the Head-Neck Taper as a Cause for Adverse Local Tissue Reactions after Total Hip Arthroplasty. *J. Bone Jt. Surg.* **2012**, *94*, 1655–1661.
58. Daniel, J.; Holland, J.; Quigley, L.; Sprague, S.; Bhandari, M. Pseudotumors Associated with Total Hip Arthroplasty. *J. Bone Jt. Surg.* **2012**, *94*, 86–93, https://doi.org/10.2106/JBJS.J.01612.
59. Pivec, R.; Meneghini, R.M.; Hozack, W.J.; Westrich, G.H.; Mont, M.A. Modular Taper Junction Corrosion and Failure: How to Approach a Recalled Total Hip Arthroplasty Implant. *J. Arthroplasty* **2014**, *29*, 1–6, https://doi.org/10.1016/j.arth.2013.08.026.
60. Jacobs, J.J.; Gilbert, J.L.; Urban, R.M. Corrosion of Metal Orthopaedic Implants. *J. Bone Jt. Surg.* **1998**, *80*, 268–282.
61. Eltit, F.; Assiri, A.; Garbuz, D.; Duncan, C.; Masri, B.; Greidanus, N.; Bell, R.; Sharma, M.; Cox, M.; Wang, R. Adverse Reactions to Metal on Polyethylene Implants: Highly Destructive Lesions Related to Elevated Concentration of Cobalt and Chromium in Synovial Fluid. *J. Biomed. Mater. Res. - Part A* **2017**, *105*, 1876–1886, https://doi.org/10.1002/jbm.a.36057.
62. Gibon, E.; Amanatullah, D.F.; Loi, F.; Pajarinen, J.; Nabeshima, A.; Yao, Z.; Hamadouche, M.; Goodman, S.B. The Biological Response to Orthopaedic Implants for Joint Replacement: Part I: Metals. *J. Biomed. Mater. Res. Part B Appl. Biomater.* **2017**, *105*, 2162–2173.

63. Kop, A.M.; Keogh, C.; Swarts, E. Proximal Component Modularity in THA - At What Cost? An Implant Retrieval Study. *Clin. Orthop. Relat. Res.* **2012**, *470*, 1885–1894, https://doi.org/10.1007/s11999-011-2155-9.
64. McGrory, B.J.; Jacobs, J.J.; Kwon, Y.M.; Fillingham, Y. Standardizing Terms for Tribocorrosion-Associated Adverse Local Tissue Reaction in Total Hip Arthroplasty. *Arthroplast. Today* **2020**, *6*, 196–200, https://doi.org/10.1016/j.artd.2020.01.008.
65. Collier, J.P.; Surprenant, V.A.; Jensen, R.E.; Mayor, M.B. Corrosion at the Interface of Cobalt-Alloy Heads on Titanium-Alloy Stems. *Clin. Orthop. Relat. Res.* **1991**, *271*, 305–312.
66. Reclaru, L.; Lerf, R.; Eschler, P.Y.; Blatter, A.; Meyer, J.M. Pitting, Crevice and Galvanic Corrosion of REX Stainless-Steel/CoCr Orthopedic Implant Material. *Biomaterials* **2002**, *23*, 3479–3485, https://doi.org/10.1016/S0142-9612(02)00055-8.
67. Hussey, D.K.; McGrory, B.J. Ten-Year Cross-Sectional Study of Mechanically Assisted Crevice Corrosion in 1352 Consecutive Patients With Metal-on-Polyethylene Total Hip Arthroplasty. *J. Arthroplasty* **2017**, *32*, 2546–2551, https://doi.org/10.1016/j.arth.2017.03.020.
68. McGrory, B.J.; MacKenzie, J.; Babikian, G. A High Prevalence of Corrosion at the Head-Neck Taper with Contemporary Zimmer Non-Cemented Femoral Hip Components. *J. Arthroplasty* **2015**, *30*, 1265–1268, https://doi.org/10.1016/j.arth.2015.02.019.
69. Brown, S.A.; Flemming, C.A.C.; Kawalec, J.S.; Placko, H.E.; Vassaux, C.; Merritt, K.; Payer, J.H.; Kraay, M.J. Fretting Corrosion Accelerates Crevice Corrosion of Modular Hip Tapers. *J. Appl. Biomater.* **1995**, *6*, 19–26, https://doi.org/10.1002/jab.770060104.
70. Goldberg, J.R.; Gilbert, J.L.; Jacobs, J.J.; Bauer, T.W.; Paprosky, W.; Leurgans, S. A Multicenter Retrieval Study of the Taper Interfaces of Modular Hip Prostheses. *Clin. Orthop. Relat. Res.* **2002**, *401*, 149–161, https://doi.org/10.1097/00003086-200208000-00018.
71. Engh, C.A. Do You Have to Remove a Corroded Femoral Stem? *J. Arthroplast.* **2014**, *29*, 139–142, https://doi.org/10.1016/j.arth.2014.03.055.
72. Hussenbocus, S.; Kosuge, D.; Solomon, L.B.; Howie, D.W.; Oskouei, R.H. Head-Neck Taper Corrosion in Hip Arthroplasty. *Biomed Res. Int.* **2015**, *2015*, 758123, https://doi.org/10.1155/2015/758123.
73. Tan, S.C.; Teeter, M.G.; Del Balso, C.; Howard, J.L.; Lanting, B.A. Effect of Taper Design on Trunnionosis in Metal on Polyethylene Total Hip Arthroplasty. *J. Arthroplasty* **2015**, *30*, 1269–1272, https://doi.org/10.1016/j.arth.2015.02.031.
74. Nam, D.; Keeney, J.A.; Nunley, R.M.; Johnson, S.R.; Clohisy, J.C.; Barrack, R.L. Metal Ion Concentrations in Young, Active Patients Following Total Hip Arthroplasty with the Use of Modern Bearing Couples. *J. Arthroplasty* **2015**, *30*, 2227–2232, https://doi.org/10.1016/j.arth.2015.06.025.
75. Mistry, J.B.; Chughtai, M.; Elmallah, R.K.; Diedrich, A.; Le, S.; Thomas, M.; Mont, M.A. Trunnionosis in Total Hip Arthroplasty: A Review. *J. Orthop. Traumatol.* **2016**, *17*, 1–6, https://doi.org/10.1007/s10195-016-0391-1.
76. Jennings, J.M.; Dennis, D.A.; Yang, C.C. Corrosion of the Head-Neck Junction After Total Hip Arthroplasty. *Biomed. Mater. Eng.* **2016**, *24*, 349–356.
77. Martin, J.R.; Camp, C.L.; Wyles, C.C.; Taunton, M.J.; Trousdale, R.T.; Lewallen, D.G. Increased Femoral Head Offset is Associated With Elevated Metal Ions in Asymptomatic Patients With Metal-on-Polyethylene Total Hip Arthroplasty. *J. Arthroplasty* **2016**, *31*, 2814–2818, https://doi.org/10.1016/j.arth.2016.05.047.
78. Hallab, N.J.; Messina, C.; Skipor, A.; Jacobs, J.J. Differences in the Fretting Corrosion of Metal-Metal and Ceramic-Metal Modular Junctions of Total Hip Replacements. *J. Orthop. Res.* **2004**, *22*, 250–259, https://doi.org/10.1016/S0736-0266(03)00186-4.
79. Kurtz, S.M.; Kocagöz, S.B.; Hanzlik, J.A.; Underwood, R.J.; Gilbert, J.L.; MacDonald, D.W.; Lee, G.-C.; Mont, M.A.; Kraay, M.J.; Klein, G.R.; et al. Do Ceramic Femoral Heads Reduce Taper Fretting Corrosion in Hip Arthroplasty? A Retrieval Study. *Clin. Orthop. Relat. Res.* **2013**, *471*, 3270–82, https://doi.org/10.1007/s11999-013-3096-2.

80. Traina, F.; Tassinari, E.; De Fine, M.; Bordini, B.; Toni, A. Revision of Ceramic Hip Replacements for Fracture of a Ceramic Component: AAOS Exhibit Selection. *JBJS* **2011**, *93*, e147, https://doi.org/10.2106/JBJS.K.00589.
81. Rahaman, M.N.; Yao, A.; Bal, B.S.; Garino, J.P.; Ries, M.D. Ceramics for Prosthetic Hip and Knee Joint Replacement. *J. Am. Ceram. Soc.* **2007**, *90*, 1965–1988, https://doi.org/10.1111/j.1551-2916.2007.01725.x.
82. Piconi, C.; Maccauro, G. Zirconia as a Ceramic Biomaterial. *Biomaterials* **1999**, *20*, 1–25.
83. Cales, B. Zirconia as a Sliding Material: Histologic, Laboratory, and Clinical Data. *Clin. Orthop. Relat. Res.* **2000**, *379*, 84–112.
84. Clarke, I.C.; Manaka, M.; Green, D.D.; Williams, P.; Pezzotti, G.; Kim, Y.-H.; Ries, M.; Sugano, N.; Sedel, L.; Delauney, C.; et al. Current Status of Zirconia Used in Total Hip Implants. *J. Bone Jt. Surg.* **2003**, *85-A*, 73–84.
85. De Aza, A.H.; Chevalier, J.; Fantozzi, G.; Schehl, M.; Torrecillas, R. Crack Growth Resistance of Alumina, Zirconia and Zirconia Toughened Alumina Ceramics for Joint Prostheses. *Biomaterials* **2002**, *23*, 937–45.
86. Chevalier, J.; Gremillard, L.; Deville, S. Low-Temperature Degradation of Zirconia and Implications for Biomedical Implants. *Annu. Rev. Mater. Res.* **2007**, *37*, 1–32, https://doi.org/10.1146/annurev.matsci.37.052506.084250.
87. Piconi, C.; Maccauro, G.; Pilloni, L.; Burger, W.; Muratori, F.; Richter, H.G. On the fracture of a Zirconia Ball Head. *J. Mater. Sci. Mater. Med.* **2006**, *17*, 289–300, https://doi.org/10.1007/s10856-006-7316-0.
88. Sheth, N.P.; Lementowski, P.; Hunter, G.; Garino, J.P. Clinical Applications of Oxidized Zirconium. *J. Surg. Orthop. Adv.* **2008**, *17*, 17–26.
89. Evangelista, G.T.; Fulkerson, E.; Kummer, F.; Di Cesare, P.E. Surface Damage to an Oxinium Femoral Head Prosthesis after Dislocation. *J. Bone Jt. Surg. - Ser. B* **2007**, *89-B*, 535–537, https://doi.org/10.1302/0301-620X.89B4.17557.
90. Jaffe, W.L.; Strauss, E.J.; Cardinale, M.; Herrera, L.; Kummer, F.J. Surface Oxidized Zirconium Total Hip Arthroplasty Head Damage Due to Closed Reduction Effects on Polyethylene Wear. *J. Arthroplasty* **2009**, *24*, 898–902, https://doi.org/10.1016/j.arth.2008.07.009.
91. Gibon, E.; Scemama, C.; David, B.; Hamadouche, M. Oxinium Femoral Head Damage Generated by a Metallic Foreign Body Within the Polyethylene Cup Following Recurrent Dislocation Episodes. *Orthop. Traumatol. Surg. Res.* **2013**, *99*, 865–869, https://doi.org/10.1016/j.otsr.2013.04.011.
92. Tribe, H.; Malek, S.; Stammers, J.; Ranawat, V.; Skinner, J.A. Advanced Wear of an Oxinium™ Femoral Head Implant Following Polyethylene Liner Dislocation. *Ann. R. Coll. Surg. Engl.* **2013**, *95*, e133-5, https://doi.org/10.1308/003588413X13629960047876.
93. Lee, R.; Essner, A.; Wang, A.; Jaffe, W.L. Scratch and Wear Performance of Prosthetic Femoral Head Components Against Crosslinked UHMWPE Sockets. *Wear* **2009**, *267*, 1915–1921, https://doi.org/10.1016/j.wear.2009.03.034.
94. Carli, A.; Koch, C.N.; Esposito, C.I.; Wright, T.M.; Padgett, D.E. Polyethylene Wear Increases in Liners Articulating with Scratched Oxidized Zirconium Femoral Heads. *Clin. Orthop. Relat. Res.* **2018**, *476*, 182–192, https://doi.org/10.1007/s11999.0000000000000040.
95. Kurtz, S.M.; Kocagöz, S.; Arnholt, C.; Huet, R.; Ueno, M.; Walter, W.L. Advances in Zirconia Toughened Alumina Biomaterials for Total Joint Replacement. *J. Mech. Behav. Biomed. Mater.* **2014**, *31*, 107 116, https://doi.org/10.1016/j.jmbbm.2013.03.022.
96. Pezzotti, G.; Saito, T.; Padeletti, G.; Cossari, P.; Yamamoto, K. Nano-Scale Topography of Bearing Surface in Advanced Alumina/Zirconia Hip Joint Before and After Severe Exposure in Water Vapor Environment. *J. Orthop. Res.* **2010**, *28*, 762–766, https://doi.org/10.1002/jor.21069.
97. Pezzotti, G.; Saito, T.; Takahashi, Y.; Fukatsu, K.; Sugano, N. Surface Topology of Advanced Alumina/Zirconia Composite Femoral Head as Compared with Commercial Femoral Heads Made of Monolithic Zirconia. *J. Am. Ceram. Soc.* **2011**, *94*, 945–950, https://doi.org/10.1111/j.1551-2916.2010.04181.x.

98. Hallan, G.; Fenstad, A.M.; Furnes, O. What Is the Frequency of Fracture of Ceramic Components in THA? Results from the Norwegian Arthroplasty Register from 1997 to 2017. *Clin. Orthop. Relat. Res.* **2020**, *478*, 1254–1261, https://doi.org/10.1097/CORR.0000000000001272.
99. Rankin, C.S.; Robinson, P.G.; Beattie, N.; Gaston, P. Fracture of a BIOLOX Delta Ceramic Femoral Head: A Case Report and Update of the Literature. *JBJS Case Connect. Connect.* **2019**, *9*, e0336, https://doi.org/10.2106/JBJS.CC.18.00336.
100. Pomeroy, E.; Rowan, F.; Masterson, E. Atraumatic Fracture of a BIOLOX Delta Ceramic Femoral Head Articulating with a Polyethylene Liner. *JBJS Case Connect.* **2015**, *5*, e112.
101. Duensing, I.M.; Stanley, S.; Bolognesi, M. Fracture of a 40-mm BIOLOX Delta Femoral Head. *Arthroplast. Today* **2021**, *10*, 144–148, https://doi.org/10.1016/j.artd.2021.06.021.
102. Dobria, P.; Patel, A.; Levine, B. Catastrophic Intraoperative Failure of a Ceramic Femoral Head. *Arthroplast. Today* **2021**, *11*, 234–238, https://doi.org/10.1016/j.artd.2021.08.017.
103. Lombardi, A. V.; Berend, K.R.; Seng, B.E.; Clarke, I.C.; Adams, J.B. Delta Ceramic-on-Alumina Ceramic Articulation in Primary THA: Prospective, Randomized FDA-IDE Study and Retrieval Analysis. *Clin. Orthop. Relat. Res.* **2010**, *468*, 367–374, https://doi.org/10.1007/s11999-009-1143-9.
104. Roy, M.E.; Noel, O.F.; Whiteside, L.A. Phase Transformation and Roughening in Artificially Aged and Retrieved Zirconia-Toughened Alumina Femoral Heads. *J. Arthroplasty* **209AD**, *34*, 772–780, https://doi.org/10.1016/j.arth.2018.12.025.
105. Takahashi, Y.; Zhu, W.; Sugano, N.; Pezzotti, G. On the Role of Oxygen Vacancies, Aliovalent Ions and Lattice Strain in the in vivo Wear Behavior of Alumina Hip Joints. *J. Mech. Behav. Biomed. Mater.* **2011**, *4*, 993–1003, https://doi.org/10.1016/j.jmbbm.2011.02.014.
106. Boffelli, M.; Doimo, A.; Marin, E.; Puppulin, L.; Zhu, W.; Sugano, N.; Clarke, I.C.; Pezzotti, G. Chemically Driven Tetragonal-to-Monoclinic Polymorphic Transformation in Retrieved ZTA Femoral Heads from Dual Mobility Hip Implants. *J. Mech. Behav. Biomed. Mater.* **2016**, *56*, 195–204, https://doi.org/10.1016/j.jmbbm.2015.11.032.
107. Bal, B.S.; Zhu, W.; Zanocco, M.; Marin, E.; Sugano, N.; McEntire, B.J.; Pezzotti, G. Reconciling in vivo and in vitro Kinetics of the Polymorphic Transformation in Zirconia-Toughened Alumina for Hip Joints: I. Phenomenology. *Mater. Sci. Eng. C* **2017**, *C72*, 252–258, https://doi.org/10.1016/j.msec.2016.11.059.
108. Pezzotti, G.; Zhu, W.; Zanocco, M.; Marin, E.; Sugano, N.; McEntire, B.J.; Bal, B.S. Reconciling in vivo and in vitro Kinetics of the Polymorphic Transformation in Zirconia-Toughened Alumina for Hip Joints: II. Theory. *Mater. Sci. Eng. C* **2017**, *71*, 446–451, https://doi.org/10.1016/j.msec.2016.10.031.
109. Pezzotti, G.; Bal, B.S.; Zanocco, M.; Marin, E.; Sugano, N.; McEntire, B.J.; Zhu, W. Reconciling in vivo and in vitro Kinetics of the Polymorphic Transformation in Zirconia-Toughened Alumina for Hip Joints: III. Molecular Scale Mechanisms. *Mater. Sci. Eng. C* **2016**, https://doi.org/10.1016/j.msec.2016.11.007.
110. Learmonth, I.D.; Young, C.; Rorabeck, C. The Operation of the Century: Total Hip Replacement. *Lancet* **2007**, *370*, 1508–1519, https://doi.org/10.1016/S0140-6736(07)60457-7.
111. Rondinella, A.; Marin, E.; McEntire, B.J.; Bock, R.M.; Bal, B.S.; Zhu, W.L.; Yamamoto, K.; Pezzotti, G. Bioceramics are Not Bioinert: The Role of Oxide and Non-Oxide Bioceramics on the Oxidation of UHMWPE Components in Artificial Joints. *Key Eng. Mater.* **2018**, *782*, 165–175, https://doi.org/10.4028/www.scientific.net/KEM.782.165.
112. Heimann, R.B. Silicon Nitride, A Close to Ideal Ceramic Material for Medical Application. *Ceramics* **2021**, *4*, 208–223.
113. Du, X.; Lee, S.S.; Blugan, G.; Ferguson, S.J. Silicon Nitride as a Biomedical Material: An Overview. *Int. J. Mol. Sci.* **2022**, *23*, 6551, https://doi.org/10.3390/ijms23126551.
114. Bozic, K.J.; Morshed, S.; Silverstein, M.D.; Rubash, H.E.; Kahn, J.G. Use of Cost-Effectiveness Analysis to Evaluate New Technologies in Orthopaedics. The Case of

Alternative Bearing Surfaces in Total Hip Arthroplasty. *J. Bone Joint Surg. Am.* **2006**, *88*, 706–14, https://doi.org/10.2106/JBJS.E.00614.
115. Johnson, D.R. *Ceramic Technology for Advanced Heat Engines Project, Semi-Annual & Annual Reports for the Periods 1988 through 1995, Prepared for the U.S. Department of Energy by Oak Ridge National Laboratory under Contract DE-AC05-840R21400*; Oak Ridge, TN 37831-6285, 1995;
116. Kaushal, T.S.; Weber, K.E. *Advanced Diesel Engine Component Development Program, Final Report Tasks 4-14, No. DOE/NASA-0329-2; NASA/CR-191203, National Aeronautics and Space Administration*; Cleveland, OH, 1994;
117. McEntire, B.J.; Wills, R.W.; Southam, R.E. *The Development and Testing of Ceramic Components in Piston Engines, Final Report, Prepared for the U.S. Department of Energy under Contract with Oak Ridge National Laboratories, Contract No. DE-AC05-840R21400*; Oak Ridge, TN 37831-6285, 1994;
118. Wang, L.; Snidle, R.W.; Gu, L. Rolling Contact Silicon Nitride Bearing Technology: A Review of Recent Research. *Wear* **2000**, *246*, 159–173, https://doi.org/10.1016/S0043-1648(00)00504-4.
119. Riley, F.L. Silicon Nitride and Related Materials. *J. Am. Ceram. Soc.* **2000**, *83*, 245–265.
120. Sorrell, C.C.; Hardcastle, P.H.; Druitt, R.K.; Howlett, C.R.; McCartney, E.R. Results of 15-Year Clinical Study of Reaction Bonded Silicon Nitride Intervertebral Spacers. *Proc. 7th World Biomater. Conf.* **2004**, 1872.
121. Mobbs, R.J.; Rao, P.J.; Phan, K.; Hardcastle, P.; Choy, W.J.; McCartney, E.R.; Druitt, R.K.; Mouatt, C.A.L.; Sorrell, C.C. Anterior Lumbar Interbody Fusion Using Reaction Bonded Silicon Nitride Implants: Long Term Case Series of the First Synthetic ALIF Spacer Implanted in Humans. *World Neurosurg.* **2018**, *120*, 256–264, https://doi.org/10.1016/j.wneu.2018.08.237.
122. Bal, B.S.; Khandkar, A.; Lakshminarayanan, R.; Clarke, I.; Hofmann, A.A.; Rahaman, M.N. Testing of Silicon Nitride Ceramic Bearings for Total Hip Arthroplasty. *J. Biomed. Mater. Res. Part B Appl. Biomater.* **2008**, *87*, 447–454, https://doi.org/10.1002/jbm.b.31123.
123. Mazzocchi, M.; Bellosi, A. On the Possibility of Silicon Nitride as a Ceramic for Structural Orthopaedic Implants. Part I: Processing, Microstructure, Mechanical Properties, Cytotoxicity. *J. Mater. Sci. Mater. Med.* **2008**, *19*, 2881–2887, https://doi.org/10.1007/s10856-008-3417-2.
124. Mazzocchi, M.; Gardini, D.; Traverso, P.L.; Faga, M.G.; Bellosi, A. On the Possibility of Silicon Nitride as a Ceramic for Structural Orthopaedic Implants. Part II: Chemical Stability and Wear Resistance in Body Environment. *J. Mater. Sci. Mater. Med.* **2008**, *19*, 2889–2901, https://doi.org/10.1007/s10856-008-3437-y.
125. Bal, B.S.; Khandkar, A.; Lakshminarayanan, R.; Clarke, I.; Hofmann, A.A.; Rahaman, M.N. Fabrication and Testing of Silicon Nitride Bearings in Total Hip Arthroplasty. *J. Arthroplasty* **2009**, *24*, 110–116, https://doi.org/10.1016/j.arth.2008.01.300.
126. McEntire, B.J.; Lakshminarayanan, R.; Ray, D.A.; Clarke, I.C.; Puppulin, L.; Pezzotti, G.; Bal, B.S.; Lakshminarayanan, A.; Bock, R. Silicon Nitride Bearings for Total Joint Arthroplasty. *Lubricants* **2016**, *4*, 1–24, https://doi.org/10.3390/lubricants4040035.
127. Yorifuji, M.; Affatato, S.; Tateiwa, T.; Takahashi, Y.; Shishido, T.; Marin, E.; Zanocco, M.; Zhu, W.; Pezzotti, G.; Yamamoto, K. Wear Simulation of Ceramic-on-Crosslinked Polyethylene Hip Prostheses: A New Non-Oxide Silicon Nitride versus the Gold Standard Composite Oxide Ceramic Femoral Heads. *Materials (Basel).* **2020**, *13*, 1–11, https://doi.org/10.3390/ma13132917.
128. Scholes, S.C.; Green, S.M.; Unsworth, A. The Wear of Metal-on-Metal Total Hip Prostheses Measured in a Hip Simulator. *Proc. Inst. Mech. Eng. Part H J. Eng. Med.* **2001**, *215*, 523–530, https://doi.org/10.1243/0954411011536118.
129. Kusaka, J.; Takashima, K.; Yamane, D.; Ikeuchi, K. Fundamental Study for All-Ceramic Artificial Hip Joint. *Wear* **1999**, *225–229*, 734–742, https://doi.org/10.1016/S0043-1648(98)00386-X.

130. Dante, R.C.; Kajdas, C.K. A Review and a Fundamental Theory of Silicon Nitride Tribochemistry. *Wear* **2012**, *288*, 27–38, https://doi.org/10.1016/j.wear.2012.03.001.
131. Mizutani, T.; Nagase, H.; Fujiwara, N.; Ogoshi, H. Silicic Acid Polymerization Catalyzed by Amines and Polyamines. *Bull. Chem. Soc. Jpn.* **1998**, *71*, 2017–2022.
132. Sui, X.; Wang, B.; Liu, Y.; Wu, H.; Li, Q.; Zhang, Q. Study on Microstructure of Polysilicic Acid in High Salinity. *J. Appl. Sci. Eng. Innov.* **2014**, *1*, 44–48.
133. Komvopoulos, K.; Saka, N.; Suh, N.P. The Mechanism of Friction in Boundary Lubrication. *J. Tribol.* **1985**, *107*, 452–462, https://doi.org/10.1115/1.3261108.
134. Persson, B.N.J. Theory of Friction and Boundary Lubrication. *Phys. Rev. B* **1993**, *48*, 18140–18158, https://doi.org/10.1103/PhysRevB.48.18140.
135. Lee, S.K.; Lawn, B.R. Contact Fatigue in Silicon Nitride. *J. Am. Ceram. Soc.* **1999**, *82*, 1281–1288.
136. Lawn, B.R.; Lee, S.K.; Peterson, I.M.; Wuttiphan, S. Model of Strength Degradation from Hertzian Contact Damage in Tough Ceramics. *J. Am. Ceram. Soc.* **1998**, *81*, 1509–1520, https://doi.org/10.1111/j.1151-2916.1998.tb02510.x.
137. Lee, K.S.; Jung, Y.; Peterson, I.M.; Lawn, B.R.; Kim, D.; Lee, S.K. Model for Cyclic Fatigue of Quasi-Plastic Ceramics in Contact with Spheres. *J. Am. Ceram. Soc.* **2000**, *83*, 2255–2262, https://doi.org/10.1111/j.1151-2916.2000.tb01544.x.
138. Shankar, S.; Prakash, L.; Kalayarasan, M. Finite Element Analysis of Different Contact Bearing Couples for Human Hip Prosthesis. *Int. J. Biomed. Eng. Technol.* **2013**, *11*, 66–80, https://doi.org/10.1504/IJBET.2013.053712.
139. Hodaei, M.; Farhang, K.; Maani, N. A Contact Model for Establishment of Hip Joint Implant Wear Metrics. *J. Biomed. Sci. Eng.* **2014**, *7*, 228–242.
140. Shankar, S.; Nithyaprakash, R. Wear Prediction on Silicon Nitride Bearing Couple in Human Hip Prosthesis using Finite Element Concepts. *Proc. Inst. Mech. Eng. Tribol.* **2014**, *0*, 1–8, https://doi.org/10.1177/1350650114526580.
141. Maskiewicz, V.K.; Williams, P.A.; Prates, S.J.; Bowsher, J.G.; Clarke, I.C. Characterization of Protein Degradation in Serum-Based Lubricants during Simulation Wear Testing of Metal-on-Metal Hip Prostheses. *J. Biomed. Mater. Res. Part B Appl. Biomater.* **2010**, *94B*, 429–440, https://doi.org/10.1002/jbm.b.31671.
142. Myant, C.; Underwood, R.; Fan, J.; Cann, P.M. Lubrication of Metal-on-Metal Hip Joints: The Effect of Protein Content and Load on Film Formation and Wear. *J. Mech. Behav. Biomed. Mater.* **2012**, *6*, 30–40, https://doi.org/10.1016/j.jmbbm.2011.09.008.
143. Bowsher, J.G.; Clarke, I.C. Thermal Conductivity of Femoral Ball Strongly Influenced UHMWPE Wear in a Hip Simulator. In Proceedings of the Transactions of the 53rd Annual Meeting of the Orthopaedic Research Society; Orthopaedic Research Society: San Diego, CA, 2007; p. 0278.
144. Tsai, S.; Salehi, A.; Aldinger, P.; Hunter, G. Heat Generation and Dissipation Behavior of Various Orthopaedic Bearing Materials. *Key Eng. Mater.* **2006**, *309–311*, 1291–1294.
145. Bal, B.S.; Rahaman, M.N. Orthopedic Applications of Silicon Nitride Ceramics. *Acta Biomater.* **2012**, *8*, 2889–2898, https://doi.org/10.1016/j.actbio.2012.04.031.
146. McEntire, B.J.; Bal, B.S.; Rahaman, M.N.; Chevalier, J.; Pezzotti, G. Ceramics and Ceramic Coatings in Orthopaedics. *J. Eur. Ceram. Soc.* **2015**, *35*, 4327–4369, https://doi.org/10.1016/j.jeurceramsoc.2015.07.034.
147. Zhou, Y.S.; Ohashi, M.; Tomita, N.; Ikeuchi, K.; Takashima, K. Study on the Possibility of Silicon Nitride—Silicon Nitride as a Material for Hip Prostheses. *Mater. Sci. Eng. C* **1997**, *5*, 125–129, https://doi.org/10.1016/S0928-4931(97)00032-5.
148. Xu, J.; Kato, K. Formation of Tribochemical Layer of Ceramics Sliding in Water and its Role for Low Friction. *Wear* **2000**, *245*, 61–75, https://doi.org/10.1016/S0043-1648(00)00466-X.
149. Shankar, S.; Nithyaprakash, R.; Sugunesh, A.P.; Selvamani, K.A.; Uddin, M.S. Experimental and Finite Element Wear Study of Silicon Nitride Against Alumina for Hip Implants with Bio-Lubricant for Various Gait Activities. *Silicon* **2021**, *13*, 633–644, https://doi.org/10.1007/s12633-020-00462-w.

150. Kong, X.; Hu, X.; Chai, W. In vitro & in vivo Investigation of the Silicon Nitride Ceramic Hip Implant's Safety and Effectiveness Evaluation. *J. Orthop. Surg. Res.* **2022**, *17*, 1–13, https://doi.org/10.1186/s13018-021-02884-7.
151. Parekh, A.; Sood, A.; Monsef, J.B.; Hamouda, M.; Hussain, A.; Gonzalez, M. Second-Generation Highly Cross-Linked Polyethylene in Total Hip Arthroplasty. *JBJS Rev.* **2021**, *9*, e20.
152. Yoon, B.-H.; Park, J.-W.; Lee, Y.-K.; Koo, K.-H.; Chang, C.B. Long-Term Wear-Related Complications of Cross-Linked Versus Conventional Polyethylene After Total Hip Arthroplasty: A Meta-Analysis. *J. Arthroplasty* **2022**, https://doi.org/10.1016/j.arth.2022.05.013.
153. Busch, A.; Jäger, M.; Klebingat, S.; Baghdadi, J.; Flörkemeier, T.; Hütter, F.; Grupp, T.M.; Haversath, M.; VITAS-Group Vitamin E-Blended Highly Cross-Linked Polyethylene Liners in Total Hip Arthroplasty: A Randomized, Multicenter Trial Using Virtual CAD-Based Wear Analysis at 5-Year Follow-Up. *Arch. Orthop. Trauma Surg.* **2020**, *140*, 1859–1866, https://doi.org/10.1007/s00402-020-03358-x.
154. Pezzotti, G.; Zhu, W.; Sugano, N.; Marin, E.; Yamamoto, K.; Nishiike, N.; Hori, T.; Rondinella, A.; McEntire, B.J.; Bock, R.; et al. Oxide Ceramic Femoral Heads Contribute to the Oxidation of Polyethylene Liners in Artificial Hip Joints. *J. Mech. Behav. Biomed. Mater.* **2018**, *82*, 168–182, https://doi.org/10.1016/j.jmbbm.2018.03.021.
155. Pezzotti, G.; Bal, B.S.; Casagrande, E.; Sugano, N.; McEntire, B.J.; Zhu, W.; Puppulin, L. On the Molecular Interaction between Femoral Heads and Polyethylene Liners in Artificial Hip Joints: Phenomenology and Molecular Scale Phenomena. *Biomed. Mater.* **2017**, *12*, 15005.
156. Khullar, P.; Zhu, D.; Gilbert, J.L. Fretting Corrosion of Si_3N_4 vs CoCrMo Femoral Heads on Ti-6Al-V Trunnions. *J. Orthop. Res.* **2020**, *38*, 1617–1626, https://doi.org/10.1002/jor.24681.
157. Mali, S.A.; Gilbert, J.L. Correlating Fretting Corrosion and Micromotions in Modular Tapers: Test Method Development and Assessment. *ASTM Spec. Tech. Publ. 1591 Implant Modul.* **2015**, 259–282, https://doi.org/10.1520/STP159120140136.
158. Kocagoz, S.B.; Underwood, R.J.; MacDonald, D.W.; Gilbert, J.L.; Kurtz, S.M. Ceramic Heads Decrease Metal Release Caused by Head-Taper Fretting and Corrosion. *Clin. Orthop. Relat. Res.* **2016**, *474*, 985–994, https://doi.org/10.1007/s11999-015-4683-1.
159. Aldinger, P.; Williams, T.; Woodard, E. Accelerated Fretting Corrosion Testing of Zironia Toughened Alumina Composite Ceramic and a New Composition of Ceramicised Metal Femoral Heads. *Annu. Meet. Orthop. Res. Soc.* **2017**.
160. Kyomoto, M.; Shoyama, Y.; Saiga, K.; Moro, T.; Ishihara, K. Reducing Fretting-Initiated Crevice Corrosion in Hip Simulator Tests Using a Zirconia-Toughened Alumina Femoral Head. *J. Biomed. Mater. Res. - Part B Appl. Biomater.* **2018**, *106*, 2815–2826, https://doi.org/10.1002/jbm.b.34062.

A Critical Review of Two Bioceramics for Total Hip Arthroplasty

Bryan J. McEntire, B. Sonny Bal, and Giuseppe Pezzotti

1 Introduction

Ceramic materials have played an important role in the evolution of total hip arthroplasty (THA). Beginning in the early 1900s, various innovations were pioneered to restore mobility to an increasingly elderly population beset with osteoarthritis, hip dysplasia, fractures, and other bone or joint maladies; but the most significant advancement in hip implant efficacy occurred around 1960 with the modular total hip arthroplasty system pioneered by Sir John Charnley [1]. It consisted of a femoral stem, ball, liner, and acetabular socket. The key articulation components within this modular design were the femoral ball, which at the time was made of a cobalt-chromium (CoCr) alloy, and a polyethylene (UHMWPE) liner. This design revolutionized THA surgery by increasing its longevity, ultimately leading to the procedure being lauded as the "Operation of the Century" [2].

In the 1970s, alumina (Al_2O_3) ceramics were introduced as alternatives for the femoral ball and acetabular liner [3, 4], mainly to address polyethylene wear. The ceramic femoral head was promoted because it promised less wear when articulating against UHMWPE. Even lower wear rates were observed when an Al_2O_3 head was coupled with an Al_2O_3 liner [3]. Among the companies that produced these Al_2O_3 components, CeramTec AG (Plochingen, Germany) became the market and

B. J. McEntire (✉)
SINTX Technologies, Inc., Salt Lake City, UT, USA
e-mail: bmcentire@sintx.com

B. S. Bal
Orthopaedic Surgeon and Attorney, Bal Consulting LLC, Columbia, MO, USA

G. Pezzotti
Ceramic Physics Laboratory, Kyoto Institute of Technology, Sakyo-ku, Matsugasaki, Kyoto, Japan

B. S. Bal et al. (eds.), *Silicon Nitride Bioceramics*,
https://doi.org/10.1007/978-3-031-67047-3_11

technology leader [5]. An improved material replaced their first-generation Al_2O_3 components in the 1980s. This second generation of devices was made using a finer raw material. It provided higher strength and overall durability. Then, proof-testing of components was added in the 1990s which further increased component reliability. Their third generation of Al_2O_3 biomedical ceramics was marketed under the tradename BIOLOX®*forte.*

Zirconia (ZrO_2 or YTZP) ceramics were the primary competitor to BIOLOX components at the time. They were touted as having higher strength, toughness, and better reliability than Al_2O_3. These improvements were achieved due to a peculiar phase-related strengthening mechanism known as transformation toughening [6]. Femoral heads were promoted and distributed by Saint Gobain Desmarquest under the tradename Prozyr®. They gained a significant market share and were in a position to challenge BIOLOX's dominance. Although reports on Prozyr® before 2000 indicated adequate clinical performance, the composition and process Desmarquest used to produce these parts was fatally flawed. The material was inherently unstable and subject to low-temperature hydrothermal degradation (LTD) [7, 8]. Reports of significant in vivo head fractures led the US FDA to recall these devices in 2001. This unfortunate event decimated the future of ZrO_2 in orthopedic devices and further solidified CeramTec's market leadership.

Yet, in responding to the Prozyr® challenge of the 1990s, a novel material was developed by CeramTec engineers. Designated as BIOLOX®*delta*, it was designed as an alumina-matrix-composite (AMC). Compositionally, *delta* is a dense ceramic consisting of 79.4 vol.% Al_2O_3 and 17.0 vol.% ZrO_2, which is partially stabilized with ~1.3 mol.% Y_2O_3, 3.0 vol.% $SrAl_2O_4$, and 0.6 vol.% Cr_2O_3 [9]. Like Prozyr, *delta*'s strength and toughness are derived from the same transformation mechanism; but unlike Prozyr, the presence of the Al_2O_3 matrix inhibits LTD and catastrophic failures. Today, BIOLOX®*delta* has a well-established record of excellent clinical outcomes dating back to its introduction in 2003, and it is currently the world's market-leading bioceramic for THA devices.

Over the past decade, persistent clinical problems, such as corrosion at modular tapers have led physicians to move away from using CoCr femoral heads articulating against polyethylene (PE) liners with *delta* gradually being the favored femoral head, and ultra-high molecular weight polyethylene (UHMWPE, mostly as a highly crosslinked formulation, HXLPE) being preferred for liners. Ceramic heads (e.g., BIOLOX®*delta*) coupled with HXLPE liners are now used in over 70% of all primary THAs in the USA (cf., Chap. 10) [10]).

Concurrent with the development of Al_2O_3, ZrO_2, and AMC bioceramics in the 1980s through 2000, multiple government agencies worldwide collaborated on the development of another high-performance ceramic, silicon nitride (Si_3N_4). Yet, the target market for this material was not implantable human prosthetics, rather, it was developed for fuel-conserving, high-temperature heat-engines [11]. Although this promise was never fully realized at the time, other investigators simultaneously began looking at Si_3N_4 as a potential bioceramic. Indeed, it was quietly employed as a spinal spacer in a clinical research trial beginning in 1986 [12, 13]. It has since set the record as being the first abiotic material to be used for this purpose, and a recent 30+ year follow-up on surviving patients indicated excellent outcomes with no adverse effects [14].

Si_3N_4's mechanical properties have been subsequently optimized, and they are comparable to *delta*, although the mechanism for Si_3N_4's toughness and strength are markedly different. Unlike *delta*, it is an inherently phase-stable material, whose grains crystallize in an elongated or fibrous structure. This type of microstructure presents a tortuous high-energy path for crack propagation, and bridging grains in the crack wake induce an arresting compressive stress at the crack tip [15, 16]. A hot-isostatically pressed version of Si_3N_4, doped with 6 wt.% Y_2O_3 and 4 wt.% Al_2O_3, was developed by SINTX Technologies, Inc. (*MC*^{2}®Si_3N_4, Salt Lake City, UT, USA) and subsequently cleared for use by the FDA in 2008 for use in spinal arthrodesis implants [17]. Like *delta*, *MC*^{2}®Si_3N_4 boasts an enviable record of successful clinical outcomes since its introduction, albeit in a different anatomical application. Even though *MC*^{2}®Si_3N_4 has yet to receive regulatory clearance for its use in hip arthroplasty, significant THA testing has been conducted. The absence of regulatory approvals has less to do with its properties and performance, and more to do with limited resources for regulatory trials and the established market dominance of BIOLOX®*delta*.

Despite significant technological innovations in materials, designs, and clinical practice over the past fifty years, considerable challenges remain in extending the lifetime of THA prostheses to the lifespans of most patients. Unfortunately, long-term survival rates for implants are no greater today than for Charnley's original low-friction cemented hip device. As separately reviewed by Callahan et al. [18] and Berry et al. [19], the 25-year survivorship of the Charnley traditional design ranged from 77% to 81%. Contrast this result with a 2019 meta-analysis by Evans et al. who evaluated the survivorship of over 13,000 contemporary hip implants [20]. Using data from national joint registries and case studies, they estimated the current 25-year survival rate at between 58% to 78%, respectively. The causative reasons for failure remarkably remain the same as in Charnley's day—fracture, dislocation, infection, polyethylene wear, osteolysis, and aseptic loosening [2]. Consequently, additional innovation is needed, particularly for younger and more physically active THA patients. This is where BIOLOX®*delta* and *MC*^{2}®Si_3N_4 may play important pivotal roles in the evolution of joint arthroplasty.

While there are other suppliers of bioceramics to the medical industry, CeramTec and SINTX are unique in that they arguably offer the two best bioceramics for THA purposes. This chapter is therefore dedicated to a comparative review of the properties and biological performance characteristics of BIOLOX®*delta* and *MC*^{2}®Si_3N_4. This review will include discussions of physical and mechanical properties, surface chemistry, wear performance, antipathogenic characteristics, mechanically assisted crevice corrosion, and thermal conductivity considerations. This chapter may contain some redundancy with other sections of the book. However, this repetition serves the purpose of maintaining continuity and facilitating a comprehensive comparison of these two significant bioceramics.

Table 1 Physical and mechanical properties of BIOLOX®*delta* and MC^2®Si_3N_4

Property	Units	Bioceramic	
		BIOLOX®*delta*	MC^2®Si_3N_4
Density	g/cc	4.37	3.26
Elastic modulus	GPa	350–358	296
Grain size	μm	0.54–0.56	0.5 x 5.0
Flexural strength	MPa	1150–1411	995–1250
Fracture toughness	$MPa{\cdot}m^{1/2}$	5.7–8.5	7.0–10.5
Hardness	GPa	17.6–19.8	14.4–15.9

2 Physical and Mechanical Properties

BIOLOX®*delta* and MC^2®Si_3N_4 are bioceramic materials well-suited for application as THA femoral heads due to their exceptional combination of physical and mechanical properties. Both materials achieve full density, thereby minimizing intrinsic porosity and enhancing their resistance to fatigue and crack propagation. Additionally, they exhibit superior fracture toughness, a critical property for withstanding impact and stress concentrations in the human body for several decades. This is complemented by high flexural strength and hardness, properties that contribute to their ability to bear substantial loads and resist wear. A comparison of their basic physical and mechanical properties is provided in Table 1 [9, 17, 21–33]. There are noteworthy differences in a few properties—particularly density, elastic modulus, flexural strength, fracture toughness, and hardness that will be reviewed in the following sub-sections.

2.1 Density

While both bioceramics achieve full volumetric density, MC^2®Si_3N_4 has a mass density ~ 34% lower than BIOLOX®*delta*. This difference is solely a function of their respective compositions. The majority of MC^2 is composed of silicon and nitrogen, among the lightest elements of the periodic table, with minor amounts of aluminum, yttrium, and oxygen. Conversely, the majority constituents of *delta* are aluminum, zirconium, and oxygen, with minor amounts of yttrium, strontium, and chromium. This density difference translates to heavier implantable *delta* components. For example, an unfinished femoral ball measuring Ø44 mm will have an MC^2 mass of ~145 g while the same ball from *delta* will be ~194 g, or 49 g heavier. In contrast, the mass of a CoCr alloy ball is about 375 g (i.e., 158% and 93% greater than MC^2 or *delta*, respectively), while a native femoral head of cortical bone is only about 67 g. Despite the lower weight of bioceramic heads compared to traditional options like CoCr, femoral head mass is an often ignored design consideration in THA. Factors such as implant geometry, surface finish, friction coefficient, patient bone structure, weight, and activity level, along with surgical implant positioning, orientation, and fixation methods have a far greater impact on overall implant stress distribution and wear performance [34]. Modern THA designs consider these factors in a complex manner to minimize the stress that is transmitted to the liner and acetabular cup. While the absolute mass of the femoral head itself may be a minor consideration, minimizing weight remains a desirable goal, especially considering the significant weight change compared to the native femoral head it replaces.

2.2 Elastic Modulus

Although *delta* has a higher modulus of elasticity than MC^2, neither material significantly impacts joint stability or performance based solely on elastic modulus. UHWMPE liners are the compliant members in this articulation couple with elastic moduli ranging from 655 to 1077 MPa [35]—at least two orders of magnitude lower in stiffness than the ceramics. Stress shielding has been raised as an issue over the use of high-modulus materials in THA, but most concerns focus on the femoral stem [36]. Few studies have been published about the stress shielding caused by the articulation force transmitted by the femoral head through the liner and cup to the underlying acetabulum. In a prospective randomized controlled clinical trial, Pitto et al. only observed moderate cortical bone density changes in the acetabulum after a one-year follow-up [37]; but more importantly, they failed to see significant differences between implants that included UHMWPE versus ceramic liners. Later, Vogel et al. performed a detailed finite element analysis of ceramic versus UHMWPE liners supported by either titanium or carbon-fiber reinforced PEEK (CFRP) cups [38]. Notably, the low modulus (15 GPa) of CFRP exerted a minimal influence on stress–strain distribution within the adjacent bone, irrespective of the liner material. These findings collectively suggest that the modulus of the acetabular cup and, to a lesser extent, the femoral head play limited roles in acetabular stress shielding. Even though the elastic moduli of MC^2 and *delta* are high compared with native bone, their influence on stress shielding and aseptic loosening appears to be inconsequential.

2.3 Flexural Strength

The flexural strength of both biomaterials was evaluated as specified in the ASTM F603 standard using the ASTM C1161 flexural test method for MC^2 and the ISO 14704 flexural strength for *delta*. These two test methods are similar but use either 3- or 4-point bending on different specimen sizes. MC^2 typically used a $3.0 \times 4.0 \times 40$ mm test geometry in 3-point bending. For *delta*, a 4-point flexure test was employed but the specific test geometry is unreported. By the standard, it could be $3.0 \times 4.0 \times 40$ mm or $3.0 \times 4.0 \times 30$ mm [23]. The question is: Are these test methods comparable? From Weibull analyses, the size/strength relationship between different test geometries can be estimated [39]:

$$(\sigma_{4pt}/\sigma_{3pt}) = (V_{E3pt}/V_{E4pt})^{1/m} \quad (1)$$

where σ_{3pt} and σ_{4pt} represent the mean or characteristic fracture strength of the ceramic for 3- and 4-point bending respectively; V_{E3pt} and V_{E4pt} are the effective volumes under stress for the two test configurations; and m = Weibull modulus. Mathematically, this equation suggests that the ratio of strengths for different test configurations can be expressed as the inverse ratio of their effective test volumes and a dimensionless factor that incorporates the influence of the Weibull modulus. This factor accounts for the variation in stress distribution due to testing geometry and the inherent variability in material flaw sizes [39, 40]. Quinn provided the necessary equations for the practical conversion of strength data between different test configurations [39]. Equation (2) is used for comparing 3- to 4-point bending with outer and inner load points of 40 mm and inner load points of 20 mm, respectively:

$$(\sigma_{4pt,40}/\sigma_{3pt,40}) = [(m+2)/2]^{1/m} \quad (2)$$

Equation (3) can be used for comparing 3- and 4-point flexural tests where the 4-point configuration has outer and inner load points of 30 and 10 mm, respectively:

$$(\sigma_{4pt,30}/\sigma_{3pt,40}) = [(m+3)/4]^{1/m} \quad (3)$$

Using actual test data for a relatively recent lot of MC^2 as shown in Fig. 1, $\sigma_{3pt} = 1253$ MPa and $m = 14.5$; and summarized data for *delta* of $\sigma_{4pt} = 1411$ MPa and $m = 14.9$ [9] (assuming 4-point 30 mm outer load points), Table 2 provides estimates of fracture strengths for alternative test configurations for both MC^2®Si_3N_4 and BIOLOX®*delta*, where $\sigma_{3,40}$ = average or characteristic strength for 3-point flexure with an outer span of 40 mm, $\sigma_{4,30}$ = average or characteristic strength for 4-point flexure with outer and inner spans of 30 and 10 mm, respectively, and $\sigma_{4,40}$ = average or characteristic strength for 4-point flexure with outer and inner spans of 40 and 20 mm, respectively.

This analysis suggests that *delta* has markedly superior flexural strength over MC^2 (~25%) regardless of flexural test configuration. However, Quinn and Morrell pointed out that the strength measured in test specimens can indicate, but not guarantee, the strength of actual components [41]. This can be due to several factors, including differences in density, microstructure, preparation methods, machining damage, and the testing methodologies between specimens and components.

To mitigate these issues, the US FDA requires manufacturers to certify the strength of their THA femoral balls using burst testing by the ISO 7206 testing standard [42]. The regulatory requirement is: "The average fracture load of at least five balls shall exceed 46 kN, and no ball shall fail at less than 20 kN." Therefore, to compare the mechanical properties of *delta* to MC^2 in a *real* component, burst testing of identical Ø28mm 12/14 + 8.5 offset femoral heads was conducted using CoCr

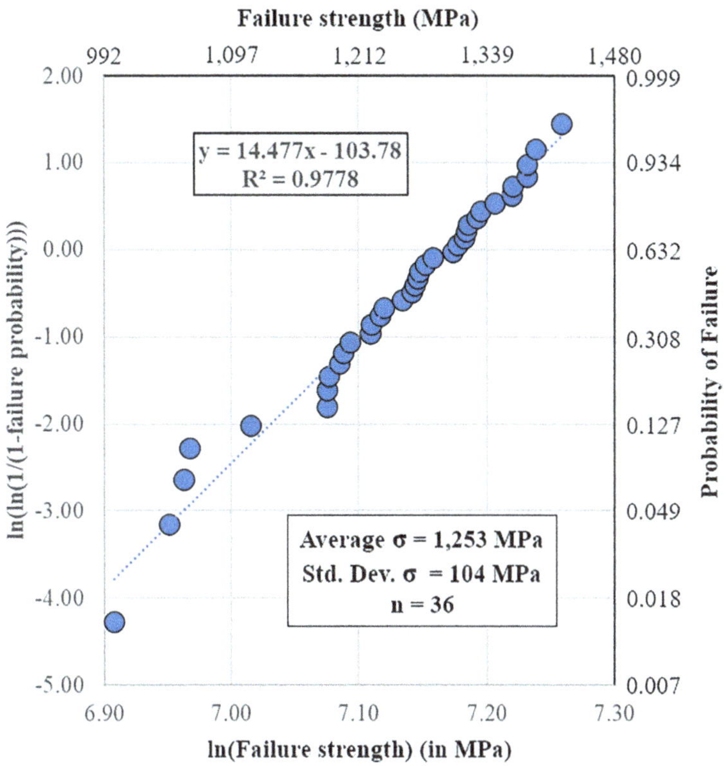

Fig. 1 Weibull strength distribution plot for *MC*^{2}Si_3N_4; PAR 2047, Courtesy SINTX Technologies, Inc.

Table 2 Sensitivity analysis of actual and predicted mean or characteristic failure stresses for *MC*^{2}®Si_3N_4 and BIOLOX®*delta* using 3-point and 4-point flexural testing configurations

Bioceramic	m	$\sigma_{3,40}$ (MPa)	$\sigma_{4,30}$ (MPa)	$\sigma_{4,40}$ (MPa)
MC^2®$Si_{i3}N_4$	14.5 Actual	1253 Actual	1132 Estimate	1083 Estimate
BIOLOX®*delta*	14.9 Actual	1561 Estimate	1411 Actual	1352 Estimate

trunnions. The +8.5 offset produces a severely elevated stress distribution within the ceramic heads. The *delta* heads were acquired from a commercial source, and the *MC*^{2} heads were produced by SINTX Technologies to exactly match the *delta* components.

Although the test was limited by the number of available heads (i.e., five for Si_3N_4 and six for *delta*), the results showed that both ceramics surpassed the 46 kN average load required by the FDA, and there were no failures less than 20 kN. The individual test values, averages, and standard deviations are presented in Table 3. A *t*-test of the two data sets resulted in a *p*-value of 0.084, indicating no significant

Table 3 Burst test results for MC^2®Si_3N_4 and BIOLOX®*delta* Ø28mm 12/14 + 8.5 offset femoral heads tested following the ISO 7206 standard

No.	Burst test results (kN)	
	MC^2®Si_3N_4	BIOLOX®*delta*
1	80.00	68.15
2	74.93	68.79
3	84.19	48.97
4	80.70	66.98
5	55.00	54.85
6		79.80
Average	74.96	64.59
Std. Dev.	11.64	11.01

differences in burst strengths for the two materials. In contrast to the flexural data presented earlier, these limited burst-test results suggest that the two materials have similar fracture strengths; but both materials are demonstrably superior in strength to biomedical Al_2O_3, which has been the traditional gold standard for THA joint prostheses [43] until the advent of BIOLOX®*delta*.

2.4 Fracture Toughness

An obvious difference in mechanical properties between the two materials is fracture toughness. The reported K_{IC} value for MC^2 is between 7.0 and 10.5 MPa·m$^{1/2}$, whereas *delta* has a range of between 5.7 and 7.0 MPa·m$^{1/2}$ (cf., Table 1). As noted earlier, the mechanisms for enhanced fracture toughness of both materials are different; but so are their testing methods. MC^2 has been traditionally evaluated using the ASTM E-399 or ASTM C-1421 standards, commonly referred to as the "Chevron notch" methods (CVN), while *delta* has been assessed using the "Single edge V-notch beam" technique (SEVNB, ISO 23146). These two methods are similar in that they both require careful specimen preparation with controlled notches. Both methods are highly recommended [44, 45], and independent studies have shown they produce reasonably comparable results for identical ceramics tested by the same laboratories [46–48]. Nevertheless, there have been no direct comparative K_{IC} tests of MC^2 and *delta* using the same method by the same laboratory. Pezzotti et al. [31] and Richter et al. [32] separately evaluated the K_{IC} of *delta* using the SEVNB method; and, Martin et al. independently measured the K_{IC} of MC^2 [22] using a double torsion (DT) test technique [49]; although DT tests are not considered to be as reliable as either the CVN or SEVNB methods [44]. Statistical comparisons of K_{IC} values for the two bioceramics are given in Table 4.

Even though the reported data for MC^2 using the DT method is lower than the CVN technique (i.e., 7.0 versus 10.5 MPa·m$^{1/2}$, respectively), a *t*-test comparison of the data in Table 4 shows *p*-values of <0.01, suggesting that the higher K_{IC} is

Table 4 Fracture toughness values (K_{IC}) of MC^2®Si_3N_4 and BIOLOX®*delta* obtained using the CVN, DT, and SEVNB test methods. NR = not reported

Bioceramic	Test method	n	K_{IC} result (MPa·m$^{1/2}$)	Ref.
MC^2®Si_3N_4	CVN	241	10.6 ± 1.3	[17]
	DT	3	7.0 ± 0.3	[22]
BIOLOX®*delta*	SEVNB	NR	6.5 ± 0.3	[26]
	SEVNB	10	5.8 ± 0.4	[32]
	SEVNB	7	5.7 ± 0.6	[31]

statistically different. Similarly, performing *t*-test comparisons between MC^2 and *delta* is also suggestive of a valid statistical difference, regardless of the test method. Indeed, the results indicate that the fracture toughness of MC^2 is superior to *delta*. This leads to an interesting paradox. The relationship between fracture toughness and flexural strength appears counterintuitive based on the Griffith equation [50, 51]:

$$\sigma_R = K_{IC}/Y(\pi a)^{1/2} \tag{4}$$

where σ_R = flexural strength, K_{IC} = critical stress intensity factor or fracture toughness, Y = *a* geometrical constant that is material dependent, and a = the flaw size. One might surmise that MC^2 should have a much higher strength than *delta*, but this is not the case. The apparent paradox arises because the relationship between fracture toughness and fracture strength is not straightforward. Other factors, such as composition, microstructural features, grain size, grain boundaries, toughening mechanism, and stress concentrations, play a role in determining the overall flexural strength of the material (e.g., the Y factor in the Griffith equation). Notably, previous studies have demonstrated that K_{IC} in Si_3N_4 can be tailored through specific additives and densification processes [52–56]. These techniques produce microstructures ranging from fine-grained and semi-isotropic to coarse-grained and anisotropic. While this allows for K_{IC} values ranging from ~5 to ≥10.0 MPa·m$^{1/2}$, larger grains also introduce potential critical flaws, partially negating the benefit of the higher $K_{IC.}$ Furushima et al. recently demonstrated the potential of convolutional neural networks (CNNs) for predicting fracture toughness in Si_3N_4 [57, 58]. Their approach utilized observations of polished Si_3N_4 microstructures obtained through CNNs and correlated them with K_{IC} values measured using the single-edge-pre-cracked-beam (SEPB) technique, spanning a range of ~4 to ~10 MPa·m$^{1/2}$. This method achieved a high correlation coefficient ($R^2 \geq 0.96$) between the predicted and experimentally measured K_{IC} values, highlighting the significant role of microstructural features in governing fracture toughness.

While silicon nitride has a greater fracture toughness, *delta* manages to achieve both high fracture toughness and impressive fracture strength due to its unique composition, and microstructure, by a different toughening mechanism than Si_3N_4. Nevertheless, in comparison to the K_{IC} of Al_2O_3 bioceramics (i.e., ~4.0 MPa·m$^{1/2}$, [30]), both MC^2 and *delta* are substantially superior.

2.5 *Hydrothermally Induced Mechanical Degradation and Subcritical Crack Growth*

An important aspect to consider regarding the mechanical integrity of *MC²*®Si_3N_4 and BIOLOX®*delta* is their susceptibility to degradation under hydrothermal conditions. The properties of both materials are affected by moisture, but these ceramics possess mechanisms to mitigate their detrimental effects. As stated previously, *delta*'s prominent improvements in fracture toughness and strength are due to transformation toughening [8]. This mechanism relies on the conversion of the ceramic's metastable tetragonal zirconia particles (*t*-ZrO_2) to their stable monoclinic form (*m*-ZrO_2) in the presence of an advancing crack. This transformation is accompanied by a ~4% volume increase, which creates a compressive stress field around the crack tip. This compressive stress resists crack propagation, resulting in an overall increase in the toughness and strength of the ceramic. However, this strengthening mechanism also has a weakness. A hydrolysis reaction coupled with external loading prematurely accelerates this phase transformation, which subsequently reduces the material's incipient fracture toughness and strength. Pezzotti et al. (2009) first demonstrated this phenomenon but also acknowledged limitations due to retarded homeostatic degradation kinetics. They suggested that any time-to-failure would likely exceed the typical lifespan of most patients [31]. Researchers from CeramTec also replicated this effect in a study published shortly afterward [9]. They showed the instability of the material by monitoring an increase in its *m*-ZrO_2 phase under increasing hydrothermal conditions and after cyclic fatigue testing. At 5 h and 100 h of aging in an autoclave (134 °C, 2 atm steam, up to 600 MPa in stress, for up to 20×10^6 cycles) the *m*-ZrO_2 content increased from ~18% to a maximum of ~47%. Despite these significant microstructural changes, they found no differences in flexural strength between pristine and fatigue-tested specimens. They concluded that increased monoclinic content at the ceramic's surface had no impact on bulk mechanical properties.

This degradation phenomenon was further extensively characterized and reviewed by Chevalier et al. as depicted in Fig. 2 [59]. In their study, the surface *m*-ZrO_2 increased from about 15% to over 60% from simple static exposure to temperature and water vapor. Note from the dotted line in the figure that a marked increase in monoclinic content is only expected to occur after about 11.4 years in vivo, but this increase appeared to be inconsequential compared to the accelerated tests conducted at elevated temperatures. The authors saw no observable decrease in bulk strength, no increase in surface roughness, or other apparent damage that might indicate concern. They, therefore, concluded that the phase transformation that occurs in BIOLOX®*delta* is, in fact, a safety feature [60].

However, despite these reassurances, concerns regarding the phase instability of *delta* persisted. A 2016 study used indentation-crack-length fracture toughness methods to assess changes in K_{IC} values as a function of crack length (i.e., *R*-curve behavior) for three bioceramics, *MC²*®Si_3N_4, BIOLOX®*forte,* and BIOLOX®*delta*. The results of this study are provided in Fig. 3a, b.

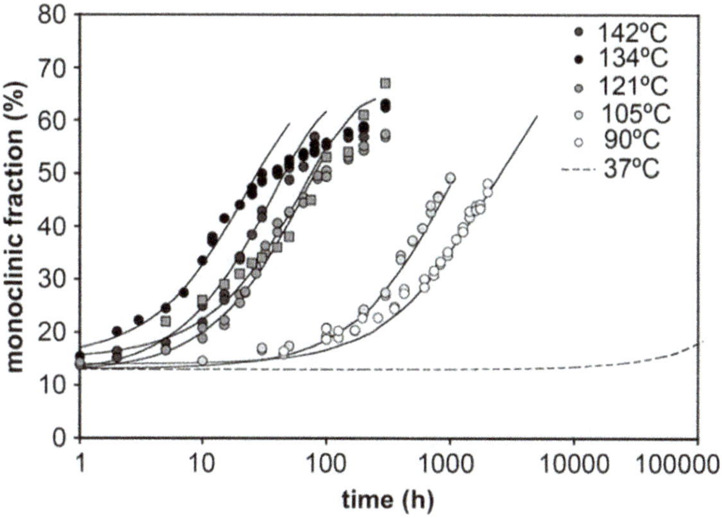

Fig. 2 Increase in surface m-ZrO_2 phase fractions on polished BIOLOX®*delta* femoral heads as a function of static exposure to a hydrothermal environment at different temperatures at 1 atm steam pressure. The dotted line represents an Arrhenius extrapolation of these data to 37 °C. Note the predicted beginning of an exponential rise in m-ZrO_2 at 10^5 h (~11.4 years in vivo). (Modified and used with permission [59])

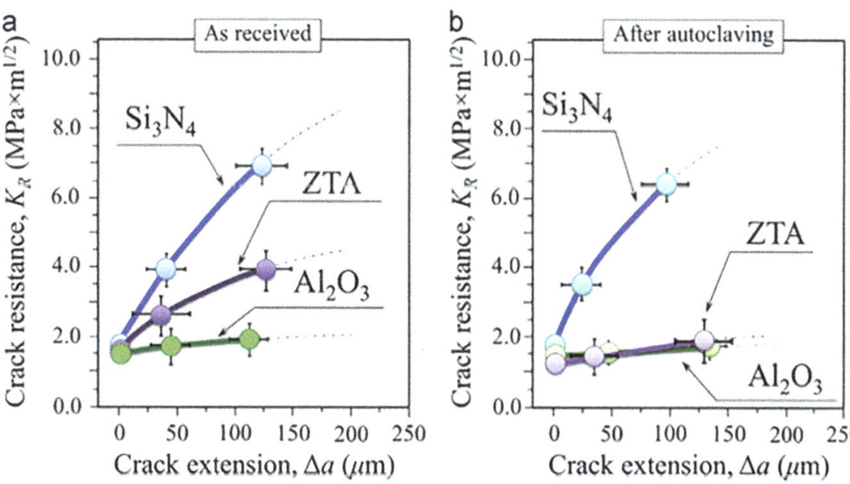

Fig. 3 Near-tip rising R-curve behavior obtained from indentation cracks for MC^2®Si_3N_4, BIOLOX®*forte* Al_2O_3, and BIOLOX®*delta* ZTA bioceramics before (**a**) and after (**b**) autoclaving. (Used with permission [61])

Figure 3a shows K_{IC} values determined on the as-received ceramics, and Fig. 3b are results after autoclaving the precracked samples for 200 h at 121 °C and 1 atm pressure. Note that the fracture toughness for MC^2 does not change with the

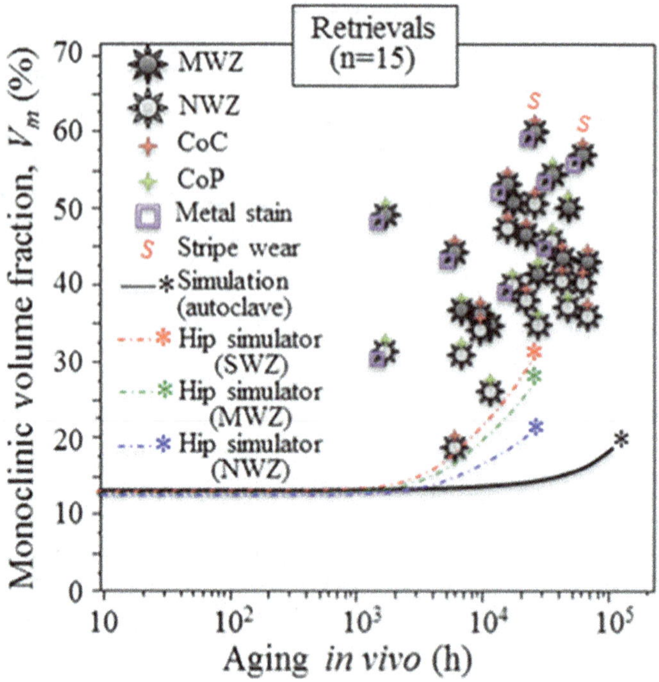

Fig. 4 Average m-ZrO_2 fractions in the main wear zone (MWZ) and non-wear zone (NWZ) on fifteen retrieved ceramic-on-polyethylene (CoP) and ceramic-on-ceramic (CoC) BIOLOX®*delta* femoral heads plotted against their service time in vivo. Additionally, data for static hydrothermal aging in an autoclave (*) and under hip simulator wear conditions are shown for the striped wear zone (SWZ, *), main wear zone (MWZ, *), and non-wear zone (NWZ, *) of pristine in vitro tested *delta* femoral heads. (Used with permission [65])

hydrothermal treatment, while *delta*'s as-received K_{IC} of ~4 MPa·m$^{1/2}$ is reduced to a comparable value for BIOLOX®*forte* (i.e., ~2 MPa·m$^{1/2}$) after autoclaving. This apparent reduction in surface toughness is a direct result of the hydrothermally induced transformation of existing t-ZrO_2 particles within the crack-propagation zone.

It was also argued that the simple Arrhenius plot for predicting time-to-failure (as shown in Fig. 2) was too simplistic because this analysis does not account for the mechanical or chemical interactions with the other modular hip components (e.g., femoral stem and acetabular cup). From a review of earlier studies by Affatato et al. [62–64], it was hypothesized and then demonstrated by Pezzotti et al. that mechanical stress and metal staining on the surface of pristine *delta* heads accelerated the polymorphic phase transformation (i.e., $t \rightarrow m$-ZrO_2) leading to reduced surface fracture toughness [65–67]. The results of this study are presented in Fig. 4.

They found that a short static autoclave treatment (only 24 h, 121 °C, 1 atm steam) on *delta* femoral heads possessing metal stains was sufficient to dramatically increase the amount of m-ZrO_2 by ≥50% compared to areas free of metal stains.

They also examined fifteen retrieved *delta* heads that had been in vivo for between two and sixteen months. The extent of transformation in these heads was greater than predicted by laboratory accelerated aging tests (cf., Fig. 2). Mechanistically, the transformation occurred due to the combination of stress and dissociation of water vapor at the metal-stained interface. An autocatalytic reaction generated an abundance of free oxygen and hydrogen ions which destabilized the *t*-ZrO_2. Their results suggested that the assessment of the surface instability of *delta* using a simplistic Arrhenius algorithm of hydrothermal activation was unrealistic [68]. These results dramatically showed increased *m*-ZrO_2 content in the near-surface regions of the ceramic at in vivo exposure times of less than one year; but it should also be pointed out that the highest stresses and greatest extent of metal contamination occurred on the internal taper of the head where it attaches to the femoral stem, not on its external articulation surface.

In 2020, Tateiwa et al. conducted a follow-up study that burst-tested a quantity of 24 Ø28mm L *delta* femoral heads after subjecting them to moderate to severe hydrothermal aging conditions [69]. The femoral heads were obtained from commercial sources in three distinct countries. Serial number analysis revealed that the samples originated from separate production batches, all manufactured during 2015. Shown in Fig. 5a are results for both pristine and treated heads using several hydrothermal aging conditions. The corresponding increase in the *m*-ZrO_2 phase content of these burst-tested heads as a function of depth is given in Fig. 5b.

Pristine femoral heads on CoCr trunnions exhibited a mean burst strength of 58 kN. Hydrothermal aging for 165 h resulted in a significant decrease (~14%) in burst strength (cf., Fig. 5a), coinciding with an increase in surface monoclinic content ranging from approximately 26% to greater than 40% (cf., Fig. 5b). Interestingly, the extent of transformation in the hydrothermally treated heads at a depth of 100 μm (20–25% *m*-ZrO_2) mirrored that of pristine heads at a depth of 0–10 μm (cf., Fig. 5b). This contrasts with the observation of only 10% *m*-ZrO_2 at 100 μm in

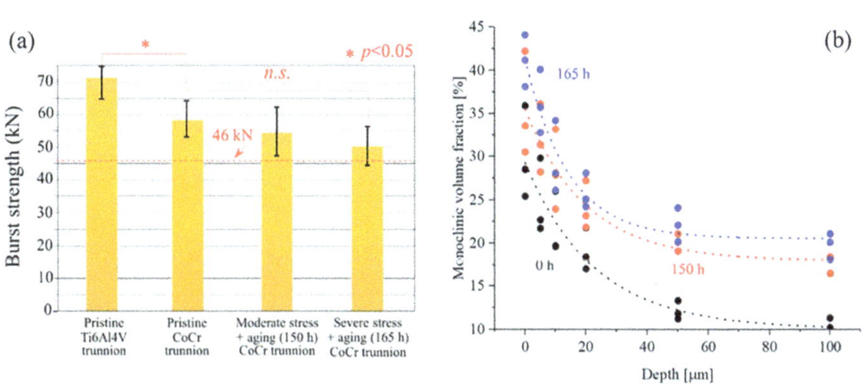

Fig. 5 (**a**) Burst-strength results obtained on a series of *delta* femoral heads; and (**b**) in-depth profile of monoclinic volume fraction as a function of hydrothermal aging, as measured by Raman spectroscopy using the Karagiri equation. (Used with permission under unrestricted access Creative Common CC BY license [69])

pristine heads, suggesting a progressive degradation mechanism that goes beyond the surface. This aligns with the understanding that BIOLOX®*delta*'s composition (designed at ~17 vol.% t-ZrO_2) sits slightly above the theoretical percolation threshold of 16 vol.%. Above this threshold, a continuous network of t-ZrO_2 grains facilitates the propagation of the $t \rightarrow m$-ZrO_2 transformation [70, 71]. To further inhibit the transformation, the size and uniformity of the t-ZrO_2 dispersoids must be carefully controlled. The presence of large dispersoids, aggregates, pores, or other inhomogeneities in the microstructure will be detrimental. However, note that even the most compromised burst-test loads surpassed the minimum FDA requirement of 46 kN (cf., Fig. 5a), highlighting the effectiveness of the processing and quality control employed during the production of this unique bioceramic.

A subsequent publication by Porporati et al. [72] contested the foregoing results. In an article titled, "Is Surface Metastability of Today's Ceramic Bearings a Clinical Issue?" they argued that the Raman spectroscopy methods overestimated the extent of the observed phase transformation. While acknowledging the occurrence of some transformation, Porporati et al. contended that the amount of m-ZrO_2 formed was negligible from a safety perspective. They further asserted that "the monoclinic content per se should not be considered as an indicator of degradation." They did not attempt to replicate the in vitro metal staining or burst-test studies. They only performed static autoclave tests and then used alternative analytical tools (i.e., XRD) and different equations to assess the m-ZrO_2 content. They also presented an analysis on only one retrieved ceramic-on-ceramic femoral head (12-y in vivo) that had evidence of metal staining. For that head, they concluded that mechanical shock was responsible for transformation as opposed to any autocatalytic effect. Furthermore, they argued that the amount of ZrO_2 that transformed in that head was $\leq 1\%$. Their answer to their article's title question was implicit: No, the surface phase transformation of *delta* is not a clinical concern. Nevertheless, it should be noted that CeramTec specifically forbids the use of steam sterilization for their *delta* components, precisely to avoid premature phase transformation and any associated possibility of early failures. As stated in their clinical FAQs, "Re-sterilization by hospitals is always prohibited (*e.g.*, especially autoclave, steam sterilization)."[1]

So, is there clinical evidence of early failures of *delta* due to hydrothermally induced degradation? A recent review of an Italian joint replacement registry for the survivorship of 29,495 *delta* hip prostheses found only two atraumatic in vivo head fractures (0.007%). While this is remarkably low, it should be noted that the average follow-up period was only 5.2 years. The authors performed detailed SEM and EDS fractographic examinations of one of the failures and expressed their concern that "these clinical outcomes suggest the occurrence of a more complex failure mode, other than failure related to concurrent or previous remarkable trauma associated with impact loading of the prosthetic joint." In fact, based on their analysis, they speculated that suboptimal seating of the ceramic head may have led to a "multi-

[1]Frequently Asked Questions, https://www.ceramtec-medical.com/en/infocenter/faq, Accessed, Oct. 12, 2023.

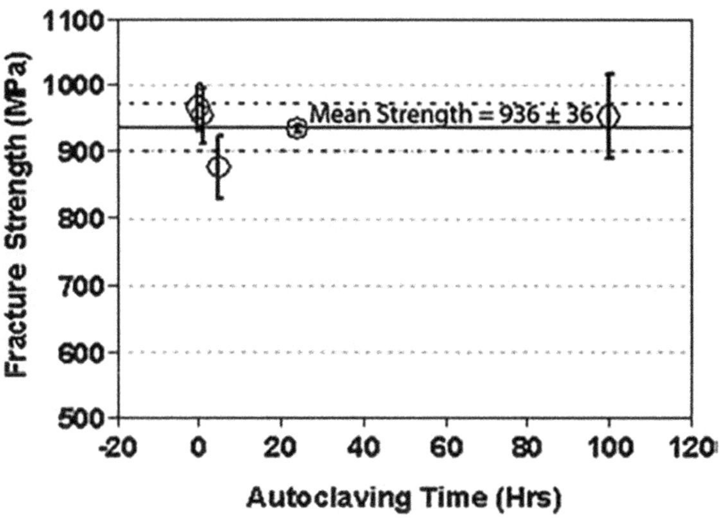

Fig. 6 Strength retention of Si_3N_4 after hydrothermal aging. (Used with permission [81])

stage crack growth mechanism," and resultant fracture [73]. Indeed, slow-crack growth over an extended period always ends in catastrophic failure. It may take 15–25 years for its occurrence, but it is inevitable (cf. dotted line in Fig. 2). Given that *delta* has only been available on the market for about 20 years, one might predict increasing numbers of atraumatic fractures in vivo, and recent reports appear to support this possibility [73–80].

In comparison to *delta*, MC^2 is a phase-stable material. It does not rely on phase transformation for its toughening mechanism. Indeed, its toughness is based on its fibrous microstructure which provides interlocking grains to hinder crack propagation. As shown in Fig. 6, the strength of Si_3N_4 is not affected by extensive static hydrothermal aging (i.e., 121 °C, 1 atm, 100 h) [81]. This is not to proclaim that Si_3N_4 is free from moisture-assisted slow-crack growth (SCG). Several studies have shown that Si_3N_4 is sensitive to low-temperature stress-induced damage associated with moist environments [82–86]. This is due primarily to an intragranular glass hydrolysis reaction. These studies found the dissolution kinetics and SCG to be most pronounced in the glass phase, but any degradation of mechanical properties was mitigated because of Si_3N_4's interlocking grain structure.

Unfortunately, there are no comparative published SCG studies on *delta* or MC^2. But in 2011, Chevalier et al. examined several biomedical ceramics in simulated body fluids [87]. Although the materials were not identified by their tradenames, the study included Al_2O_3, ZrO_2, a conventional zirconia-toughened alumina, ZTA (similar, if not identical to *delta*), and an experimental ZTA. The results of their investigation are shown in Fig. 7a. They found that Al_2O_3 and ZrO_2 were inferior to the conventional (microcomposite) and experimental (nanocomposite) ZTAs. The threshold values for SCG initiation were about 2.5 and 2.8 $MPa{\cdot}m^{1/2}$ for Al_2O_3 and

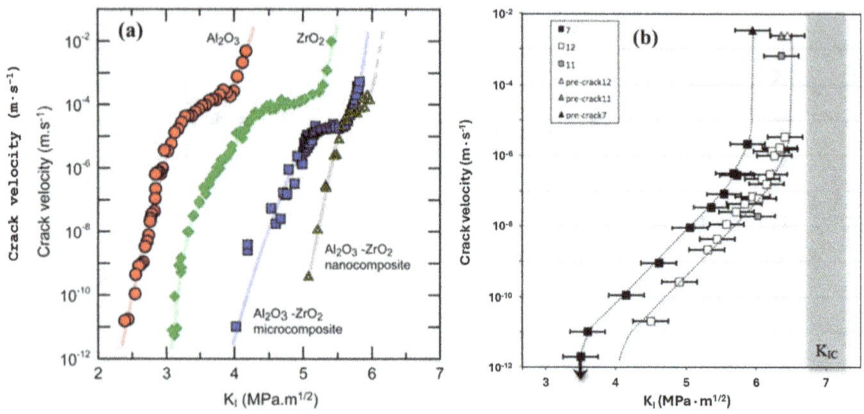

Fig. 7 Slow crack growth (V-K) diagrams for biomedical ceramics including (**a**) Al_2O_3, ZrO_2, a ZTA (microcomposite, similar to BIOLOX®*delta*) and an experimental ZTA (nanocomposite) [87]; and (**b**) *MC*^{2}®Si_3N_4, [22]. (Figures modified, combined, and used with permission)

ZrO_2, respectively. The conventional and experimental ZTAs had values of ~4.0 and ~5.0 MPa·m$^{1/2}$, respectively. The SCG fast fracture toughness limits of the four ceramics were about 4.2, 5.5, 6.0, and 6.2 MPa·m$^{1/2}$, respectively [87]. Thus, the ZTAs were more than twice as resistant to SCG as Al_2O_3 or ZrO_2.

Using the identical method, Martin et al. completed a dynamic fatigue test of *MC*^{2}®Si_3N_4 specimens in simulated body fluid for approximately one month [22]. The results of their work are presented in Fig. 7b. The data suggested that Si_3N_4 had comparable SCG characteristics to the microcomposite ZTA from Chevalier's 2011 publication [87]; but threshold values for initiation of SCG in *MC*^{2} were slightly lower than the microcomposite (i.e., 3.5 to 4.0 MPa·m$^{1/2}$), while the SCG fast fracture limit was ~6.5 MPa·m$^{1/2}$ (i.e., slightly higher than the micro- or nano-composite ZTAs). These results show that *MC*^{2} and *delta* exhibit similar SCG characteristics but by different mechanisms. SCG in Si_3N_4 occurs in the intergranular glass due to a moisture-assisted chemical attack of Si-O and Si-N bonds. Crack propagation is mechanically restrained by the ceramic's interlocking fibrous microstructure. Under increasing stress, SCG proceeds by debonding and fracturing the bridging grains [86, 88, 89]. In *delta*, intergranular SCG also occurs. It is initially constrained by the presence of *t*-ZrO_2 particles which induce localized compressive stresses on pre-existing microcracks. However, these microcracks provide channels for moisture diffusion. Water radicals can penetrate, react, and destabilize the *t*-ZrO_2 lattice. Its transformation to *m*-ZrO_2 results in a volume expansion that generates additional localized microcracks which provide pathways for water diffusion and further transformation. The amount, size, and dispersion uniformity of the *t*-ZrO_2 particles within its Al_2O_3 matrix govern the material's stability [7, 8, 70].

Although there is no evidence to support the clinical effectiveness of Si_3N_4 in THA, *delta* continues to gather significant patient statistics. A recent review of Norwegian hip replacement patients found that *delta* femoral heads were much less likely to fracture than BIOLOX®*forte* (Al_2O_3) [90]. Between 2005 and 2017, there was one *delta* fracture in 7901 implantations (0.0127%), compared to 34 *forte* fractures in 25,678 implantations (0.132%). The median follow-up time was 6.3 years. The study found that *forte* femoral heads were 12.2 times more likely to fail than *delta* heads (95% confidence: 4.0 to 37.6, $p < 0.001$). This is a commendable improvement, but not surprising given their respective K_I values for SGC initiation (cf., Fig. 7a). However, the researchers concluded their report with this cautionary statement: "…the long-term clinical … performance of *delta* bearings has not been extensively studied and should be studied further" [90]. Hopefully, this same recommendation will eventually be applied to MC^2®Si_3N_4 bearings.

2.6 Hardness

BIOLOX®*delta* and MC^2®Si_3N_4 differ in their measured hardness values (i.e., 17.6 to 19.8 GPa versus 14.4–15.9 GPa, respectively, cf., Table 1). While the major reason for this difference is their intrinsic chemical compositions and composite structures, there are minor contributions due to grain size and toughness. For *delta*, the Al_2O_3/Cr_2O_3 matrix primarily controls hardness. It comprises ~80 vol.% of the microstructure with *t*-ZrO_2 and $SrAl_2O_4$ at 17 vol.% and 3 vol.%, respectively. It has been speculated that the minor addition of Cr_2O_3 in *delta* increases its hardness, but Kuntz et al. completed a parametric compositional study showing that less than 0.5 wt.% Cr_2O_3 had no effect [33]. However, they noted a measurable reduction in hardness (~1.9 GPa) associated with an increase in grain size as shown in Fig. 8a. The range of observed hardness values for standard production of

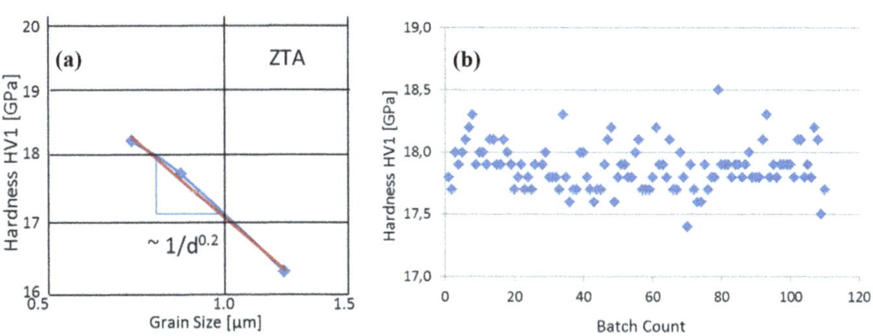

Fig. 8 Hardness measurements of BIOLOX®*delta*: (**a**) The effect of grain size on hardness with curve fitting; and (**b**) Hardness values obtained from routine quality control of regular powder batches in 2011–2014, $n \approx 110$. (Used with permission by Creative Commons CC-BY license [33])

BIOLOX®*delta* are reproduced in Fig. 8b. They demonstrate a tight distribution and excellent process control over the material.

For MC²®Si_3N_4, its ß-phase crystal structure is responsible for its hardness. It comprises about 93.6 vol.% of the microstructure. The glassy grain-boundary phase is only about 6.4 vol.%. Similar to *delta*, the hardness of Si_3N_4 varies with grain size [91, 92]. Mukhopadhyay et al. developed the following empirical linear relationship for hardness versus grain size from several hot-pressed and sintered Si_3N_4 and SiAlON ceramics [91]:

$$H = \left[7.2 + 8.9(\mathrm{GS})^{-0.5}\right] \pm 0.14\ \mathrm{GPa} \tag{5}$$

where H = hardness and GS = average grain size. For grain sizes in the range of 1.5 μm to 4 μm, their analysis indicated that doubling the average grain size resulted in a ~14% reduction in hardness. This translated to a decrease from 14.3 GPa to 12.1 GPa ($\Delta H = -2.2$ GPa) for a grain size increase of 1.6 μm to 3.2 μm, respectively. Notably, this change in hardness is similar in magnitude to the observed reduction in the hardness for *delta* under a comparable grain size increase (cf., Fig. 8a). Production batches of *MC²*®Si_3N_4 have typical hardness values of 15.12 ± 0.25 GPa ($n = 132$) for a 6σ statistical range of between 14.37 to 15.87 GPa (cf. Table 1).

Finally, the material's toughening mechanisms also influence its hardness. Hardness values are obtained through standardized testing methods like ASTM C1327 or ISO 14705. These techniques involve Vickers or Knoop indentation on polished sample surfaces, followed by an analysis of the penetration depth area. In toughened ceramics, such as BIOLOX®*delta* and *MC²*®Si_3N_4 composites, the presence of "plastic process zones" and R-curve behavior around the indenters restricts their penetration depth. This limitation leads to higher measured hardness values compared to predictions based on a simple rule of mixtures approach [61, 93].

Does the observed difference in hardness between *delta* and *MC²* have implications for their clinical performance in THA? Higher hardness has been assumed to translate to a smoother articulation surface, improved wettability, and better scratch resistance, reducing the frictional wear of the mating HXLPE liner. Compared to traditional CoCr alloys used in hip arthroplasty with a hardness of 3–4 GPa [43], both *delta* (17.6–19.8 GPa) and *MC²* (14.4–15.9 GPa) are significantly harder. Early in vitro simulator studies suggested that ceramics perform better than CoCr in minimizing polyethylene wear [94, 95], and recent rigorous systematic reviews and meta-analyses appear to confirm this finding [96, 97].[2] These results suggest

[2]The review by Mertz et al. is particularly instructive [97]. It included 36 studies of patients that collectively received 1657 CoCr or 659 BIOLOX®*delta* heads along with HXLPE liners at a mean follow-up of 7 years. Using a pooled weighted mean analysis method, the HXLPE wear rates were 0.063 mm/year for CoCr versus 0.047 mm/year for *delta* ($p < 0.01$). A sub-group meta-analysis of four included studies that directly compared CoCr and *delta* demonstrated 0.029 mm/year more wear for CoCr couples, although this result was not statistically significant.

that femoral head hardness is a factor in HXLPE wear, but the difference between *delta* and MC^2 is minimal compared to the difference between *delta* and CoCr. Indeed, a hip simulator study directly comparing *delta* and MC^2 heads articulating against HXLPE liners showed no statistically significant difference in wear [98] (cf., Sect. 4).

Enhanced scratch resistance is the more important outcome of using high-hardness ceramics. In a hip simulator test, Lee et al. demonstrated that femoral heads made from BIOLOX®*delta* were superior to CoCr. They found that *delta* reduced femoral head damage by 97%, with a corresponding equivalent reduction in XLPE wear [99]. In vivo, femoral heads can be scratched due to subluxation, dislocation, or hard foreign third-body particles within the artificial joint (e.g., bone, cement, or metal) [100, 101]. Consequently, in qualifying new prosthetic joint materials, the US FDA requires third-body wear testing of articulation couples in a hip simulator according to ISO 14242-1,2.[3]

Although there are no direct comparative scratch tolerance tests on BIOLOX®*delta* and MC^2®Si_3N_4, there are established predictive models based on material properties. Evans and Wilshaw [102], and later, Evans and Marshall [103, 104] proposed the following relationships between damage induced on the surface of a ceramic material and its intrinsic hardness and fracture toughness, respectively:

$$\ddot{V}_i \propto \left(P_i^{5/4}\right)/\left[\left(H_v^{1/2}\right)\left(K_{IC}^{3/4}\right)\right] \tag{6}$$

$$\ddot{V}_i \propto \left(P_i^{9/8}\right)(E/H_v)^{4/5}/\left[\left(K_{IC}^{1/2}\right)\left(H_v^{5/8}\right)\right] \tag{7}$$

where $\ddot{V}_i$ = the volume of material removed from the surface under an abrasive particle, P_i = load applied to the abrasive particle, H_v = Vickers hardness, K_{IC} = Fracture toughness, and E = Elastic modulus. Substituting the range of reported property values for *delta* and MC^2 (cf., Table 1) into these equations and performing sensitivity analyses suggests no differences in damage tolerance between *delta* and MC^2. Both materials should perform equally well under the conditions of third-body wear.

2.7 Other Comparative Property Assessments

It is worth noting that other investigators have completed physical and mechanical property comparative assessments of various Si_3N_4 materials (albeit not

[3]Personal communication between T. Shepherd (FDA) and B. J. McEntire (SINTX Technologies), dated Aug. 13, 2015.

Table 5 Mechanical and wear properties of three different Si_3N_4 ceramics compared to reference values obtained from BIOLOX®*delta*

Property	Material			
	Si_3N_4 – A	Si_3N_4 – B	Si_3N_4 – C	BIOLOX®*delta*
Flexural strength (MPa)	1104 ± 95	965 ± 60	943 ± 74	1390
Fracture toughness (MPa·m$^{1/2}$)	5.6 ± 0.0	5.5 ± 0.1	5.7 ± 0.1	6.4
Vickers hardness (GPa)	15.6 ± 0.2	18.9 ± 1.0	15.3 ± 0.5	19.1
Wear (mm^3)	135.72 ± 7.17	~100 ± 15*	~100 ± 30[a]	0.05 ± 0.01

[a]Estimated from graphs. Numerical statistics were not provided. Si_3N_4-A, B, and C were optimized for strength, hardness, and toughness, respectively [23]

MC^2®Si_3N_4) against BIOLOX®*delta*. For instance, Häußler et al. examined three Si_3N_4 compositions that were optimized for either strength, hardness, or toughness [23]. After evaluating their basic mechanical properties, they completed a 100 h ring-on-disc wear test (i.e., ceramic-on-ceramic) of these compositions matched against identical BIOLOX®*delta* samples. The results of these tests are given in Table 5. Although the investigators acknowledged that the three Si_3N_4 materials achieved good results in terms of mechanical properties, the Si_3N_4 data were inferior to BIOLOX®*delta*. The wear test data also showed *delta*'s superiority, although the Si_3N_4 samples exhibited significant edge chipping during the study, which may have been specific to this test set-up. Nevertheless, all of the *delta* components were free of this type of defect. Without providing a complete set of statistical details, the investigators claimed that *delta* was superior to the three Si_3N_4 materials in every aspect ($p < 0.05$).

A group of Chinese researchers recently completed a comprehensive in vitro and in vivo (animal model) investigation of Si_3N_4 for arthroplasty purposes [105], and compared their results to the BIOLOX®*delta* material standard (i.e., ISO 6474-2.2). The Si_3N_4 they used in their investigation was produced by Black Ceramic® Medical Technology Co., Ltd. (BCM Si_3N_4, Suzhou, China). Nine in vitro and five in vivo tests were conducted including an evaluation of mechanical properties and a biocompatibility assessment per ISO 10993. They also conducted a three-month hip implantation study in minipigs. Results of their comparative mechanical properties assessment are provided in Table 6.

Their data indicated that the BCM Si_3N_4 had superior mechanical performance to BIOLOX®*delta* Type S and was only inferior to Type X on flexural strength. Additionally, each of the ISO 10993 tests demonstrated acceptable biocompatibility. The minipig hip implantation test was also successful. The animals were ambulated and functioned well during the three observational months. CT scans showed the implants remained well positioned, blood tests were normal, and pathologies were negative. The researchers concluded that "silicon nitride has the right combination of high strength, high hardness, excellent fracture toughness, lower density, better wear resistance, good biocompatibility, inherent stability, corrosion resistance and bioactivity, and bone integration capability, all of which are desirable in medical implants" [105].

Table 6 Comparative mechanical property assessment of BCM Si_3N_4 and the BIOLOX®*delta* ISO 6474-2 Standard

Property or observation	BCM Si_3N_4	BIOLOX®*delta* per ISO 6474-2 - Type S	BIOLOX®*delta* per ISO 6474-2 - Type X
Elastic Modulus per ISO 17561 (GPa)	299 ± 3	≥ 320	≥ 320
Flexural strength per ISO 14704 (MPa)	796 ± 53	≥ 750	≥ 1000
Flexural strength after accelerated Aging per ISO 14704 (MPa)	788 ± 44	≥ 750	≥ 1000
Fracture toughness per ISO 21618 ($MPa{\cdot}m^{1/2}$)	6.25 ± 0.10	≥ 3.5	≥ 4.0
Fracture toughness after accelerated aging per ISO 21618 ($MPa{\cdot}m^{1/2}$)	6.26 ± 0.07	N.R.	N.R.
Hardness per ISO 14705 (GPa)	14.51 ± 0.13	≥ 15.5	≥ 16.0
Hardness after accelerated aging per ISO 14705 (GPa)	14.43 ± 0.17	N.R.	N.R.
Cyclic fatigue per ISO 22214 for 10 MM Cycles	No failures for n = 5 @ 417 MPa	No failures for n ≥ 5 @ 300 MPa	No failures for n ≥ 5 @ 400 MPa
Cyclic fatigue after accelerated aging per ISO 22214 for 10 MM cycles	No failures for n = 5 @ 417 MPa	No failures for n ≥ 5 @ 240 MPa	No failures for n ≥ 5 @ 320 MPa

N.R. Not reported or not required [105]

3 Surface Chemistry

The concept of a truly bioinert ceramic remains elusive and oxymoronic [106]. While all materials elicit some degree of response from the immune system, the level and type of interaction at the implant interface differ. The immune system prioritizes removing or isolating foreign objects, with dissolution being the preferred outcome, followed by encapsulation. Only if deemed immunocompatible, will integration into tissues occur [43]. Consequently, the interfacial chemistry of a bioceramic plays a critical role in determining its success, influencing both cellular response and implant fate. This principle applies to all bioceramics, including BIOLOX®*delta* and *MC*^{2}®Si_3N_4.

For instance, when *delta* comes into contact with bodily fluids, hydrolysis forms various aluminum moieties. Depending on their concentration, they induce an inflammatory response from macrophages which in turn generate reactive oxygen species (ROS) [107]. While the biological intent of ROS is to destroy the foreign material, it reduces local pH and can damage adjacent tissue [108]. Si_3N_4 also hydrolyzes, but it releases reactive nitrogen species (RNS) instead of ROS and facilitates the formation of nitric oxide (NO) [109]. While experimental evidence suggests that NO has a catabolic function in the joint, there is also evidence that NO and its derivatives may play a beneficial role in chondrocytes. Additionally, NO has favorable functions in other tissues, and these activities could also potentially occur in chondrocytes [108]. RNS can scavenge ROS from the local environment and neutralize hydrons (protons). This is a unique therapeutic property of Si_3N_4 that does not exist with oxide bioceramics [110]. Si_3N_4's ability to eliminate ROS, neutralize hydrons, and enhance local pH, makes it a potential therapeutic material for osteoarthritis and other conditions caused by oxidative stress [111].

The surface chemistry of biomaterials like *MC*^{2} and *delta* can influence prosthetic joint function through several key factors, including inflammatory response, solubility, wear resistance, and bacterial adhesion (cf., Chaps. 4, 5, and 6 for additional details). Diseased joints are characterized by acidic synovial fluid compared to the homeostatic pH of healthy joints [112, 113]. This is attributed to the release of reactive oxygen species (ROS) and acids by macrophages during inflammation, a common feature of joint disease. ROS are known to have detrimental effects on cells and contribute to chondrocyte death [108]. Si_3N_4 exhibits beneficial effects by buffering the synovial fluid's pH, whereas Al_2O_3 (e.g., *delta*) tends to promote a more acidic environment. Consequently, Si_3N_4 may mitigate inflammation, while Al_2O_3 could potentially make it worse.

Additionally, $Al(OH)_3$, the primary surface dissolution product of Al_2O_3 at physiological pH, is essentially insoluble. $Al(OH)_3$ accumulates within the joint capsule alongside mechanically generated wear debris throughout the implant's lifespan. The presence of these combined foreign materials triggers the aforementioned immune response, potentially leading to osteolysis and aseptic loosening [114]. While all prosthetic components generate wear particles, Si_3N_4 may offer an advantage. At physiological pH, its dissolution products and wear debris convert into soluble and resorbable moieties, namely NH_4^+ and $Si(OH)_4$ [115–117]. The body can eliminate excess amounts through regular renal function.

4 Wear Performance

The surface chemistry of the femoral head can also significantly influence the wear performance of hip prostheses, particularly when paired with polyethylene liners. This effect was demonstrated in laboratory studies where different oxide and *MC*^{2} ceramic femoral heads were evaluated in combination with advanced polyethylene liners [118, 119]. The testing involved wear simulation after hydrothermal aging in an autoclave. Notably, zirconia-toughened-alumina (ZTA, i.e., *delta*) induced a substantial increase in both polymer crystallinity and oxidation. These combined chemical and physical alterations contribute to brittleness within the polymer, potentially leading to accelerated wear over time. Conversely, *MC*^{2} exhibited a protective effect on the polyethylene liner by scavenging free oxygen from the tribological interface.

Subsequent research provided further evidence to substantiate this observed effect [120]. Figure 9a–d summarize a basic in vitro experiment designed to assess

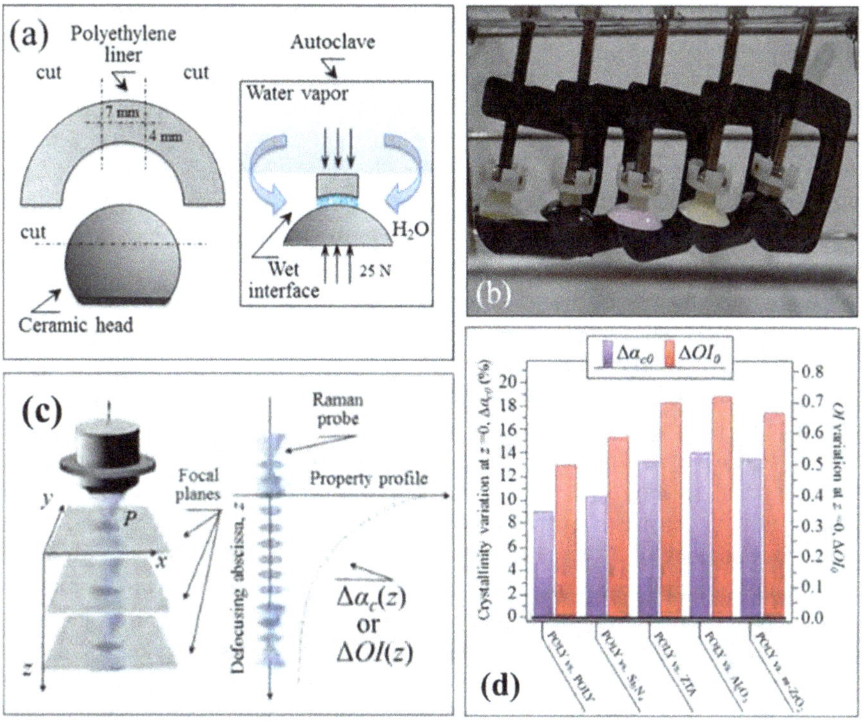

Fig. 9 (**a**) Schematic of the sample preparation procedure including sectioning and coupling of UHMWPE/ceramic couples in an autoclave operating in a steam environment; (**b**) a picture of a set of coupled samples as assembled before insertion into the autoclave; (**c**) in-depth confocal scanning procedure for characterizing the sub-surface profiles in the UHMWPE after hydrothermal testing; and (**d**) crystallinity and oxidation values from the UHMWPE samples coupled with different ceramic counter faces. (Used with permission [120])

the impact of coupling different ceramic femoral head materials with an advanced, highly crosslinked polyethylene liner (X3™, Stryker Orthopedics). Sections of alumina (BIOLOX®*forte*), zirconia-toughened-alumina (BIOLOX®*delta*), *m*-ZrO_2 (OXINIUM™), and Si_3N_4 (MC^2) femoral heads were paired with the liner under a 25 N static load. These samples were then subjected to steam autoclave conditions (121 °C, 1 atm, 24 h). Following drying, the polyethylene samples were analyzed using Raman spectroscopy to detect detailed chemical and physical alterations. The findings from this brief in vitro test revealed that the most significant polyethylene oxidation and embrittlement occurred with the oxide ceramic heads, while the MC^2 group exhibited the least degradation, comparable to the control samples. This short-term experiment offered insight into the potential long-term in vivo service of these articulation couples. The results suggest that MC^2 may be a more favorable femoral head material for extending the functional lifespan of advanced polyethylene liners.

These observations may explain the slightly lower wear rates observed for advanced polyethylene liners paired with MC^2 and *delta* in the only available hip simulator study directly comparing the two bioceramics [98]. *MC*^{2} consistently generated lower polyethylene wear throughout the study, with the difference becoming more pronounced after approximately 750,000 cycles. While the wear rate disparity between *MC*^{2} and *delta* did not reach statistical significance, it aligns with the trends observed in the aforementioned short-term laboratory tests. Collectively, these findings suggest that *MC*^{2} femoral heads exhibit at least equivalent wear performance to *delta* in the short term. Yet, its potential to mitigate polyethylene oxidation and embrittlement suggests *MC*^{2} may provide lower wear in the long term. (cf., Chap. 10 for additional details.)

The performance of *MC*^{2} and *delta* in ceramic-on-ceramic (CoC) bearings presents another crucial consideration. While *delta* holds a well-established market position, concerns exist. A retrospective review reported a 6.2% revision surgery rate for BIOLOX®*forte* and *delta* CoC hips from various causes, with a 3% incidence of bearing noise (squeaking) at 10 years [121]. A contemporary study comparing CoC and ceramic-on-polyethylene (CoP) devices over 13 years found similar overall revision rates, but a 6% squeaking incidence in CoC bearings, and significantly higher patient satisfaction with CoP [122]. This finding contrasts with initial expectations of CoC superiority. However, higher costs, stricter surgical implantation requirements, and noise susceptibility of CoC bearings have contributed to a recent shift towards CoP prostheses. Bearing noise, traditionally reported at 0.7% to 20.9%, is believed to be underreported, and a major source of patient dissatisfaction [123]. Supporting this trend, the Orthopedic Network News annual report indicates that "hard-on-hard bearings" (including CoC) comprised less than 1% of all cases in 2020, compared to a peak of 43% in 2007 [124].

MC^2®Si_3N_4 is not recommended for use in CoC bearings [125], and BIOLOX®*delta* shouldn't be used for these devices either. However, it is not for the obvious reasons of cost, detailed surgical protocols, or noise. This is because the surfaces of both materials hydrolyze in vivo (cf., Sect. 3 and Chaps. 4 and 5). These hydrolysis reactions consume water by the following basic equations:

$$Si_3N_4 + 16H_2O \rightarrow 3Si(OH)_4 + 4NH_4OH, \Delta G_{298^\circ K} = -\ 1203.3\ \text{kJ/mol} \tag{8}$$

$$SiO_2 + 2H_2O \rightarrow Si(OH)_4, \Delta G_{298^\circ K} = 792.7\ \text{kJ/mol} \tag{9}$$

$$Al_2O_3 + 3H_2O \rightarrow 2Al(OH)_3, \Delta G_{298^\circ K} = 260.9\ \text{kJ/mol} \tag{10}$$

The hydrolysis of Si_3N_4 occurs spontaneously once its protective oxide coating (SiO_2) is removed. Its removal in CoC bearings is due to the cyclic breakdown of fluid film, the resumption of boundary lubrication, and abrasive wear between the mating components during the start–stop sequence which is part of normal human gait. At the immediate point of contact between the mating parts, a locally water-starved tribochemical reaction occurs, and its observable manifestations are stick-slip motion and squeaking [125].

An analogous tribochemical reaction occurs for self-mated Al_2O_3 (*forte*) or ZTA (*delta*) bearings, although frictional heating is required; yet the localized water consumption by the Al_2O_3 couple is less than for Si_3N_4. This is why Al_2O_3 couples outperformed *MC*^{2} bearings in a hip simulator under a non-standard stop-dwell-start sequence (cf., Chap. 10) [125]. This study modified the standard 24/7 continuous Paul orbital duty cycle by including a static dwell at maximum load. This change was hypothesized to be more clinically relevant because patients function quite differently, with intermittent, multidirectional, and varied motion. It was envisaged that a dwell at constant load would expel serum lubricant from the contact zone between head and liner pairs requiring subsequent acceleration to overcome unlubricated static friction, with the combined effect of adverse wear and noise for these hard-on-hard bearings. The results confirmed the hypothesis. During the continuous motion portion of the test— up to 3 million cycles—no significant differences in wear were noted between the bearing pairs. However, when the stop-dwell-start segment was introduced, all three bearing pairs showed increased wear, but the two *MC*^{2} pairs were worse than the Al_2O_3 couple.

5 Antipathogenic Performance

The surface chemistry of bioceramics plays a vital role in their resistance to pathogens. To date, approximately fifty studies have confirmed that *MC*^{2}®Si_3N_4 lyses or inactivates nosocomial bacteria [126–135], fungi [136, 137], and viruses [138], including several variants of SARS-CoV-2 [139–141], while three pertinent studies cite the antipathogenicity of BIOLOX®*delta* [142–144].

Piconi et al. examined three ceramics for their resistance to biofilm formation—BIOLOX®*forte*, BIOLOX®*delta*, and an unspecified Si_3N_4 against two common bacterial species—*S. epidermidis* and *E. coli* [142]. All three materials exhibited significantly reduced colonization compared to polystyrene controls (i.e., between 30% and 60% lower). Among the three bioceramics, only *forte* demonstrated significantly lower *S. epidermidis* biofilm formation ($p < 0.005$) compared to *delta* and Si_3N_4, both of which had similar behavior; and no significant differences in *E. coli* biofilm formation were observed between any of the ceramics.

Trampuz et al. subsequently conducted a multicenter retrieval study on patients undergoing revision surgery for either infection or aseptic loosening [143]. Retrieved implants underwent qualitative and quantitative microbiological analysis to determine the presence of retained bacteria. Notably, ceramic components (e.g., BIOLOX®*forte* and *delta*) exhibited significantly lower bacterial colony forming units (CFU) compared to retrieved metal and polymer components.

A study sponsored by CeramTec investigated *delta*, *forte*, a CoCr alloy, and an XLPE polyethylene for their ability to resist bacterial adhesion [144]. Two pathogens were selected, *S. aureus*, and *S. epidermidis*. For control, a lysing solution consisting of 10 vol.% sodium dodecyl sulfate and 50 vol.% dimethylformamide in distilled water was applied to each sample series. Measurements were made via optical density (OD). The viability index was obtained by normalizing OD data from the experimental material to those obtained from control specimens. Results are

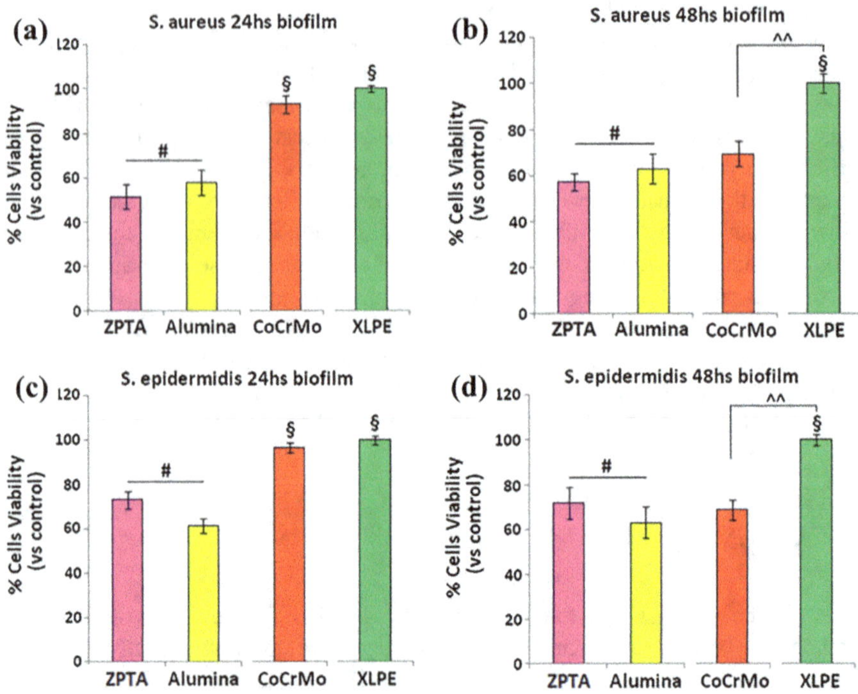

Fig. 10 *S. aureus* biofilm adhesion to orthopedic materials for (**a**) 24 h and (**b**) 48 h; and *S. epidermidis* biofilm adhesion to orthopedic materials for (**c**) 24 h and (**d**) 48 h. Percent viability is in comparison to lysed controls. ZPTA = BIOLOX®*delta*, Alumina = BIOLOX®*forte*, CoCrMo = Cobalt chromium alloy, and XLPE = Crosslinked polyethylene. (Used with permission [144])

presented in Fig. 10. Although d*elta* and *forte* supported lower quantities of viable cells than CoCr or XLPE, their amounts were by no means inconsequential, ranging from about 50% to 75% of the control groups. XLPE proved to be the worst biomaterial for biofilm formation with up to 100% viability for both pathogens. Unfortunately, this study did not include MC^2®Si_3N_4.

A separate contemporary publication compared the antibacterial properties of *delta* and MC^2 using *S. epidermidis*, with polyetheretherketone (PEEK) serving as a control group. Fluorescence imaging was used to assess absolute bacterial viability. Time points in the study included 8, 12, 24, and 48 h [131, 132]. The results of these tests are provided in Fig. 11. These data showed that MC^2 consistently and significantly resisted the formation of biofilms throughout the study in contrast to *delta* and PEEK. At 8, 12, and 24 h, the amount of observed biofilm on Si_3N_4 was about two orders of magnitude lower than *delta* (i.e., < 99%), whereas a three-order of magnitude difference was seen at 48 h (i.e., < 99.9%). Unlike the prior comparative studies, these findings suggest a reduced efficacy of *delta* in inhibiting biofilm formation compared to MC^2. However, *delta* still demonstrated superior performance to plastics and metals.

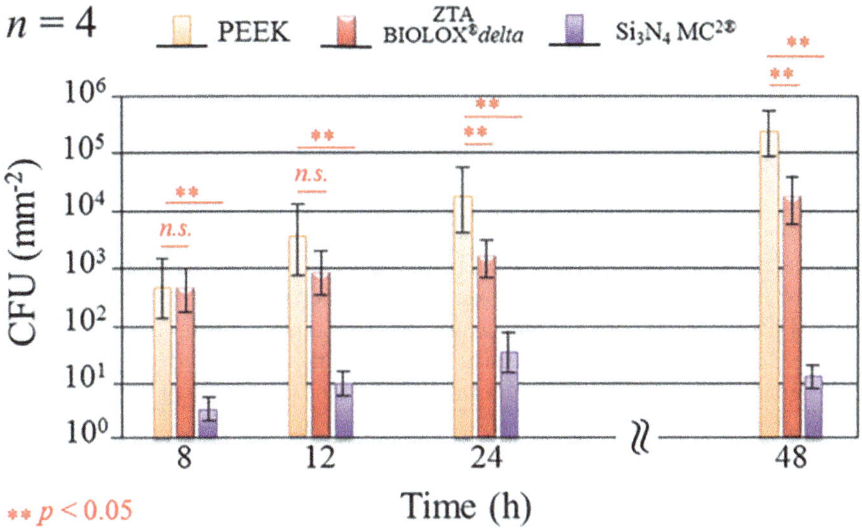

Fig. 11 Quantitative assessment of viable bacteria and its statistical validation based on the analysis of fluorescence micrographs covering the entire samples ($n = 4$ per substrate type). (Used with permission from the Royal Society of Chemistry under Creative Commons CC BY License [131, 132])

Periprosthetic joint infections (PJIs) are a serious complication after joint replacement surgery and can cause significant patient morbidity. More than 25% of joint replacement revisions are due to PJIs, and their incidence is expected to rise. This increase is partly due to an aging population, obesity, diabetes, other comorbiditics, and antibiotic resistance. One study suggests that the cost of treating PJIs will exceed $1.85 billion in the USA annually by the end of the current decade [145]. The contemporary therapy for PJIs typically involves a two-stage reoperation, including: (1) removal of the implant, debridement of infected tissue, application of antibiotics to the wound, systemically administered antibiotics, and implantation of a temporary antibiotic-loaded implant; and (2) replacement of the temporary implant with a new permanent device once the infection has been eradicated [146]. However, one of the major concerns with this protocol is increased antibiotic resistance.

Recognizing this challenge, MC^2®Si_3N_4 may have an impact on reducing cost and minimizing antibiotic resistance; but more importantly, its use in joint arthroplasty may improve patient recovery, preserve lifestyle, and increase longevity. While there are no clinical data to support its antipathogenic effectiveness in hip arthroplasty, its use as a spinal arthrodesis device may have predictive value [147]. The reported incidence of infection of MC^2®Si_3N_4 implanted devices up through 2020 is only 0.03%,[4] while the total incidence of periprosthetic spinal

[4]Personal Communication with Mr. Jason Parberry, Sr. Director for Quality Assurance and Regulatory Affairs, SINTX Technologies, Inc., Salt Lake City, UT USA, Oct. 27, 2023.

infections is typically between 1% and 4%, with some reviews indicating that it may be as high as 20% [148]. This contrasts with the reported annual incidence of PJIs in the United States of approximately 1–2% for total knee arthroplasty (TKA) and 0.5–1% for total hip arthroplasty (THA) [149].

Mechanistically, the same surface chemistry that was presented earlier (cf., Sect. 3) is responsible for *MC*^{2}®Si_3N_4's antipathogenic properties. Specifically, minute dissolution of Si_3N_4 releases nitrogen moieties—ammonia (NH_3) and ammonium (NH_4^+)—into the local environment, which are enzymatically converted to various reactive nitrogen species (RNS), including nitric oxide (NO) and peroxynitrite ($^-$OONO), which damage prokaryotic membranes and cellular DNA [131, 132]. Because this is a chemically based lytic mechanism, it is expected to reduce the chance of bacterial resistance. Given the growing demand for joint replacements by an aging population and the increased prevalence of joint infections along with antibiotic resistance, *MC*^{2}®Si_3N_4 is poised to provide cost and patient benefits related to its antipathogenic properties that may challenge the capabilities of BIOLOX®*delta*.

6 Mechanically Assisted Crevice Corrosion (MACC)

MACC is a phenomenon that occurs in modular total hip replacements. It is caused by corrosion at the taper junction of the stem and femoral head. It results in the release of metal ions and particulate debris into the surrounding joint space. MACC is a serious complication resulting in (1) osteolysis and implant loosening due to the necrosis of supporting bone structures, (2) adverse local tissue reactions (ALTR) which affect the soft tissues leading to pain, swelling, and the formation of pseudotumors, and (3) metallosis, which is the release of high concentrations of metal ions into the bloodstream causing heart, liver, and kidney comorbidities. This iatrogenic disease (also known as trunnionosis) was described and analyzed in detail by Gilbert et al. in 1993 [150]. Unfortunately, it can be asymptomatically present in all patients with Ti and CoCr stems or sleeves coupled to CoCr femoral heads. One study, at 10-year follow-up, indicated that 22% of patients had serum cobalt levels >1 ppb, which is consistent with MACC, and ALTR was symptomatically present in half of these individuals. Furthermore, 59% of those with documented MACC underwent revision surgeries [151].

Kurtz et al. (2013) investigated the occurrence of MACC in total hip arthroplasty using retrieved femoral heads [152]. Their study employed a matched cohort design, analyzing fifty metal and ceramic femoral heads each paired with either CoCr or Ti stems. Key results from their research are presented in Fig. 12a, b.

The ceramic explants of this study included five first-generation BIOLOX®, 30 second-generation BIOLOX®*forte*, and 15 third-generation BIOLOX®*delta* heads. The average implantation time for the ceramic heads was just 3.3 ± 3.7 years (range, 0.5 to 18 y), which was matched to the metal heads of 3.2 ± 3.8 years (range, 0.5 to 17 y), with no significant difference. There were no

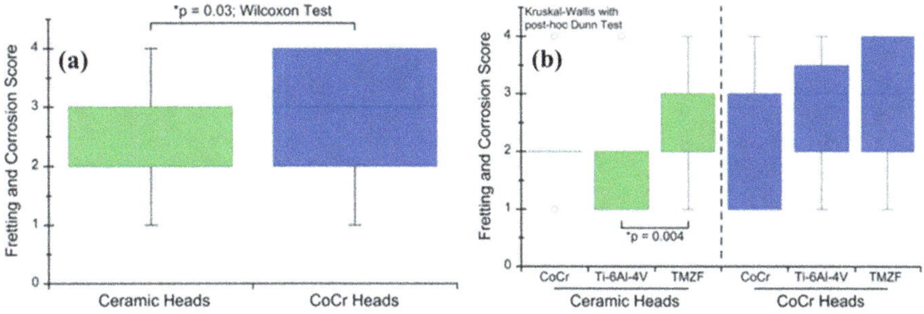

Fig. 12 (**a**) Femoral stem taper fretting and corrosion scores for matched ceramic and CoCr head cohorts. Although damage scores were significantly lower for the ceramic cohort ($p = 0.03$), they were not eliminated; and (**b**) box plot showing femoral stem taper fretting and corrosion score versus stem alloy for the ceramic and metal head cohorts. Titanium stems were worse than CoCr, and a proprietary Ti alloy (TMZF) was worse in both cohorts. (Used with permission [152])

other differences in patient characteristics except for age. Patients receiving ceramic heads were younger (52 ± 10 y) than those implanted with metal heads (57 ± 14 y, $p = 0.03$).

The extent of MACC was evaluated using a previously developed four-point scale. A score of 1 denoted minimal damage (fretting on less than 10% of the surface and no corrosion). Scores of 2, 3, and 4 indicated progressively more severe damage, ranging from mild (fretting on greater than 10% of the surface and/or localized corrosion) to moderate (fretting exceeding 30% of the surface and/or aggressive corrosion with debris) to severe (fretting on over 50% of the surface with extensive corrosion and debris). The investigators concluded that ceramic femoral heads were superior to metal ($p = 0.03$).

Using the matched cohorts of CoCr stems and ceramic femoral heads from the foregoing study, a quantitative method was developed to estimate volumetric material loss from each head and stem. Then, the amount of corroded metal released by each group was normalized to their short in vivo service lives [153]. As shown in Fig. 13, the cumulative volumetric material loss for the ceramic cohort had a median of 0.0 mm^3 per year (range, 0.0–0.4 mm^3) and the CoCr cohort had a median of 0.1 mm^3/year (range 0.0–9 mm^3). In their publication, the only factor that was found to be associated with decreasing material loss from taper junctions was femoral head material. The researchers concluded that there is "evidence that ceramic heads do not appear to have the same level of stem or bore fretting-corrosion concerns as with metallic heads" [153].

A subsequent in vivo study by Eichler et al. (2021) examined fretting corrosion of Ti stems assembled to large-diameter BIOLOX®*delta* heads (Ø36 to Ø48 mm) paired with *delta* acetabular liners [154]. For femoral heads greater than Ø44 mm, an additional Ti adapter sleeve was used between the stem and head (i.e., BIOLOX®*Option*). The investigators monitored serum Ti, Co, and Cr levels as markers of fretting corrosion between the components in a cohort of 57 patients over five years who underwent primary THA at a single institution. They found no

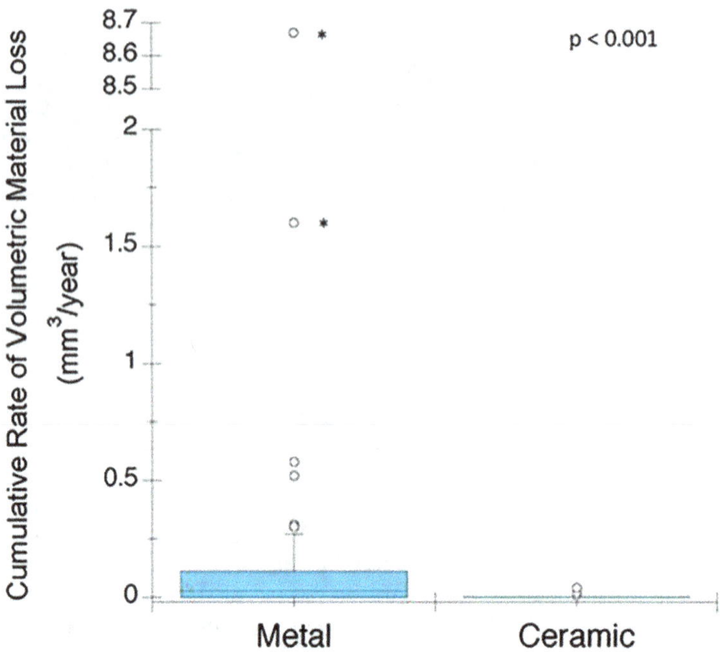

Fig. 13 Box plot showing the rate of material loss from the metal and ceramic cohorts. The median and the maximum values seen for the CoCr cohort (median = 0.1 mm^3, maximum = 9 mm^3) are an order of magnitude greater compared with the ceramic cohort (median = 0.0 mm^3, maximum = 0.4 mm^3). (Used with permission under CC BY Open Access license [153])

changes in Co or Cr serum levels and all patients had Ti concentrations below the recommended safety level of 10 μg/L. Indeed, in a subgroup analysis of 25 patients who had previous measurements, they recorded a decline in Ti levels between the 20th and 78th month. Furthermore, there were no statistical differences in Ti concentrations related to bearing size or the presence or absence of the Ti adapter sleeve. The investigators concluded that "there is no important wear and corrosion at the taper junction," although they conceded that trunnionosis was likely only a CoCr disease. Several other limited retrieval studies have confirmed these findings [155–157]. These results demonstrate that Al_2O_3-based femoral heads have a reduced incidence of MACC and ALTR compared to metal heads. Additionally, the selection of the stem and sleeve materials is crucial in mitigating this disease. Evidence suggests that Ti stems and sleeves are preferable interfacial components when paired with BIOLOX®*delta* heads.

There are no in vivo MACC studies on Si_3N_4, and no in vitro studies directly comparing *MC*^{2} and *delta*. There have been only two in-depth in vitro studies—one for each material—and the tests are not directly comparable. However, the results may provide an interesting perspective. A 2012 fretting corrosion study examined the effect of adding Ti sleeves (i.e., BIOLOX®*Option*) to large diameter *delta* heads

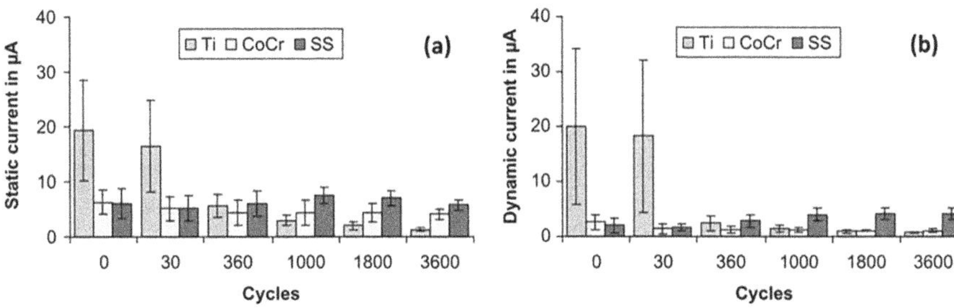

Fig. 14 (**a**) Static and (**b**) dynamic currents of fretting corrosion as a function of the testing cycle for Ø60 mm BIOLOX®*delta* femoral heads affixed to Ti sleeves (BIOLOX®*Option*) and Ti, CoCr, and SS trunnions. (Used with permission [158])

(i.e., up to Ø60 mm) which were affixed to either Ti, CoCr, or steel trunnions [158]. There were no separate materials used as controls in this study, and they did not examine the fretting of the trunnion materials directly against *delta* heads. However, they did measure static and dynamic fretting currents for the three trunnion types after seating the heads at a compression load of 4.5 kN and exposing them to a simulated corrosive salt solution buffered with lactic acid to a pH value of 4.5. Loads during testing varied between zero and 3 kN. The results of these tests are provided in Fig. 14a, b. These data indicate that there were high initial corrosion currents for the Ti trunnions (i.e., up to 20 µA), but much lower values for the CoCr and SS trunnions (ranging between ~1 and ~5 µA for static or dynamic loads). The high currents are the result of the initial repassivation of the oxide coating on the Ti trunnions that was disrupted during seating. The SS trunnions started with a low current, but it increased over time, due to the build-up of acidic pockets that accelerated corrosion. The investigators recommended the use of titanium alloy adapter sleeves because they exhibited only small fretting corrosion and wear; and they concluded that "our results demonstrate that large ceramic heads … with a metal adapter sleeve do not affect corrosion of modular taper connections" [158].

For *MC*^{2}®Si_3N_4, a single 2020 study compared *MC*^{2} and CoCr femoral heads affixed to Ti trunnions without sleeves [159]. In a similar electrochemical setup as the foregoing *delta* study, researchers measured short- and long-term fretting currents along with onset loads for corrosion. Matched Ø32 mm +8 mm offset 12/14 taper femoral heads were affixed onto Ti trunnions under 4 kN loads, then assessed under static loads at up to 4 kN, and dynamic loads of 3.2 kN at 15 Hz in phosphate-buffered saline for one million cycles. Results are presented in Fig. 15a–e. The applied loads at the onset of fretting are given in Fig. 15a, b for the *MC*^{2} and CoCr femoral heads, respectively. This occurred at about 2.2 kN and 1.8 kN for the two different heads. The Si_3N_4 heads had a higher onset load, indicating that they were more resistant to oxide disruption. Additionally, once fretting began for both groups, the rate of increase in fretting current was lower for the Si_3N_4 heads, as evidenced by the smaller slope of the current versus load.

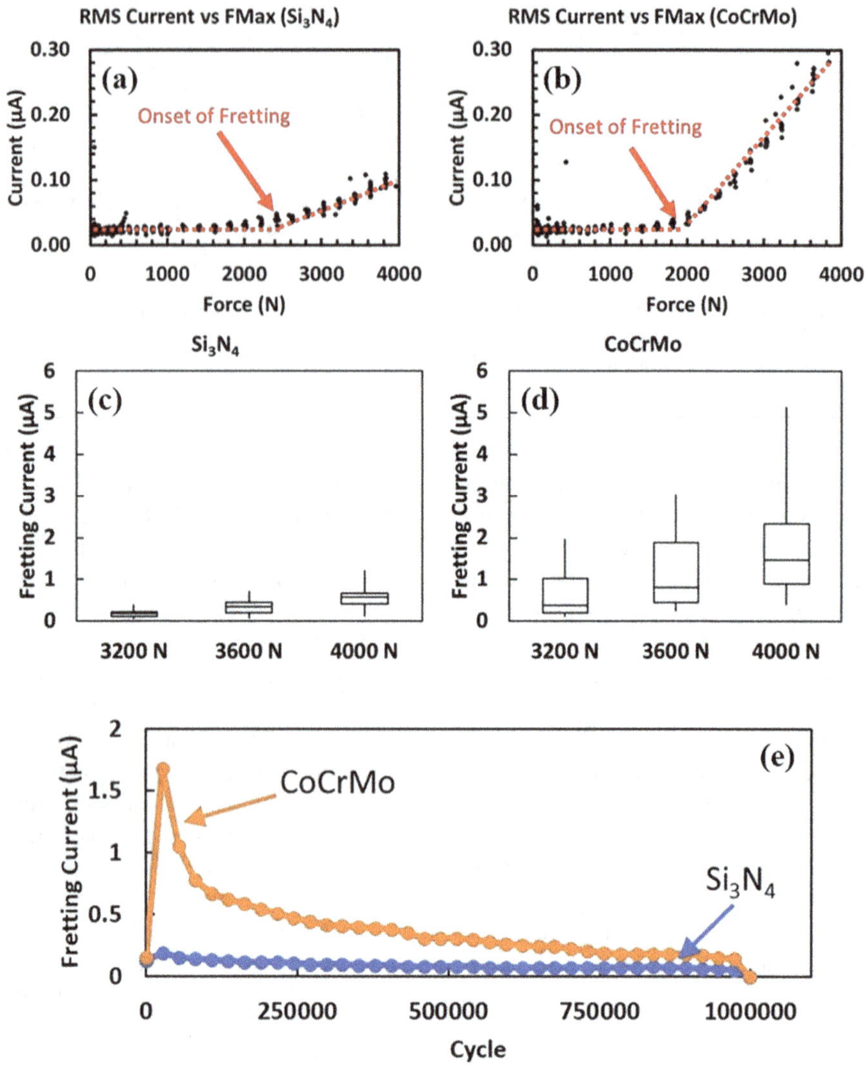

Fig. 15 RMS fretting current versus maximum cyclic load for (**a**) Si_3N_4 and (**b**) CoCr femoral heads. The abrupt change in slope indicates the onset of fretting. RMS = root mean square; box plots showing fretting currents at three static loads for (**c**) Si_3N_4 and (**d**) CoCr femoral heads. The Si_3N_4 heads had statistically lower average fretting current ($p < 0.05$, $n = 10$ for each group) at each load when compared to CoCr heads; and (**e**) Fretting currents versus cycles at 3200 N and 15 Hz for both Si_3N_4 and CoCr femoral heads during one million cycle fretting tests. (Used with permission [159])

Observed fretting currents versus three static loads are given in Fig. 15c, d. Although the data is only presented for three applied loads, the CoCr heads had a higher average fretting current than the Si_3N_4 heads at all applied loads above the onset ($p < 0.05$). Higher cyclic loads led to higher fretting current in both groups, indicating that oxide disruption increased with increasing load. However, the statistically lower average fretting currents for the MC^2 heads indicated that they were more corrosion-resistant. Note that the range of fretting currents for the static loads was always under 0.1 μA for the Si_3N_4/Ti couples versus about 0.1 to 5.0 μA for the CoCr/Ti couples. The CoCr/Ti values from this test are remarkably similar to the results obtained for the sleeved BIOLOX®*delta* static test for Ti trunnions presented in Fig. 14a; and yet, the Si_3N_4/Ti static values are significantly lower than all the observed data for the trunnions used in the *delta* study, (cf., Figs. 14a and 15c).

The fretting current under cyclic loading of the MC^2/Ti and CoCr/Ti couples during the one-million cycle test is shown in Fig. 15e. A significant spike in fretting current (~1.8 μA) was observed for the CoCr/Ti couple early in the test (~25 k cycles), followed by a slow decline over the entire test period. Note a similar spike in current occurred for the sleeved *delta* study using Ti trunnions, although the increase in current was an order of magnitude larger (i.e., 10–20 μA versus 1 to 2 μA, cf., Figs. 14b and 15e). Conversely, the Si_3N_4/Ti couples showed only a minor rise in amperage during the start-up phase (i.e., ≤ 0.2 μA), and then maintained a stable low current for the duration of the test (cf., Fig. 15e).

Do these data suggest that MC^2/Ti couples could have better corrosion characteristics than *delta*/Ti couples? While the test set-ups and loads were similar between the two studies, a direct comparison of the two materials is difficult because the *delta* heads were larger, and they had an intermediary Ti sleeve. However, from a surface chemistry perspective, there are fundamental reasons to believe that Si_3N_4/Ti couples may be an improvement. The chemistry behind MACC involves an ever-increasing acidic nature of the trapped fluid in the crevices between adjoining modular components [160]. The dominant electrochemical mechanism driving increased acidity is oxygen reduction. This leads to the breakdown of the protective passive coating on the Ti or CoCr stem or trunnion, followed by a corrosive attack of the underlying metal. It was demonstrated earlier, that when Al_2O_3-based ceramics are in contact with biological fluids, a surface reaction leads to acidification of the local aqueous environment. Consequently, *delta* may exacerbate MACC due to a concomitant reduction in pH. Conversely, MC^2 has the opposite effect. The elution of nitrogen from Si_3N_4 and its conversion to ammonium (NH_4^+) may improve corrosion resistance. Ammonium and nitrate ions (NO_3^-) have been shown to buffer the pH of crevice solutions, which inhibits their growth. Also, during the early stages of crevice formation, NH_4^+ or ammonium hydroxide (NH_4OH) in the crevice helps in the immediate repassivation of the metal, thus preventing crevice initiation and growth [161].

7 Thermal Conductivity Considerations

Frictional heating in total hip implants can be a severe problem. Excessive heat damages surrounding soft and hard tissue which can result in aseptic loosening and implant failure. Higher temperatures degrade synovial fluid and increase the risks of oxidation, creep, and wear of polyethylene liners. Prolonged or repetitive temperature spikes can also create patient discomfort or pain. The amount of frictional heat generated is a function of several factors including implant design, weight and activity of the patient, quality of the synovial fluid, the presence of prosthetic wear or other debris, and the thermal properties of the articulating materials [162]. Several in vivo and multiple in vitro studies have characterized frictional heating based on various implant and patient parameters [163–168]. Measurements suggest that temperatures can rapidly exceed 43 °C to 45 °C for heavy patients during extended walking periods exceeding 70 minutes or for highly active younger patients in shorter timeframes [169, 170]. While most studies focused on implant design, kinematics, and patient characteristics, few accounted for the thermal conductivity of the prosthetic materials. However, appropriately selected components can reduce thermal spikes. In a hip simulator, Bowsher and Clarke were among the first researchers to demonstrate that differences in the thermal conductivity of femoral heads had a profound impact on the dissipation of heat within the joint space [171]. In turn, they found that heat degraded the quality of the joint lubricant. Protein precipitation occurred on femoral heads having lower thermal conductivities. Higher prosthesis temperatures cause several problems including increased wear of the polyethylene liner, increased risk of osteolysis, and discomfort for active lifestyle patients. The extent of this problem is underreported due to the other more common symptomatic causes of joint malfunction and failure. Table 7 provides a list of thermal conductivities for materials utilized in prosthetic hip joints [43, 172]. Note that MC^2 has more than twice the thermal conductivity of *delta*. Although no studies compare MC^2 to *delta* regarding prosthetic joint temperatures, the preferential use of MC^2 over *delta* as a femoral head material may reduce temperature spikes and improve prosthetic joint longevity.

Table 7 Thermal conductivities of prosthetic materials used in total hip arthroplasty

Material	Thermal conductivity (W/m°K)
Polyetheretherketone (PEEK)	0.29
Magnesia partially stabilized zirconia (PSZ)	2
Yttria tetragonal zirconia polycrystals (YTZP)	2–3
Ultra-high molecular weight polyethylene (UHMWPE)	5–10
Titanium alloy (Ti6Al4V)	6.7–7
Cobalt-chromium alloy (CoCr)	12.7
BIOLOX®*delta* (Alumina matrix composite, AMC)	17
BIOLOX®*forte* (Alumina, Al_2O_3)	30
MC^2 silicon nitride (Si_3N_4)	30–50

8 Summary and Conclusions

Both BIOLOX®*delta* and *MC*^{2}®Si_3N_4 are considered premier bioceramics for total hip arthroplasty applications. While *delta* has an excellent short- to mid-term in vivo record, a comparison of each material's functional characteristics may suggest that *MC*^{2}®Si_3N_4 should perform equally well in the short- to mid-term. Yet, given the surface chemistry differences of the two materials, *MC*^{2}®Si_3N_4 may have advantages in the long term. Based on the information provided in this review, given in Table 8 is a comparison of features and benefits associated with the use of BIOLOX®*delta* or *MC*^{2}®Si_3N_4 as femoral head materials. Following the table, a bulleted list highlights the key points of this review for further consideration.

- Physical and mechanical properties: *MC*^{2}®Si_3N_4 and BIOLOX®*delta* have comparable strength, fracture toughness, slow-crack-growth characteristics, and hardness, which are superior to CoCr, Al_2O_3, and ZrO_2. *MC*^{2} has slightly higher fracture toughness which is unaffected by hydrothermal conditions. Conversely, *delta* has demonstrably superior fracture strength based on its metastable toughening mechanism. The slow-crack-growth characteristics of both materials are similar, but the result of different mechanisms. Hardness values between the two materials are different, but not enough to affect scratch or wear performance.
- Surface chemistry: Compared to Al_2O_3 (*delta*), Si_3N_4 (*MC*^{2}) exhibits favorable surface chemistry properties. In physiological fluids, Si_3N_4 releases minimal quantities of soluble and resorbable nitrogen and silicon species, contributing to a slightly basic local joint environment. This may help counteract the acidic synovial fluid observed in arthritic or diseased joints. Furthermore, the released moieties might potentially scavenge reactive oxygen species (ROS) within the joint capsule, potentially mitigating inflammation. Conversely, the surface chemistry of Al_2O_3 promotes an acidic environment. It does not exhibit true bioinertness, but rather exacerbates joint inflammation and generates insoluble wear debris.
- Wear performance: Surface chemistry also plays a significant role in bearing wear performance. For CoP-bearing couples, *MC*^{2} appears to protect polyethylene liners. In short-term simulator and static contact studies, it produced lower oxidation and embrittlement of polyethylene than Al_2O_3, ZrO_2, and *delta*. In a standard five million cycle hip simulator study using the most advanced vitamin E stabilized polyethylene, the wear performance of the *MC*^{2} polyethylene couples was consistently better than matched *delta* pairs, although the results were not statistically significant. From these studies, one can conclude that *MC*^{2} is at least equivalent to *delta* for CoP bearings in the short term (i.e., $\leq$10–15 years); but *MC*^{2} may have an advantage of increased implant longevity. For CoC bearings, *delta* is better, but it suffers from a similar tribochemical reaction that creates bearing noise and wear. However, this may now be a moot point because practitioners are currently avoiding implanting CoC devices.

Table 8 Summary of the properties, features, and benefits of BIOLOX®*delta* and *MC*²®Si_3N_4

	Properties, features, and benefits	
Functional characteristics	BIOLOX®*delta*	*MC*²®Si_3N_4
Mechanical properties	High toughness: 5.7–8.5 $MPa{\cdot}m^{1/2}$ High strength: 1150–1411 MPa Weibull modulus: 14.9 Burst test (Ø28 mm): 64.6 ± 11.0 kN Offered in 3 Std. offsets: −3.5, 0, +3.5 mm; Ti sleeve used for larger offsets. Metastable *m*-ZrO_2 phase, susceptible to hydrothermal destabilization and reduced surface fracture toughness. Slow-crack growth (SCG) start: ~ 4.0 $MPa{\cdot}m^{1/2}$, SCG fast fracture: ~ 6.0 $MPa{\cdot}m^{1/2}$	High toughness: 4.4–15 $MPa{\cdot}m^{1/2}$ High strength: 995–1250 MPa Weibull modulus: 14.5 Burst test (Ø28 mm): 75.0 ± 11,6 kN Potential for use with larger offsets with or without Ti sleeve. Stable ß-Si_3N_4 phase, surface, and bulk toughness are unaffected by the hydrothermal environment. Slow-crack growth (SCG) start: ~ 3.5 $MPa{\cdot}m^{1/2}$, SCG fast fracture: ~ 6.5 $MPa{\cdot}m^{1/2}$
Surface chemistry	*Delta* is not bioinert. Surface reactions lead to a local acidic environment. Reactive nitrogen species (ROS) are released which can be harmful to local tissue. *Delta* acidifies synovial fluid and releases insoluble $Al(OH)_3$.	Si_3N_4 is bioactive. Surface reactions lead to a local basic environment. Reactive nitrogen species (RNS) are released and may have a beneficial effect on local tissue. Si_3N_4 buffers the synovial fluid and releases resorbable NH_4^+ and $Si(OH)_4$.
Wear performance	*Delta* increases oxidation and crystallinity in polyethylene liners by releasing oxygen. Advanced polyethylene liners consistently showed slightly greater wear against *delta* during *a* five million cycle hip simulator test. *Delta* CoC bearings have a 3% to 6% chance of squeaking. Patients prefer CoP couples. Now, CoC couples account for <1% of implant cases.	Si_3N_4 exerts a protective role towards polyethylene liners by scavenging oxygen. Advanced polyethylene liners consistently showed slightly lower wear in a direct comparative five million cycle hip simulator test. Si_3N_4 is not recommended for use in CoC bearings due to its hydrolysis which can create increased wear and noise during normal ambulation.
Antipathogenic characteristics	*Delta* has lower bacterial pathogenicity than *forte* or polyethylene, but it is not considered to be an antipathogenic biomaterial.	Si_3N_4 showed two orders of magnitude lower bacterial pathogenicity than *delta* at 24 h (99%) at 48 h, it was 99.9% lower than *delta*.

Mechanically Assisted Crevice Corrosion (MACC)	MACC is significantly lower using *delta* heads than CoCr, but it was not eliminated. Long-term viability may still be problematic. The surface chemistry of *delta* may exacerbate crevice corrosion because of its acidic nature, thereby shortening prosthesis life.	Si_3N_4 heads showed significantly lower fretting currents than CoCr heads and may be at least equivalent to *delta* in corrosion resistance. The surface chemistry of Si_3N_4 may buffer the acidic nature of crevice pH, thereby improving prosthesis life.
Thermal conductivity considerations	The thermal conductivity of *delta* is about 17 W/m°K. *Delta* may promote temperature spikes within the joint space for heavy or physically active patients.	The thermal conductivity of Si_3N_4 is 30 to 50 W/m°K. Si_3N_4 will have lower temperature spikes within the joint space of heavy or physically active patients.

- Antipathogenic performance: While both BIOLOX®*delta* and *MC*^{2}®Si_3N_4 demonstrate reduced bacterial adhesion and biofilm formation on their surfaces compared to CoCr and UHMWPE, *MC*^{2}®Si_3N_4 offers a distinct advantage. Available research suggests that *MC*^{2}®Si_3N_4 possesses inherent bactericidal, fungicidal, and virucidal activity. This translates to the potential for the material to not only impede bacterial colonization but also actively lyse and inactivate a broader spectrum of pathogens on the implant surface.
- Mechanically assisted crevice corrosion (MACC): Available research demonstrates that ceramic femoral heads (i.e., *delta* or *MC*^{2}) significantly reduce MACC compared to CoCr heads, regardless of the metal trunnion material. The overall reduction can be as much as an order of magnitude based on short-term retrieval studies [153]. While no studies have been conducted directly comparing BIOLOX®*delta* to *MC*^{2}®Si_3N_4, two independent reports suggest that Si_3N_4 may be at least equivalent to *delta* in corrosion resistance. However, *delta* may contribute to accelerated corrosion due to a reduction in pH in crevice pockets; whereas *MC*^{2} buffers pH which restrains corrosion, and it aids in re-passivation of any damage to the protective metallic-oxide layer.
- Thermal conductivity considerations: *MC*^{2}®Si_3N_4 possesses a higher thermal conductivity compared to BIOLOX®*delta*. This property may be advantageous in reducing transient temperature spikes within the joint space. Temperature fluctuations around the implant are a concern, particularly for active patients or those with increased body mass index (BMI). By facilitating more efficient heat dissipation, *MC*^{2}®Si_3N_4 has the potential to mitigate these thermal spikes and potentially improve implant stability and periprosthetic tissue health.

This review has comparatively evaluated BIOLOX®*delta* and *MC*^{2}®Si_3N_4, two advanced bioceramics for use in human implants. Both materials have received regulatory approval for implant applications, and both boast a strong performance history over the past two decades. This review has highlighted their shared advantages, such as excellent biocompatibility, strength, toughness, and wear resistance, while also detailing their material property differences. While BIOLOX®*delta* enjoys a well-established track record as the leading bioceramic for THA, *MC*^{2}®Si_3N_4 presents potential advantages in areas like phase stability, surface chemistry, antipathogenic properties, and thermal conductivity. Further in vitro and especially clinical studies are warranted to directly compare the long-term performance of these materials in the context of THA. Such research will be crucial for guiding future advancements and elongating prosthetic lifetimes to match patient lifespans.

References

1. Charnley, J. Arthroplasty of the Hip: A New Operation. *Lancet* **1961**, 1129–1132.
2. Learmonth, I.D.; Young, C.; Rorabeck, C. The Operation of the Century: Total Hip Replacement. *Lancet* **2007**, *370*, 1508–1519, doi:https://doi.org/10.1016/S0140-6736(07)60457-7.

3. Bal, B.S.; Garino, J.; Ries, M.; Rahaman, M.N. Ceramic Materials in Total Joint Arthroplasty. *Semin. Arthroplasty* **2006**, *17*, 94–101, doi:https://doi.org/10.1053/j.sart.2006.09.002.
4. Rahaman, M.N.; Yao, A.; Bal, B.S.; Garino, J.P.; Ries, M.D. Ceramics for Prosthetic Hip and Knee Joint Replacement. *J. Am. Ceram. Soc.* **2007**, *90*, 1965–1988, doi:https://doi.org/10.1111/j.1551-2916.2007.01725.x.
5. Willmann, G. The Evolution of Ceramics in Total Hip Replacement. *Hip Int.* **2000**, *10*, 193–203.
6. Heuer, A.H. Transformation Toughening in ZrO_2-Containing Ceramics. *J. Am. Ceram. Soc.* **1987**, *70*, 689–698, doi:https://doi.org/10.1111/j.1151-2916.1987.tb04865.x.
7. Chevalier, J.; Gremillard, L.; Deville, S. Low-Temperature Degradation of Zirconia and Implications for Biomedical Implants. *Annu. Rev. Mater. Res.* **2007**, *37*, 1–32, doi:https://doi.org/10.1146/annurev.matsci.37.052506.084250.
8. Chevalier, J.; Gremillard, L.; Virkar, A. V.; Clarke, D.R. The Tetragonal-Monoclinic Transformation in Zirconia: Lessons Learned and Future Trends. *J. Am. Ceram. Soc.* **2009**, *92*, 1901–1920, doi:https://doi.org/10.1111/j.1551-2916.2009.03278.x.
9. Kuntz, M.; Masson, B.; Pandorf, T. Current State of the Art of the Ceramic Composite Material BIOLOX(R)Delta. In *Strength of Materials*; Mendes, G., Lago, B., Eds.; Nova Science Publishers, Inc, 2009; pp. 133–155 ISBN 978-1-60741-500-8.
10. American Academy of Orthopedic Surgeons, A.J.R.R. *The Eighth Annual Report of the AJRR on Hip and Knee Arthroplasty*; Rosemont, IL USA, **2021**;
11. McEntire, B.J.; Hengst, R.R.; Collins, W.T.; Taglialavore, A.P.; Yeckley, R.L. Ceramic Component Processing Development for Advanced Gas Turbine Engines. *J. Eng. Gas Turbines Power* **1993**, *115*, 1–8.
12. Mouatt, C.A.L. The Development of a Silicon Nitride Spinal Implant. *Mater. Sci. Forum* **1988**, *34–36*, 403–406, doi:https://doi.org/10.4028/www.scientific.net/MSF.34-36.403.
13. Sorrell, C.C.; Hardcastle, P.H.; Druitt, R.K.; Howlett, C.R.; McCartney, E.R. Results of 15-Year Clinical Study of Reaction Bonded Silicon Nitride Intervertebral Spacers. *Proc. 7th World Biomater. Conf.* **2004**, 1872.
14. Mobbs, R.J.; Rao, P.J.; Phan, K.; Hardcastle, P.; Choy, W.J.; McCartney, E.R.; Druitt, R.K.; Mouatt, C.A.L.; Sorrell, C.C. Anterior Lumbar Interbody Fusion Using Reaction Bonded Silicon Nitride Implants: Long Term Case Series of the First Synthetic ALIF Spacer Implanted in Humans. *World Neurosurg.* **2018**, *120*, 256–264, doi: https://doi.org/10.1016/j.wneu.2018.08.237.
15. Pezzotti, G.; Muraki, N.; Maeda, N.; Satou, K.; Nishida, T. In situ Measurement of Bridging Stresses in Toughened Silicon Nitride Using Raman Microprobe Spectroscopy. *J. Am. Ceram. Soc.* **1999**, *82*, 1249–1256, doi:https://doi.org/10.1111/j.1151-2916.1999.tb01903.x.
16. Pezzotti, G.; Enomoto, Y.; Zhu, W.; Boffelli, M.; Marin, E.; McEntire, B.J. Surface Toughness of Silicon Nitride Bioceramics: I, Raman Spectroscopy-Assisted Micromechanics. *J. Mech. Behav. Biomed. Mater.* **2015**, *54*, 328–345.
17. McEntire, B.J.; Lakshminarayanan, R.; Thirugnanasambandam, P.; Seitz-Sampson, J.; Bock, R.; O'Brien, D. Processing and Characterization of Silicon Nitride Bioceramics. *Bioceram. Dev. Appl.* **2016**, *6*, 1000093, doi:https://doi.org/10.4172/2090-5025.100093.
18. Callaghan, J.J.; Albright, J.C.; Goetz, D.D.; Olejniczak, J.P.; Johnston, R.C. Charnley Total Hip Arthroplasty with Cement: Minimum Twenty-Five-Year Follow-Up. *J. Jt. Bone Surg.* **2000**, *82*, 487.
19. Berry, D.J.; Harmsen, W.S.; Cabanela, M.E.; Morrey, B.F. Twenty-Five-Year Survivorship of Two Thousand Consecutive Primary Charnley Total Hip Replacements: Factors Affecting Survivorship of Acetabular and Femoral Components. *J. Bone Jt. Surg.* **2002**, *84*, 171–177.
20. Evans, J.T.; Evans, J.P.; Walker, R.W.; Blom, A.W.; Whitehouse, M.R.; Sayers, A. How Long Does a Hip Replacement Last? A Systematic Review and Meta-Analysis of Case Series and National Registry Reports With More Than 15 Years of Follow-Up. *Lancet* **2019**, *393*, 647–654, doi:https://doi.org/10.1016/S0140-6736(18)31665-9.

21. Masson, B.; Kuntz, M. Long-Term Stability of Ceramic Composite in Total Hip Arthroplasty. In *Total Hip Arthroplasty.*; Knahr, K., Ed.; Springer: Berlin, Heidelberg, 2012; pp. 145–153 ISBN 978-3-642-27360-5.
22. Martin, L.; Chevalier, J.; Gremillard, L. Slow Crack Growth Behavior of Silicon Nitride Using the Double Torsion Test Method. *SINTX Res. Rep.* **2012**, 1–5.
23. Häußler, K.L.; Preuß, R.; Streicher, R.M. Wear of Large Ceramic-on-Ceramic Bearings for Total Hip Arthroplasty and the Mechanical and Tribological Properties of Silicon Nitrides. In *Tribology in Total Hip and Knee Arthroplasty*; Knahr, K., Ed.; Springer Berlin Heidelberg: Berlin, Heidelberg, 2014; pp. 85–94 ISBN 978-3-642-45265-9.
24. Piconi, C. 5 - Ceramics for Joint Replacement: Design and Application of Commercial Bearings. In *Advances in Ceramic Biomaterials*; Palmero, P., Cambier, F., De Barra, E.B. T.-A. in C.B., Eds.; Woodhead Publishing, 2017; pp. 129–179 ISBN 978-0-08-100881-2.
25. *SINTX Technical Data Sheet: MC2 Silicon Nitride Typical Properties and Specifications.* SINTX Technologies, Inc.: Salt Lake City, UT 84119 2024, p. 1.
26. *CeramTec Technical Brochure: BIOLOX$^{(R)}$delta Scientific Information and Performace Data.* CeramTec, AG: Plochingen, Germany 2006, pp. 1–15.
27. Burger, W.; Richter, H.G. High Strength and Toughness Alumina Matrix Composites by Transformation Toughening and "*in situ*" Platelet Reinforcement (ZPTA) - The New Generation of Bioceramics. *Key Eng. Mater.* **2001**, *192–195*, 545–548.
28. Heimke, G.; Leyen, S.; Willmann, G. Knee Arthroplasty: Recently Developed Ceramics Offer New Solutions. *Biomaterials* **2002**, *23*, 1539–51.
29. Merkert, P. Next Generation Ceramic Bearings. In Proceedings of the Bioceramics in Joint Arthroplasty; Steinkopff, 2003; pp. 123–125.
30. Kuntz, M. Validation of a New High-Performance Alumina Matrix Composite for Use in Total Joint Replacement. *Semin. Arthroplasty* **2006**, *17*, 141–145, doi:https://doi.org/10.1053/j.sart.2006.09.008.
31. Pezzotti, G.; Yamada, K.; Porporati, A.A.; Kuntz, M.; Yamamoto, K. Fracture Toughness Analysis of Advanced Ceramic Composite for Hip Prosthesis. *J. Am. Ceram. Soc.* **2009**, *92*, 1817–1822, doi:https://doi.org/10.1111/j.1551-2916.2009.03114.x.
32. Richter, B.I.; Kellner, S.; Menzel, H.; Behrens, P.; Denkena, B.; Ostermeier, S.; Hurschler, C. Mechanical Characterization of Nacre as an Ideal-Model for Innovative New Endoprosthesis Materials. *Arch. Orthop. Trauma Surg.* **2011**, *131*, 191–196, doi:https://doi.org/10.1007/s00402-010-1118-z.
33. Kuntz, M.; Krüger, R. The Effect of Microstructure and Chromia Content on the Properties of Zirconia Toughened Alumina. *Ceram. Int.* **2018**, *44*, 2011–2020, doi:https://doi.org/10.1016/j.ceramint.2017.10.146.
34. Brown, T.D.; Bartel, D.L. What Design Factors Influence Wear Behavior at the Bearing Surfaces in Total Joint Replacements? *J. Am. Acad. Orthop. Surg.* **2008**, *16 Suppl 1*, 101–106, doi:https://doi.org/10.5435/00124635-200800001-00020.
35. Malito, L.G.; Arevalo, S.; Kozak, A.; Spiegelberg, S.; Bellare, A.; Pruitt, L. Material Properties of Ultra-High Molecular Weight Polyethylene: Comparison of Tension, Compression, Nanomechanics and Microstructure Across Clinical Formulations. *J. Mech. Behav. Biomed. Mater.* **2018**, *83*, 9–19, doi:https://doi.org/10.1016/j.jmbbm.2018.03.029.
36. Savio, D.; Bagno, A. When the Total Hip Replacement Fails: A Review on the Stress-Shielding Effect. *Processes* **2022**, *10*, doi:https://doi.org/10.3390/pr10030612.
37. Pitto, R.P.; Bhargava, A.; Pandit, S.; Munro, J.T. Retroacetabular Stress-Shielding in THA. *Clin. Orthop. Relat. Res.* **2008**, *466*, 353–358, doi:https://doi.org/10.1007/s11999-007-0043-0.
38. Vogel, D.; Klimek, M.; Saemann, M.; Bader, R. Influence of the Acetabular Cup Material on the Shell Deformation and Strain Distribution in the Adjacent Bone-A Finite Element Analysis. *Materials (Basel).* **2020**, *13*, doi:https://doi.org/10.3390/ma13061372.
39. Quinn, G.D. Weibull Strength Scaling for Standardized Rectangular Flexure Specimens. *J. Am. Ceram. Soc.* **2003**, *86*, 508–510.

40. Quinn, G.D.; Quinn, J.B. A Practical and Systematic Review of Weibull Statistics for Reporting Strengths of Dental Materials. *Dent. Materaisl* **2010**, *26*, 135–147, doi:https://doi.org/10.1016/j.dental.2009.09.006.A.
41. Quinn, G.D.; Morrell, R. Design Data for Engineering Ceramics: A Review of the Flexure Test. *J. Am. Ceram. Soc.* **1991**, *74*, 2037–2066, doi:https://doi.org/10.1111/j.1151-2916.1991.tb08259.x.
42. Guidance Document for the Preparation of Premarket Notifications for Ceramic Ball Hip Systems. *US Food Drug Adm. Med. Devices* **2009**, 5–3.
43. McEntire, B.J.; Bal, B.S.; Rahaman, M.N.; Chevalier, J.; Pezzotti, G. Ceramics and Ceramic Coatings in Orthopaedics. *J. Eur. Ceram. Soc.* **2015**, *35*, 4327–4369, doi:https://doi.org/10.1016/j.jeurceramsoc.2015.07.034.
44. Morrell, R. Fracture Toughness Testing for Advanced Technical Ceramics: Internationally Agreed Good Practice. *Adv. Appl. Ceram.* **2006**, *105*, 88–98, doi:https://doi.org/10.1179/174367606X84422.
45. Cesar, P.F.; Della Bona, A.; Scherrer, S.S.; Tholey, M.; van Noort, R.; Vichi, A.; Kelly, R.; Lohbauer, U. ADM Guidance—Ceramics: Fracture Toughness Testing and Method Selection. *Dent. Mater.* **2017**, *33*, 575–584, doi:https://doi.org/10.1016/j.dental.2017.03.006.
46. Boccaccini, A.R.; Rawlings, R.D.; Dlouhý, I. Reliability of the Chevron-Notch Technique for Fracture Toughness Determination in Glass. *Mater. Sci. Eng. A* **2003**, *347*, 102–108, doi: https://doi.org/10.1016/S0921-5093(02)00612-3.
47. Kailer, A.; Stephan, M. On the Feasibility of the Chevron Notch Beam Method to Measure Fracture Toughness of Fine-Grained Zirconia Ceramics. *Dent. Mater.* **2016**, *32*, 1256–1262, doi:https://doi.org/10.1016/j.dental.2016.07.011.
48. Mizuno, M.; Okuda, H. VAMAS Round Robin on Fracture Toughness of Silicon Nitride. *J. Am. Ceram. Soc.* **1995**, *78*, 1793–1801, doi:https://doi.org/10.1111/j.1151-2916.1995.tb08891.x.
49. Chevalier, J.; Saadaoui, M.; Olagnon, C.; Fantozzi, G. Double-Torsion Testing a 3Y-TZP Ceramic. *Ceram. Int.* **1996**, *22*, 171–177, doi: https://doi.org/10.1016/0272-8842(95)00076-3.
50. Wiederhorn, S. Brittle Fracture and Toughening Mechanisms in Ceramics. *Annu. Rev. Mater. Sci.* **1984**, *14*, 373–403.
51. Lube, T.; Danzer, R. *Mechanical Properties and Reliability of Advanced Ceramics*; Elsevier Inc., 2014; ISBN 9780123946195.
52. Kawashima, T.; Okamoto, H.; Yamamoto, H.; Kitamura, A. Grain Size Dependence of the Fracture Toughness of Silicon Nitride Ceramics. *J. Ceram. Soc. Japan* **1991**, *99*, 320–323.
53. Hirosaki, N.; Akimune, Y.; Mitomo, M. Effect of Grain Growth of ß-Silicon Nitride on Strength, Weibull Modulus, and Fracture Toughness. *J. Am. Ceram. Soc.* **1993**, *76*, 1892–1894, doi: https://doi.org/10.1111/j.1151-2916.1993.tb06670.x.
54. Becher, P.F.; Sun, E.Y.; Plucknett, K.P.; Alexander, K.B.; Hsueh, C.; Lin, H.; Waters, S.B.; Westmoreland, C.G. Microstructural Design of Silicon Nitride with Improved Fracture Toughness: I, Effects of Grain Shape and Size. *J. Am. Ceram. Soc.* **1998**, *81*, 2821–2830.
55. Sun, E.Y.; Becher, P.F.; Plucknett, K.P.; Hsueh, C.-H.; Alexander, K.B.; Waters, S.B. Microstructural Design of Silicon Nitride with Improved Fracture Toughness: II, Effects of Yttria and Alumina Additives. *J. Am. Ceram. Soc.* **1998**, *81*, 2831–2840.
56. Yang, J.; Sekino, T.; Niihara, K. Effect of Grain Growth and Measurement on Fracture Toughness of Silicon Nitride Ceramics. *J. Mater. Sci.* **1999**, *34*, 5543–5548, doi:https://doi.org/10.1023/A:1004724814778.
57. Furushima, R.; Maruyama, Y.; Nakashima, Y.; Ngo, M.C.; Ohji, T.; Fukushima, M. Fracture Toughness Evaluation of Silicon Nitride from Microstructures via Convolutional Neural Network. *J. Am. Ceram. Soc.* **2023**, *106*, 817–821, doi: https://doi.org/10.1111/jace.18795.
58. Furushima, R.; Nakashima, Y.; Maruyama, Y.; Hirao, K.; Ohji, T.; Fukushima, M. Artificial Intelligence-Based Determination of Fracture Toughness and Bending Strength of Silicon Nitride Ceramics. *J. Am. Ceram. Soc.* **2023**, *106*, 4944–4954, doi: https://doi.org/10.1111/jace.19147.

59. Chevalier, J.; Grandjean, S.; Kuntz, M.; Pezzotti, G. On the Kinetics and Impact of Tetragonal to Monoclinic Transformation in an Alumina/Zirconia Composite for Arthroplasty Applications. *Biomaterials* **2009**, *30*, 5279–82, doi:https://doi.org/10.1016/j.biomaterials.2009.06.022.
60. Kuntz, M.; Shneider, N.; Heros, R. Controlled Zirconia Phase Transformation in BIOLOX®*delta* - A Feature of Safety. In *Bioceramics and Alternative Bearings in Joint Arthroplasty*; Steinkopff: New York, 2005; pp. 79–84.
61. McEntire, B.J.; Enomoto, Y.; Zhu, W.; Boffelli, M.; Marin, E.; Pezzotti, G. Surface Toughness of Silicon Nitride Bioceramics: II, Comparison with Commercial Oxide Materials. *J. Mech. Behav. Biomed. Mater.* **2016**, *54*, 346–359, doi:https://doi.org/10.1016/j.jmbbm.2015.08.044.
62. Affatato, S.; Modena, E.; Toni, A.; Taddei, P. Retrieval Analysis of Three Generations of BIOLOX® Femoral Heads: Spectroscopic and SEM Characterisation. *J. Mech. Behav. Biomed. Mater.* **2012**, *13*, 118–28, doi:https://doi.org/10.1016/j.jmbbm.2012.04.003.
63. Taddei, P.; Ruggiero, A.; Pavoni, E.; Affatato, S. Transfer of Metallic Debris After *in vitro* Ceramic-on-Metal Simulation: Wear and Degradation in Biolox® Delta Composite Femoral Heads. *Compos. Part B Eng.* **2016**.
64. Affatato, S.; Ruggiero, A.; De Mattia, J.S.; Taddei, P. Does Metal Transfer Affect the Tribological Behaviour of Femoral Heads? Roughness and Phase Transformation Analyses on Retrieved Zirconia and Biolox® Delta Composites. *Compos. Part B Eng.* **2016**, *92*, 290–298.
65. Bal, B.S.; Zhu, W.; Zanocco, M.; Marin, E.; Sugano, N.; McEntire, B.J.; Pezzotti, G. Reconciling *in vivo* and *in vitro* Kinetics of the Polymorphic Transformation in Zirconia-Toughened Alumina for Hip Joints: I. Phenomenology. *Mater. Sci. Eng. C* **2017**, *C72*, 252–258, doi:https://doi.org/10.1016/j.msec.2016.11.059.
66. Pezzotti, G.; Zhu, W.; Zanocco, M.; Marin, E.; Sugano, N.; McEntire, B.J.; Bal, B.S. Reconciling *in vivo* and *in vitro* Kinetics of the Polymorphic Transformation in Zirconia-Toughened Alumina for Hip Joints: II. Theory. *Mater. Sci. Eng. C* **2017**, *71*, 446–451, doi:https://doi.org/10.1016/j.msec.2016.10.031.
67. Pezzotti, G.; Zhu, W.; Zanocco, M.; Marin, E.; Sugano, N.; McEntire, B.J.; Bal, B.S.; Zanocco, M.; Marin, E.; Sugano, N.; *et al.* Reconciling *in vivo* and *in vitro* Kinetics of the Polymorphic Transformation in Zirconia-Toughened Alumina for Hip Joints: III. Molecular Scale Mechanisms. *Mater. Sci. Eng. C* **2017**, *C71*, 552–557, doi:https://doi.org/10.1016/j.msec.2016.11.007.
68. Pezzotti, G.; Affatato, S.; Rondinella, A.; Yorifuji, M.; Marin, E.; Zhu, W.; McEntire, B.; Bal, S.; Yamamoto, K. *In Vitro* versus *In Vivo* Phase Instability of Zirconia-Toughened Alumina Femoral Heads: A Critical Comparative Assessment. *Materials (Basel).* **2017**, *10*, 466, doi: https://doi.org/10.3390/ma10050466.
69. Tateiwa, T.; Marin, E.; Rondinella, A.; Ciniglio, M.; Zhu, W.; Saverio, A.; Pezzotti, G.; Bock, R.M.; McEntire, B.J.; Bal, B.S.; *et al.* Burst Strength of BIOLOX®*delta* Femoral Heads and its Dependence on Low-Temperature Environmental Degradation. *Materials (Basel).* **2020**, *13*, 350, doi:https://doi.org/10.3390/ma13020350.
70. Pecharromán, C.; Bartolomé, J.F.; Requena, J.; Moya, J.S.; Deville, S.; Chevalier, J.; Fantozzi, G.; Torrecillas, R. Percolative Mechanism of Aging in Zirconia-Containing Ceramics for Medical Applications. *Adv. Mater.* **2003**, *15*, 507–511, doi: https://doi.org/10.1002/adma.200390117.
71. Deville, S.; Chevalier, J.; Dauvergne, C.; Fantozzi, G.; Bartolome, J.F.; Moya, J.S.; Torrecillas, R. Microstructural Investigation of the Aging Behavior of (3Y-TZP)-Al_2O_3 Composites. *J. Am. Ceram. Soc.* **2005**, *88*, 1273–1280, doi:https://doi.org/10.1111/j.1551-2916.2005.00221.x.
72. Porporati, A.A.; Gremillard, L.; Chevalier, J.; Pitto, R.; Deluca, M. Is Surface Metastability of Today's Ceramic Bearings a Clinical Issue? *J. Compos. Sci.* 2021, *5*.
73. Lucchini, S.; Baleani, M.; Giardina, F.; Martelli, A.; Castagnini, F.; Bordini, B.; Traina, F. A Case-Driven Hypothesis for Multi-Stage Crack Growth Mechanism in Fourth-Generation

Ceramic Head Fracture. *J. Orthop. Surg. Res.* **2022**, *17*, 293, doi:https://doi.org/10.1186/s13018-022-03190-6.
74. Dobria, P.; Patel, A.; Levine, B. Catastrophic Intraoperative Failure of a Ceramic Femoral Head. *Arthroplast. Today* **2021**, *11*, 234–238, doi:https://doi.org/10.1016/j.artd.2021.08.017.
75. Pomeroy, E.; Rowan, F.; Masterson, E. Atraumatic Fracture of a BIOLOX Delta Ceramic Femoral Head Articulating with a Polyethylene Liner. *JBJS Case Connect.* **2015**, *5*, e112.
76. Patetta, M.J.; Kayupov, E.; Karam, J.A.; Chmell, S.J. Fourth-Generation Ceramic Head Fracture in Dual Mobility Total Hip Arthroplasty: A Case Report. *J. Orthop. Case Reports* **2022**, *12*, 98–101, doi:https://doi.org/10.13107/jocr.2022.v12.i01.2638.
77. Rankin, C.S.; Robinson, P.G.; Beattie, N.; Gaston, P. Fracture of a BIOLOX Delta Ceramic Femoral Head: A Case Report and Update of the Literature. *JBJS Case Connect. Connect.* **2019**, *9*, e0336, doi:https://doi.org/10.2106/JBJS.CC.18.00336.
78. Duensing, I.M.; Stanley, S.; Bolognesi, M. Fracture of a 40-mm BIOLOX Delta Femoral Head. *Arthroplast. Today* **2021**, *10*, 144–148, doi:https://doi.org/10.1016/j.artd.2021.06.021.
79. Oak Salvatore J.; Rueda, Carlos A. Higuera; Klika, Alison; Molloy, Robert, S.R.. F. Atraumatic Delta Head Fracture in a Ceramic-On-Polyethylene Total Hip Arthroplasty. *J. Hip Surg.* **2017**, *01*, 118–120, doi:https://doi.org/10.1055/s-0037-1603628.
80. Heiner, A.D.; Mahoney, C.R. Fracture of a BIOLOX Delta Ceramic Femoral Head Articulating Against a Polyethylene Liner: A Case Report. *JBJS Case Connect.* **2014**, *4*, 1–6.
81. Bal, B.S.; Khandkar, A.; Lakshminarayanan, R.; Clarke, I.; Hofmann, A.A.; Rahaman, M.N. Fabrication and Testing of Silicon Nitride Bearings in Total Hip Arthroplasty. *J. Arthroplasty* **2009**, *24*, 110–116, doi:https://doi.org/10.1016/j.arth.2008.01.300.
82. Kawakubo, T.; Komeya, K. Static and Cyclic Fatigue Behavior of a Sintered Silicon Nitride at Room Temperature. *J. Am. Ceram. Soc.* **1987**, *70*, 400–405, doi:https://doi.org/10.1111/j.1151-2916.1987.tb05659.x.
83. Cornelissen, B.E.; Dauskardt, R.H.; Ritchie, R.O.; Thomas, G. Cyclic Fatigue Behavior and Fracture Toughness of Silicon Nitride Ceramics Sintered with Rare-Earth Oxides. *Acta Metall. Mater.* **1994**, *42*, 3055–3064, doi:https://doi.org/10.1016/0956-7151(94)90402-2.
84. Ishihara, S.; McEvily, A.J.; Goshima, T. Effect of Atmospheric Humidity on the Fatigue Crack Propagation Behavior of Short Cracks in Silicon Nitride. *J. Am. Ceram. Soc.* **2000**, *83*, 571–577.
85. Jacobs, D.S.; Chen, I.-W. Mechanical and Environmental Factors in the Cyclic and Static Fatigue of Silicon Nitride. *J. Am. Ceram. Soc.* **1994**, *77*, 1153–1161.
86. Bhatnagar, A.; Hoffman, M.J.; Dauskardt, R.H. Fracture and Subcritical Crack Growth Behavior of Y-Si-Al-O-N Glasses and Si_3N_4 Ceramics. *J. Am. Ceram. Soc.* **2000**, *83*, 585–596.
87. Chevalier, J.; Taddei, P.; Gremillard, L.; Deville, S.; Fantozzi, G.; Bartolomé, J.F.; Pecharroman, C.; Moya, J.S.; Diaz, L.A.; Torrecillas, R.; *et al.* Reliability Assessment in Advanced Nanocomposite Materials for Orthopaedic Applications. *J. Mech. Behav. Biomed. Mater.* **2011**, *4*, 303–314, doi:https://doi.org/10.1016/j.jmbbm.2010.10.010.
88. Gilbert, C.J.; Dauskardt, R.H.; Ritchie, R.O. Behavior of Cyclic Fatigue Cracks in Monolithic Silicon Nitride. *J. Am. Ceram. Soc.* **1995**, *78*, 2291–2300.
89. Gilbert, C.J.; Dauskardt, R.H.; Ritchie, R.O. Microstructural Mechanisms of Cyclic Fatigue-Crack Propagation in Grain-Bridging Ceramics. *Ceram. Int.* **1997**, *23*, 413–418, doi:https://doi.org/10.1016/S0272-8842(96)00048-X.
90. Hallan, G.; Fenstad, A.M.; Furnes, O. What Is the Frequency of Fracture of Ceramic Components in THA? Results from the Norwegian Arthroplasty Register from 1997 to 2017. *Clin. Orthop. Relat. Res.* **2020**, *478*, 1254–1261, doi:https://doi.org/10.1097/CORR.0000000000001272.
91. Mukhopadhyay, A.K.; Datta, S.K.; Chakraborty, D. Hardness of Silicon Nitride and Sialon. *Ceram. Int.* **1991**, *17*, 121–127, doi:https://doi.org/10.1016/0272-8842(91)90040-7.
92. Zutshi, A.; Haber, R.A.; Niesz, D.E.; Adams, J.W.; Wachtman, J.B.; Ferber, M.K.; Hsu, S.M. Processing, Microstructure, and Wear Behavior of Silicon Nitride Hot-Pressed with

Alumina and Yttria. *J. Am. Ceram. Soc.* **1994**, *77*, 883–890, doi:https://doi.org/10.1111/j.1151-2916.1994.tb07243.x.
93. Hannink, R.H.J.; Kelly, P.M.; Muddle, B.C. Transformation Toughening in Zirconia-Containing Ceramics. *J. Am. Ceram. Soc.* **2000**, *83*, 461–487.
94. Fisher, J.; Jennings, L.M.; Galvin, A.L. Wear of Highly Crosslinked Polyethylene Against Cobalt Chrome and Ceramic Femoral Heads. In Proceedings of the Bioceramics and Alternative Bearings in Joint Arthroplasty; Benazzo, F., Falez, F., Dietrich, M., Eds.; Steinkopff: Darmstadt, 2006; pp. 185–188.
95. Galvin, A.L.; Jennings, L.M.; Tipper, J.L.; Ingham, E.; Fisher, J. Wear and Creep of Highly Crosslinked Polyethylene against Cobalt Chrome and Ceramic Femoral Heads. *Proc. Inst. Mech. Eng. Part H J. Eng. Med.* **2010**, *224*, 1175–1183.
96. Merfort, R.; Maffulli, N.; Hofmann, U.K.; Hildebrand, F.; Simeone, F.; Eschweiler, J.; Migliorini, F. Head, Acetabular Liner Composition, and Rate of Revision and Wear in Total Hip Arthroplasty: A Bayesian Network Meta-Analysis. *Sci. Rep.* **2023**, *13*, 1–23, doi:https://doi.org/10.1038/s41598-023-47670-z.
97. Mertz, K.C.; Yang, J.W.; Chung, B.C.; Chen, X.; Mayfield, C.K.; Heckmann, N.D. Ceramic Femoral Heads Exhibit Lower Wear Rates Compared to Cobalt Chrome: A Meta-Analysis. *J. Arthroplasty* **2023**, *38*, 397–405, doi:https://doi.org/10.1016/j.arth.2022.09.008.
98. Yorifuji, M.; Affatato, S.; Tateiwa, T.; Takahashi, Y.; Shishido, T.; Marin, E.; Zanocco, M.; Zhu, W.; Pezzotti, G.; Yamamoto, K. Wear Simulation of Ceramic-on-Crosslinked Polyethylene Hip Prostheses: A New Non-Oxide Silicon Nitride versus the Gold Standard Composite Oxide Ceramic Femoral Heads. *Materials (Basel).* **2020**, *13*, 1–11, doi:https://doi.org/10.3390/ma13132917.
99. Lee, R.; Essner, A.; Wang, A.; Jaffe, W.L. Scratch and Wear Performance of Prosthetic Femoral Head Components Against Crosslinked UHMWPE Sockets. *Wear* **2009**, *267*, 1915–1921, doi:https://doi.org/10.1016/j.wear.2009.03.034.
100. Najjar, D.; Bigerelle, M.; Migaud, H.; Iost, A. Identification of Scratch Mechanisms on a Retrieved Metallic Femoral Head. *Wear* **2005**, *258*, 240–250, doi:https://doi.org/10.1016/j.wear.2004.09.035.
101. Kruger, K.M. Computational and Experimental Biomechanics of Total Hip Wear Increase Due to Femoral Head Damage, The University of Iowa, 2014.
102. Evans, A.G.; Wilshaw, T.R. Quasi-Static Solid Particle Damage in Brittle Solids—I. Observations Analysis and Implications. *Acta Metall.* **1976**, *24*, 939–956, doi:https://doi.org/10.1016/0001-6160(76)90042-0.
103. Evans, A.G.; Marshall, D.B. Wear Mechanisms in Ceramics. In *Fundamentals Friction and Wear of Materials*; American Society of Metals Press: Metals Park, OH, USA, 1981; p. 439.
104. Hsu, S.M.; Shen, M. Wear Prediction of Ceramics. *Wear* **2004**, *256*, 867–878, doi:https://doi.org/10.1016/j.wear.2003.11.002.
105. Kong, X.; Hu, X.; Chai, W. *In vitro* & *in vivo* Investigation of the Silicon Nitride Ceramic Hip Implant's Safety and Effectiveness Evaluation. *J. Orthop. Surg. Res.* **2022**, *17*, 1–13, doi:https://doi.org/10.1186/s13018-021-02884-7.
106. Pezzotti, G. Bioceramics are not Bioinert. *Mater. Today* **2017**, *20*, 395–398, doi:https://doi.org/10.1016/j.mattod.2017.06.008.
107. Mestres, G.; Espanol, M.; Xia, W.; Tenje, M.; Ott, M. Evaluation of Biocompatibility and Release of Reactive Oxygen Species of Aluminum Oxide-Coated Materials. *ACS Omega* **2016**, *1*, 706–713, doi:https://doi.org/10.1021/acsomega.6b00198.
108. Abramson, S.B. Nitric Oxide in Inflammation and Pain Associated with Osteoarthritis. *Arthritis Res. Ther.* **2008**, *10*, 1–7, doi:https://doi.org/10.1186/ar2463.
109. McEntire, B.J.; Bal, B.S.; Pezzotti, G. Antimicrobial Nitric Oxide Releasing Compounds and Scaffolds. In *Antimicrobial Combination Devices*; Urish, K.L., Mihalko, W.M., Eds.; ASTM International: West Conshohocken, PA, USA, 2020; pp. 105–137.
110. Pezzotti, G. Silicon Nitride as a Biomaterial. *J. Ceram. Soc. Japan* **2023**, *131*, 398–428, doi:https://doi.org/10.2109/jcersj2.23057.

111. Rahal, A.; Kumar, A.; Singh, V.; Yadav, B.; Tiwari, R.; Chakraborty, S.; Dhama, K. Oxidative Stress, Prooxidants, and Antioxidants: The Interplay. *Biomed Res. Int.* **2014**, *2014*, 761264, doi:https://doi.org/10.1155/2014/761264.
112. Cummings, N.A.; Nordby, G.L. Measurement of Synovial Fluid pH in Normal and Arthritic Knees. *Arthritis Rheum.* **1966**, *9*, 47–56, doi: https://doi.org/10.1002/art.1780090106.
113. Milošev, I.; Levašič, V.; Vidmar, J.; Kovač, S.; Trebše, R. pH and Metal Concentration of Synovial Fluid of Osteoarthritic Joints and Joints with Metal Replacements. *J. Biomed. Mater. Res. Part B Appl. Biomater.* **2017**, *105*, 2507–2515, doi: https://doi.org/10.1002/jbm.b.33793.
114. Sargeant, A.; Goswami, T. Hip Implants: Paper V. Physiological Effects. *Mater. Des.* **2006**, *27*, 287–307, doi: https://doi.org/10.1016/j.matdes.2004.10.028.
115. Pettersson, M.; Skjoldebrand, C.; Filho, L.; Engqvist, H.; Persson, C. Morphology and Dissolution Rate of Wear Debris from Silicon Nitride Coatings. *ACS Biomater. Sci. Eng.* **2016**, 1–28.
116. Martin, K.R. The Chemistry of Silica and its Potential Health Benefits. *J. Nutr. Heal. Aging* **2007**, *11*, 94–98.
117. Adeva, M.M.; Souto, G.; Blanco, N.; Donapetry, C. Ammonium Metabolism in Humans. *Metabolism* **2012**, *61*, 1495–1511, doi: https://doi.org/10.1016/j.metabol.2012.07.007.
118. Pezzotti, G.; Zhu, W.; Sugano, N.; Marin, E.; Yamamoto, K.; Nishiike, N.; Hori, T.; Rondinella, A.; McEntire, B.J.; Bock, R.; *et al.* Oxide Ceramic Femoral Heads Contribute to the Oxidation of Polyethylene Liners in Artificial Hip Joints. *J. Mech. Behav. Biomed. Mater.* **2018**, *82*, 168–182, doi:https://doi.org/10.1016/j.jmbbm.2018.03.021.
119. Rondinella, A.; Marin, E.; McEntire, B.J.; Bock, R.M.; Bal, B.S.; Zhu, W.L.; Yamamoto, K.; Pezzotti, G. Bioceramics are Not Bioinert: The Role of Oxide and Non-Oxide Bioceramics on the Oxidation of UHMWPE Components in Artificial Joints. *Key Eng. Mater.* **2018**, *782*, 165–175, doi:https://doi.org/10.4028/www.scientific.net/KEM.782.165.
120. Pezzotti, G.; Bal, B.S.; Casagrande, E.; Sugano, N.; McEntire, B.J.; Zhu, W.; Puppulin, L. On the Molecular Interaction between Femoral Heads and Polyethylene Liners in Artificial Hip Joints: Phenomenology and Molecular Scale Phenomena. *Biomed. Mater.* **2017**, *12*, 15005.
121. McCarthy, C.J.; Mahon, J.; Sheridan, G.A.; Welch-Phillips, A.; O'Byrne, J.M.; Kenny, P.J. Third and Fourth Generation Ceramic-on-Ceramic Total Hip Arthroplasty at a Minimum of Ten Years. *J. Clin. Orthop. Trauma* **2022**, *31*, 101942, doi: https://doi.org/10.1016/j.jcot.2022.101942.
122. Giuseppe, M.; Mattia, B.; Nadia, B.; Raffaele, V.; Pasquale, R.; Stefano, D.A.; Mattia, S.; Vincenzo, D.S.; Giulio, M. Ceramic-on-Ceramic Versus Ceramic-on-Polyethylene in Total Hip Arthroplasty: A Comparative Study at a Minimum of 13 Years Follow-Up. *BMC Musculoskelet. Disord.* **2021**, *22*, 1–9, doi:https://doi.org/10.1186/s12891-021-04950-x.
123. Mai, K.; Verioti, C.; Ezzet, K. a; Copp, S.N.; Walker, R.H.; Colwell, C.W. Incidence of "Squeaking" after Ceramic-on-Ceramic Total Hip Arthroplasty. *Clin. Orthop. Relat. Res.* **2010**, *468*, 413–7, doi:https://doi.org/10.1007/s11999-009-1083-4.
124. Mendenhall, S. Orthopedic Network News. *Orthop. Netw. News* **2020**, *31*, 1–24.
125. McEntire, B.J.; Lakshminarayanan, R.; Ray, D.A.; Clarke, I.C.; Puppulin, L.; Pezzotti, G.; Bal, B.S.; Lakshminarayanan, A.; Bock, R. Silicon Nitride Bearings for Total Joint Arthroplasty. *Lubricants* **2016**, *4*, 1–24, doi:https://doi.org/10.3390/lubricants4040035.
126. Gorth, D.J.; Puckett, S.; Ercan, B.; Webster, T.J.; Rahaman, M.; Bal, B.S. Decreased Bacteria Activity on Si_3N_4 Surfaces Compared with PEEK or Titanium. *Int. J. Nanomedicine* **2012**, *7*, 4829–4840, doi:https://doi.org/10.2147/IJN.S35190.
127. Webster, T.J.; Patel, A.A.; Rahaman, M.N.; Bal, B.S. Anti-Infective and Osteointegration Properties of Silicon Nitride, Poly (Ether Ether Ketone), and Titanium Implants. *Acta Biomater.* **2012**, *8*, 4447–4454.
128. Pezzotti, G.; Bock, R.M.; McEntire, B.J.; Jones, E.; Boffelli, M.; Zhu, W.; Baggio, G.; Boschetto, F.; Puppulin, L.; Adachi, T.; *et al.* Silicon Nitride Bioceramics Induce Chemically Driven Lysis in Porphyromonas Gingivalis. *Langmuir* **2016**, *32*, 3024–3035, doi:https://doi.org/10.1021/acs.langmuir.6b00393.

129. Bock, R.M.; Jones, E.N.; Ray, D.A.; Bal, B.S.; Pezzotti, G.; McEntire, B.J. Bacteriostatic Behavior of Surface-Modulated Silicon Nitride in Comparison to Polyetheretherketone and Titanium. *J. Biomed. Mater. Res. Part A* **2017**, *105*, 1521–1534, doi:https://doi.org/10.1002/jbm.a.35987.
130. Ishikawa, M.; Bentley, K.L. de M.; McEntire, B.J.; Bal, B.S.; Schwarz, E.M.; Xie, C.; Avenue, E.; McEntire, B.J.; Schwarz, E.M.; Xie, C. Surface Topography of Silicon Nitride Affects Antimicrobial and Osseointegrative Properties of Tibial Implants in a Murine Model. *J. Biomed. Mater. Res. A* **2017**, *105*, 3413–3421, doi:https://doi.org/10.1002/jbm.a.36189.
131. Pezzotti, G.; Bock, R.M.; McEntire, B.J.; Adachi, T.; Marin, E.; Boschetto, F.; Zhu, W.; Mazda, O.; Bal, B.S. *In vitro* Antibacterial Activity of Oxide and Non-Oxide Bioceramics for Arthroplastic Devices: I. In situ Time-Lapse Raman Spectroscopy. *Analyst* **2018**, *143*, 3708–3721, doi:https://doi.org/10.1039/C8AN00233A.
132. Boschetto, F.; Toyama, N.; Horiguchi, S.; Bock, R.M.; McEntire, B.J.; Adachi, T.; Marin, E.; Zhu, W.; Mazda, O.; Bal, B.S.; *et al. In vitro* Antibacterial Activity of Oxide and Non-Oxide Bioceramics for Arthroplastic Devices: II. *In situ* Time-Lapse Fourier Transform Infrared Spectroscopy. *Analyst* **2018**, *143*, 2128–2140, doi:https://doi.org/10.1039/C8AN00234G.
133. Pezzotti, G. A Spontaneous Solid-State NO Donor to Fight Antibiotic Resistant Bacteria. *Mater. Today Chem.* **2018**, *9*, 80–90, doi:https://doi.org/10.1016/j.mtchem.2018.05.004.
134. Boschetto, F.; Adachi, T.; Horiguchi, S.; Fainozzi, D.; Parmigiani, F.; Marin, E.; Zhu, W.; McEntire, B.J.; Yamaoto, T.; Kanamura, N.; *et al.* Monitoring Metabolic Reactions in *Staphylococcus epidermidis* Exposed to Silicon Nitride using *in situ* Time-Lapse Raman Spectroscopy. *J. Biomed. Opt.* **2018**, *23*, 1–10, doi:https://doi.org/10.1117/1.JBO.23.5.056002.
135. Pezzotti, G. Silicon Nitride: A Bioceramic with a Gift. *ACS Appl. Mater. Interfaces* **2019**, *11*, 26619–26636, doi:https://doi.org/10.1021/acsami.9b07997.
136. Pezzotti, G.; Fujita, Y.; Boschetto, F.; Zhu, W.; Marin, E.; Vandelle, E.; McEntire, B.J.; Bal, B.S.; Giarola, M.; Makimura, K.; *et al.* Activity and Mechanism of Action of the Bioceramic Silicon Nitride as an Environmentally Friendly Alternative for the Control of the Grapevine Downy Mildew Pathogen Plasmopara viticola. *Front. Microbiol.* **2020**, *11*, 1–17, doi:https://doi.org/10.3389/fmicb.2020.610211.
137. Pezzotti, G.; Asai, T.; Adachi, T.; Ogitani, E.; Yamamoto, T.; Kanamura, N.; Boschetto, F.; Zhu, W.; Zanocco, M.; Marin, E.; *et al.* Antifungal Activity of Polymethyl Methacrylate/Si_3N_4 Composites against Candida Albicans. *Acta Biomater.* **2021**, *126*, 259–276, doi:https://doi.org/10.1016/j.actbio.2021.03.023.
138. Pezzotti, G.; Boschetto, F.; Ogitani, E.; Fujita, Y.; Zhu, W.; Marin, E.; McEntire, B.J.; Bal, B.S.; Mazda, O. Silicon Nitride: A Potent Solid-State Bioceramic Inactivator of ssRNA Viruses. *Sci. Rep.* **2021**, *11*, 2977, doi:https://doi.org/10.1038/s41598-021-82608-3.
139. Pezzotti, G. Rapid Inactivation of SARS-CoV-2 by Silicon Nitride, Copper, and Aluminum Nitride. *BioRxiv Prepr.* **2020**, 1–16, doi:https://doi.org/10.1101/2020.06.19.159970.
140. Pezzotti, G.; Ohigitani, E.; Shin-Ya, M.; Adachi, T.; Marin, E.; Boshetto, F.; Zhu, W.; Mazda, O. Instantaneous "Catch-and-Kill" Inactivation of SARS-CoV-2 by Nitride Ceramics. *Clin. Transl. Med.* **2020**, 1–4, doi:https://doi.org/10.1002/ctm2.212.
141. Pezzotti, G.; Boschetto, F.; Ohgitani, E.; Fujita, Y.; Shin-Ya, M.; Adachi, T.; Yamamoto, T.; Kanamura, N.; Marin, E.; Zhu, W.; *et al.* Raman Molecular Fingerprints of SARS-CoV-2 British Variant and the Concept of Raman Barcode. *Adv. Sci.* **2021**, *2103287*, 1–10, doi:https://doi.org/10.1002/advs.202103287.
142. Piconi, C.; Ionescu, A.C.; Cochis, A.; Iasi, E.; Brambilla, E.; Rimondini, L. Bioceramic Materials Show Reduced Pathological Biofilm Formation. *Key Eng. Mater.* **2015**, *631*, 448–453, doi:https://doi.org/10.4028/www.scientific.net/KEM.631.448.
143. Trampuz, A.; Maiolo, E.M.; Winkler, T.; Perka, C. Biofilm Formation on Ceramic, Metal, and Polyethylene Bearing Components From Hip Joint Replacement Systems. *Orthop. Proc.* **2016**, *98-B*, 80, doi:doi:https://doi.org/10.1302/1358-992X.98BSUPP_10.ISTA2015-080.

144. Sorrentino, R.; Cochis, A.; Azzimonti, B.; Caravaca, C.; Chevalier, J.; Kuntz, M.; Porporati, A.A.; Streicher, R.M.; Rimondini, L. Reduced Bacterial Adhesion on Ceramics used for Arthroplasty Applications. *J. Eur. Ceram. Soc.* **2018**, *38*, 963–970, doi:https://doi.org/10.1016/j.jeurceramsoc.2017.10.008.
145. Naufal, E.; Wouthuyzen-Bakker, M.; Babazadeh, S.; Stevens, J.; Choong, P.F.M.; Dowsey, M.M. Methodological Challenges in Predicting Periprosthetic Joint Infection Treatment Outcomes: A Narrative Review. *Front. Rehabil. Sci.* **2022**, *3*, 1–7, doi:https://doi.org/10.3389/fresc.2022.824281.
146. Mian, H.M.; Lyons, J.G.; Perrin, J.; Froehle, A.W.; Krishnamurthy, A.B. A Review of Current Practices in Periprosthetic Joint Infection Debridement and Revision Arthroplasty. *Arthroplasty* **2022**, *4*, 1–12, doi:https://doi.org/10.1186/s42836-022-00136-5.
147. Rambo, W.M. Treatment of Lumbar Discitis using Silicon Nitride Spinal Spacers: A Case Series and Literature Review. *Int. J. Surg. Case Rep.* **2018**, *43*, 61–68, doi:https://doi.org/10.1016/j.ijscr.2018.02.009.
148. Karczewski, D.; Pumberger, M.; Müller, M.; Andronic, O.; Perka, C.; Winkler, T. Implications for Diagnosis and Treatment of Peri-Spinal Implant Infections from Experiences in Periprosthetic Joint Infections—A Literature Comparison and Review. *J. Spine Surgery; Vol 6, No 4 (December 17, 2020) J. Spine Surg.* **2020**, *6*, 800–813.
149. Premkumar, A.; Kolin, D.A.; Farley, K.X.; Wilson, J.M.; McLawhorn, A.S.; Cross, M.B.; Sculco, P.K. Projected Economic Burden of Periprosthetic Joint Infection of the Hip and Knee in the United States. *J. Arthroplasty* **2021**, *36*, 1484-1489.e3, doi: https://doi.org/10.1016/j.arth.2020.12.005.
150. Gilbert, J.L.; Buckley, C.A.; Jacobs, J.J. In vivo Corrosion of Modular Hip Prosthesis Components in Mixed and Similar Metal Combinations. The Effect of Crevice, Stress, Motion, and Alloy Coupling. *J. Biomed. Mater. Res.* **1993**, *27*, 1533–1544, doi: https://doi.org/10.1002/jbm.820271210.
151. McGrory, B.J. High Incidence of Mechanically Assisted Crevice Corrosion at 10 Years in Non-Cemented, Non-Recalled, Contemporary Total Hip Arthroplasties, *J. Arthroplasty* **2022**, *37*, S941–S946, doi: https://doi.org/10.1016/j.arth.2021.11.026.
152. Kurtz, S.M.; Kocagöz, S.B.; Hanzlik, J.A.; Underwood, R.J.; Gilbert, J.L.; MacDonald, D.W.; Lee, G. C.; Mont, M.A.; Kraay, M.J.; Klein, G.R.; *et al.* Do Ceramic Femoral Heads Reduce Taper Fretting Corrosion in Hip Arthroplasty? A Retrieval Study. *Clin. Orthop. Relat. Res.* **2013**, *471*, 3270–82, doi:https://doi.org/10.1007/s11999-013-3096-2.
153. Kocagoz, S.B.; Underwood, R.J.; MacDonald, D.W.; Gilbert, J.L.; Kurtz, S.M. Ceramic Heads Decrease Metal Release Caused by Head-Taper Fretting and Corrosion. *Clin. Orthop. Relat. Res.* **2016**, *474*, 985–994, doi:https://doi.org/10.1007/s11999-015-4683-1.
154. Eichler, D.; Barry, J.; Lavigne, M.; Massé, V.; Vendittoli, P.A. No Radiological and Biological Sign of Trunnionosis with Large Diameter Head Ceramic Bearing Total Hip Arthroplasty after 5 Years. *Orthop. Traumatol. Surg. Res.* **2021**, *107*, 102543, doi:https://doi.org/10.1016/j.otsr.2019.12.015.
155. Cooper, H.J.; Urban, R.M.; Wixson, R.L.; Meneghini, R.M.; Jacobs, J.J. Adverse Local Tissue Reaction Arising from Corrosion at the Femoral Neck-Body Junction in a Dual-Taper Stem with a Cobalt-Chromium Modular Neck. *J. Bone Jt. Surg.* **2013**, *95 Am*, 865–72, doi:https://doi.org/10.2106/JBJS.L.01042.
156. Koch, C.N.; Figgie, M.; Figgie, M.P.; Elpers, M.E.; Wright, T.M.; Padgett, D.E. Ceramic Bearings with Titanium Adapter Sleeves Implanted During Revision Hip Arthroplasty Show Minimal Fretting or Corrosion: a Retrieval Analysis. *HSS J.* **2017**, *13*, 241–247, doi:https://doi.org/10.1007/s11420-017-9566-4.
157. Wyles, C.C.; Kolz, J.M.; Van Citters, D.W.; Berry, D.J.; Trousdale, R.T. *In Vivo* Corrosion of Sleeved Ceramic Femoral Heads: A Retrieval Study. *J. Arthroplasty* **2021**, *36*, 1133–1137, doi: https://doi.org/10.1016/j.arth.2020.10.005.

158. Preuss, R.; Lars Haeussler, K.; Flohr, M.; Streicher, R.M. Fretting Corrosion and Trunnion Wear-Is it also a Problem for Sleeved Ceramic Heads? *Semin. Arthroplasty* **2012**, *23*, 251–257, doi:https://doi.org/10.1053/j.sart.2013.01.008.
159. Khullar, P.; Zhu, D.; Gilbert, J.L. Fretting Corrosion of Si_3N_4 vs CoCrMo Femoral Heads on Ti-6Al-V Trunnions. *J. Orthop. Res.* **2020**, *38*, 1617–1626, doi:https://doi.org/10.1002/jor.24681.
160. Betts, A.J.; Boulton, L.H. Crevice Corrosion: Review of Mechanisms, Modelling, and Mitigation. *Br. Corros. J.* **1993**, *28*, 279–295, doi:https://doi.org/10.1179/000705993799156299.
161. Kamachimudali, U.; Sridhar, T.M.; Raj, B. Corrosion of Bio Implants. *Sadhana* **2003**, *28*, 601–637, doi:https://doi.org/10.1007/BF02706450.
162. Uddin, M.S.; Majewski, P. Frictional Heating in Hip Implants—A Review. *Procedia Eng.* **2013**, *56*, 725–730, doi: https://doi.org/10.1016/j.proeng.2013.03.185.
163. Bergmann, G.; Graichen, F.; Rohlmann, A.; Verdonschot, N.; van Lenthe, G.H. Frictional Heating of Total Hip Implants, Part 1: Measurements in Patients. *J. Biomech.* **2001**, *34*, 421–428, doi: https://doi.org/10.1016/S0021-9290(00)00188-3.
164. Bergmann, G.; Graichen, F.; Rohlmann, A.; Verdonschot, N.; van Lenthe, G.H. Frictional Heating of Total Hip Implants. Part 2: Finite Element Study. *J. Biomech.* **2001**, *34*, 429–435, doi: https://doi.org/10.1016/S0021-9290(00)00234-7.
165. Damm, P.; Bender, A.; Duda, G.; Bergmann, G. In vivo Measured Joint Friction in Hip Implants During Walking After a Short Rest. *PLoS One* **2017**, *12*, e0174788.
166. Sagbas, B.; Durakbasa, M.N.; Sagbas, M.; Koyun, A. Measurement and Theoretical Determination of Frictional Temperature Rise Between Sliding Surfaces of Artificial Hip Joints. *Measurement* **2014**, *51*, 411–419, doi: https://doi.org/10.1016/j.measurement.2013.12.020.
167. Tsai, S.; Salehi, A.; Aldinger, P.; Hunter, G. Heat Generation and Dissipation Behavior of Various Orthopaedic Bearing Materials. *Key Eng. Mater.* **2006**, *309–311*, 1291–1294.
168. Sagbas, B.; Durakbasa, M.N. Optimization of Frictional Heating Parameters of Artificial Joint Materials by Taguchi Approach. In Proceedings of the 12th IMEKO TC10 Workshop on Technical Diagnostics; Florence, Italy, 2013; pp. 96–101.
169. Mattei, L.; Di Puccio, F.; Piccigallo, B.; Ciulli, E. Lubrication and Wear Modelling of Artificial Hip Joints: A Review. *Tribol. Int.* **2011**, *44*, 532–549, doi:https://doi.org/10.1016/j.triboint.2010.06.010.
170. Krismer, M. Sports Activities After Total Hip Arthroplasty. *EFORT Open Rev.* **2017**, *2*, 189–194, doi:https://doi.org/10.1302/2058-5241.2.160059.
171. Bowsher, J.G.; Clarke, I.C. Thermal Conductivity of Femoral Ball Strongly Influenced UHMWPE Wear in a Hip Simulator. In Proceedings of the Transactions of the 53rd Annual Meeting of the Orthopaedic Research Society; Orthopaedic Research Society: San Diego, CA, 2007; p. 0278.
172. Candadai, A.A.; Weibel, J.A.; Marconnet, A.M. Thermal Conductivity of Ultrahigh Molecular Weight Polyethylene: From Fibers to Fabrics. *ACS Appl. Polym. Mater.* **2020**, *2*, 437–447, doi:https://doi.org/10.1021/acsapm.9b00900.

Silicon Nitride for Dental Applications

Ichiro Nishimura, Giuseppe Pezzotti, and Bryan J. McEntire

1 Introduction

The oral cavity is not only subjected to physical, chemical, and thermal injuries [1] but also more significantly carries the largest variety of microbial colonies [2, 3]. The exposed surface of the tooth is composed of enamel tissue, the hardest structure in the entire body. However, the fermentation of carbohydrates by oral microorganisms leads to the production of an acidic environment which can demineralize the enamel layer [4]. The tooth-supporting gingiva is one of the most active barrier tissues, which may overreact to the pathological oral environment resulting in chronic inflammation [5]. Imbalances in the oral homeostatic environment and dysbiosis of oral microbes can cause oral maladies such as dental caries and periodontal disease [4]. Furthermore, numerous reports have repeatedly established that dental diseases impactfully affect the whole human body [6–8]. Therefore, the maintenance of oral health is critically important. The World Health Organization (WHO) proposes that the best approach to address chronic noncommunicable diseases is comprehensive prevention and control. The WHO further proposes four levels of prevention, corresponding to different phases of the development of diseases as illustrated in Fig. 1 [9].

I. Nishimura (✉)
The Weintraub Center for Reconstructive Biotechnology, UCLA School of Dentistry, Los Angeles, CA, USA
e-mail: inishimura@dentistry.ucla.edu

G. Pezzotti
Ceramic Physics Laboratory, Kyoto Institute of Technology, Sakyo-ku, Matsugasaki, Kyoto, Japan

B. J. McEntire
Emeritus Scientific Officer, SINTX Technologies, Inc., Salt Lake City, UT, USA

B. S. Bal et al. (eds.), *Silicon Nitride Bioceramics*,
https://doi.org/10.1007/978-3-031-67047-3_12

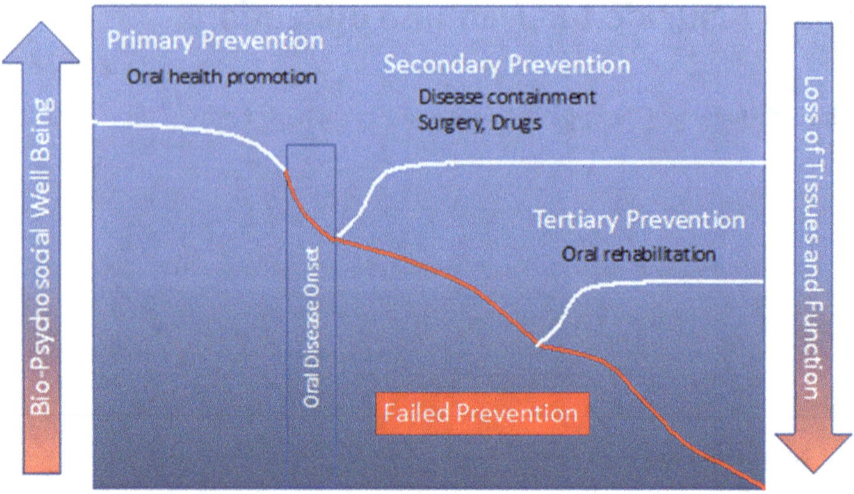

Fig. 1 Diagram depicting the dynamic transition of disease onset to unmanaged progression. The WHO proposes providing specialized measures to prevent the loss of tissues and function and to maintain bio-psychosocial well-being. (Original figure based on Ref. [9])

Primordial prevention aims to establish and maintain a healthy environment through public health policy and health promotion. Dental and oral primordial prevention includes oral hygiene promotion and fluoridation of community water. *Primary prevention* aims to reduce the incidence of disease in selected groups and individuals at elevated risk. For example, periodic fluoride application by a dental healthcare provider and fisher sealant for pediatric patients are used to prevent dental caries. *Secondary prevention* targets individuals with initial stages of established diseases and reduces the prevalence by shortening the disease duration. Interventional therapies for dental caries and periodontal diseases are considered secondary prevention. *Tertiary prevention* aims to reduce the impact of complications for late-stage diseases. The goal of edentulism treatment is consistent with tertiary prevention and is achieved through oral rehabilitation to obtain new bio-psychosocial stability. Dental biomaterials have been used as critical elements to achieve oral rehabilitation. The therapeutic goal in each stage of the prevention model is to stabilize the declining bio-psychosocial well-being of the patient and to minimize the loss of tissue and function.

As the demand for metal-free and biocompatible solutions in dentistry grows, silicon nitride (Si_3N_4) is emerging as a captivating contender. Although it is renowned for its exceptional strength, toughness, and wear resistance [10], Si_3N_4 offers a unique combination of features that could revolutionize dental applications. Beyond its mechanical benefits, it is proven to be exceptionally biocompatible; and it has inherent osteoconductive and antibacterial properties, attributes that are also highly sought-after in dental restorations and implants [11–16]. Compared to metal implants, Si_3N_4 offers inherent X-ray radiolucency, enabling precise intraosseous placement and visualization of adjacent bone and soft tissues during surgical procedures [17, 18]. Studies have also confirmed that it can be used as a filler or a porous scaffold to strengthen and harden dental resins [19–21]. While its intrinsic light-gray coloration and visible-light translucency currently fall short of glass-ceramic veneers or zirconia (ZrO_2), research efforts are actively pursuing advancements in adaptive coloration and translucency to bridge this gap [22–26]. Because of these attributes, recent review articles have explored the potential of this innovative material for the oral environment [27–30]; and prototype Si_3N_4 implants with excellent biocompatibility and mechanical properties have been produced from CAD/CAM blocks and by 3D-printing [31, 32]. Consequently, this chapter delves into the enthusiasm for Si_3N_4 as a dental prosthesis by comprehensively exploring its key facets, including: (1) mechanical stability—strength, fracture toughness, hardness, and wear resistance are crucial factors for enduring the demanding oral environment; (2) osseous integration—appropriate interactions with skeletal hard tissues are essential for long-term implant success; and (3) antimicrobial characteristics—Si_3N_4's unique antipathogenic chemistry leads to enhanced implant hygiene and reduced peri-implant inflammation.

2 Mechanical Stability

The major goals of dental treatments are to stabilize and maintain oral functions supporting the overall health and welfare of patients. One of the tasks of the jaw is to masticate food producing crushed small pieces for easy deglutition or swallowing. Masticatory muscles generate a maximum occlusal force of over 1000 N in males and nearly 700 N in females [33]. However, it has been reported that the compressive forces required to chew modern food such as beef jerky, carrots, and bread are 300 N, 250 N, and 50 N, respectively [34]. The average effective occlusal area of the maxillary and mandibular first molars is approximately 17 mm^2 [35] and thus the required withstanding force at the first molar may be 40 ~ 60 MPa. Historically, dental materials such as precious and nonprecious alloys, silver amalgam, composite resins, and ceramics have been applied for the restoration of damaged dentition. Because these restorative biomaterials must withstand the occlusal force, their mechanical properties have been the primary focus of dental biomaterials research. Mechanical stability is also a critical issue for dental implant systems.

On the one hand, commercially available pure titanium (cp Ti) exhibits tensile strengths of between 330–860 MPa and a fracture toughness >70 $MPa{\cdot}m^{1/2}$ [36], while yttria-doped tetragonal zirconia polycrystals (YTZP or simply ZrO_2) reportedly have a range of flexural strengths of 700–1500 MPa and fracture toughness of 4.5–20 $MPa{\cdot}m^{1/2}$ [37]. Both materials have properties well above the required strength to withstand occlusal forces. Polyetheretherketone (PEEK) and carbon-fiber-reinforced polyetheretherketone (CFR-PEEK) have also been touted to be appropriate dental biomaterials. Even though their compressive and flexural strength, fracture toughness, and hardness are substantially lower than Ti or ZrO_2,

Table 1 Comparative properties for titanium (cp Ti), a Ti alloy (Ti6Al4V), zirconia (YTZP ZrO_2), polyetheretherketone (PEEK), carbon-fiber-reinforced PEEK (CFR-PEEK), silicon nitride (Si_3N_4), and cortical bone [36–40]

Property	Units	Material				
		Ti and Ti alloy	YTZP ZrO_2	Si_3N_4	PEEK and CFR-PEEK	Cortical bone
Density	g/cc	4.43–4.50	6.00–6.05	3.22–3.35	1.3–1.4	1.5–2.0
Elastic modulus	GPa	105–120	210–223	296–313	3.5–18.0	7.5–25.8
Poisson's ratio	NA	0.31–0.34	0.30–0.33	0.27	0.3–0.4	0.19–0.48
Compressive strength	MPa	800–970	2000–2200	4000	118–210	70–260
Flexural or tensile strength	MPa	330–970	700–1500	800–1100	100–337	90–228
Fracture toughness	$MPa{\cdot}m^{1/2}$	46.3–93.3	4.5–20.0	4.40–15.0	5.5–10.4[a]	1.0–20.0
Hardness	GPa	2.8–3.3	11.0–12.5	15.0	99–105[b]	0.68–0.78

Additional physical and mechanical property data for PEEK and CFR-PEEK obtained from https://invibio.com/en/materials-solutions. (Accessed on 12 Feb 2024)
[a]Notched Izod Impact Strength in kJ/m^2; [b]Rockwell M Scale

they appear adequate for occlusal forces. Ti, ZrO_2, PEEK, and CFR-PEEK have US regulatory approval for use as various dental prostheses, including permanent implants.

On the other hand, Si_3N_4, a nonoxide structural ceramic, also appears to have appropriate properties for dental appliances, but it has yet to receive US regulatory clearance for this purpose. Its flexural strength and fracture toughness of 800–1100 MPa and 4.4–15 $MPa{\cdot}m^{1/2}$, respectively, are comparable to ZrO_2; thus, it may be considered a new material candidate for dentistry and implant systems. Comparative properties of these materials are provided in Table 1. A depiction of the implants is shown in Fig. 2.

While Ti and its alloys are widely used in dental implants due to their good mechanical properties and biocompatibility, corrosion concerns in the oral environment persist. Microorganisms can collect along subgingival fissures and pockets triggering electrochemical corrosion of Ti, leading to the release of cytotoxic Ti^{4+} ions. This can result in the local generation of pro-inflammatory cytokines, and osteoclastic action, leading to gingiva and bone recession. This situation is exacerbated in individuals who have metal allergies, creating systemic nonlocalized hypersensitive reactions [41]. To address these limitations, patients, dental practitioners, and researchers actively pursue the mantra, "no metal in the mouth." Nonmetallic implants (e.g., ceramics, polymers), appear to provide superior functionality while possessing desirable aesthetics and enhanced resistance to the harsh oral environment.

Despite being a relatively recent addition to the dental material landscape, ZrO_2 is now well accepted for implants, abutments, crowns, and bridges because of its

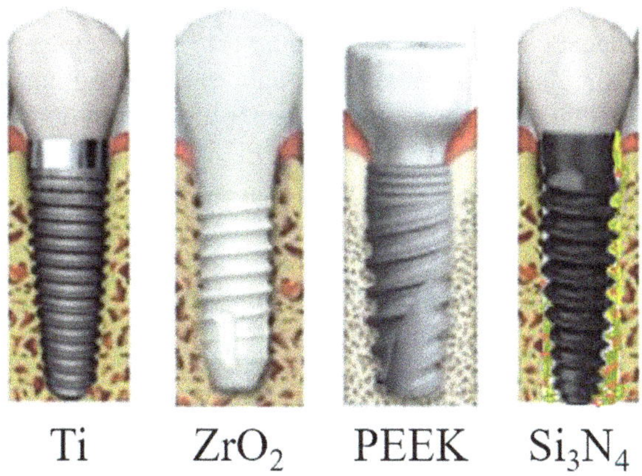

Fig. 2 Dental implant materials. (Modified and used with permission under Creative Commons Attribution License 4.0 [30])

outstanding mechanical and esthetic properties [42]. As indicated in Table 1, its compressive and flexural strength are superior to Ti materials. Although lower than Ti, its fracture toughness is adequate and basically equivalent to cortical bone. Additionally, its high hardness aligns with its inherent nature as an oxide ceramic. Nevertheless, there are mechanical concerns with ZrO_2. Its impressive strength stems from its metastable tetragonal (t-ZrO_2) phase chemistry [43]. Component fracture is ostensibly arrested when a propagating crack is placed under compression by the localized transformation of t-ZrO_2 to its stable monoclinic form, m-ZrO_2. However, it is well known that premature transformation can also spontaneously occur due to the hydrothermal aging conditions present within the oral cavity. Consequently, the toughening mechanism responsible for YTZP ZrO_2's improved strength, can, over time, be progressively reversed under the combined influence of cyclic loading and chemical interactions [44]. Still, this issue remains controversial because recent in vitro and retrieval studies suggest that, although the transformation of the implant's surface occurs, it is not necessarily detrimental to mechanical strength or long-term survivability [45, 46]. Also, since the commercial introduction of ZrO_2 for dentistry in 2004, compositional modifications and glaze applications have been developed to inhibit the $t \rightarrow m$ phase transformation [47]; and recent systematic reviews and meta-analyses place the 1 to 5-year survival rate for ZrO_2 implants at between 92% and 97.2% [48, 49]. Although ZrO_2 implants do not suffer from problems of metal sensitivity and microbial-induced metallic corrosion, their osseous integration is not markedly superior to Ti implants, and failures associated with fractures are higher. Nevertheless, their short-term survivability (i.e., minimum 12-month follow-up) was deemed to be statistically equivalent to titanium implants [50].

The mechanical property profiles of Ti and its alloys, ZrO_2 and Si_3N_4 are superior to PEEK materials; yet both bioceramics along with Ti have higher elastic moduli than the PEEK compounds. Having an elastic modulus similar to native bone is touted to be one of the significant advantages of PEEK because it reduces the possibility of stress shielding, which purportedly leads to bone resorption, implant loosening, and eventual prosthesis failure [40]. Yet, there is a paucity of clinical evidence supporting this hypothesis [51, 52]. Implant success is likely influenced by a combination of factors, and understanding the interplay between material properties, biomechanics, and patient-specific conditions remains an ongoing area of research [40]. Nevertheless, PEEK is growing in popularity as a dental prosthesis because of its adequate mechanical properties, ease of manufacturability (e.g., adaptability to CAD/CAM systems and 3D printing), radio transparency, metal-free nature, and aesthetics. Unfortunately, PEEK is also completely bioinert and hydrophobic, which limits its ability to adequately integrate with bone; but several engineering strategies are being explored to overcome this deficiency [17, 53, 54].

Although Si_3N_4 has yet to receive regulatory approval for use in dentistry, the data presented in Table 1 indicate that it has the requisite mechanical stability. It is a superior choice over PEEK materials in all mechanical attributes except elastic modulus. It has the highest stiffness of any dental ceramic. While this may be concerning because of stress shielding, it is important to note that osseous integration is not simply a function of implant stiffness. Indeed, it is a multifactorial phenomenon that not only includes elastic modulus, but also must account for implant design and placement, interfacial loading conditions, implant surface chemistry, and manufacturing techniques [55]. In comparison with Ti, its alloys, and ZrO_2, Si_3N_4 has similar flexural and higher compressive strengths while its fracture toughness is comparable to ZrO_2 and cortical bone. Some publications may suggest ZrO_2 has higher hardness than Si_3N_4 (i.e., $H > 17$ GPa) [56], but this finding is an artifact of ZrO_2's transformation toughening mechanism which limits indenter penetration. Si_3N_4 has the highest hardness of any of the biomaterials considered for dental applications with the possible exception of alumina (Al_2O_3, $H = 18.0 \sim 23.0$ GPa) [37]. Si_3N_4 is also an inherently mechanically stable bioceramic. Its toughness and high strength are the result of its fibrous microstructure which forces a propagating crack to follow a high-energy tortuous path, while bridging grains in the crack-wake place the crack-tip in compression [57, 58]. Unlike ZrO_2, its phase chemistry and mechanical properties remain unaffected even under severe hydrothermal conditions [59]. In summary, Si_3N_4 is emerging as a promising bioceramic for dental applications, combining strength, toughness, and corrosion resistance to oppose environmental challenges.

3 Osseous Integration

Dental implant use has witnessed a remarkable surge in popularity, with prevalence jumping from 0.7% in 1999–2000 to 5.7% in 2015–2016, representing an almost ninefold increase in just over a decade. Notably, the largest absolute increase (12.9%) occurred within the 65–74 age group, while the most dramatic relative increase (~1000%) was observed among individuals aged 55–64. Looking ahead, projections suggest continued growth, with dental implant prevalence reaching an estimated 23% by 2026 [60]. The success of dental implants hinges on a critical process known as osseointegration, where surgically placed implants directly fuse with alveolar bone. Initially defined as a purely structural connection between bone and implants, the concept of osseointegration has evolved to encompass both structural and functional aspects, emphasizing its clinical significance. This paradigm shift stemmed from the pioneering work of Brånemark in the 1950s, who accidentally discovered the remarkable stability achieved by titanium fixtures within bone tissue. Before this transformative discovery, traditional stainless-steel blade-vent implants relied on fibrous tissue encapsulation, offering a less robust means of integration [61]. The clinical success of blade implants was debated, and the risk of failure was suspected to be caused, in part, by the fibrous integration responding to a prolonged inflammatory response to stainless steel materials.

3.1 Titanium, Zirconia, and PEEK

Ti-based biomaterials have revolutionized osseointegration implant systems, achieving remarkable clinical success with 10-year survival rates ranging from 80.5% to 97.9% [62]. This exceptional performance is attributed, in part, to the biofriendly nature of Ti, which minimizes inflammatory responses compared to materials like stainless steel. Furthermore, Ti exhibits minimal influence on bone metabolism, promoting a more favorable osseointegration environment. Experimental evaluation of osseointegration for Ti implants relies heavily on histological observation, specifically the measurement of bone-to-implant contact (BIC) area. Preclinical studies have consistently demonstrated a positive correlation between increased surface roughness of Ti implants and improved BIC values. This relationship was further validated in a human study where micro-Ti implants with varying surface topographies (machined/smooth vs. sandblasted and acid-etched) were retrieved after a two-month healing period under unloaded conditions. The study revealed a statistically significant increase in mean BIC, rising from 23.1% ± 12.0% for smooth surfaces to 42.8% ± 9.8% for roughened surfaces, highlighting the impact of micro-topography on osseointegration success [63].

Extensive research has delved into the mechanisms underlying osseointegration with rough-surface implants. Initially, the increased surface area created by roughness was hypothesized to enhance mechanical interlocking between bone tissue and the implant, thereby bolstering osseointegration [64]. Finite element analysis simulations of rough-surface implants correlated with in vivo data, lending initial support to this theory. However, further investigation revealed discrepancies between the modeled and actual data, suggesting that purely mechanical interlocking cannot fully explain this complex phenomenon [65].

Ogawa and colleagues uncovered compelling evidence for altered intrinsic biomechanical properties of bone tissue formed in close contact with rough-surface implants, both in vitro [66, 67] and in vivo [68]. Notably, bone tissue associated with acid-etched implants exhibited three times greater hardness compared to that integrated with machined implants, approaching the mechanical properties of cortical bone. This intriguing observation prompted the exploration of the underlying mechanisms. It is well established that the mechanical properties of bone tissue are directly influenced by the biological components synthesized by osteoblasts. In the context of osteogenesis imperfecta (OI), a group of genetic disorders characterized by softer and brittle bone, mutations primarily affect genes encoding bone extracellular matrix molecules like type I collagen [69]. However, collagen cross-linking enzymes may also play a significant role in influencing bone's mechanical properties. Utilizing a differential gene expression platform, Ogawa and Nishimura identified the upregulation of specific genes, such as prolyl 4-hydroxylase (P4H), in bone tissue associated with rough-surface implants [70]. Given P4H's crucial role in facilitating collagen cross-linking, the authors postulated that rough implant surface-induced P4H upregulation might contribute to the observed increase in surrounding bone hardness, warranting further research into this intriguing potential mechanism.

Numerous in vitro studies have employed osteoblastic cells to investigate the biological effects of surface-treated Ti biomaterials [71–74]. These studies have consistently utilized established assays to demonstrate enhanced or accelerated osteogenic differentiation of the cultured cells on rough-surface Ti compared to smoother counterparts. This observation is further corroborated by quantitative reverse-transcription polymerase chain reaction (qRT-PCR) analyses, which reveal upregulated expression of key osteoblastic genes [75, 76]. Collectively, this compelling evidence suggests that rough-surface Ti biomaterials exhibit demonstrably improved osteoconductive properties, promoting bone formation and integration [77]. However, caution is warranted when interpreting qRT-PCR results, as inherent limitations can lead to false positives [78]. This underlines the importance of employing more comprehensive gene expression analyses for Ti biomaterials. Fortunately, recent advancements in unbiased genome-wide platforms like microarrays and single-cell RNA sequencing offer deeper insights. Interestingly, such studies have revealed that rough-surface Ti biomaterials do not significantly influence the expression of previously identified osteogenic genes [79]. Conversely, these materials were found to significantly modulate unexpected gene clusters like circadian clock genes [80, 81]. These findings challenge the initial assumption that rough-surface Ti biomaterials possess inherent osteoconductive properties, necessitating further research to fully characterize their osseointegration mechanisms [82, 83].

Bone formation is an energy-consuming biological process. Mitochondria are cytosolic organelles that uptake glucose-derived molecules, fatty acids, and other fuels for oxidative phosphorylation to generate adenosine triphosphate (ATP). Osteoblasts are metabolically active and require high levels of energy in the form of ATP [84, 85]. Targeted deletion of a core subunit composing mitochondrial complex I in mouse skeletal progenitor cells resulted in deficient bone embryogenesis and spontaneous bone fracture [86]. Increasing evidence supports the critical role of mitochondria in osteogenic cell proliferation and differentiation [87]. Cyclophilin D knockout mice, a genetic model of high oxidative metabolism, showed greater osseointegration in the spinal fusion site [88]. The role of implant materials in regulating mitochondrial activity in the peri-implant microenvironment suitable for osseointegration and bone regeneration has gained recent attention [89].

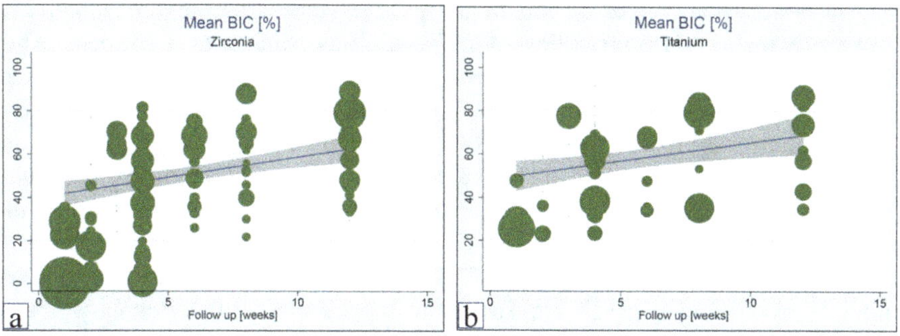

Fig. 3 Mean %BIC of implants installed in animals (**a**): zirconia, (**b**): titanium after different healing periods [weeks]. The green spots represent single observations gathered from the included studies. Spot sizes are relative to the weighting of the observations. (Reconfigured and used with permission [90])

As previously indicated, the 5-year survivability and performance of ZrO_2 implants were found to be statistically equivalent to Ti implants [49, 50]; so it could be assumed that the osseous integration mechanism for ZrO_2 might be the same as for Ti devices. Studies have sought to elucidate osseointegration differences between the two materials. Tests in laboratory animals have been conducted for this purpose, using monkeys, dogs, pigs, rats, rabbits, and sheep as models [90]. Pieralli et al. compiled a systematic review and meta-analysis of 54 animal studies and combined data irrespective of model. Osseous integration was determined by several factors including bone-to-implant-contact (BIC), removal torque analysis (RTQ), and push-in tests. The comparative BIC results for ZrO_2 and Ti implants are shown in Fig. 3a, b, respectively. The %BIC following implantation varied significantly between materials, follow-up duration, and animal model. For instance, after one week in pig maxillae, BIC was minimal (0%), while BIC reached up to 89.1% in rabbit femurs after 90 weeks. Data from 879 animals and 2916 implants were analyzed based on 249 observations. Overall, titanium implants achieved a mean BIC of 60.7% at the mean follow-up time. In comparison, ZrO_2 implants demonstrated a lower mean BIC of 57.2%. The weighted mean average for the ZrO_2 implants was particularly low for early osseointegration times (i.e., < 5 weeks). However, after >10 weeks, statistical analysis revealed no significant difference in BIC between the two implant materials ($p > 0.31$) [90]. This same conclusion was found for the RTQ and push-out tests.

Although there was a broad range of observed %BICs, the authors concluded that, on balance, there were no significant differences between animal types ($p > 0.75$); and the location of the implant also did not affect %BIC outcomes. However, their analyses suggested significant differences based on surface topography. Compared to implants with additive topographical changes (i.e., injection molded microtopography), subtractive modifications (i.e., sandblasting, acid etching) resulted in a statistically significant reduction in %BIC of -9.1% ($p = 0.03$). Furthermore, implants with smooth surfaces displayed the largest decrease in BIC, reaching −13.2% ($p = 0.02$). These findings suggest that surface texturization plays a key role in the osseointegration of ZrO_2 implants, similar to Ti devices. This conclusion is perhaps not surprising since both materials have similar reported ranges for isoelectric points ($ZrO_2 \approx 4.7 \sim 7.0$, [91, 92]; Ti $\approx 3.5 \sim 6.3$, [93, 94]), both spontaneously form a hydrated oxide layer on their surfaces when exposed to water or biological fluids [95], and both exhibit similar hydrophilicity characteristics [96]. Several studies have investigated differences in osteoblast interaction with zirconia and titanium surfaces [71, 97]. Although they reported subtle differences in early proliferation and adhesion of osteoblast cells, both materials exhibited similar protein synthesis, ultrastructural appearance of osteoblasts, and osteointegration effectiveness. These types of observations led one reviewer of the two materials to conclude that, "although surface characteristics, such as topography and chemistry, have often been discussed independently of each other, these characteristics are virtually inseparable" [98].

It should be noted that the osseous integration and antimicrobial characteristics of both Ti- and ZrO_2-based biomaterials can be improved by functionalizing their surfaces with bioactive Si_3N_4 powders. For instance, Pezzotti and coworkers were the first to demonstrate that Si_3N_4 powder can be embedded into titanium surfaces using laser-cladding. The resulting surface composite exhibited greater osteogenic activity than pure Ti alone [99]. (A review of this process and its outcomes is provided earlier in this compilation, cf., Chap. 7.) Later, Guo et al. used micro-arc oxidation (MAO) to deposit different concentrations of Si_3N_4 nanoparticles onto Ti substrates. Their study also demonstrated a remarkable improvement in cell proliferation, osteoblastic differentiation, and bactericidal efficacy [100].

Laser-cladding of Si_3N_4 powder onto the surface of dental ZrO_2 was first conducted by Marin et al. [101, 102]. The laser-cladding operation produced a uniformly coarse coating of partially melted Si_3N_4 powder, along with amorphous and nano-crystalline silicon from the decomposition of Si_3N_4 with a thickness of 15 ± 5 μm. The in vitro osteogenic viability and activity of SaOS-2 osteosarcoma cells were monitored using WST assays, fluorescence microscopy, and Raman spectroscopy. Four comparative materials were included in the study: (1) Monolithic Si_3N_4 (control), (2) ZrO_2 with a Si_3N_4 laser-cladded coating, (3) Roughened ZrO_2, and (4) Polished ZrO_2. The key result from this investigation is shown in Fig. 4. Using fluorescence microscopy, the amount and viability of the SaOS-2 adhering to the surface of each sample were analyzed by counting their living nuclei, and the presence of osteopontin and osteocalcin, two matricellular proteins that are markers of osteogenic activity. Consistent with reports on other oxide ceramics like alumina [15], cell proliferation was significantly higher on monolithic Si_3N_4 compared to either of the ZrO_2 materials ($p < 0.05$). Roughness generally promoted cellular adhesion, but only the coated ZrO_2 samples exhibited a homogeneous distribution of cell nuclei across their surfaces. Furthermore, both osteocalcin and osteopontin levels were significantly higher on coated ZrO_2 samples compared to both the smooth and rough ZrO_2 specimens ($p < 0.05$). Interestingly, the coated ZrO_2 samples displayed lower cellular proliferation and, consequently, lower osteocalcin and osteopontin levels compared to stoichiometric Si_3N_4 ($p < 0.05$). These findings

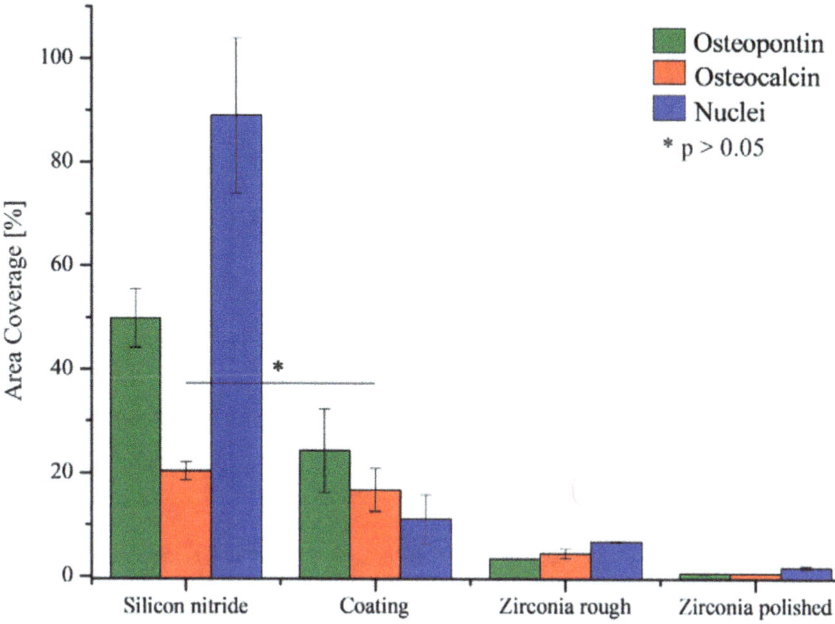

Fig. 4 Quantification of SaOS-2 cell proliferation and bone formation based on fluorescence microscopy. Direct cell nuclei (stained blue), osteopontin (stained red), and osteocalcin (stained green) were counted in the images for four substrates (1) Monolithic Si_3N_4, (2) ZrO_2 with a Si_3N_4 coating by laser-cladding, (3) Surface roughened ZrO_2, and (4) Polished ZrO_2. (Used with permission [102])

confirm that the positive control (monolithic Si_3N_4) supported the highest cell number, correlating with the most significant matrix protein synthesis. Quantification of osteocalcin and osteopontin also demonstrated a significant increase in the ZrO_2-coated samples compared to both smooth and rough ZrO_2 ($p < 0.05$). Remarkably, the smooth ZrO_2 specimens exhibited the lowest levels of mineralized matrix formation.

PEEK has emerged as a promising alternative to traditional metal implants in diverse fields like orthopedics, craniofacial surgery, spine surgery, and dentistry [103, 104]. Its biocompatibility, physical and chemical stability, and favorable mechanical properties make it a compelling candidate for such applications. However, a critical limitation of unmodified PEEK is its inherent hydrophobicity and bioinert nature, which significantly hinders its ability to achieve osseointegration. This obstacle remains the primary impediment to PEEK's broader clinical adoption across multiple medical disciplines. A systematic review by AlOtaibi et al. highlights the limitations in existing evidence for PEEK's osseointegration [105]. They identified a scarcity of randomized controlled trials and a lack of preclinical data from animal models demonstrating efficacy. Based on their analysis, surface functionalization approaches, either physical or chemical modifications, coatings, or composites appear to be crucial for enhancing PEEK's performance and

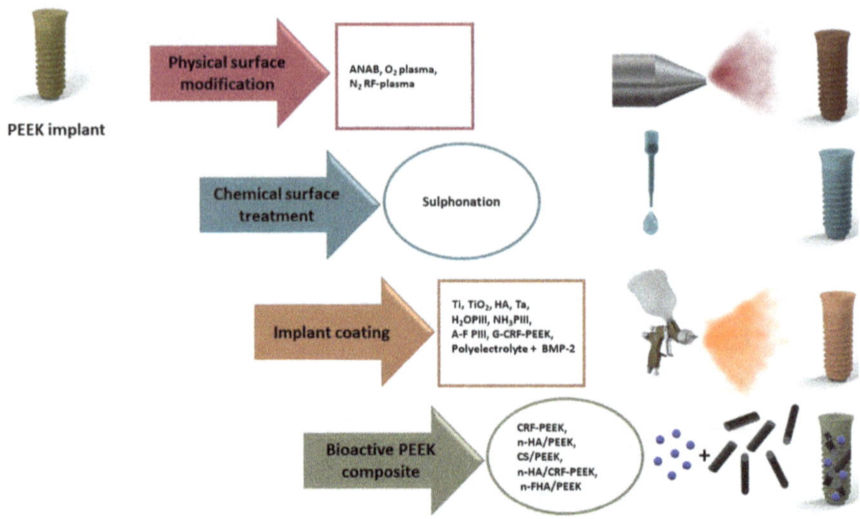

Fig. 5 The scheme of current strategies to modify the bioactivity of PEEK implants. (Used with permission under Creative Commons Attribution Noncommercial Share Alike License [105])

acceptance as a reliable bone prosthesis. Their proposed methods for functionalizing PEEK are shown in Fig. 5. Each of these techniques can be independently applied or synergistically combined to improve PEEK's osseointegration ability.

Dondani et al. [106] and Pidhatika et al. [107] completed contemporary independent systematic reviews of surface functionalization methods for PEEK. Physical treatments included the use of an accelerated neutron atom beam (ANAB) or various reactive plasmas like nitrogen, oxygen, argon, water vapor, or combinations. Some methods linked a plasma pretreatment with subsequent exposure to hydrofluoric acid or hydrogen peroxide, while others utilized the plasma to prepare the surface for a polymeric or antimicrobial coating. Each treatment favorably increased PEEK's wettability, adhesion, and differentiation of osteoblasts.

Traditional sandblasting and laser ablation have also been employed to preroughen implants in preparation for the application of a bioactive coating. In terms of coatings, sulfonation of the surface was found to be particularly effective in improving osseous integration, and numerous chemical recipes have been reported using sulfuric acid in combination with nitric acid or other agents. Sulfonation has also been utilized with plasma treatments. Other inorganic chemical treatments have been evaluated including phosphonation and silanization. Particulate coatings are also popular, including hydroxyapatite (HAp) or Ti. Polymer coatings have been explored, such as dexamethasone in combination with plasma treatments or with the inclusion of inorganic complexes (e.g., Mg and Zn polymers).

The authors did not address the application of Si_3N_4 powder coatings embedded in PEEK, which have promising in vitro potential for promoting osseous integration, as evidenced by findings from several research groups [108–111]. Dondani et al.

also reported that PEEK composites have been developed to improve their osseous integration ability, including mixtures of PEEK with Ti, titanium dioxide (TiO_2), HAp, carbon fibers, and calcium phosphate. Lastly, Si_3N_4 has also been incorporated into PEEK to form improved osseous-friendly composites [17, 112, 113]. Nevertheless, each of these technological approaches is at an early development stage, and significant additional in vitro, in vivo animal studies, and human clinical data will be required before surface functionalized PEEK implants become common in modern dentistry.

3.2 *Silicon Nitride*

Si_3N_4 has demonstrated promising properties for biomedical applications, with a notable track record of clinical effectiveness as an orthopedic implant, particularly in spinal fusion procedures [114–120]. This success motivates exploration of its potential in the oral cavity, where biocompatibility and long-term efficacy are crucial. However, direct extrapolation from orthopedic applications to the oral environment requires cautious consideration due to the distinct physiological and mechanical demands of this unique niche. Si_3N_4's biocompatibility and potential efficacy in oral applications must take into account factors such as osseous integration, tissue interactions, wear resistance, susceptibility to oral fluids, and bacterial colonization. Nevertheless, Si_3N_4 bioceramics are being increasingly considered as the next-generation biomaterial for both orthopedic and dental implants due to their osteoconductive activity, as demonstrated by several recent reviews [10, 121, 122].

While Si_3N_4 has yet to be cleared by relevant regulatory authorities for use in dentistry, significant in vitro, and in vivo animal studies have been conducted to demonstrate its osseous integration potential. For instance, in an early in vitro study, Si_3N_4 was compared with Ti6Al4V and PEEK for in vitro bone formation by MC3T3-E1 preosteoclasts [123]. The study found that Si_3N_4 caused a significant increase in in vitro mineralization, and Raman spectroscopy revealed that the mineralized nodules on Si_3N_4 closely resembled natural bone. A similar, but more recent study, found that Si_3N_4 also promoted cell proliferation and mineralization of MC3T3-E1 preosteoclasts [124]. In this experiment, Si_3N_4 particles were embedded in gelatin/chitosan cryogels, where Si_3N_4 chemistry, rather than surface topography, was isolated as the primary osteopromotive factor.

Several comparative in vitro investigations also explored the impact of metal debris on implant failure. Nanoscale metal particles, particularly in orthopedic and dental applications, can trigger severe inflammatory responses, potentially contributing to implant failure [125, 126]. In particular, Ti nanoparticles induce a time-dependent decline in cell viability within both mesenchymal stem cells (MSC) and fibroblast populations, accompanied by increased reactive oxygen species (ROS). Tissue analysis demonstrates oxidative stress, elevated extracellular and intracellular Ti levels, and disruption in bone turnover balance. These findings suggest that Ti particles may trigger the recruitment of abnormally high numbers of neutrophils. The neutrophils, in turn, likely release elevated levels of metalloproteinases, resulting in collagen fiber degradation. This cascade of events could negatively impact MSC commitment and ultimately hinder proper bone regeneration [126]. A recent donor study compared human peripheral blood mononuclear cells (PBMC) inflammatory responses to Si_3N_4, Ti6Al4V, and CoCr nanoparticles. This investigation revealed a significant association between particle size, surface chemistry, and immune response. Interestingly, the researchers found that Si_3N_4 did not cause any adverse reactions whereas CoCr particles caused donor-dependent cytotoxicity, TNF-α cytokine release, oxidative stress, and DNA damage in PBMNCs; and Ti6Al4V particles induced significantly elevated levels of cytokines in one of the donors [127]. This observation, coupled with the similar biological responses observed between Si_3N_4 nanoparticles and microparticles, suggests the potential of Si_3N_4 as an alternative to Ti with reduced inflammation risk.

One of the first Si_3N_4 in vivo studies was performed in a minipig model. Neumann et al. manufactured and implanted prototype Si_3N_4 mini-fixation systems within facial frontal bone defects created in test animals to determine its osteosynthesis efficacy. The researchers employed histological analysis, computed tomography (CT), and magnetic resonance imaging (MRI) to evaluate the implant's integration [128]. Following the three-month implantation period, the study revealed no instances of implant loss, displacement, or fracture. Complete osteosynthesis was achieved, evidenced by full bone healing. New bone formation was observed in direct contact with the implants, indicating favorable biocompatibility and seamless mucosal attachment. This was further corroborated by the partial osseous overgrowth of the implanted plates. The implants generated no artifacts during radiological imaging, and no adverse interactions with the surrounding bone tissues were detected. Another early in vivo study that investigated the potential of Si_3N_4 as an osseous implant in rabbit tibias also showed promising results [129]. Subsequent histological analysis revealed favorable bone integration surrounding the implants, indicating good biocompatibility comparable to Ti implants. Similar in vivo studies were subsequently performed in ovine, rat, and caprine models with comparable outcomes [130–132].

In clinical settings, successful osseointegration is traditionally assessed based on its ability to withstand mechanical forces. This assessment considers two key stages of implant stability: (1) initial stability—established at the time of implant placement, this depends primarily on the mechanical locking of the implant fixture within the jawbone. Bone quality and quantity play a dominant role in determining initial stability; and (2) secondary stability—arising over time through bone regeneration and remodeling, this reflects the true achievement of osseointegration. This stage is critical for long-term implant success [133]. Jawbone quality and quantity are crucial factors for the lasting viability of dental implants. The edentulous posterior maxilla, characterized by a scarcity of cortical bone and a predominance of lower-density, weaker trabecular bone (type IV), presents a particular challenge [134]. A 5-year analysis of machined surface implants placed in type IV bone reported high rates of implant failure [135]. More recently, a retrospective study of over 12,000 implants found significantly lower survival rates for type IV bone compared to type I bone

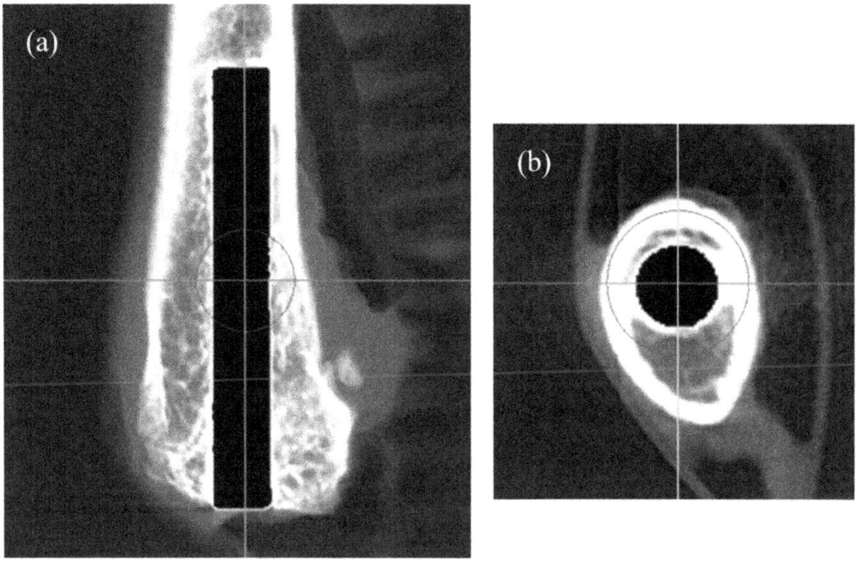

Fig. 6 MicroCT images of an experimental Si_3N_4 implant placed in a mouse femur: (**a**) Longitudinal; and (**b**) Axial views. The results indicate uniform bone formation around the Si_3N_4 implant. (Gonzales, Khade, and Nishimura, unpublished study)

(97.6% vs. 88.8%) [136]. This highlights the potential impact of bone quality on implant prognosis. To alleviate the disadvantageous host factors represented by type IV bone, new implant designs were investigated. One approach involved incorporating bone morphogenetic proteins (BMPs) into implant systems, aiming to enhance their osteoconductive properties and stimulate bone formation [137]. More recently, research investigated the potential of mRNA delivery protocols to provide sustained BMP signaling within biomaterials [138–140]. This represents a promising avenue for overcoming the limitations of conventional BMP delivery methods.

As an alternative therapy, Si_3N_4 dental implants have the potential to upregulate osteogenesis without exogenous BMP. This has been shown in several orthopedic studies [141–143], and a recent preliminary experiment at the UCLA School of Dentistry demonstrated excellent BIC ratios for the implantation of an experimental Si_3N_4 miniature rod in a murine femur as shown in Fig. 6a, b.

Adjunct to this research, RNA sequencing of human bone marrow stem cells (BMSC) was performed on samples cultured on Si_3N_4, machined Ti (mTi), and acid-etched Ti (aeTi). Unexpectedly, Si_3N_4 was found to induce angiogenesis through the BMSCs. Immunohistochemistry was then conducted on murine femur implants for Si_3N_4, mTi, and aeTi using VEGF receptor 3, CD31, and CD34 antibodies. It was discovered that bone tissue induced by Si_3N_4 in the mouse femur was rich in VEGF receptor 3 positive microvasculature (Gonzales, Khade, and Nishimura, unpublished). It has been published that VEGF receptor 3 positive CD31 supports vascular endothelial cells and bone formation [144, 145]. Thus, it appears that one of

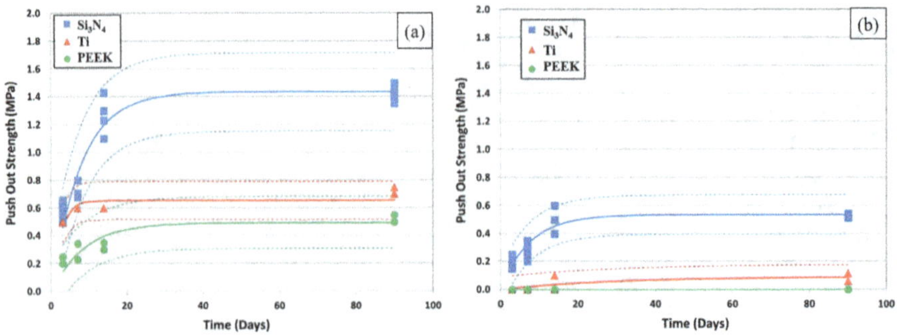

Fig. 7 Implant push-out strengths: (**a**) without bacterial inoculation; and (**b**) with 1×10^4 *S. epidermidis* bacterial inoculation during surgical intervention. Nonlinear regression analysis using the simplified Boltzmann equation. Dashed lines represent a 95% confidence interval. (Reconfigured and used with permission [131])

Si_3N_4's biological advantages is to stimulate angiogenesis, which in turn induces bone formation. Other investigators have also found that Si_3N_4 may be effective in upregulating endothelial tissues for the development of new blood vessels [13, 146, 147].

While no clinical data exist concerning the effectiveness of Si_3N_4 for osseous integration within the oral cavity, a foundational study evaluated its properties in the parietal crania of mature Wistar rats [131]. This in vivo investigation compared the osseointegration and anti-infective characteristics of three dense biomaterials relevant for oral applications—Si_3N_4, Ti, and PEEK. Ninety-six skeletally mature Wistar rats were used in the study. Following anesthesia, elliptical defects encompassing the nasofrontal area to the external occipital protuberance were created in the mid-sagittal plane, exposing the calvarium. The trephined bone was meticulously removed to preserve the dura mater. Each defect was randomly filled with either Ti, ($n = 24$), PEEK, ($n = 24$), or Si_3N_4, ($n = 48$) test coupons. Subsequently, half of the animals within each biomaterial group were randomly inoculated with a standardized aliquot of 1×10^4 *S. epidermidis* for bacterial challenge, while the remaining half received a matching volume of sterile saline as a control. The animals were cared for and then euthanized at 3, 7, 14 days, and 3 months post-operatively. Block sections of the calvarium surrounding the defect and biomaterial were harvested from each animal. Samples were either cryogenically frozen or dehydrated and cured. The frozen samples were later thawed and subjected to push-out strength tests, while the dehydrated samples were embedded in resin and prepared for histomorphometry. Push-out strength tests were used to assess osseous integration—the higher the strength, the greater the integration of the implant. The results of these strength tests under nonbacterial inoculation conditions are presented in Fig. 7a.

Si_3N_4 consistently exhibited the highest push-out strength among all implant materials at each investigated time point. At the 90-day observation period, the strength of Si_3N_4 was demonstrably more than double that of the other two implant

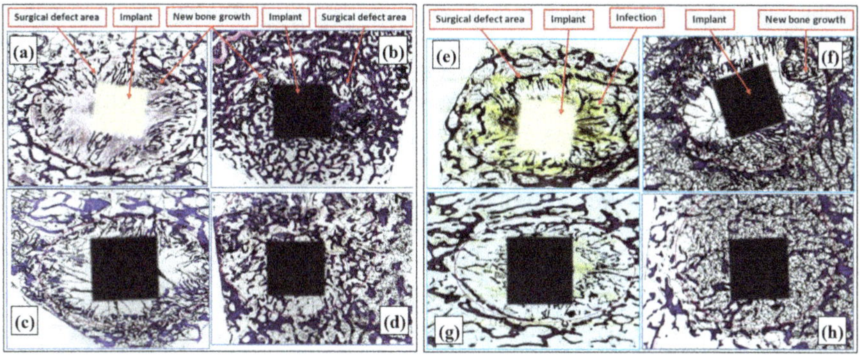

Fig. 8 Wistar rat calvaria histology 3 months after implantation without bacterial inoculation: (**a**) PEEK, 8% new bone at implant interface, 24% new bone in surgical area; (**b**) Si_3N_4, 65% new bone at implant interface, 71% new bone in surgical area; (**c**) Ti, 19% new bone at implant interface, 36% new bone in surgical area; (**d**) Si_3N_4, 52% new bone at implant interface, 66% new bone in surgical area; and with 1×10^4 *S. epidermidis* bacterial inoculation; (**e**) PEEK, 5% new bone at implant interface, 95% bacteria at implant interface, 21% new bone growth in surgical area, 88% bacterial growth in surgical area; (**f**) Si_3N_4, 25% new bone at implant interface, 0% bacteria at implant interface, 39% new bone growth in surgical area, 0% bacteria in surgical area; (**g**) Ti, 9% bone at implant interface, 67% bacteria at implant interface, 26% new bone growth in surgical area, 21% bacteria in surgical area; (**h**) Si_3N_4, 21% new bone at implant interface, 0% bacteria at implant interface, 42% new bone growth in surgical area, 0% bacteria in surgical area. (Reconfigured and used with permission [131])

materials. Statistical analysis revealed significant differences ($p < 0.01$) in push-out strength between Si_3N_4 and the other materials at both the 14- and 90-day time points. While the push-out strength of Ti remained superior to PEEK at 90 days post-surgery, it was demonstrably lower compared to Si_3N_4.

Under bacterial inoculation, the push-out strengths of all biomaterials markedly declined as shown in Fig. 7b. Even so, this testing revealed significantly greater push-out strengths for Si_3N_4 implants compared to both PEEK and Ti implants. Remarkably, the push-out strength of Si_3N_4 at 90 days post-implantation was over five times higher than the values observed for both Ti and PEEK ($p < 0.01$). While the push-out strength of Ti implants was statistically greater than that of PEEK at 90 days post-surgery, this difference was not statistically significant ($p = 0.10$).

Histological analyses of representative implants are shown in Fig. 8a–h. The noninoculated implants at 90 days post-surgery revealed favorable in vivo behavior and biocompatibility for all three materials, with minimal inflammatory response and no signs of infection. Bone growth into the calvarial defects was observed for all implants, but it was substantially greater for Si_3N_4. Although attempts were made to evaluate inoculated implants at 3 and 7 days, instability prevented meaningful histological analysis at these time points. Consequently, histological studies were only performed at 14- and 90-day post-surgery.

PEEK and Ti implants inoculated with bacteria exhibited minimal or no evidence of osteointegration at both time points. In contrast, Si_3N_4 consistently displayed significantly greater new bone formation within the calvarial defects compared to both Ti and PEEK at all implantation periods. Furthermore, the Si_3N_4 implants displayed considerably reduced bacterial biofilm formation compared to PEEK and Ti. These findings corroborate the reported bacteriostatic properties of Si_3N_4 and its potential to promote new bone formation [142, 148].

Overall, these studies suggest Si_3N_4 has potential as a biocompatible material for dental applications. However, further research is required to definitively assess its osteoconductive properties in vivo. While these findings are encouraging, future human clinical studies focusing on bone remodeling and growth around Si_3N_4 implants are needed, particularly in jawbones with low-density trabecular bone (type IV). Only then can it be determined if Si_3N_4 will fulfill its promise as a highly effective solution for overcoming the osseous integration challenges associated with dental implantology.

3.3 Silicon Nitride and Endodontics

Perhaps an overlooked area of osteointegration in dentistry is related to dental caries. This chronic disease is globally prevalent affecting both children and adults. Promoting the formation of restorative dentin in carious lesions is an important clinical objective. The dental pulp contributes significantly to the production of dental hard tissues, including restorative dentin, in response to both physiological and pathological stimuli. Pulp tissue cells, derived from mesenchymal stem cells, possess multipotent differentiation potential. Under appropriate stimulation, these cells differentiate into odontoblasts, responsible for secreting the dentin matrix. Consequently, the success of pulp capping treatments is critically dependent on maintaining pulp tissue vitality and facilitating the formation of dentin bridges. As an osteopromotive compound, Si_3N_4 has been examined for its therapeutic effectiveness in upregulating odontoblasts and the formation of dentin bridges [149]. In this one study, four types of Si_3N_4 granules were cultured in vitro for up to 14 days with harvested rat dental pulp cells (rDPCs) in an osteogenic medium. The four Si_3N_4 granule types included pristine as-synthesized (P-Si_3N_4), glacial acetic acid treated (A-Si_3N_4), thermally oxidized at 200 °C (T-Si_3N_4) and sintered with 3 wt.% Y_2O_3 (S-Si_3N_4). The four different Si_3N_4 compositions represented a range of surface chemistries and pH. It was postulated that modulating the surface chemistry and ammonia (NH_3) elution of Si_3N_4 might provide an optimum environment for upregulating odontoblasts while preventing cytotoxicity. The use of acetic acid on A-Si_3N_4 removed surface silanol groups, making this composition more reactive during the initial stages of hydrolysis whereas the thermal treatment of T-Si_3N_4 inhibited the release of NH_3 by the formation of a thin SiO_2 surface layer. It was believed that the S-Si_3N_4 might provide an intermediate result, attenuating NH_3 release and optimizing pH buffering for cell proliferation. Cell viability and proliferation were determined using live/dead staining in culture wells containing 0.1 mg and 1 mg of the various Si_3N_4 compositions per well. Polystyrene was utilized as a positive control. Results are presented in Fig. 9a, b.

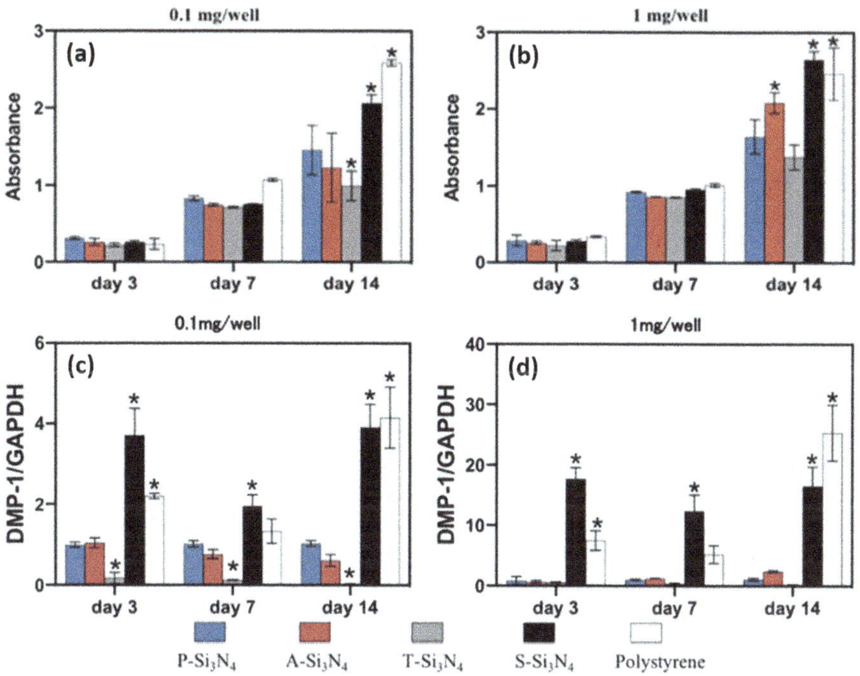

Fig. 9 Evaluation of cytotoxicity and proliferation of rDPCs in vitro. Cells incubated with (**a**) 0.1 and (**b**) 1 mg/well of Si_3N_4 for 3, 7, and 14 days, and analyzed by Cell Counting Kit-8 assays. In (**c**) and (**d**) Real-time qPCR to detect the expression of odontoblast-specific Dentin Matrix Acidic Phosphoprotein (DMP-1) gene expression at 3, 7, and 14 days on plates with P-, A-, T-, S-Si_3N_4 at 0.1 mg and 1 mg doses, respectively. Data are presented as the mean ± standard deviation ($n = 3$). The differences in mean values among the test group were evaluated a one-way analysis of variance and Tukey's multiple comparisons test. * $p < 0.05$: vs. the P-Si_3N_4 group. (Modified and used with permission under Creative Common CC BY license [149])

Following the 14-day cell culture period, the well containing S-Si_3N_4 exhibited a statistically significant increase in total cell number compared to wells with other Si_3N_4 variants. Remarkably, the number of cells on P-Si_3N_4, A-Si_3N_4, and T-Si_3N_4 substrates decreased over time relative to polystyrene controls, while the cell population on S-Si_3N_4 remained statistically similar. Real-time quantitative polymerase chain reaction (qRT-PCR) was employed to quantify Dentin Matrix Acidic Phosphoprotein 1 (DMP-1) expression, a marker associated with odontogenic differentiation. DMP-1 can initiate hydroxyapatite formation through its ability to bind calcium ions. As observed in Fig. 9c, d, the S-Si_3N_4 granular bed promoted early differentiation of rDPCs, implying that Si_3N_4, when prepared under optimal conditions, possesses the potential to stimulate odontogenic differentiation in rDPCs. The mechanism governing the enhanced rDPC activation by S-Si_3N_4 remains to be fully elucidated. However, byproducts from the hydrolytic degradation of Si_3N_4, including silanols and nitrogenous species like NH_3 and nitric oxide (NO), may have contributed to the observed proliferation and differentiation of rDPCs. While the in vivo effects of S-Si_3N_4 on dentin formation warrant further investigation, these findings demonstrate the efficacy of various treatments in modifying the surface properties of Si_3N_4 to improve its bioactivity and promote differentiation. This could be an alternative clinical therapy to the use of calcium hydroxide in the formation of dentin bridges and calcified barriers.

4 Antimicrobial Characteristics

Despite significant advancements in dental biomaterials over the past few decades, the complex interface between these materials and the diverse oral microbiome remains a significant challenge to long-term implant success and overall oral health. Recognizing this critical interplay, future developments in dental biomaterials must prioritize strategies that effectively interface with the oral microbiome. This can be achieved by focusing on two key areas: (1) integrating inherent antimicrobial functionalities into biomaterials to directly combat microbial colonization and biofilm formation, reducing the risk of caries, peri-implant infections, and associated complications; and (2) optimizing biomaterial surface characteristics to minimize bacterial adhesion and biofilm formation while promoting beneficial host cell interactions. This can involve manipulating surface chemistry and topography to discourage microbial attachment and promote tissue integration.

A classic 1993 survey from the UCLA School of Dentistry reported a secondary caries prevalence exceeding 6% based on dental radiographs [150]. More recently, a 6-month recall study by the University of Leuven & Dentistry, University Hospitals Leuven, Belgium, found a lower prevalence of 3.6% in both composite and silver amalgam restorations [151]. Regardless of these variations, the development of secondary caries necessitates the removal of the affected tooth structure along with the failed biomaterial to halt disease progression. This often leads to larger tooth defects, potentially exceeding restorable limits and ultimately resulting in tooth extraction and edentulism. Similarly, periodontitis can develop around dental materials, severely damaging the supporting tissues and ultimately leading to tooth loss. Secondary dental diseases do not originate from the dental materials themselves. Instead, a gap at the interface zone creates an atypical micro-environment within the oral cavity, harboring disproportionately copious amounts of oral microbial biofilm. This biofilm, dominated by cariogenic bacteria like *Streptococcus mutans*, can lead to demineralization of the enamel and the development of secondary caries. Unfortunately, current biomaterials lack specific design features to inhibit carious lesions and prevent periodontitis; and several strategies have been explored to resist caries formation and improve restorative materials, as evidenced by research studies [152–155]. However, there are reasons to believe that Si_3N_4 may be well positioned to address these important pathogenic issues.

4.1 Anticariogenic Properties

Within the indigenous oral flora, *S. mutans* is considered to be the major bacterium linked to dental caries [156]. The cariogenicity of this species is generally held in check by an antagonism between it and *S. sanguinis* [157]. However, dysbiotic proportions of the two bacteria and other commensal strains coupled with (or as a result of) poor diet, inadequate oral care, or heredity allow *S. mutans* to become opportunistic [158]. In a nutrient-rich environment, *S. mutans* rather than *S. sanguinis* can sustain rapid sugar metabolism deep within dental plaque [159] and *S. mutans* reportedly is highly efficient in colonizing multiple abiotic restorative materials [160]. The virulence of *S. mutans* results from its ability to produce large quantities of lactic acid by metabolizing carbohydrates while concurrently surviving in a self-produced low-pH environment [161]. It also can synthesize homopolymers from sucrose which serve as tenacious adhesions to pellicles [156]. Being a facultative anaerobe, *S. mutans* can utilize aerobic respiration if oxygen is present, or switch to anaerobic respiration under hypoxic conditions when embedded within a biofilm. While there are a large number of bacterial species within the complex commensal community that cause carious lesions, *S. mutans* holds a dominant position and it typically comprises 40–60% of oral plaque [162].

Assessments of the susceptibility of oral biomaterials to harbor bacteria and form plaque have been an ongoing objective within the dental community. For example, several oral biomaterials were recently evaluated in two in vivo studies [163, 164]. Testing involved human volunteers who were outfitted with oral appliances containing multiple small discs or squares of the test materials. After wearing the appliances for up to 24 h, the samples were removed, cleaned, stained, and analyzed via fluorescence microscopy. In the first of these studies, gold- (Au) and silver-based (Ag) noble metal alloys were compared to ZrO_2, polyetherketoneketone (PEKK, i.e., similar to PEEK), and a titanium zirconium alloy (TiZr, control) [163]. All materials had statistically equivalent surface finishes. Unsurprisingly, the lowest amount of plaque and the fewest CFU counts were observed for the Ag-alloy. The antibacterial activity of Ag compounds is well known [165, 166]; and it is attributed to the inactivation of essential bacterial proteins and inhibition of DNA replication. The gold-based noble metal had the next lowest plaque and CFU counts (1.5× greater than the Ag alloy). This was followed by PEKK (1.7× Ag), TiZr (2.2× Ag), and ZrO_2 (2.4× Ag). In the second study, four biomaterials were examined including conventional Ti and ZrO_2, and two varieties of PEEK made by different manufacturers (BioHPP and VestaKeep) [164]. The biofilm results of this study are presented in Fig. 10. In this study, ZrO_2 presented with the lowest amount of biofilm plaque, amounting to <10% surface coverage, whereas the two PEEK materials showed the highest concentrations and were statistically equivalent. The conventional Ti was modestly higher than the ZrO_2 but below the PEEK compounds. Although the results of this study appear contradictory to the prior in vivo experiment, the authors pointed out that due to the various modifications of PAEK materials, the results of their study may not be generalizable, particularly

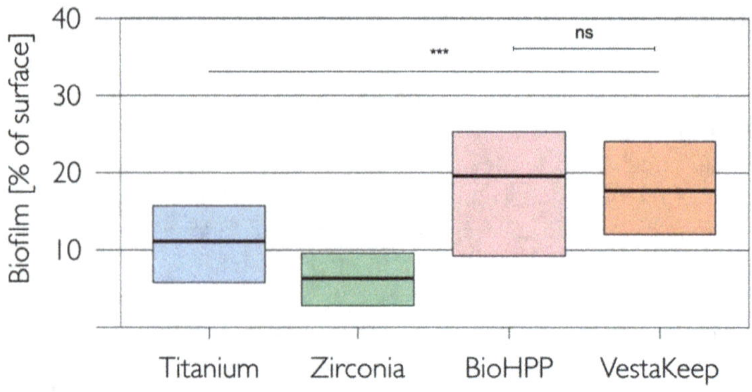

Fig. 10 Percentage coverage of in vivo biofilm after 24 h on four test implant abutment materials. Data are presented as medians and interquartile ranges (25th/75th percentiles). ns, not significant; *, $p < 0.001$. (Used with permission under Creative Commons Attribution (CC BY) license [164])

when using Ti and ZrO_2 as comparisons. Nevertheless, both studies are instructive in that demonstrate the range of bacterial survivability on these biomaterials in the oral environment.

The oral flora has evolved over millennia to survive within a severe environment—one that sees remarkable variations in the composition and availability of nutrients, biological fluids, temperature, and pH [167]. It is within this context that therapeutic research has focused on methods to dislodge, disrupt, remove, and destroy dental plaque. While the current "gold standard" continues to be consistent personal oral care coupled with regular dental office visits for scaling, root planing, and polishing, various chemical treatments have also been envisioned. In particular, the use of nitrate/nitrite oral rinses [168, 169] or nitric oxide (NO) eluting siliceous dentifrices [170] have been proposed as novel methods of preventing caries. This research is pertinent because the bacteriolysis mechanism for these oral therapies is identical to the one operating for Si_3N_4.

Indeed, an unpublished in vitro oral biofilm study has been conducted comparing Si_3N_4 to chemically pure titanium (CP-Ti) and surface functionalized titanium (SF-Ti), and ZrO_2 in disc form (Ø12.7 × 2 mm).[1] While the CP-Ti discs had an as-machined surface, the SF-Ti was functionalized by a proprietary process generally described by Smeets et al. [171] (i.e., grit-blasting and HF acid-etching [172, 173]). The Si_3N_4 samples were left in their "as-fired" state. A commensal strain of bacteria (i.e., *Streptococcus mutans* [160], *Streptococcus sanguis* [174], *Porphyromonas gingivalis* [175], *Lactobacillus salivarius* [176], *Staphylococcus aureus* [177], *Actinomyces oris* [178], and *L. salivarius* [179]), was utilized within a multi-strain testing protocol known as the "Zurich Model" [180, 181]. After incubation in test wells for periods of 12 h and 168 h under anaerobic conditions,

[1] "Antibacterial Assessment of Silicon Nitride for Potential Use as a Dental Implant," Unpublished Research Report Prepared by SINTX Technologies Corporation, March 20, 2019.

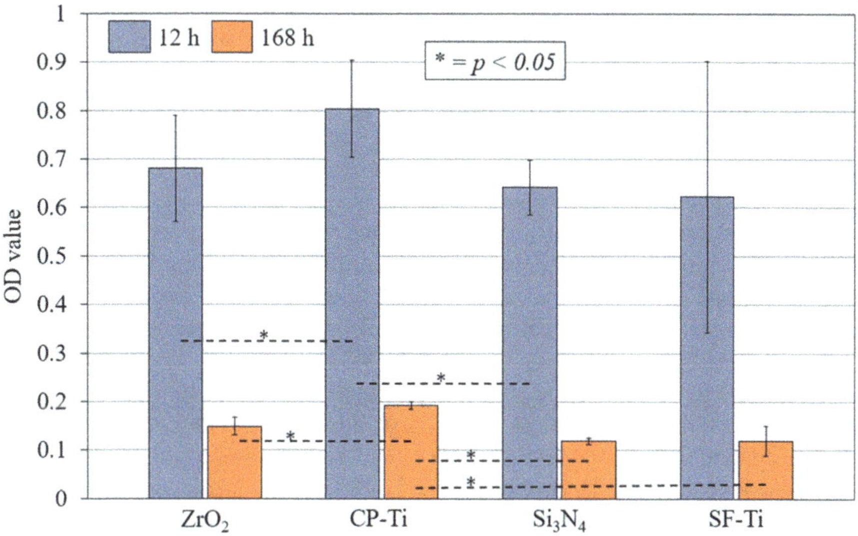

Fig. 11 WST-1 assay optical density (OD_{600}) results for live bacteria on the various biomaterials after 12 h and 168 h of incubation. (Unpublished study, Courtesy SINTX Technologies, Inc.)

the samples were evaluated for bacterial proliferation and viability using standard WST-1 assays. The key result from this internal study is provided in Fig. 11.

After 12 h of incubation, the optical density for living bacteria on CP-Ti was the largest of any of the biomaterials, and statistically higher than ZrO_2 and Si_3N_4. Although SF-Ti showed the lowest average amount of viable bacteria at 12 h, it was statistically indistinguishable from the remaining biomaterials. At 168 h, the optical density values for living bacteria on all of the biomaterials had substantially declined. However, the CP-Ti retained the highest number of living bacteria in comparison to all of the other biomaterials ($p < 0.05$). Furthermore, the WST-1 optical density results for Si_3N_4 were the lowest among all samples, suggesting that it may be the most effective biomaterial in inactivating the commensal biofilm but the statistical differences between it and ZrO_2 or SF-Ti were not significant. However, the similarity of these results to the previously reviewed in vivo studies suggests that Si_3N_4 is at least as effective as the current biomaterial standard (i.e., SF-Ti) for endosseous implants. Similar studies have shown that other nitride-based materials and coatings (i.e., TiN, ZrN, and (Ti1-xZrx)N) exhibit antimicrobial benefits—likely by the same mechanism as Si_3N_4 [182].

4.2 *Periodontitis and Peri-Implantitis*

Periodontitis is a common worldwide oral disease, but estimates of its prevalence vary widely due to methodological and geographical factors, ranging from less than

1.6% to greater than 11.2% [183, 184]; while apical periodontitis is suspected to be present in at least half of the world's population [185]. Periodontal diseases are polymicrobial infections resulting from a complex interaction between numerous bacterial pathogens. These bacteria engage in a dysbiotic interplay with host tissues, ultimately leading to the destruction of the periodontium. This includes the alveolar bone, periodontal ligament, and other structures supporting the teeth. Over fifty bacterial pathogens are implicated, with *Porphyromonas gingivalis*, *Tannerella forsythia*, and *Treponema denticola* being among the most prominent [184].

Peri-implantitis is a progressive inflammatory disease affecting the peri-implant tissues, including the gingiva, subgingival tissue, and bone surrounding dental implants and abutments. Its etiology is essentially identical to periodontitis, principally involving the same dysbiotic microbiota; but inflammatory conditions can also be causal due to an innate reaction to foreign bodies (e.g., titanium particles, ions, and dental adhesives). Regardless of the cause, this disease manifests with progressive inflammation, mucosal recession (i.e., mucositis), sub-gingival pocket formation, bone loss, and ultimately implant failure [186]. A systematic review and meta-analysis reported a mean prevalence of peri-implantitis of 19.5% (95% CI, 12.9–26.2%) at the patient level and 12.5% (95% CI, 11.7–13.4%) at the implant level [187]. This surprisingly high incidence rate highlights the importance of biomaterial selection, proper care, and maintenance for long-term implant success.

As presented and discussed in Sect. 3.1, Ti and ZrO_2 essentially have the same bone and tissue integration. This appears to be true because both implant materials undergo similar modifications to their surfaces by either machining, grit-blasting, acid-etching, or laser ablation. However, there are subtle and unique differences between these materials and their interactions with the oral microbiota. ZrO_2 tends to have reduced biofilm formation in comparison to Ti, particularly with regard to the predominant periodontitis pathogens. Conversely, analysis of immunoinflammatory markers within the mucosal tissues surrounding both ZrO_2 and Ti implants, such as *T*- and *B*-lymphocytes, suggests no significant material-dependent differences in the host inflammatory response elicited by bacterial biofilms. Additionally, inflammation leading to peri-implantitis for Ti could also be due to an allergic reaction or metallosis [188]. While this has not been observed for ZrO_2, current follow-up studies (up to 7.8 years) suggest that ZrO_2 implants exhibit peri-implantitis with a similar clinical presentation as periodontitis to natural teeth [189]. Supporting the notion of minimal material-based disparity, several recent systematic reviews and meta-analyses comparing Ti and ZrO_2 implants or abutments reported few or negligible differences in peri-implantitis susceptibility or implant survival rates between the two materials [49, 50, 190–192]. This observed similarity in their response to bacterial colonization and biofilm formation aligns with the established properties of both biomaterials. Neither Ti nor ZrO_2 readily undergoes hydrolysis, nor do they release significant quantities of antimicrobial ions. These inherent characteristics likely contribute to a comparable tissue response at the implant interface, regardless of the material employed.

On the other hand, despite its aesthetically pleasing color and shock-absorbing capabilities, PEEK has a poor record of in vitro and in vivo microbial resistance [131, 148, 193, 194]. This vulnerability can be attributed, in large part, to the inherent hydrophobicity and bioinertness of PEEK's surface chemistry. Studies have demonstrated a strong correlation between increased surface hydrophobicity and enhanced pathogen attachment [195]. Poor wetting behavior, characteristic of hydrophobic surfaces, is considered a primary factor promoting biofilm development. Generally, negatively charged bacterial membranes tend to adhere less favorably to negatively charged biomaterial surfaces [196]. This phenomenon is further accentuated by the presence of a hydrated or hydroxylated surface, which creates an energetically unfavorable environment for bacterial attachment. Conversely, hydrophobic surfaces lack such inherent barriers, facilitating microbial colonization. Several strategies have been explored to address the surface limitations of PEEK and enhance its functionality, as detailed in Sect. 3.1. These strategies have not only shown promise in improving the material's osseous integration ability but also its resistance to microbial colonization. The incorporation of Si_3N_4 particles as an embedded outer layer on PEEK devices has been particularly effective. In vitro studies have demonstrated several orders of magnitude reduction in microbial colonization [108, 111–113]. 3D-printing of Si_3N_4 surface-coated prototype implants has recently demonstrated that this technology is progressing toward clinical implementation [197].

Si_3N_4's bactericidal efficacy has been demonstrated against several pathogens including *Escherichia coli*, *Staphylococcus epidermidis*, *Pseudomonas aeruginosa*, *Enterococcus*, and two strains of *Staphylococcus aureus* including methicillin-resistant (MRSA) [17, 131, 148, 193, 198–200] and at least one oral species—*Porphyromonas gingivalis* (*P. gingivalis*) [201]. This last study is particularly instructive because it highlights Si_3N_4's ability to effectively lyse the primary causative pathogen for periodontal disease and peri-implantitis. Using Raman spectroscopy, researchers were able to show that contact of living *P. gingivalis* with monolithic Si_3N_4 in a brain-heart infusion culture under anaerobic conditions led to progressive structural changes to the pathogen's cellular nucleic acid, proteins, and lipids. Mechanistically, this is depicted in the schematic diagram of Fig. 12.

Within the biological milieu, a hydrolytic reaction at the surface of Si_3N_4 releases silanols $SiH_{4-n}(OH)_n$ (n = 1 to 4) and ammonia (NH_3). The presence of these moieties increases the local pH which stabilizes NH_3 over ammonium (NH_4^+) [15]. However, *P. gingivalis* is a unique pathogen as it can metabolize nitrogenous compounds, unlike most bacteria which rely on sugars for energy. This process generates a microenvironment rich in ammonia and other byproducts [202]. The conversion of nitrate to nitrite and ultimately ammonia also involves the production of nitric oxide (NO) through nitrite reductions [203–205]. Despite being a freely diffusible, highly reactive, and cytotoxic gas, *P. gingivalis* possesses mechanisms to efficiently scavenge NO under normal metabolic conditions. Yet, the presence of silanols, which are highly unstable themselves, generates reactive oxygen species

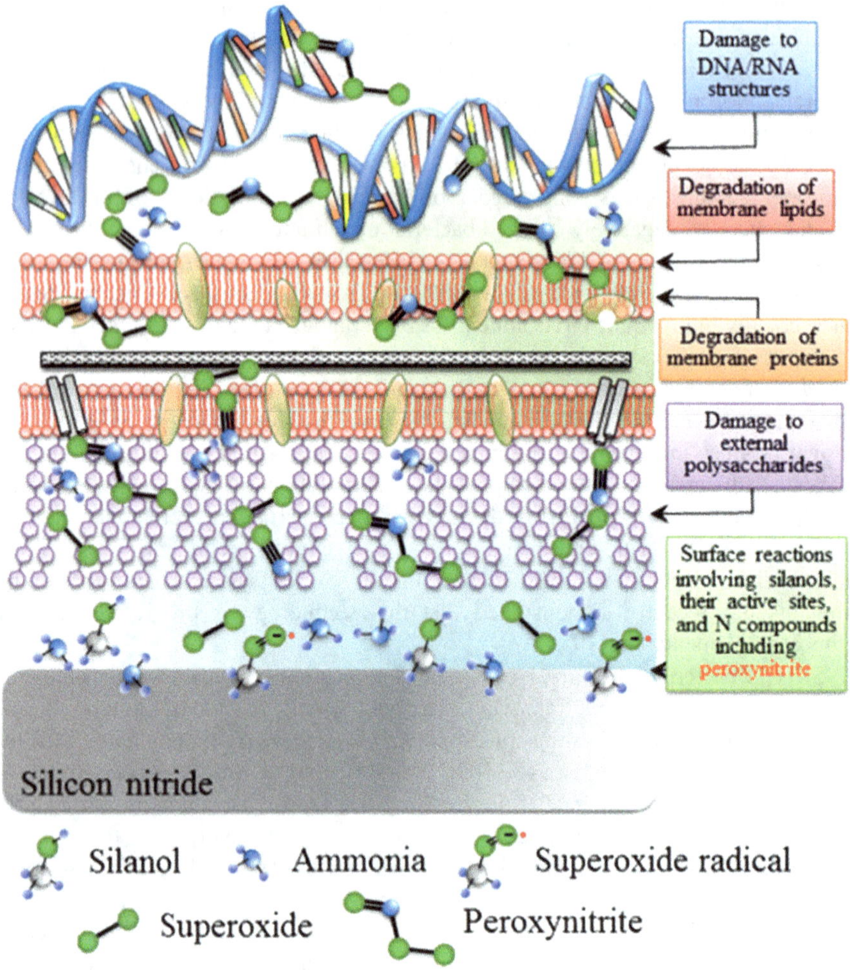

Fig. 12 Schematic representation of the proposed mechanism for Si_3N_4-mediated feeding and poisoning of *P. gingivalis*. The diagram depicts the interface between solid Si_3N_4 and *P. gingivalis*, highlighting the key chemical species formed at the ceramic/bacterium interface. (Used with permission [201])

(ROS) near NO. Under these conditions, the interaction between Si_3N_4-generated superoxide radicals and the NO produced by *P. gingivalis* metabolism leads to the rapid formation of peroxynitrite ($O{=}N{-}O{-}O^-$), an oxidizer and potent bactericidal agent [206]. Its presence damages external polysaccharides, degrades membrane proteins and lipids, and incapacitates the DNA and RNA of the prokaryote. Essentially, *P. gingivalis* is simultaneously fed and poisoned by an abundance of nitrous compounds.

5 Summary and Conclusions

This comprehensive review evaluated the mechanical properties, osseointegration potential, and antimicrobial efficacy of leading biomaterials currently employed or under investigation for dental applications. Titanium (Ti), with its long history of successful clinical use, remains the benchmark for dental implants and abutments. However, dental zirconia (ZrO_2) is gaining significant traction due to its excellent biocompatibility and aesthetic appeal, catering to the growing demand for "metal-free" restorations. Polyetheretherketone (PEEK) is a newer biomaterial with regulatory approval for dental prostheses. While its inherent mechanical properties, particularly the carbon-fiber-reinforced variant, are deemed acceptable, they fall short of the strength and toughness offered by Ti or ZrO_2. All three materials lack substantial bioactivity. Both titanium and zirconia exhibit comparable properties in terms of osseointegration and microbial resistance. Conversely, PEEK was shown to be hydrophobic and wholly bioinert, lacking both intrinsic osteointegration and antimicrobial features. Various surface modification techniques are being explored as the means of imparting these necessary functionalities onto PEEK.

Silicon nitride (Si_3N_4) has emerged as a promising candidate for dental biomaterials. It boasts mechanical properties comparable to Ti and ZrO_2. More importantly, numerous studies have demonstrated its superior capability for integration with bone and soft tissues. Si_3N_4 demonstrably upregulates the activity of osteoblasts and odontoblasts, and recent research confirms its angiogenic properties. Additionally, it exhibits at least equivalent efficacy to titanium in suppressing cariogenic bacteria, while surpassing it in eliminating periodontal pathogens. While Si_3N_4 awaits regulatory approval for oral applications, its advantageous properties hold significant promise for future clinical validation and widespread adoption.

References

1. Levin, L.; Day, P.F.; Hicks, L.; O'Connell, A.; Fouad, A.F.; Bourguignon, C.; Abbott, P. V. International Association of Dental Traumatology Guidelines for the Management of Traumatic Dental Injuries: General introduction. *Dent. Traumatol.* **2020**, *36*, 309–313, https://doi.org/10.1111/edt.12574.
2. Arweiler, N.B.; Netuschil, L. The Oral Microbiota. In *Microbiota of the Human Body*; Schwiertz, A., Ed.; Springer International Publishing: Cham, 2016; pp. 45 60 ISBN 978-3-319-31248-4.
3. Verma, D.; Garg, P.K.; Dubey, A.K. Insights into the Human Oral Microbiome. *Arch. Microbiol.* **2018**, *200*, 525–540, https://doi.org/10.1007/s00203-018-1505-3.
4. Lamont, R.J.; Koo, H.; Hajishengallis, G. The Oral Microbiota: Dynamic Communities and Host Interactions. *Nat. Rev. Microbiol.* **2018**, *16*, 745–759, https://doi.org/10.1038/s41579-018-0089-x.
5. Moutsopoulos, N.M.; Konkel, J.E. Tissue-Specific Immunity at the Oral Mucosal Barrier. *Trends Immunol.* **2018**, *39*, 276–287, https://doi.org/10.1016/j.it.2017.08.005.

6. Carrizales-Sepúlveda, E.F.; Ordaz-Farías, A.; Vera-Pineda, R.; Flores-Ramírez, R. Periodontal Disease, Systemic Inflammation and the Risk of Cardiovascular Disease. *Hear. Lung Circ.* **2018**, *27*, 1327–1334, https://doi.org/10.1016/j.hlc.2018.05.102.
7. Sanz, M.; Marco del Castillo, A.; Jepsen, S.; Gonzalez-Juanatey, J.R.; D'Aiuto, F.; Bouchard, P.; Chapple, I.; Dietrich, T.; Gotsman, I.; Graziani, F.; et al. Periodontitis and Cardiovascular Diseases: Consensus Report. *J. Clin. Periodontol.* **2020**, *47*, 268–288, https://doi.org/10.1111/jcpe.13189.
8. Sabharwal, A.; Stellrecht, E.; Scannapieco, F.A. Associations Between Dental Caries and Systemic Diseases: A Scoping Review. *BMC Oral Health* **2021**, *21*, 472, https://doi.org/10.1186/s12903-021-01803-w.
9. Bonita, R.; Beaglehole, R.; Kjellström, T. *Basic Epidemiology*; World Health Organization, 2006; ISBN 9241547073.
10. Du, X.; Lee, S.S.; Blugan, G.; Ferguson, S.J. Silicon Nitride as a Biomedical Material: An Overview. *Int. J. Mol. Sci.* **2022**, *23*, 6551, https://doi.org/10.3390/ijms23126551.
11. Heimann, R.B. Silicon Nitride, A Close to Ideal Ceramic Material for Medical Application. *Ceramics* **2021**, *4*, 208–223.
12. Wu, J.; Liu, Y.; Zhang, H.; Wu, Y.; Chu, Z.; Wu, Q.; Lu, M.; Tang, C. Silicon Nitride as a Potential Candidate for Dental Implants: Osteogenic Activities and Antibacterial Properties. *J. Mater. Res.* **2021**, *36*, 1866–1882, https://doi.org/10.1557/s43578-021-00249-8.
13. Liu, Z.; Wang, R.; Liu, W.; Liu, Y.; Feng, X.; Zhao, F.; Chen, P.; Shao, L.; Rong, M. Recent Advances in the Application and Biological Mechanism of Silicon Nitride Osteogenic Properties: A Review. *Biomater. Sci.* **2023**, Advance Article, https://doi.org/10.1039/D3BM00877K.
14. He, J.; Liu, Y.; Zeng, X.; Tong, Y.; Liu, R.; Wang, K.; Shangguan, X.; Qiu, G.; Sipaut, C.S. Silicon Nitride Bioceramics Sintered by Microwave Exhibit Excellent Mechanical Properties, Cytocompatibility *In Vitro*, and Anti-Bacterial Properties. *J. Funct. Biomater.* 2023, *14*.
15. Pezzotti, G. Silicon Nitride as a Biomaterial. *J. Ceram. Soc. Japan* **2023**, *131*, 398–428, https://doi.org/10.2109/jcersj2.23057.
16. Mohammadi, H.; Beddu, S.; Petrů, M.; Sepantafar, M.; Ebadi, M.; Yap, B.K.; Bang, L.T.; Yong, T.C.; Ramesh, S.; Rahimian Koloor, S.S.; et al. Advances in Silicon Nitride Ceramic Biomaterials for Dental Applications—A Review. *J. Mater. Res. Technol.* **2024**, *28*, 2778–2791, https://doi.org/10.1016/j.jmrt.2023.12.186.
17. Pezzotti, G.; Marin, E.; Adachi, T.; Lerussi, F.; Rondinella, A.; Boschetto, F.; Zhu, W.; Kitajima, T.; Inada, K.; McEntire, B.J.; et al. Incorporating Si_3N_4 into PEEK to Produce Antibacterial, Osteoconductive, and Radiolucent Spinal Implants. *Macromol. Biosci.* **2018**, *18*, 1080033, https://doi.org/10.1002/mabi.201800033.
18. Heimann, R.B. Silicon Nitride Ceramics: Structure, Synthesis, Properties, and Biomedical Applications. *Materials (Basel).* **2023**, *16*, 1–16, https://doi.org/10.3390/ma16145142.
19. Pezzotti, G.; Asai, T.; Adachi, T.; Ogitani, E.; Yamamoto, T.; Kanamura, N.; Boschetto, F.; Zhu, W.; Zanocco, M.; Marin, E.; et al. Antifungal Activity of Polymethyl Methacrylate/Si_3N_4 Composites against Candida Albicans. *Acta Biomater.* **2021**, *126*, 259–276, https://doi.org/10.1016/j.actbio.2021.03.023.
20. Marin, E.; Boschetto, F.; Honma, T.; Yang, Q.; Ajunwa, O.M.; Marsili, E.; Zhu, W.; Pezzotti, G. Particulate Reinforced Bone Cements: The Potential Antibacteric Role of Nitrides. *Materialia* **2023**, *28*, 101748, https://doi.org/10.1016/j.mtla.2023.101748.
21. Wang, F.; Guo, J.; Li, K.; Sun, J.; Zeng, Y.; Ning, C. High Strength Polymer/Silicon Nitride Composites for Dental Restorations. *Dent. Mater.* **2019**, *35*, 1–10, https://doi.org/10.1016/j.dental.2019.05.022.
22. Wananuruksawong, R.; Jinawath, S.; Padipatvuthikul, P.; Wasanapiarnpong, T. Fabrication of Silicon Nitride Dental Core Ceramics with Borosilicate Veneering Material. *IOP Conf. Ser. Mater. Sci. Eng.* **2011**, *18*, 192010, https://doi.org/10.1088/1757-899X/18/19/192010.

23. Wasanapiarnpong, T.; Cherdtham, N.; Padipatvuthikul, P.; Mongkolkachit, C.; Wananuruksawong, R.; Jinawath, S. Fabrication of Dental Ceramics from Silicon Nitride Core with Borosilicate Glass Veneer. In Proceedings of the Advanced Materials Research; 2012; Vol. 506, pp. 493–496.
24. Wananuruksawong, R.; Wasanapiarnpong, T.; Dhanesuan, N.; Didron, P.P. Microhardness and Biocompatibility of Silicon Nitride Ceramic Developed for Dental Applications. *Mater. Sci. Appl.* **2014**, *5*, 1034, https://doi.org/10.4236/msa.2014.514106.
25. Özdoğan, M.S.; Güngörmüş, M.; Çelİk, A.; Topateş, G. Silicon Nitride Ceramic for All-Ceramic Dental Restorations. *Dent. Mater. J.* **2020**, 1–7, https://doi.org/10.4012/dmj.2020-134.
26. Jiang, L.; Wang, T.; Yin, S.; Luo, T.; Fang, X.; Yuan, J.; Tang, X.; Wang, Y.; Li, Q.; Yang, J. Preparation and Properties of Cordierite Glass-Ceramic as Veneer Porcelain for Silicon Nitride Dental Ceramics. *J. Eur. Ceram. Soc.* **2022**, *42*, 6182–6191, https://doi.org/10.1016/j.jeurceramsoc.2022.06.072.
27. Mussano, F.; Genova, T.; Munaron, L.; Faga, M.G.; Carossa, S. Ceramic Biomaterials for Dental Implants : Current Use and Future Perspectives. In *Dental Implantology and Biomaterial*; INTECH Open Access, 2016; pp. 63–90.
28. Badran, Z.; Strullou, X.; Hughes, F.J.; Soueidan, A.; Hoornaert, A.; Ide, M. Silicon Nitride (Si_3N_4) Implants: The Future of Dental Implantology? *J. Oral Implantol.* **2017**, *43*, 240–244.
29. Raza, S.M.; Khurshid, Z.; Zafar, M.S.; Najeeb, S.; Ul Yaqin, S.A. 15—Silicon Nitride (SiN): An Emerging Material for Dental Implant Applications. In *Woodhead Publishing Series in Biomaterials*; Zafar, M.S., Khurshid, Z., Khan, A.S., Najeeb, S., Sefat, F.B.T.-D.I., Eds.; Woodhead Publishing, 2020; pp. 287–299 ISBN 978-0-12-819586-4.
30. Aparicio-razo, M.; Mongalo-vázquez, J.L.J.; Ramos, J.A.Y.; Navarro-zárate, A.; Santos-enríquez, V.H.; Vivanco-pérez, I.; Méndez, J.F.; Paredes-juárez, G.A. Comparison of Material Properties for Dental Implants: Titanium, Polyetheretherketone, Zirconium and Silicon Nitride. *Adv. Mater. Lett.* **2022**, *13*, 2022.011684; 1–13.
31. Krajangta, N.; Sarinnaphakorn, L.; Didron, P.P.; Wasanapiarnpong, T. Development of Silicon Nitride Ceramic for CAD/CAM Restoration. *Dent. Mater. J.* **2020**, *3*, 1–6, https://doi.org/10.4012/dmj.2019-119.
32. Zou, R.; Bi, L.; Huang, Y.; Wang, Y.; Wang, Y.; Li, L.; Liu, J.; Feng, L.; Jiang, X., Deng, B. A Biocompatible Silicon Nitride Dental Implant Material Prepared by Digital Light Processing Technology. *J. Mech. Behav. Biomed. Mater.* **2023**, *141*, 105756, https://doi.org/10.1016/j.jmbbm.2023.105756.
33. Hatch, J.P.; Shinkai, R.S.A.; Sakai, S.; Rugh, J.D.; Paunovich, E.D. Determinants of Masticatory Performance in Dentate Adults. *Arch. Oral Biol.* **2001**, *46*, 641–648, https://doi.org/10.1016/S0003-9969(01)00023-1.
34. Cho, Y.-E.; Park, E.-J.; Koak, J.-Y.; Kim, S.-K.; Heo, S.-J.; Park, J.-M. Strain Gauge Analysis of Occlusal Forces on Implant Prostheses at Various Occlusal Heights. *Int. J. Oral Maxillofac. Implants* **2014**, *29*, 1034–1041, https://doi.org/10.11607/jomi.3040.
35. Yurkstas, A.; Manly, R.S. Measurement of Occlusal Contact Area Effective in Mastication. *Am. J. Orthod.* **1949**, *35*, 185–195, https://doi.org/10.1016/0002-9416(49)90028-7.
36. Liu, X.; Chen, S.; Tsoi, J.K.H.; Matinlinna, J.P. Binary Titanium Alloys as Dental Implant Materials—A Review. *Regen. Biomater.* **2017**, *4*, 315–323, https://doi.org/10.1093/rb/rbx027.
37. McEntire, B.J.; Bal, B.S.; Rahaman, M.N.; Chevalier, J.; Pezzotti, G. Ceramics and Ceramic Coatings in Orthopaedics. *J. Eur. Ceram. Soc.* **2015**, *35*, 4327–4369, https://doi.org/10.1016/j.jeurceramsoc.2015.07.034.
38. Han, X.; Yang, D.; Yang, C.; Spintzyk, S.; Scheideler, L.; Li, P.; Li, D.; Geis-Gerstorfer, J.; Rupp, F. Carbon Fiber Reinforced PEEK Composites Based on 3D-Printing Technology for Orthopedic and Dental Applications. *J. Clin. Med.* **2019**, *8*, https://doi.org/10.3390/jcm8020240.

39. Steinberg, E.L.; Rath, E.; Shlaifer, A.; Chechik, O.; Maman, E.; Salai, M. Carbon Fiber Reinforced PEEK Optima—A Composite Material Biomechanical Properties and Wear/Debris Characteristics of CF-PEEK Composites for Orthopedic Trauma Implants. *J. Mech. Behav. Biomed. Mater.* **2013**, *17*, 221–228, https://doi.org/10.1016/j.jmbbm.2012.09.013.
40. Najeeb, S.; Zafar, M.S.; Khurshid, Z.; Siddiqui, F. Applications of Polyetheretherketone (PEEK) in Oral Implantology and Prosthodontics. *J. Prosthodont. Res.* **2016**, *60*, 12–19, https://doi.org/10.1016/j.jpor.2015.10.001.
41. Nagay, B.E.; Cordeiro, J.M.; Barao, V.A.R. Insight Into Corrosion of Dental Implants: From Biochemical Mechanisms to Designing Corrosion-Resistant Materials. *Curr. Oral Heal. Reports* **2022**, *9*, 7–21, https://doi.org/10.1007/s40496-022-00306-z.
42. Zhang, Y.; Lawn, B.R. Novel Zirconia Materials in Dentistry. *J. Dent. Res.* **2018**, *97*, 140–147, https://doi.org/10.1177/0022034517737483.
43. Chevalier, J.; Gremillard, L.; Virkar, A. V.; Clarke, D.R. The Tetragonal-Monoclinic Transformation in Zirconia: Lessons Learned and Future Trends. *J. Am. Ceram. Soc.* **2009**, *92*, 1901–1920, https://doi.org/10.1111/j.1551-2916.2009.03278.x.
44. Keuper, M.; Berthold, C.; Nickel, K.G. Long-time Aging in 3 mol.% Yttria-Stabilized Tetragonal Zirconia Polycrystals at Human Body Temperature. *Acta Biomater.* **2014**, *10*, 951–9, https://doi.org/10.1016/j.actbio.2013.09.033.
45. Kohal, R.J.; Trinkner, A.; Burkhardt, F.; Patzelt, S.B.M.; Vach, K.; Kušter, M.; Abram, A.; Kocjan, A.; Nold, J. Long-Term Stability of Hydrothermally Aged and/or Dynamically Loaded One-Piece Diameter Reduced Zirconia Oral Implants. *J. Funct. Biomater.* **2023**, *14*, https://doi.org/10.3390/jfb14030123.
46. Gil, J.; Delgado-García-Menocal, J.A.; Velasco-Ortega, E.; Bosch, B.; Delgado, L.; Pérez-Antoñanzas, R.; Fernández-Fairén, M. Comparison of Zirconia Degradation in Dental implants and Femoral Balls: An X-ray Diffraction and Nanoindentation Study. *Int. J. Implant Dent.* **2021**, *7*, https://doi.org/10.1186/s40729-021-00383-2.
47. Camposilvan, E.; Leone, R.; Gremillard, L.; Sorrentino, R.; Zarone, F.; Ferrari, M.; Chevalier, J. Aging Resistance, Mechanical Properties and Translucency of Different Yttria-Stabilized Zirconia Ceramics for Monolithic Dental Crown Applications. *Dent. Mater.* **2018**, *34*, 879–890, https://doi.org/10.1016/j.dental.2018.03.006.
48. Hashim, D.; Cionca, N.; Courvoisier, D.S.; Mombelli, A. A Systematic Review of the Clinical Survival of Zirconia Implants. *Clin. Oral Investig.* **2016**, *20*, 1403–1417, https://doi.org/10.1007/s00784-016-1853-9.
49. Roehling, S.; Gahlert, M.; Bacevic, M.; Woelfler, H.; Laleman, I. Clinical and Radiographic Outcomes of Zirconia Dental Implants—A Systematic Review and Meta-Analysis. *Clin. Oral Implants Res.* **2023**, *34*, 112–124, https://doi.org/10.1111/clr.14133.
50. Padhye, N.M.; Calciolari, E.; Zuercher, A.N.; Tagliaferri, S.; Donos, N. Survival and Success of Zirconia Compared with Titanium Implants: A Systematic Review and Meta-Analysis. *Clin. Oral Investig.* **2023**, *27*, 6279–6290, https://doi.org/10.1007/s00784-023-05242-5.
51. Verma, V.; Hazari, P.; Verma, P. Do Implants Made of Polyetheretherketone and its Composites Have Reduced Stress Shielding Effects Compared to Other Dental Implant Materials? A Systematic Review. *Evid. Based. Dent.* **2023**, *24*, 193–194, https://doi.org/10.1038/s41432-023-00935-y.
52. Korabi, R.; Shemtov-Yona, K.; Rittel, D. On Stress/Strain Shielding and the Material Stiffness Paradigm for Dental Implants. *Clin. Implant Dent. Relat. Res.* **2017**, *19*, 935–943, https://doi.org/10.1111/cid.12509.
53. Panayotov, I.V.; Orti, V.; Cuisinier, F.; Yachouh, J. Polyetheretherketone (PEEK) for Medical Applications. *J. Mater. Sci. Mater. Med.* **2016**, *27*, https://doi.org/10.1007/s10856-016-5731-4.
54. Qin, L.; Yao, S.; Zhao, J.; Zhou, C.; Oates, T.W.; Weir, M.D.; Wu, J.; Xu, H.H.K. Review on Development and Dental Applications of Polyetheretherketone-Based Biomaterials and Restorations. *Materials (Basel).* **2021**, *14*, 1–25, https://doi.org/10.3390/ma14020408.

55. Dias Corpa Tardelli, J.; Monteiro de Barros Ciribelli Alves, B.; Lima da Costa Valente, M.; Cândido dos Reis, A. Influence of the Modulus of Elasticity of Dental Implants on the Distribution of Stresses in the Alveolar Bone by the Finite Element Nethod: A Systematic Review. *J. Oral Maxillofac. Surgery, Med. Pathol.* **2023**, *35*, 383–389, https://doi.org/10.1016/j.ajoms.2023.02.008.
56. Majić Renjo, M.; Ćurković, L.; Štefančić, S.; Ćorić, D. Indentation Size Effect of Y-TZP Dental Ceramics. *Dent. Mater.* **2014**, *30*, e371–e376, https://doi.org/10.1016/j.dental.2014.08.367.
57. Pezzotti, G.; Enomoto, Y.; Zhu, W.; Boffelli, M.; Marin, E.; McEntire, B.J. Surface Toughness of Silicon Nitride Bioceramics: I, Raman Spectroscopy-Assisted Micromechanics. *J. Mech. Behav. Biomed. Mater.* **2015**, *54*, 328–345.
58. McEntire, B.J.; Enomoto, Y.; Zhu, W.; Boffelli, M.; Marin, E.; Pezzotti, G. Surface Toughness of Silicon Nitride Bioceramics: II, Comparison with Commercial Oxide Materials. *J. Mech. Behav. Biomed. Mater.* **2016**, *54*, 346–359, https://doi.org/10.1016/j.jmbbm.2015.08.044.
59. Bal, B.S.; Khandkar, A.; Lakshminarayanan, R.; Clarke, I.; Hofmann, A.A.; Rahaman, M.N. Fabrication and Testing of Silicon Nitride Bearings in Total Hip Arthroplasty. *J. Arthroplasty* **2009**, *24*, 110–116, https://doi.org/10.1016/j.arth.2008.01.300.
60. Elani, H.W.; Starr, J.R.; Da Silva, J.D.; Gallucci, G.O. Trends in Dental Implant Use in the U.S., 1999–2016, and Projections to 2026. *J. Dent. Res.* **2018**, *97*, 1424–1430, https://doi.org/10.1177/0022034518792567.
61. Schneider, P.; Makek, M.; Feigel, A.; Ochsenbein, H.; Lutz, F.; Mühlemann, H.R. Histologic Studies of Endosseous Blade-Vent Implants of a Human Mandible: Report of an Autopsy Case. *J. Oral Implantol.* **1978**, *8*, 31–40.
62. Howe, M.-S.; Keys, W.; Richards, D. Long-Term (10-year) Dental Implant Survival: A Systematic Review and Sensitivity Meta-Analysis. *J. Dent.* **2019**, *84*, 9–21, https://doi.org/10.1016/j.jdent.2019.03.008.
63. Grassi, S.; Piattelli, A.; de Figueiredo, L.C.; Feres, M.; de Melo, L.; Iezzi, G.; Alba Jr., R.C.; Shibli, J.A. Histologic Evaluation of Early Human Bone Response to Different Implant Surfaces. *J. Periodontol.* **2006**, *77*, 1736–1743, https://doi.org/10.1902/jop.2006.050325.
64. Halldin, A.; Ander, M.; Jacobsson, M.; Hansson, S. Simulation of the Mechanical Interlocking Capacity of a Rough Bone Implant Surface During Healing. *Biomed. Eng. Online* **2015**, *14*, 45, https://doi.org/10.1186/s12938-015-0038-0.
65. Butz, F.; Ogawa, T.; Nishimura, I. Interfacial Shear Strength of Endosseous Implants. *Int. J. Oral Maxillofac. Implants* **2011**, *26*.
66. Saruwatari, L.; Aita, H.; Butz, F.; Nakamura, H.K.; Ouyang, J.; Yang, Y.; Chiou, W.-A.; Ogawa, T. Osteoblasts Generate Harder, Stiffer, and More Delamination-Resistant Mineralized Tissue on Titanium Than on Polystyrene, Associated With Distinct Tissue Micro- and Ultrastructure. *J. Bone Miner. Res.* **2005**, *20*, 2002–2016, https://doi.org/10.1359/JBMR.050703.
67. Att, W.; Kubo, K.; Yamada, M.; Maeda, H.; Ogawa, T. Biomechanical Properties of Jaw Periosteum-Derived Mineralized Culture on Different Titanium Topography. *Int. J. Oral Maxillofac. Implants* **2009**, *24*.
68. Butz, F.; Aita, H.; Wang, C.J.; Ogawa, T. Harder and Stiffer Bone Osseointegrated to Roughened Titanium. *J. Dent. Res.* **2006**, *85*, 560–565, https://doi.org/10.1177/154405910608500616.
69. Marini, J.C.; Cabral, W.A. Osteogenesis Imperfecta. In *Genetics of Bone Biology and Skeletal Disease*; Thakker, R. V, Whyte, M.P., Eisman, J.A., Igarashi, T.B.T.-G. of B.B. and S.D. (Second E., Eds.; Academic Press: Tokyo, Japan, 2018; pp. 397–420. ISBN 978-0-12-804182-6.
70. Ogawa, T.; Nishimura, I. Genes Differentially Expressed in Titanium Implant Healing. *J. Dent. Res.* **2006**, *85*, 566–570, https://doi.org/10.1177/154405910608500617.

71. Depprich, R.; Ommerborn, M.; Zipprich, H.; Naujoks, C.; Handschel, J.; Wiesmann, H.-P.; Kübler, N.R.; Meyer, U. Behavior of Osteoblastic Cells Cultured on Titanium and Structured Zirconia Surfaces. *Head Face Med.* **2008**, *4*, 29, https://doi.org/10.1186/1746-160X-4-29.
72. Le Guehennec, L.; Lopez-Heredia, M.-A.; Enkel, B.; Weiss, P.; Amouriq, Y.; Layrolle, P. Osteoblastic Cell Behaviour on Different Titanium Implant Surfaces. *Acta Biomater.* **2008**, *4*, 535–543, https://doi.org/10.1016/j.actbio.2007.12.002.
73. He, P.; Wang, X.; Ning, C.; Liu, X.; Li, M.; Xu, H.; Guo, G.; Mao, G.; Liu, G.; Xu, B.; et al. Regulation of Osteoblast Functions on Titanium Surfaces with Different Micro/Nano Topographies and Compositions. *Sci. China Technol. Sci.* **2019**, *62*, 559–568, https://doi.org/10.1007/s11431-018-9313-3.
74. Le Guéhennec, L.; Soueidan, A.; Layrolle, P.; Amouriq, Y. Surface Treatments of Titanium Dental Implants for Rapid Osseointegration. *Dent. Mater.* **2007**, *23*, 844–854, https://doi.org/10.1016/j.dental.2006.06.025.
75. Ogawa, T.; Ozawa, S.; Shih, J.-H.; Ryu, K.H.; Sukotjo, C.; Yang, J.-M.; Nishimura, I. Biomechanical Evaluation of Osseous Implants Having Different Surface Topographies in Rats. *J. Dent. Res.* **2000**, *79*, 1857–1863, https://doi.org/10.1177/00220345000790110701.
76. Ozawa, S.; Ogawa, T.; Iida, K.; Sukotjo, C.; Hasegawa, H.; Nishimura, R.D.; Nishimura, I. Ovariectomy Hinders the Early Stage of Bone-Implant Integration: Histomorphometric, Biomechanical, and Molecular Analyses. *Bone* **2002**, *30*, 137–143, https://doi.org/10.1016/S8756-3282(01)00646-9.
77. Schwartz, Z.; Nasazky, E.; Boyan, B.D. Surface Microtopography Regulates Osteointegration: The Role of Implant Surface Microtopography in Osteointegration. *Alpha Omegan* **2005**, *98*, 9–19.
78. Bilgrau, A.E.; Falgreen, S.; Petersen, A.; Kjeldsen, M.K.; Bødker, J.S.; Johnsen, H.E.; Dybkær, K.; Bøgsted, M. Unaccounted Uncertainty from qPCR Efficiency Estimates Entails Uncontrolled False Positive Rates. *BMC Bioinformatics* **2016**, *17*, 159, https://doi.org/10.1186/s12859-016-0997-6.
79. Nishimura, I. Genetic Networks in Osseointegration. *J. Dent. Res.* **2013**, *92*, 109S-118S, https://doi.org/10.1177/0022034513504928.
80. Mengatto, C.M.; Mussano, F.; Honda, Y.; Colwell, C.S.; Nishimura, I. Circadian Rhythm and Cartilage Extracellular Matrix Genes in Osseointegration: A Genome-Wide Screening of Implant Failure by Vitamin D DFeficiency. *PLoS One* **2011**, *6*, e15848, https://doi.org/10.1371/journal.pone.0015848.
81. Okawa, H.; Egusa, H.; Nishimura, I. Implications of the Circadian Clock in Implant Dentistry. *Dent. Mater. J.* **2020**, *39*, 173–180, https://doi.org/10.4012/dmj.2019-291.
82. Hassan, N.; McCarville, K.; Morinaga, K.; Mengatto, C.M.; Langfelder, P.; Hokugo, A.; Tahara, Y.; Colwell, C.S.; Nishimura, I. Titanium Biomaterials with Complex Surfaces Induced Aberrant Peripheral Circadian Rhythms in Bone Marrow Mesenchymal Stromal Cells. *PLoS One* **2017**, *12*, 1–22, https://doi.org/10.1371/journal.pone.0183359.
83. Morinaga, K.; Sasaki, H.; Park, S.; Hokugo, A.; Okawa, H.; Tahara, Y.; Colwell, C.S.; Nishimura, I. Neuronal PAS Domain 2 (Npas2) Facilitated Osseointegration of Titanium Implant with Rough Surface through a Neuroskeletal Mechanism. *Biomaterials* **2019**, *192*, 62–74, https://doi.org/10.1016/j.biomaterials.2018.11.003.
84. Sautchuk, R.; Eliseev, R.A. Cell Energy Metabolism and Bone Formation. *Bone Reports* **2022**, *16*, 101594, https://doi.org/10.1016/j.bonr.2022.101594.
85. Sabini, E.; Arboit, L.; Khan, M.P.; Lanzolla, G.; Schipani, E. Oxidative Phosphorylation in Bone Cells. *Bone Reports* **2023**, *18*, 101688, https://doi.org/10.1016/j.bonr.2023.101688.
86. Lin, C.; Yang, Q.; Guo, D.; Xie, J.; Yang, Y.-S.; Chaugule, S.; DeSouza, N.; Oh, W.-T.; Li, R.; Chen, Z.; et al. Impaired Mitochondrial Oxidative Metabolism in Skeletal Progenitor Cells Leads to Musculoskeletal Disintegration. *Nat. Commun.* **2022**, *13*, 6869, https://doi.org/10.1038/s41467-022-34694-8.

87. Li, Q.; Gao, Z.; Chen, Y.; Guan, M.-X. The Role of Mitochondria in Osteogenic, Adipogenic and Chondrogenic Differentiation of Mesenchymal Stem Cells. *Protein Cell* **2017**, *8*, 439–445, https://doi.org/10.1007/s13238-017-0385-7.
88. Shum, L.C.; Hollenberg, A.M.; Baldwin, A.L.; Kalicharan, B.H.; Maqsoodi, N.; Rubery, P.T.; Mesfin, A.; Eliseev, R.A. Role of Oxidative Metabolism in Osseointegration During Spinal Fusion. *PLoS One* **2020**, *15*, e0241998.
89. Zhang, X.; Sun, J.; Zhou, M.; Li, C.; Zhu, Z.; Gan, X. The Role of Mitochondria in the Peri-Implant Microenvironment. *Exp. Physiol.* **2023**, *108*, 398–411, https://doi.org/10.1113/EP090988.
90. Pieralli, S.; Kohal, R.-J.J.; Lopez Hernandez, E.; Doerken, S.; Spies, B.C. Osseointegration of Zirconia Dental Implants in Animal Investigations: A Systematic Review and Meta-Analysis. *Dent. Mater.* **2018**, *34*, 171–182, https://doi.org/10.1016/j.dental.2017.10.008.
91. Leong, Y.-K.; Scales, P.J.; Healy, T.W.; Boger, D. V Effect of Particle Size on Colloidal Zirconia Rheology at the Isoelectric Point. *J. Am. Ceram. Soc.* **1995**, *78*, 2209–2212, https://doi.org/10.1111/j.1151-2916.1995.tb08638.x.
92. Jachimska, B.; Adamczyk, Z. Characterization of Rheological Properties of Colloidal Zirconia. *J. Eur. Ceram. Soc.* **2007**, *27*, 2209–2215, https://doi.org/10.1016/j.jeurceramsoc.2006.07.013.
93. Müller, C.; Lüders, A.; Hoth-Hannig, W.; Hannig, M.; Ziegler, C. Initial Bioadhesion on Dental Materials as a Function of Contact Time, pH, Surface Wettability, and Isoelectric point. *Langmuir* **2010**, *26*, 4136–4141, https://doi.org/10.1021/la903299y.
94. Borghi, F.; Vyas, V.; Podestà, A.; Milani, P. Nanoscale Roughness and Morphology Affect the IsoElectric Point of Titania Surfaces. *PLoS One* **2013**, *8*, 1–14, https://doi.org/10.1371/journal.pone.0068655.
95. Hanawa, T. Zirconia Versus Titanium in Dentistry: A Review. *Dent. Mater. J.* **2020**, *39*, 24–36, https://doi.org/10.4012/dmj.2019-172.
96. Aragoneses, J.M.J.; Valverde, N.L.; Fernandez-Dominguez, M.; Mena-Alvarez, J.; Rodriguez, C.; Gil, J.; Aragoneses, J.M.J. Relevant Aspects of Titanium and Zirconia Dental Implants for Their Fatigue and Osseointegration Behaviors. *Materials (Basel).* 2022, *15*.
97. Kohal, R.J.; Bächle, M.; Att, W.; Chaar, S.; Altmann, B.; Renz, A.; Butz, F. Osteoblast and Bone Tissue Response to Surface Modified Zirconia and Titanium Implant Materials. *Dent. Mater.* **2013**, *29*, 763–776, https://doi.org/10.1016/j.dental.2013.04.003.
98. Bosshardt, D.D.; Chappuis, V.; Buser, D. Osseointegration of Titanium, Titanium Alloy and Zirconia Dental Implants: Current Knowledge and Open Questions. *Periodontol. 2000* **2017**, *73*, 22–40, https://doi.org/10.1111/prd.12179.
99. Zanocco, M.; Boschetto, F.; Zhu, W.; Marin, E.; McEntire, B.J.; Bal, B.S.; Adachi, T.; Yamamoto, T.; Kanamura, N.; Ohgitani, E.; et al. 3D-Additive Deposition of an Antibacterial and Osteogenic Silicon Nitride Coating on Orthopaedic Titanium Substrate. *J. Mech. Behav. Biomed. Mater.* **2020**, *103*, 103557, https://doi.org/10.1016/j.jmbbm.2019.103557.
100. Guo, L.; Gao, C.; Wang, F.; Wei, J.; Hu, J.; Xu, Y. Influence of Content of Silicon Nitride Nanoparticles into Micro-Arc Oxidation Coating of Titanium on Bactericidal Capability and Osteoblastic Differentiation. *Surf. Coatings Technol.* **2023**, *458*, 129346, https://doi.org/10.1016/j.surfcoat.2023.129346.
101. Marin, E.; Zanocco, M.; Boschetto, F.; Yamamoto, T.; Kanamura, N.; Zhu, W.; McEntire, B. J.; Bal, B.S.; Ashida, R.; Mazda, O.; et al. *In Vitro* Comparison of Bioactive Silicon Nitride Laser Claddings on Different Substrates. *Appl. Sci.* **2020**, *10*, 9039, https://doi.org/10.3390/app10249039.
102. Marin, E.; Zanocco, M.; Boschetto, F.; Santini, M.; Zhu, W.; Adachi, T.; Ohgitani, E.; McEntire, B.J.; Bal, B.S.; Pezzotti, G. Silicon Nitride Laser Cladding: A Feasible Technique to Improve the Biological Response of Zirconia. *Mater. Des.* **2020**, *191*, 108649, https://doi.org/10.1016/j.matdes.2020.108649.
103. Kurtz, S.M.; Devine, J.N. PEEK Biomaterials in Trauma, Orthopedic, and Spinal Implants. *Biomaterials* **2007**, *28*, 4845–4869.

104. Verma, S.; Sharma, N.; Kango, S.; Sharma, S. Developments of PEEK (Polyetheretherketone) as a Biomedical Material: A Focused Review. *Eur. Polym. J.* **2021**, *147*, 110295, https://doi.org/10.1016/j.eurpolymj.2021.110295.
105. AlOtaibi, N.; Naudi, K.; Conway, D.; Ayoub, A. The Current State of PEEK Implant Osseointegration and Future Perspectives: A Systematic Review. *Eur. Cell. Mater.* **2020**, *40*, 1–20, https://doi.org/10.22203/eCM.v040a01.
106. Dondani, J.R.; Iyer, J.; Tran, S.D. Surface Treatments of PEEK for Osseointegration to Bone. *Biomolecules* **2023**, *13*, 464, https://doi.org/10.3390/biom13030464.
107. Pidhatika, B.; Widyaya, V.T.; Nalam, P.C.; Swasono, Y.A.; Ardhani, R. Surface Modifications of High-Performance Polymer Polyetheretherketone (PEEK) to Improve Its Biological Performance in Dentistry. *Polymers (Basel).* **2022**, *14*, 1–38, https://doi.org/10.3390/polym14245526.
108. Xu, Z.; Wu, H.; Wang, F.; Kaewmanee, R.; Pan, Y.; Wang, D.; Qu, P.; Wang, Z.; Hu, G.; Zhao, J.; et al. A Hierarchical Nanostructural Coating of Amorphous Silicon Nitride on Polyetheretherketone with Antibacterial Activity and Promoting Responses of rBMSCs for Orthopedic Applications. *J. Mater. Chem. B* **2019**, *7*, 6035–6047, https://doi.org/10.1039/c9tb01565e.
109. Dai, Y.; Chu, L.; Luo, Z.; Tang, T.; Wu, H.; Wang, F.; Mei, S.; Wei, J.; Wang, X.; Shang, X. Effects of a Coating of Nano Silicon Nitride on Porous Polyetheretherketone on Behaviors of MC3T3-E1 Cells *in vitro* and Vascularization and Osteogenesis *in Vivo*. *ACS Biomater. Sci. Eng.* **2019**, *5*, 6425–6435, https://doi.org/10.1021/acsbiomaterials.9b00605.
110. Dai, Y.; Guo, H.; Chu, L.; He, Z.; Wang, M.; Zhang, S.; Shang, X. Promoting Osteoblasts Responses *in vitro* and Improving Osteointegration *in vivo* through Bioactive Coating of Nanosilicon Nitride on Polyetheretherketone. *J. Orthop. Transl.* **2020**, *24*, 198–208, https://doi.org/10.1016/j.jot.2019.10.011.
111. Boschetto, F.; Marin, E.; Ohgitani, E.; Adachi, T.; Zanocco, M.; Horiguchi, S.; Zhu, W.; McEntire, B.J.; Mazda, O.; Bal, B.S.; et al. Surface Functionalization of PEEK with Silicon Nitride. *Biomed. Mater.* **2020**, *16*, 015015, https://doi.org/10.2139/ssrn.3447526.
112. Wu, H.; Yang, L.; Qian, J.; Wang, D.; Pan, Y.; Wang, X.; Nabanita, S.; Tang, T.; Zhao, J.; Wei, J. Microporous Coatings of PEKK/SN Composites Integration with PEKK Exhibiting Antibacterial and Osteogenic Activity, and Promotion of Osseointegration for Bone Substitutes. *ACS Biomater. Sci. Eng.* **2019**, acsbiomaterials.8b01508, https://doi.org/10.1021/acsbiomaterials.8b01508.
113. Hu, X.; Mei, S.; Wang, F.; Tang, S.; Xie, D.; Ding, C.; Du, W.; Zhao, J.; Yang, L.; Wu, Z.; et al. A Microporous Surface Containing Si_3N_4/Ta Microparticles of PEKK Exhibits both Antibacterial and Osteogenic Activity for Inducing Cellular Response and Improving Osseointegration. *Bioact. Mater.* **2021**, *6*, 3136–3149, https://doi.org/10.1016/j.bioactmat.2021.02.027.
114. Mobbs, R.J.; Rao, P.J.; Phan, K.; Hardcastle, P.; Choy, W.J.; McCartney, E.R.; Druitt, R.K.; Mouatt, C.A.L.; Sorrell, C.C. Anterior Lumbar Interbody Fusion Using Reaction Bonded Silicon Nitride Implants: Long Term Case Series of the First Synthetic ALIF Spacer Implanted in Humans. *World Neurosurg.* **2018**, *120*, 256–264, https://doi.org/10.1016/j.wneu.2018.08.237.
115. Ball, H.T.; McEntire, B.J.; Bal, B.S. Accelerated Cervical Fusion of Silicon Nitride versus PEEK Spacers: A Comparative Clinical Study. *J. Spine* **2017**, *6*, 1000396, https://doi.org/10.4172/2165-7939.1000396.
116. Rambo, W.M. Treatment of Lumbar Discitis using Silicon Nitride Spinal Spacers: A Case Series and Literature Review. *Int. J. Surg. Case Rep.* **2018**, *43*, 61–68, https://doi.org/10.1016/j.ijscr.2018.02.009.
117. Smith, M.W.; Romano, D.R.; McEntire, B.J.; Bal, B.S. A Single Center Retrospective Clinical Evaluation of Anterior Cervical Discectomy and Fusion Comparing Allograft Spacers to Silicon Nitride Cages. *J. Spine Surg.* **2018**, *4*, 349–360, https://doi.org/10.21037/jss.2018.06.02.

118. Calvert, G.C.; Huffmon III, G. V; Rambo Jr, W.M.; Smith, M.W.; McEntire, B.J.; Bal, B.S. Clinical Outcomes for Anterior Cervical Discectomy and Fusion with Silicon Nitride Spine Cages: A Multicenter Study. *J. Spine Surg.* **2019**, *5*, 504–519, https://doi.org/10.21037/jss.2019.11.17.
119. Calvert, G.C.; Huffmon, G.V.B.I.; Rambo, W.M.J.; Smith, M.W.; McEntire, B.J.; Bal, B.S. Clinical Outcomes for Lumbar Fusion using Silicon Nitride versus other Biomaterials. *J. Spine Surg.* **2020**, *6*, 33–48, https://doi.org/10.21037/jss.2019.12.11.
120. Gray, M.T.; Davis, K.P.; McEntire, B.J.; Bal, B.S.; Smith, M.W. Transforaminal Lumbar Interbody Fusion with a Silicon Nitride Cage Demonstrates Early Radiographic Fusion. *J. Spine Surg.* **2022**, *8*, 29–43, https://doi.org/10.21037/jss-21-115.
121. Pezzotti, G. Silicon Nitride: A Bioceramic with a Gift. *ACS Appl. Mater. Interfaces* **2019**, *11*, 26619–26636, https://doi.org/10.1021/acsami.9b07997.
122. Fiani, B.; Jarrah, R.; Shields, J.; Sekhon, M. Enhanced Biomaterials: Systematic Review of Alternatives to Supplement Spine Fusion Including Silicon Nitride, Bioactive Glass, Amino Peptide Bone Graft, and Tantalum. *Neurosurg. Focus* **2021**, *50*, 1–9, https://doi.org/10.3171/2021.3.FOCUS201044.
123. Ahuja, N.; Awad, K.R.; Brotto, M.; Aswath, P.B.; Varanasi, V. A Comparative Study on Silicon Nitride, Titanium and Polyether Ether Ketone on Mouse Pre-osteoblast Cells. *Med. Devices Sensors* **2021**, *4*, 1–12, https://doi.org/10.1002/mds3.10139.
124. Lee, S.S.; Laganenka, L.; Du, X.; Hardt, W.D.; Ferguson, S.J. Silicon Nitride, a Bioceramic for Bone Tissue Engineering: A Reinforced Cryogel System With Antibiofilm and Osteogenic Effects. *Front. Bioeng. Biotechnol.* **2021**, *9*, 1–18, https://doi.org/10.3389/fbioe.2021.794586.
125. Bitar, D.; Parvizi, J. Biological Response to Prosthetic Debris. *World J. Orthop.* **2015**, *6*, 172–189, https://doi.org/10.5312/wjo.v6.i2.172.
126. Loss, P.B.; Bressan, E.; Ferroni, L.; Gardin, C.; Bellin, G.; Sbricoli, L.; Sivolella, S.; Brunello, G.; Schwartz-arad, D.; Mijiritsky, E.; et al. Metal Nanoparticles Released from Dental Implant Surfaces: Potential Contribution to Chronic Inflammation and Peri-Implant Bone Loss. *Materials (Basel).* **2019**, *12*, 2036.
127. Lal, S.; Caseley, E.; Tipper, J.L. Biological Impact of Silicon Nitride for Orthopaedic Applications: Role of Particle Size, Surface Composition, and Donor Variation. **2018**, 1–12, https://doi.org/10.1038/s41598 018 27494-y.
128. Neumann, A.; Unkel, C.; Werry, C.; Herborn, C.U.; Maier, H.R.; Ragoss, C.; Jahnke, K. Prototype of a Silicon Nitride Ceramic-Based Miniplate Osteofixation System for the Midface. *Otolaryngol. Neck Surg.* **2006**, *134*, 923 30, https://doi.org/10.1016/j.otohns.2006.01.022.
129. Guedes e Silva, C.C.; Konig, B.; Carbonari, M.J.; Yoshimoto, M.; Allegrini, S.; Bressiani, J.C. Tissue Response Around Silicon Nitride Implants in Rabbits. *J. Biomed. Mater. Res.* **2008**, *84A*, 337–343, https://doi.org/10.1002/jbm.a.
130. Anderson, M.C.; Olsen, R. Bone Ingrowth into Porous Silicon Nitride. *J. Biomed. Mater. Res.* **2010**, *92A*, 1598–1605, https://doi.org/10.1002/jbm.a.32498.
131. Webster, T.J.; Patel, A.A.; Rahaman, M.N.; Bal, B.S. Anti-Infective and Osteointegration Properties of Silicon Nitride, Poly (Ether Ether Ketone), and Titanium Implants. *Acta Biomater.* **2012**, *8*, 4447–4454.
132. Kersten, R.F.M.R.; Wu, G.; Pouran, B.; van der Veen, A.J.; Weinans, H.H.; de Gast, A.; Öner, F.C.; van Gaalen, S.M. Comparison of Polyetheretherketone versus Silicon Nitride Intervertebral Spinal Spacers in a Caprine Model. *J. Biomed. Mater. Res. Part B Appl. Biomater.* **2019**, *107*, 688–699, https://doi.org/10.1002/jbm.b.34162.
133. Atsumi, M.; Park, S.; Wang, H.-L. Methods Used to Assess Implant Stability: Current Status. *Int. J. Oral Maxillofac. Implants* **2007**, *22*, 743–754.
134. Shemtov-Yona, K. Quantitative Assessment of the Jawbone Quality cCassification: A Meta-Analysis Study. *PLoS One* **2021**, *16*, e0253283.
135. Jaffin, R.A.; Berman, C.L. The Excessive Loss of Branemark Fixtures in Type IV Bone: A 5-Year Analysis. *J. Periodontol.* **1991**, *62*, 2–4, https://doi.org/10.1902/jop.1991.62.1.2.

136. Goiato, M.C.; dos Santos, D.M.; Jr. Santiago, J.F.; Moreno, A.; Pellizzer, E.P. Longevity of Dental Implants in Type IV Bone: A Systematic Review. *Int. J. Oral Maxillofac. Surg.* **2014**, *43*, 1108–1116, https://doi.org/10.1016/j.ijom.2014.02.016.
137. Haimov, H.; Yosupov, N.; Pinchasov, G.; Juodzbalys, G. Bone Morphogenetic Protein Coating on Titanium Implant Surface: a Systematic Review. *J. Oral Maxillofac. Res.* **2017**, *8*, 1–12, https://doi.org/10.5037/jomr.2017.8201.
138. Badieyan, Z.S.; Berezhanskyy, T.; Utzinger, M.; Aneja, M.K.; Emrich, D.; Erben, R.; Schüler, C.; Altpeter, P.; Ferizi, M.; Hasenpusch, G.; et al. Transcript-Activated Collagen Matrix as Sustained mRNA Delivery System for Bone Regeneration. *J. Control. Release* **2016**, *239*, 137–148, https://doi.org/10.1016/j.jconrel.2016.08.037.
139. Rajendran, A.K.; Amirthalingam, S.; Hwang, N.S. A Brief Review of mRNA Therapeutics and Delivery for Bone Tissue Engineering. *RSC Adv.* **2022**, *12*, 8889–8900, https://doi.org/10.1039/d2ra00713d.
140. Balmayor, E.R.; Geiger, J.P.; Koch, C.; Aneja, M.K.; van Griensven, M.; Rudolph, C.; Plank, C. Modified mRNA for BMP-2 in Combination with Biomaterials Serves as a Transcript-Activated Matrix for Effectively Inducing Osteogenic Pathways in Stem Cells. *Stem Cells Dev.* **2016**, *26*, 25–34, https://doi.org/10.1089/scd.2016.0171.
141. Pezzotti, G.; McEntire, B.J.; Bock, R.; Boffelli, M.; Zhu, W.; Vitale, E.; Puppulin, L.; Adachi, T.; Yamamoto, T.; Kanamura, N.; et al. Silicon Nitride: A Synthetic Mineral for Vertebrate Biology. *Sci. Rep.* **2016**, *6*, 31717, https://doi.org/10.1038/srep11384.
142. Pezzotti, G.; Marin, E.; Adachi, T.; Rondinella, A.; Boschetto, F.; Zhu, W.-L.; Sugano, N.; Bock, R.M.; McEntire, B.J.; Bal, B.S. Bioactive Silicon Nitride: A New Therapeutic Material for Osteoarthropathy. *Sci. Rep.* **2017**, *7*, 44848.
143. Pezzotti, G.; Oba, N.; Zhu, W.; Marin, E.; Rondinella, A.; Boschetto, F.; McEntire, B.J.; Yamamoto, K.; Bal, B.S. Human Osteoblasts Grow Transitional Si/N Apatite in Quickly Osteointegrated Si_3N_4 Cervical Insert. *Acta Biomater.* **2017**, *64*, 411–420.
144. Taipale, J.; Makinen, T.; Arighi, E.; Kukk, E.; Karkkainen, M.; Alitalo, K. Vascular Endothelial Growth Factor Receptor-3. In *Vascular Growth Factors and Angiogenesis*; Claesson-Welsh, L., Ed.; Springer Berlin Heidelberg: Berlin, Heidelberg, 1999; pp. 85–96. ISBN 978-3-642-59953-8.
145. Navone, S.E.; Marfia, G.; Invernici, G.; Cristini, S.; Nava, S.; Balbi, S.; Sangiorgi, S.; Ciusani, E.; Bosutti, A.; Alessandri, G.; et al. Isolation and Expansion of Human and Mouse Brain Microvascular Endothelial Cells. *Nat. Protoc.* **2013**, *8*, 1680–1693, https://doi.org/10.1038/nprot.2013.107.
146. Monte, F.A. do; Awad, K.R.; Ahuja, N.; Kim, H.K.W.; Aswath, P.; Brotto, M.; Varanasi, V.G. Amorphous Silicon Oxynitrophosphide-Coated Implants Boost Angiogenic Activity of Endothelial Cells. *Tissue Eng. Part A* **2019**, *26*, 15–27, https://doi.org/10.1089/ten.tea.2019.0051.
147. Shen, Y.; Fang, K.; Xiang, Y.; Xu, K.; Yu, L.; Chen, J.; Ma, P.; Cai, K.; Shen, X.; Liu, J. Improvement in Osteogenesis, Vascularization, and Corrosion Resistance of Titanium with Silicon-Nitride Doped Micro-Arc Oxidation Coatings. *Front. Bioeng. Biotechnol.* **2022**, *10*, 1023032.
148. Bock, R.M.; Jones, E.N.; Ray, D.A.; Bal, B.S.; Pezzotti, G.; McEntire, B.J. Bacteriostatic Behavior of Surface-Modulated Silicon Nitride in Comparison to Polyetheretherketone and Titanium. *J. Biomed. Mater. Res. Part A* **2017**, *105*, 1521–1534, https://doi.org/10.1002/jbm.a.35987.
149. Gong, Y.; Honda, Y.; Adachi, T.; Marin, E.; Yoshikawa, K.; Pezzotti, G.; Yamamoto, K. Tailoring Silicon Nitride Surface Chemistry for Facilitating Odontogenic Differentiation of Rat Dental Pulp Cells. *Int. J. Mol. Sci.* **2021**, *22*, 13130, https://doi.org/10.3390/ijms222313130.
150. Hewlett, E.R.; Atchison, K.A.; White, S.C.; Flack, V. Radiographic Secondary Caries Prevalence in Teeth with Clinically Defective Restorations. *J. Dent. Res.* **1993**, *72*, 1604–1608, https://doi.org/10.1177/00220345930720121301.

151. Nedeljkovic, I.; De Munck, J.; Vanloy, A.; Declerck, D.; Lambrechts, P.; Peumans, M.; Teughels, W.; Van Meerbeek, B.; Van Landuyt, K.L. Secondary Caries: Prevalence, Characteristics, and Approach. *Clin. Oral Investig.* **2020**, *24*, 683–691, https://doi.org/10.1007/s00784-019-02894-0.
152. Cury, J.A.; de Oliveira, B.H.; dos Santos, A.P.P.; Tenuta, L.M.A. Are Fluoride Releasing Dental Materials Clinically Effective on Caries Control? *Dent. Mater.* **2016**, *32*, 323–333, https://doi.org/10.1016/j.dental.2015.12.002.
153. Elango, A.V.; Vasudevan, S.; Shanmugam, K.; Solomon, A.P.; Neelakantan, P. Exploring the Anti-Caries Properties of Baicalin Against Streptococcus mutans: An *in vitro* Study. *Biofouling* **2021**, *37*, 267–275, https://doi.org/10.1080/08927014.2021.1897789.
154. Balhaddad, A.A.; Kansara, A.A.; Hidan, D.; Weir, M.D.; Xu, H.H.K.; Melo, M.A.S. Toward Dental Caries: Exploring Nanoparticle-Based Platforms and Calcium Phosphate Compounds for Dental Restorative Materials. *Bioact. Mater.* **2019**, *4*, 43–55, https://doi.org/10.1016/j.bioactmat.2018.12.002.
155. Zhang, N.; Zhang, K.; Xie, X.; Dai, Z.; Zhao, Z.; Imazato, S.; Al-Dulaijan, Y.A.; Al-Qarni, F. D.; Weir, M.D.; Reynolds, M.A.; et al. Nanostructured Polymeric Materials with Protein-Repellent and Anti-Caries Properties for Dental Applications. *Nanomaterials* 2018, *8*.
156. Krzyściak, W.; Jurczak, A.; Kościelniak, D.; Bystrowska, B.; Skalniak, A. The Virulence of Streptococcus mutans and the Ability to Form Biofilms. *Eur. J. Clin. Microbiol. Infect. Dis.* **2014**, *33*, 499–515, https://doi.org/10.1007/s10096-013-1993-7.
157. Kreth J., Merritt J., Shi W., Q.F. Competition and Coexistence between Streptococcus mutans and Streptococcus sanguinis in the Dental Biofilm. *J. Bacteriol.* **2005**, *187*, 7193–7203, https://doi.org/10.1128/JB.187.21.7193.
158. Duchin, S.; Van Houte, J. Colonization of Teeth in Humans by Streptococcus mutans as related to its Concentration in Saliva and Host Age. *Infect. Immun.* **1978**, *20*, 120–125.
159. Yamada, T.; Abbe, K. Effects of Oxygen on Pyruvate Formate-Lyase In Situ and Sugar Metabolism of Streptococcus mutans and Streptococcus sanguis. **1985**, *47*, 129–134.
160. Montanaro, L.; Campoccia, D.; Rizzi, S.; Donati, M.E.; Breschi, L.; Prati, C.; Arciola, C.R. Evaluation of Bacterial Adhesion of Streptococcus mutans on Dental Restorative Materials. *Biomaterials* **2004**, *25*, 4457–4463, https://doi.org/10.1016/j.biomaterials.2003.11.031.
161. Lemos, J.A.; Quivey, R.G.; Koo, H.; Abranches, J. Streptococcus mutans: A new Gram-Positive Paradigm? *Microbiology* **2013**, *159*, 436–445, https://doi.org/10.1099/mic.0.066134-0.
162. Zhang, Y.; Wang, X.; Li, H.; Ni, C.; Du, Z.; Yan, F. Human Oral Microbiota and its Modulation for Oral Health. *Biomed. Pharmacother.* **2018**, *99*, 883–893, https://doi.org/10.1016/j.biopha.2018.01.146.
163. Zeller, B.; Stöckli, S.; Zaugg, L.K.; Astasov-Frauenhoffer, M.; Hauser-Gerspach, I.; Waltimo, T.; Zitzmann, N.U. Biofilm Formation on Metal Alloys, Zirconia and Polyetherketoneketone as Implant Materials *in vivo*. *Clin. Oral Implants Res.* **2020**, *31*, 1078–1086, https://doi.org/10.1111/clr.13654.
164. Wiessner, A.; Wassmann, T.; Wiessner, J.M.; Schubert, A.; Wiechens, B.; Hampe, T.; Bürgers, R. *In Vivo* Biofilm Formation on Novel PEEK, Titanium, and Zirconia Implant Abutment Materials. *Int. J. Mol. Sci.* **2023**, *24*, 1779, https://doi.org/10.3390/ijms24021779.
165. Yin, I.X.; Zhang, J.; Zhao, I.S.; Mei, M.L.; Li, Q.; Chu, C.H. The Antibacterial Mechanism of Silver Nanoparticles and its Application in Dentistry. *Int. J. Nanomedicine* **2020**, *15*, 2555–2562, https://doi.org/10.2147/IJN.S246764.
166. Fakhruddin, K.S.; Egusa, H.; Ngo, H.C.; Panduwawala, C.; Pesee, S.; Samaranayake, L.P. Clinical Efficacy and the Antimicrobial Potential of Silver Formulations in Arresting Dental Caries: A Systematic Review. *BMC Oral Health* **2020**, *20*, 1–13, https://doi.org/10.1186/s12903-020-01133-3.
167. Wade, W.G. The Oral Microbiome in Health and Disease. *Pharmacol. Res.* **2013**, *69*, 137–143, https://doi.org/10.1016/j.phrs.2012.11.006.

168. Radcliffe, C.E.; Akram, N.C.; Hurrell, F.; Drucker, D.B. Effects of Nitrite and Nitrate on the Growth and Acidogenicity of Streptococcus mutans. *J. Dent.* **2002**, *30*, 325–331, https://doi.org/10.1016/S0300-5712(02)00046-5.
169. Xia, D.; Liu, Y.; Zhang, C.; Yang, S.; Wang, S. Antimicrobial Effect of Acidified Nitrate and Nitrite on Six Common Oral Pathogens *in vitro*. *Chin. Med. J. (Engl).* **2006**, *119*, 1904–1909, https://doi.org/10.1111/1467-8411.00038.
170. Backlund, C.J.; Sergesketter, A.R.; Offenbacher, S.; Schoenfisch, M.H. Antibacterial Efficacy of Exogenous Nitric Oxide on Periodontal Pathogens. *J. Dent. Res.* **2014**, *93*, 1089–1094, https://doi.org/10.1177/0022034514529974.
171. Smeets, R.; Stadlinger, B.; Schwarz, F.; Beck-Broichsitter, B.; Jung, O.; Precht, C.; Kloss, F.; Gröbe, A.; Heiland, M.; Ebker, T. Impact of Dental Implant Surface Modifications on Osseointegration. *Biomed Res. Int.* **2016**, *2016*, 6285620, https://doi.org/10.1155/2016/6285620.
172. Dohan Ehrenfest, D.M.; Vazquez, L.; Park, Y.-J.; Sammartino, G.; Bernard, J.-P. Identification Card and Codification of the Chemical and Morphological Characteristics of 14 Dental Implant Surfaces. *J. Oral Implantol.* **2011**, *37*, 525–542, https://doi.org/10.1563/AAID-JOI-D-11-00080.
173. Coelho, P.G.; Jimbo, R.; Tovar, N.; Bonfante, E.A. Osseointegration: Hierarchical Designing Encompassing the Macrometer, Micrometer, and Nanometer Length Scales. *Dent. Mater.* **2015**, *31*, 37–52, https://doi.org/10.1016/j.dental.2014.10.007.
174. Da Silva, C.H.F.P.; Vidigal, G.M.; De Uzeda, M.; De Almeida Soares, G. Influence of Titanium Surface Roughness on Attachment of Streptococcus sanguis: An *in vitro* Study. *Implant Dent.* **2005**, *14*, 88–93, https://doi.org/10.1097/01.id.0000154793.84609.2c.
175. Buechel, F.F.; Helbig, T.E.; Pappas, M.J.; Trier, K.; Dean, S.W. 31 Year Evolution of the Rotating-Platform Total Knee Replacment: Coping With "Spinout" and Wear. *J. ASTM Int.* **2012**, *9*, 1–14, https://doi.org/10.1520/JAI103329.
176. Godoy-Gallardo, M.; Rodríguez-Hernández, A.G.; Delgado, L.M.; Manero, J.M.; Javier Gil, F.; Rodríguez, D. Silver Deposition on Titanium Surface by Electrochemical Anodizing Process Reduces Bacterial Adhesion of Streptococcus sanguinis and Lactobacillus salivarius. *Clin. Oral Implants Res.* **2015**, *26*, 1170–1179, https://doi.org/10.1111/clr.12422.
177. Harris, L.G.; Tosatti, S.; Wieland, M.; Textor, M.; Richards, R.G. Staphylococcus aureus Adhesion to Titanium Oxide Surfaces Coated with Non-Functionalized and Peptide-Functionalized Poly(L-lysine)-Grafted-Poly(ethylene glycol) Copolymers. *Biomaterials* **2004**, *25*, 4135–4148, https://doi.org/10.1016/j.biomaterials.2003.11.033.
178. Zhang, H.; Komasa, S.; Mashimo, C.; Sekino, T.; Okazaki, J. Effect of Ultraviolet Treatment on Bacterial Attachment and Osteogenic Activity to Alkali-Treated Titanium with Nanonetwork Structures. *Int. J. Nanomedicine* **2017**, *12*, 4633–4646, https://doi.org/10.2147/IJN.S136273.
179. Fernández, L.; Langa, S.; Martín, V.; Maldonado, A.; Jiménez, E.; Martín, R.; Rodríguez, J.M. The Human Milk Microbiota: Origin and Potential Roles in Health and Disease. *Pharmacol. Res.* **2013**, *69*, 1–10, https://doi.org/10.1016/j.phrs.2012.09.001.
180. Guggenheim, B.; Giertsen, E.; Schupbach, P.; Shapiro, S. Validation of an *in vitro* Biofilm Model of Supragingival Plaque. *J. Dent. Res.* **2001**, *80*, 363–370, https://doi.org/10.1177/00220345010800011201.
181. Darrene, L.N.; Cecile, B. Experimental Models of Oral Biofilms Developed on Inert Substrates: A Review of the Literature. *Biomed Res. Int.* **2016**, *2016*, 7461047, https://doi.org/10.1155/2016/7461047.
182. Ji, M.; Park, S.; Lee, K.; Kang, I.; Yun, K.; Kim, H.; Lim, H.-P. Evaluation of Antibacterial Activity and Osteoblast-Like Cell Viability of TiN, ZrN and (Ti1-xZrx)N Coating on Titanium. *J. Adv. Prosthodont.* **2015**, *7*, 166–71, https://doi.org/10.4047/jap.2015.7.2.166.
183. Bouziane, A.; Hamdoun, R.; Abouqal, R.; Ennibi, O. Global Prevalence of Aggressive Periodontitis: A Systematic Review and Meta-Analysis. *J. Clin. Periodontol.* **2020**, *47*, 406–428, https://doi.org/10.1111/jcpe.13266.

184. Patini, R.; Staderini, E.; Lajolo, C.; Lopetuso, L.; Mohammed, H.; Rimondini, L.; Rocchetti, V.; Franceschi, F.; Cordaro, M.; Gallenzi, P. Relationship between Oral Microbiota and Periodontal Disease: A Systematic Review. *Eur. Rev. Med. Pharmacol. Sci.* **2018**, *22*, 5775–5788.
185. Tibúrcio-Machado, C.S.; Michelon, C.; Zanatta, F.B.; Gomes, M.S.; Marin, J.A.; Bier, C.A. The Global Prevalence of Apical Periodontitis: A Systematic Review and Meta-Analysis. *Int. Endod. J.* **2021**, *54*, 712–735, https://doi.org/10.1111/iej.13467.
186. Fragkioudakis, I.; Tseleki, G.; Doufexi, A.E.; Sakellari, D. Current Concepts on the Pathogenesis of Peri-implantitis: A Narrative Review. *Eur. J. Dent.* **2021**, *15*, 379–387, https://doi.org/10.1055/s-0040-1721903.
187. Diaz, P.; Gonzalo, E.; Villagra, L.J.G.; Miegimolle, B.; Suarez, M.J. What is the Prevalence of Peri-Implantitis? A Systematic Review and Meta-Analysis. *BMC Oral Health* **2022**, *22*, 1–13, https://doi.org/10.1186/s12903-022-02493-8.
188. Nemec, M.; Behm, C.; Maierhofer, V.; Gau, J.; Kolba, A.; Jonke, E.; Rausch-Fan, X.; Andrukhov, O. Effect of Titanium and Zirconia Nanoparticles on Human Gingival Mesenchymal Stromal Cells. *Int. J. Mol. Sci.* **2022**, *23*, https://doi.org/10.3390/ijms231710022.
189. Webber, L.P.; Chan, H.L.; Wang, H.L. Will Zirconia Implants Replace Titanium Implants? *Appl. Sci.* **2021**, *11*, https://doi.org/10.3390/app11156776.
190. Fernandes, P.R.E.; Otero, A.I.P.; Fernandes, J.C.H.; Nassani, L.M.; Castilho, R.M.; de Oliveira Fernandes, G.V. Clinical Performance Comparing Titanium and Titanium–Zirconium or Zirconia Dental Implants: A Systematic Review of Randomized Controlled Trials. *Dent. J.* **2022**, *10*, 0–11, https://doi.org/10.3390/dj10050083.
191. Halim, F.C.; Pesce, P.; De Angelis, N.; Benedicenti, S.; Menini, M. Comparison of the Clinical Outcomes of Titanium and Zirconia Implant Abutments: A Systematic Review of Systematic Reviews. *J. Clin. Med.* **2022**, *11*, 5052, https://doi.org/10.3390/jcm11175052.
192. Laleman, I.; Lambert, F.; Gahlert, M.; Bacevic, M.; Woelfler, H.; Roehling, S. The Effect of Different Abutment Materials on Peri-implant Tissues—A Systematic Review and Meta-Analysis. *Clin. Oral Implants Res.* **2023**, *34*, 125–142, https://doi.org/10.1111/clr.14159.
193. Gorth, D.J.; Puckett, S.; Ercan, B.; Webster, T.J.; Rahaman, M.; Bal, B.S. Decreased Bacteria Activity on Si_3N_4 Surfaces Compared with PEEK or Titanium. *Int. J. Nanomedicine* **2012**, *7*, 1829–1840, https://doi.org/10.2147/IJN.S35190.
194. Khonsari, R.H.; Berthier, P.; Rouillon, T.; Perrin, J.-P.; Corre, P. Severe Infectious Complications after PEEK-Derived Implant Placement: Report of Three Cases. *J. Oral Maxillofac. Surgery, Med. Pathol.* **2014**, *26*, 477–482, https://doi.org/10.1016/j.ajoms.2013.04.006.
195. Gittens, R.A.; Scheideler, L.; Rupp, F.; Hyzy, S.L.; Geis-Gerstorfer, J.; Schwartz, Z.; Boyan, B.D. A Review on the Wettability of Dental Implant Surfaces II: Biological and Clinical Aspects. *Acta Biomater.* **2014**, *10*, 2907–2918, https://doi.org/10.1016/j.surg.2006.10.010.Use.
196. Gallo, J.; Holinka, M.; Moucha, C.S. Antibacterial Surface Treatment for Orthopaedic Implants. *Int. J. Mol. Sci.* **2014**, *15*, 13849–13880, https://doi.org/10.3390/ijms150813849.
197. Basgul, C.; DeSantis, P.; Derr, T.; Hickok, N.; Bock, R.; Kurtz, S. Exploring the Mechanical Strength, Antimicrobial Performance, and Bioactivity of 3D-Printed Silicon Nitride-PEEK Composites in Cervical Spinal Cages. *Int. J. Bioprinting* **2024**, *X*, 1, https://doi.org/10.36922/ijb.2124.
198. Ishikawa, M.; Bentley, K.L. de M.; McEntire, B.J.; Bal, B.S.; Schwarz, E.M.; Xie, C.; Avenue, E.; McEntire, B.J.; Schwarz, E.M.; Xie, C. Surface Topography of Silicon Nitride Affects Antimicrobial and Osseointegrative Properties of Tibial Implants in a Murine Model. *J. Biomed. Mater. Res. A* **2017**, *105*, 3413–3421, https://doi.org/10.1002/jbm.a.36189.
199. Pezzotti, G.; Bock, R.M.; McEntire, B.J.; Adachi, T.; Marin, E.; Boschetto, F.; Zhu, W.; Mazda, O.; Bal, B.S. *In vitro* Antibacterial Activity of Oxide and Non-Oxide Bioceramics for Arthroplastic Devices: I. In situ Time-Lapse Raman Spectroscopy. *Analyst* **2018**, *143*, 3708–3721, https://doi.org/10.1039/C8AN00233A.

200. Boschetto, F.; Toyama, N.; Horiguchi, S.; Bock, R.M.; McEntire, B.J.; Adachi, T.; Marin, E.; Zhu, W.; Mazda, O.; Bal, B.S.; et al. *In vitro* Antibacterial Activity of Oxide and Non-Oxide Bioceramics for Arthroplastic Devices: II. In situ Time-Lapse Fourier Transform Infrared Spectroscopy. *Analyst* **2018**, *143*, 2128–2140, https://doi.org/10.1039/C8AN00234G.
201. Pezzotti, G.; Bock, R.M.; McEntire, B.J.; Jones, E.; Boffelli, M.; Zhu, W.; Baggio, G.; Boschetto, F.; Puppulin, L.; Adachi, T.; et al. Silicon Nitride Bioceramics Induce Chemically Driven Lysis in Porphyromonas Gingivalis. *Langmuir* **2016**, *32*, 3024–3035, https://doi.org/10.1021/acs.langmuir.6b00393.
202. Boutrin, M.-C.; Wang, C.; Aruni, W.; Li, X.; Fletcher, H.M. Nitric Oxide Stress Resistance in Porphyromonas Gingivalis is Mediated by a Putative Hydroxylamine Reductase. *J. Bacteriol.* **2012**, *194*, 1582–1592.
203. Lundberg, J.O.; Weitzberg, E.; Gladwin, M.T. The Nitrate-Nitrite-Nitric Oxide Pathway in Physiology and Therapeutics. *Nat. Rev. Drug Discov.* **2008**, *7*, 156–167.
204. Rodionov, D.A.; Dubchak, I.L.; Arkin, A.P.; Alm, E.J.; Gelfand, M.S. Dissimilatory Metabolism of Nitrogen Oxides in Bacteria: Comparative Reconstruction of Transcriptional Networks. *PLoS Comput. Biol.* **2005**, *1*, e55–e55.
205. Ye, R.W.; Averill, B.A.; Tiedje, J.M. Denitrification: Production and Consumption of Nitric Oxide. *Appl. Environ. Microbiol.* **1994**, *60*, 1053–1058.
206. Radi, R.; Beckman, J.S.; Bush, K.M.; Freeman, B.A. Peroxynitrite-Induced Membrane Lipid Peroxidation: The Cytotoxic Potential of Superoxide and Nitric Oxide. *Arch. Biochem. Biophys.* **1991**, *288*, 481–487, https://doi.org/10.1016/0003-9861(91)90224-7.

The Future of Silicon Nitride: Biomedical Applications and Beyond

Giuseppe Pezzotti and Bryan J. McEntire

1 Introduction

As extensively described by various authors in this book, silicon nitride (Si_3N_4) possesses several key properties relevant to biomedical applications. Although not yet fully exploited in medical devices compared to other biomaterials, Si_3N_4 is attracting increasing attention for several biomedical purposes due to its unique combination of properties [1, 2]. Its biocompatibility, osteoconductivity, and concurrent antibacterial behavior make it well-suited for spinal prosthetic implants. Additionally, its superior mechanical properties, including high-fracture toughness, strength, hardness, wear resistance, and low-friction coefficient benefit artificial hip, knee, and shoulder joints. Articulation devices require these properties to ensure long prosthetic life and to mimic the patient's natural movements [3, 4]. Furthermore, Si_3N_4's radiolucency facilitates clear visualization of the implant site for accurate surgical placement and postimplantation monitoring by conventional medical imaging [5]. This combination of key properties is not fully available in any other presently known biomaterial for implant applications. For instance, titanium alloys exhibit excellent mechanical properties, but they lack radiolucency, osteoconductivity, and antibacterial characteristics. Biocompatible polymers, like polyetheretherketone (PEEK), possess high resilience and biocompatibility, but they are not antibacterial and are radiotransparent instead of radiolucent. Despite exhibiting promising properties for various biomedical applications, the clinical adoption of Si_3N_4 lags behind that of conventional biomaterials such as titanium,

G. Pezzotti (✉)
Ceramic Physics Laboratory, Kyoto Institute of Technology, Sakyo-ku, Matsugasaki, Kyoto, Japan
e-mail: pezzotti@kit.ac.jp

B. J. McEntire
SINTX Technologies, Inc., Salt Lake City, UT, USA

B. S. Bal et al. (eds.), *Silicon Nitride Bioceramics*,
https://doi.org/10.1007/978-3-031-67047-3_13

stainless steel, alumina, zirconia, and PEEK, or other established biomedical polymers [6]. As researchers and manufacturers strive to develop advanced biomaterials tailored to specific medical needs, Si_3N_4 appears to be a promising candidate in this continuous exploration.

Perhaps the most unique property of Si_3N_4 is its surface chemistry, which has been described as a "gift" for biomedical applications [7]. In an aqueous environment, Si_3N_4's surface undergoes a slow but continuous process of hydrolysis, eluting nitrogen and silicon moieties as ammonium/ammonia and silanols, respectively. This creates a localized alkaline environment; and the release of ammonia, whose amount depends on pH, remains within concentrations that can be metabolized by eukaryotic cells, but not by bacteria or viruses. Its presence disrupts the metabolic processes of prokaryotes, which in turn inhibits their growth and survival. This inherent characteristic makes Si_3N_4 particularly attractive for medical implants, where preventing bacterial colonization is crucial. As shown in this book, Si_3N_4 has been extensively studied for its potential antibacterial and antipathogenic effects, and the release of ammonia is one aspect that undoubtedly will play a role in future biomedical applications. Additionally, research efforts are underway to explore surface modifications of Si_3N_4 to further modulate its interactions with various eukaryotic cells and microorganisms [8–12]. For example, changes in surface chemistry have been found to affect the adhesion of bacteria to the material, making it more challenging for pathogens to attach and form biofilms. Experiments on several different ssRNA viruses, including *Influenza A/B* and *Enteroviruses* [13], *Herpes* viruses [14], and different *Coronavirus* variants [15, 16], have demonstrated the antiviral efficacy of Si_3N_4 powders. This effect is attributed to a "catch and kill" mechanism, where the viruses are electrostatically attracted to the particle surface followed by disruption mediated by released ammonia [16, 17]. Ongoing research in this area focuses on modulating chemical interactions at the biomaterial interface to broaden the applicability of Si_3N_4 in medical devices and implants.

The scope of this final chapter is twofold: (1) The present and future biomedical applications will be briefly reviewed, and (2) new possibilities and future applications in bioscience and engineering will be explored. As research in materials science and engineering progresses, Si_3N_4 is likely to find even more diverse applications. Collaborative efforts between researchers, engineers, and industry stakeholders will be crucial in addressing challenges and advancing the translational science of Si_3N_4 across multiple fields.

2 Medical Applications of Silicon Nitride

2.1 Orthopedic and Dental Implants

Driven by its biocompatibility, radiolucency, anti-infectivity, and osteoconductivity, Si_3N_4's primary application in orthopedics has been in spinal implants [18–20]. Examination of a spinal implant retrieved after short-term in vivo implantation confirmed rapid bone ingrowth, starting from the internal surface of a Si_3N_4 fusion cage [21]. Other case reports have demonstrated improved antipathogenicity [22]. Recent reviews have discussed similar positive results for several other biomedical applications [23, 24]. Si_3N_4's high strength and wear resistance make it a candidate for joint replacement components, like hip and knee implants. Its low-friction coefficient may contribute to reduced wear and improved durability [25]. In contrast to oxide-based ceramics, Si_3N_4's surface exhibits a high affinity for oxygen. This unique chemical feature is believed to be protective to polyethylene sliding counterparts in artificial joints [25–27]. Consequently, Si_3N_4 is a promising candidate for extending the longevity of innovative joint replacements, potentially reaching the current target in orthopedics of exceeding 30 years [28, 29].

Silicon nitride's hardness and wear resistance may also make it suitable for dental crowns and bridges, where durability and resistance to wear are essential [30, 31]. Due to its favorable combination of biocompatibility, osteoconductive properties, and antibacterial activity, Si_3N_4, in both bulk form and as a coating on metallic components, holds promise as a material for dental implants and abutments, which interface with bone and mucosal tissue [32]. Silicon nitride coatings on traditional implant materials (e.g., titanium) may enhance both their bioactivity and antibacterial properties [33]. The application of these coatings has the potential to enhance osseointegration and mitigate the risk of infection-associated complications, thereby contributing to improved implant longevity, a current challenge in dentistry. Polymethyl methacrylate matrix-silicon nitride composites have also been proposed as an intrinsically antifungal material for dentures [34]. Silicon nitride may also be used as a bone-graft substitute [35] or as a filler in bioresorbable polymers due to its excellent biocompatibility [36]. It provides a stable framework for bone ingrowth and has the potential to support fast mandibular bone regeneration. Indeed, dentistry appears to be a promising field for new applications in Si_3N_4 bioceramic implants.

Finally, the mechanical properties of Si_3N_4, particularly its high hardness and wear resistance, make it a suitable candidate for orthopedic and dental surgical instruments, potentially contributing to enhanced instrument longevity and performance [37]. The hardness of Si_3N_4 can be advantageous in the development of drill bits and cutting tools used in orthopedic and dental procedures. Silicon nitride ceramic balls possess a combination of properties that make them highly suitable for use in high-speed dental drill bearings. These properties include exceptional hardness and a low-centrifugal force due to their reduced weight. Additionally, they exhibit superior resistance to corrosion, friction, and wear compared to conventional steel balls. Their thermal expansion coefficient is significantly lower (approximately 0.25 times that of steel), minimizing thermal mismatch issues that can arise during operation. Remarkably, Si_3N_4 balls can operate effectively in harsh environments and without lubrication, while generating minimal heat and requiring less demanding cooling systems. Due to this compelling combination of characteristics, Si_3N_4 bearings are the preferred material for high-speed dental drills.[1]

Despite the promising properties of Si_3N_4, its application in orthopedic and dental implants remains an area of active investigation. Ongoing research focuses on optimizing its material characteristics for targeted applications and elucidating its long-term behavior within the physiological environment. These advancements in material science and technology are expected to facilitate the broader adoption of Si_3N_4 within the orthopedic and dental fields.

2.2 Neurosurgical and Drug Delivery

Another future medical application in which Si_3N_4 holds promise is neurosurgery, particularly in the areas of brain-computer interfaces (BCIs), neural prosthetics, and bone flap replacements [38, 39]. Silicon nitride's electrical insulating properties and biocompatibility make it a potential candidate for manufacturing electrode arrays used in BCIs [40]. These arrays can be implanted in the brain to record neural signals or stimulate neurons for therapeutic purposes. The chemical properties of Si_3N_4 may contribute to the long-term stability of implantable devices used in BCIs. Its ability to resist corrosion in physiological environments is also crucial for maintaining the functionality of neural interfaces over time. As an additional future possibility, Si_3N_4's biocompatibility and chemical properties could be beneficial for encapsulating neural prosthetic devices [41–43]. The bioceramic protects the internal components from the surrounding biological environment while minimizing the risk of adverse reactions. As highlighted by Shen et al. [41], achieving an in vivo functional lifespan for neural implants that matches that of a healthy patient remains an unmet challenge. Current neural implant designs utilize titanium enclosures with brazed ceramic inserts for electrode feedthrough. However, this limits miniaturization, increased channel density, and wireless communication capabilities. Silicon nitride emerges as a potential solution for encapsulation due to its superior dielectric properties, excellent corrosion resistance, biocompatibility, and high-mechanical strength. Moreover, Si_3N_4's thin films enable the deposition and patterning of structures with tailored conductivities. This unique property allows for the creation of integrated conductors and insulators on the same implant, promoting miniaturization and improved functionality. Additionally, the inherent flexibility achievable with thin films opens doors for the development of novel and flexible neural implant substrates.

The compelling combination of mechanical strength, low weight, biocompatibility, and antibacterial properties positions Si_3N_4 as a promising candidate for bone flap replacements in neurosurgery. Notably, its versatility extends to both bulk Si_3N_4 and polymer composites (i.e., dispersion of Si_3N_4 powder within a biocompatible

[1] https://www.tsubaki-nakashima.com/en/products/ceramic_balls/

polymer) [34, 44, 45]. These implants could provide structural support, and their biocompatibility would contribute to the healing process. Si_3N_4's radiolucency could also be a valuable property for postoperative imaging, thereby facilitating postoperative monitoring of the healing process.

Finally, biodegradable Si_3N_4 nanoparticles or microparticles could be used for targeted drug delivery to the brain and to the posterior segment of the eye [46]. In these locations, because of its chemical stability and biocompatibility, Si_3N_4 may be advantageously used as a carrier for therapeutic agents.

While the potential of Si_3N_4 in neurosurgery holds promise, further materials science research is necessary to optimize its properties for specific applications. The intricate and fragile nature of neural tissue demands meticulous attention to biocompatibility, long-term stability, and performance. Continued collaboration between materials scientists, engineers, and neurosurgeons is paramount to drive the development and clinical integration of Si_3N_4. As technology progresses, Si_3N_4 is poised to play a significant role in expanding the capabilities and improving the outcomes of neurosurgical procedures.

2.3 Nosocomial Infections and Public Health

The antibacterial, antiviral, and antifungal properties of Si_3N_4 can potentially be harnessed to combat healthcare-associated infections (HAIs) and to improve public health. As discussed above, incorporating Si_3N_4 into the surfaces of medical implants, such as orthopedic implants, catheters, and other prosthetics, may help prevent bacterial adhesion and biofilm formation [47], especially for devices that come into direct contact with bodily fluids or tissues, where infections are more likely to occur. Moreover, due to its antibacterial properties, Si_3N_4 biomaterials could also be used in wound dressings to prevent or treat infections in wounds and surgical incisions [48]. Similarly, Si_3N_4 foams could also be combined with creams or ointments for topical applications to combat fungal infections [49].

From a different perspective, Si_3N_4 filters with antiviral properties could be incorporated into air filtration systems in healthcare facilities. This could help reduce the spread of airborne diseases and improve air quality within hospital environments. Note that Si_3N_4 has already been applied as powder dispersion in facemask filters to impart antiviral properties [50, 51]. This approach could help in reducing the viability and transmission of viruses, providing an additional layer of protection against respiratory diseases. Similarly, Si_3N_4-based materials with antibacterial properties could be used in water purification systems within healthcare facilities to prevent waterborne infections. Wright et al. [52] successfully reported the development of microslit Si_3N_4 membranes with isoporous sub-micrometer rectangular prisms (porosity ~33%). Compared to traditional sterile polymer filters, this is a dramatically different structure that provides superior filtration efficacy. This opens a door to its future use in the production of nonprotein biological therapeutics and similar applications.

Finally, applying Si_3N_4 coatings to high-touch surfaces in healthcare settings, such as door handles, bedrails, and medical equipment, will likely minimize the transmission of infectious agents [53]. Similarly, antipathogenic Si_3N_4 coatings could be applied to personal protective equipment, such as gowns and drapes to provide an additional layer of protection against viral disease transmission.

2.4 Biosensing Technologies

Silicon nitride is used in biosensing technologies and has gained attention for its applications in various biosensors [54, 55]. Again, its combination of properties makes it suitable for developing biosensing devices. Silicon nitride surfaces can be modified and functionalized with specific biomolecules, antibodies, or receptors. This allows for the selective detection of target analytes, such as proteins, DNA, or other biomarkers, through specific binding interactions. Silicon nitride field-effect transistors (Si_3N_4 FETs) are already used as biosensors. The binding of biomolecules to the sensor surface induces changes in charge distribution or conductivity, which can be measured and correlated with the concentration of the target analytes.

Silicon nitride waveguides and resonators are finding application in photonic biosensors. In a pioneering work, Bryan et al. [56] demonstrated the integration of a silicon nitride microring resonator biosensor with an on-chip microring filter bank spectrometer for label-free detection of C-reactive proteins, an inflammatory biomarker, in serum. This approach utilizes the sensitive detection of changes in the resonant wavelength caused by biomolecular interactions altering the surrounding medium's refractive index. This innovation paves the way for real-time, label-free biomolecule detections. In microfluidic devices for biosensing applications, the surface chemistry of silicon nitride can also be exploited to manipulate small volumes of fluids containing biological samples, enhancing the efficiency of sample preparation and analysis.

Similarly, silicon nitride microcantilevers or resonators can be used as mass-sensitive biosensors [57]. The binding of molecules to the surface causes changes in the resonant frequency, allowing for the detection of mass changes associated with biomolecular interactions.

Silicon nitride has also been explored as a substrate for DNA sequencing platforms [58–62]. It can support the immobilization of DNA strands, enabling the detection and analysis of DNA sequences. Notably, the biocompatibility of Si_3N_4 makes it suitable for implantable biosensors. These devices can monitor specific biomarkers or physiological parameters within the body and transmit data for remote monitoring.

In summary, the combination of Si_3N_4 with microfabrication techniques allows for the integration of multiple sensing elements, enhancing the sensitivity and specificity of biosensors. Silicon nitride's amenability to surface modifications, coupled with its compatibility with established microfabrication techniques, underpins its potential utilization in biosensing applications across diverse fields, including medical diagnostics and environmental monitoring.

3 Beyond Medicine: Other Applications for Silicon Nitride

Besides healthcare, Si_3N_4 is well known for its usage in many industrial, aerospace, automotive, and electronics applications. While these areas have been developed for decades, they are expected to experience continued innovation and growth. Indeed, the global market for Si_3N_4 was estimated to be $1.2 Billion in 2021. With an annual growth rate of 7.2%, its total market value should reach $2.2 Billion by 2030.[2] The largest segment will continue to be industrial with individual products such as bearings, cutting tools, pump components, welding nozzles, wear liners, and refractories. Aerospace and transportation segments will also experience strong growth. Dominant products in this segment include bearings, turbine blades, vanes, heat shields, nozzles, landing gear components, flap actuators, turbochargers, cams, wear pads, and other structural elements. The electronics segment is expected to have the fastest growth rate during the next decade due in part to products such as semiconductor packaging, power modules, LED lighting, high-frequency circuits, electrical insulators, and thermal management devices. However, there are new uses for Si_3N_4 on the horizon that may significantly increase its growth rate. Several of these include its use in agriculture, solar energy generation, and quantum computing.

3.1 Uses in Agriculture

While silicon nitride has demonstrated antipathogenic properties against bacteria, viruses, and fungi in certain applications, its use in preventing plant diseases, such as downy mildew in grapevines or controlling other harmful plant diseases, is yet in its infancy [63]. The antipathogenic mechanisms in agriculture are the same as those observed in mammalian applications. The presence of moisture results in the elution of ammonia and an increase in pH. However, in plant science, silicon is a beneficial nutrient that has an important role in enhancing yields and inhibiting disease [64–67]. Silicon is an abundant element that some plants absorb directly into their tissues. Its presence has been observed to contribute to antipathogenicity and pest resistance [68–70]. Plants require nitrogen not only as a nutrient for growth and development but also as a modulator of their adaptation to abiotic stresses. The ability to metabolize ammonium or ammonia molecules is crucial for nitrogen utilization [71, 72]. To exploit Si_3N_4 as a plant disease control agent, several key considerations

[2] SkyQuest Global Market Research, https://www.skyquestt.com/report/silicon-nitride-market, accessed Mar. 20, 2024.

warrant investigation: (1) compatibility across diverse plant species must be assessed, as potential adverse effects on growth and development can vary and (2) the efficacy of silicon nitride against specific plant pathogens (e.g., downy mildew) necessitates its thorough evaluation under both in vitro and in vivo conditions. An understanding of the susceptibility of different pathogens to the antipathogenic properties of Si_3N_4 is of crucial importance. While Si_3N_4 may potentially offer a lower environmental impact compared to conventional copper-based pesticides, a comprehensive evaluation is necessary. Specifically, experiments are needed to determine if the release of Si_3N_4 particles poses ecological risks. Regulatory compliance, rigorous safety, and efficacy testing are essential prerequisites for any agricultural application of Si_3N_4. In summary, the antipathogenic properties of Si_3N_4 in plants are not yet fully elucidated, highlighting a gap in current knowledge that warrants further investigation. However, this area holds significant promise for future research.

3.2 Solar Energy

Silicon nitride has become a well-established material in solar energy technology, particularly for the fabrication of solar cells and photovoltaic modules [73–77]. In particular, Si_3N_4 is applied to various components of solar cells to enhance their performance and durability [73], and its thin films are often used as antireflection coatings on the surface of solar cells [76]. These coatings minimize sunlight reflection, allowing more light to penetrate the cell and be converted into electricity [77]. Consequently, the antireflection properties of Si_3N_4 contribute to overall solar cell efficiency. Silicon nitride is also employed as a passivation layer on the surface of solar cells and as an encapsulation material in the construction of photovoltaic modules. Passivation helps reduce surface recombination of charge carriers (electrons and holes), which also improves the overall efficiency of solar cells. Hence, passivation layers contribute to better electrical performance and stability over time, while encapsulation helps protect the solar cells from environmental degradation factors (e.g., moisture, UV radiation, and temperature variations), thereby increasing the longevity and reliability of the solar module.

Another application of Si_3N_4 layers in solar cells is that of a material for backside reflectors [78–81]. These reflectors redirect photons that have not been initially absorbed by the front surface of the solar cell, increasing the overall absorption of sunlight and enhancing cell efficiency. Silicon nitride coatings could also be applied to modify the surface texture of solar cells. Moreover, Si_3N_4 coatings on surface-textured silicon devices have been applied to enhance light trapping, since it increases the absorption of sunlight and improves the overall efficiency of the solar cell [82]. Silicon nitride coatings have also been used to protect the surfaces of silicon wafers during manufacturing. This protective layer prevents contamination and damage to the silicon substrate [83–91]. The continued use of Si_3N_4 in solar cell technology is expected to contribute to advancements in efficiency and durability.

This can lead to the development of more cost-competitive solar panels, potentially accelerating the adoption of solar power as a viable and sustainable energy source.

3.3 *Quantum Computing*

Silicon nitride has also been explored for its potential applications in quantum computing [92–96]. The realization of quantum computers necessitates materials with specific properties to enable the control and manipulation of quantum information units, known as qubits. Silicon nitride has emerged as a promising material due to its advantageous optical characteristics, which is capable of performing the necessary manipulation [92, 97]. Its low-optical losses and compatibility with established silicon microfabrication techniques, make it a compelling material for photonic quantum information processing. This material enables the construction of integrated photonic circuits, where quantum information encoded in photonic states can be manipulated and transmitted using Si_3N_4 waveguides and resonators [98]. Indeed, the successful development of various quantum photonics devices, including on-chip sources, detectors, and modulators, further underscores the potential of Si_3N_4 in this field [94, 99, 100]. These devices are essential components of quantum photonic circuits. They allow for the development of hybrid quantum systems once they are integrated with traditional silicon electronics. Achieving seamless integration is crucial for interfacing quantum processors with classical control and readout electronics.

Beyond its role in photonic circuits, Si_3N_4 is also under investigation for its potential application in quantum memory devices. These devices are critical components in specific quantum computing architectures, as they enable the storage and retrieval of quantum information. Gritsenko et al. employed quantum chemical simulations and experimental evaluations of thermal ionization energies to identify Si–Si bonds within the silicon nitride lattice as responsible for trapping electrons and holes, laying the groundwork for potential applications in quantum memory [101]. The aforementioned properties of silicon nitride position it as a front-runner for various applications in quantum computing. Its low-optical losses, compatibility with established microfabrication techniques, and potential for integration with classical electronics make it a highly attractive material for the development of photonic quantum circuits. Furthermore, ongoing research into its application in quantum memory devices underscores its versatility within the realm of quantum information processing. As the field of quantum computing continues its rapid evolution, advancements in material science and device engineering are likely to unlock the full potential of Si_3N_4, paving the way for groundbreaking discoveries and the realization of scalable quantum computers.

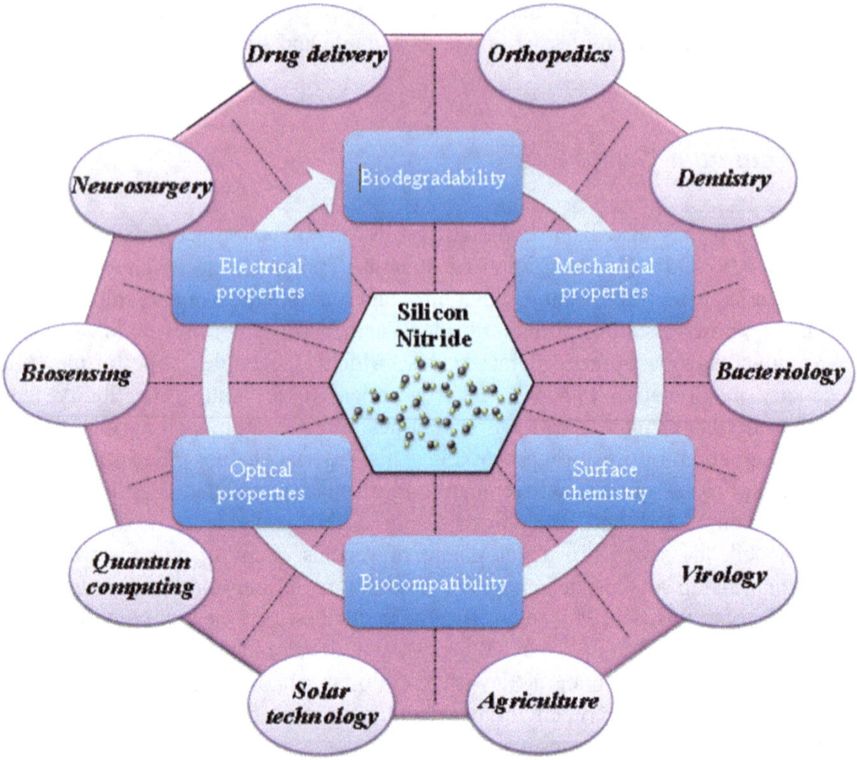

Fig. 1 Properties of Si_3N_4 and their relationships to different technological fields

4 Conclusion and Future Outlook

Extensive research has established that biomolecular interactions at the Si_3N_4 biointerface exhibit a fascinating duality: positively sensed by mammalian cells and negatively impacting a broad spectrum of pathogens, including bacteria, fungi, viruses, and plant diseases. These localized chemical reactions trigger specific cellular responses that govern cell fate and activate antipathogenic mechanisms. While macroscopically resembling many other ceramics in terms of structural stability, silicon nitride exhibits a unique surface chemistry on the microscopic scale, leading to strong interactions with microorganisms at the biointerface. This phenomenon aligns with the concept of far-field/solid-surface dualism. Notably, silicon nitride harnesses this duality to simultaneously support mammalian cell metabolism and exhibit selective toxicity toward pathogens. This remarkable property holds immense potential for groundbreaking discoveries not only in the medical field but also across various scientific and practical disciplines.

This chapter has endeavored to showcase the promising potential of Si_3N_4 across diverse scientific and technological fields. Fig. 1 visually summarizes the alignment between Si_3N_4's properties and its suitability for various technological applications. These are promising research areas, and they present exciting prospects for the continued development, clinical translation, and industrial adoption of this remarkable material. Our aim in this closing chapter is to inspire further in-depth investigations into a multitude of new challenges presented by this unique biomaterial. Undoubtedly, those who do the necessary research and development will be rewarded with groundbreaking discoveries and scientific advancements.

References

1. McEntire, B.J.; Bal, B.S.; Rahaman, M.N.; Chevalier, J.; Pezzotti, G. Ceramics and Ceramic Coatings in Orthopaedics. *J. Eur. Ceram. Soc.***2015**, *35*, 4327–4369, doi:https://doi.org/10.1016/j.jeurceramsoc.2015.07.034.
2. Krstic, Z.; Krstic, V.D. Silicon Nitride: The Engineering Material of the Future. *J. Mater. Sci.***2011**, *47*, 535–552, doi:https://doi.org/10.1007/s10853-011-5942-5.
3. Merola, M.; Affatato, S. Materials for Hip Prostheses: A Review of Wear and Loading Considerations. *Materials (Basel).***2019**, *12*, doi:https://doi.org/10.3390/ma12030495.
4. McEntire, B.J.; Lakshminarayanan, R.; Thirugnanasambandam, P.; Seitz-Sampson, J.; Bock, R.; O'Brien, D. Processing and Characterization of Silicon Nitride Bioceramics. *Bioceram. Dev. Appl.***2016**, *6*, 1000093, doi:https://doi.org/10.4172/2090-5025.100093.
5. Ghalme, S.; Falak, Y. Review on Evolution of Silicon Nitride in the Field of Orthopedics. *Curr. Trends Biomed. Eng. Biosci.***2018**, *11*, 1–3, doi:https://doi.org/10.19080/CTBEB.2018.11.555824.
6. Du, X.; Lee, S.S.; Blugan, G.; Ferguson, S.J. Silicon Nitride as a Biomedical Material: An Overview. *Int. J. Mol. Sci.***2022**, *23*, 6551, doi:https://doi.org/10.3390/ijms23126551.
7. Pezzotti, G. Silicon Nitride: A Bioceramic with a Gift. *ACS Appl. Mater. Interfaces***2019**, *11*, 26619–26636, doi:https://doi.org/10.1021/acsami.9b07997.
8. Pezzotti, G. Silicon Nitride as a Biomaterial. *J. Ceram. Soc. Japan***2023**, *131*, 398–428, doi: https://doi.org/10.2109/jcersj2.23057.
9. Pezzotti, G.; Bock, R.M.; Adachi, T.; Rondinella, A.; Boschetto, F.; Zhu, W.; Marin, E.; McEntire, B.; Bal, B.S.; Mazda, O. Silicon Nitride Surface Chemistry: A Potent Regulator of Mesenchymal Progenitor Cell Activity in Bone Formation. *Appl. Mater. Today***2017**, *9*, 82–95, doi:https://doi.org/10.1016/j.apmt.2017.05.005.
10. Pezzotti, G.; Bock, R.M.; McEntire, B.J.; Jones, E.; Boffelli, M.; Zhu, W.; Baggio, G.; Boschetto, F.; Puppulin, L.; Adachi, T.; et al. Silicon Nitride Bioceramics Induce Chemically Driven Lysis in Porphyromonas Gingivalis. *Langmuir***2016**, *32*, 3024–3035, doi:https://doi.org/10.1021/acs.langmuir.6b00393.
11. Bock, R.M.; Jones, E.N.; Ray, D.A.; Bal, B.S.; Pezzotti, G.; McEntire, B.J. Bacteriostatic Behavior of Surface-Modulated Silicon Nitride in Comparison to Polyetheretherketone and Titanium. *J. Biomed. Mater. Res. Part A***2017**, *105*, 1521–1534, doi:10.1002/jbm.a.35987.
12. Pezzotti, G.; Marin, E.; Adachi, T.; Rondinella, A.; Boschetto, F.; Zhu, W.-L.; Sugano, N.; Bock, R.M.; McEntire, B.J.; Bal, B.S. Bioactive Silicon Nitride: A New Therapeutic Material for Osteoarthropathy. *Sci. Rep.***2017**, *7*, 44848.
13. Pezzotti, G.; Boschetto, F.; Ogitani, E.; Fujita, Y.; Zhu, W.; Marin, E.; McEntire, B.J.; Bal, B. S.; Mazda, O. Silicon Nitride: A Potent Solid-State Bioceramic Inactivator of ssRNA Viruses. *Sci. Rep.***2021**, *11*, 2977, doi:https://doi.org/10.1038/s41598-021-82608-3.
14. Pezzotti, G.; Ohgitani, E.; Ikegami, S.; Shin-ya, M.; Adachi, T. Instantaneous Inactivation of Herpes Simplex Virus by Silicon Nitride Bioceramics. *J. Mol. Sci.***2023**, *24*, 12657, doi:https://doi.org/10.3390/ijms241612657.

15. Pezzotti, G.; Boschetto, F.; Ohgitani, E.; Fujita, Y.; Shin-Ya, M.; Adachi, T.; Yamamoto, T.; Kanamura, N.; Marin, E.; Zhu, W.; et al. Mechanisms of Instantaneous Inactivation of SARS-CoV-2 by Silicon Nitride Bioceramic. *Mater. Today Bio***2021**, 100144, doi:https://doi.org/10.1016/j.mtbio.2021.100144.
16. Pezzotti, G.; Ohigitani, E.; Shin-Ya, M.; Adachi, T.; Marin, E.; Boshetto, F.; Zhu, W.; Mazda, O. Instantaneous "Catch-and-Kill" Inactivation of SARS-CoV-2 by Nitride Ceramics. *Clin. Transl. Med.***2020**, 1–4, doi:https://doi.org/10.1002/ctm2.212.
17. Pezzotti, G. A Spontaneous Solid-State NO Donor to Fight Antibiotic Resistant Bacteria. *Mater. Today Chem.***2018**, *9*, 80–90, doi:https://doi.org/10.1016/j.mtchem.2018.05.004.
18. Katsaros, I.; Zhou, Y.; Welch, K.; Xia, W.; Persson, C. Bioactive Silicon Nitride Implant Surfaces with Maintained Antibacterial Properties. *J. Funct. Biomater.***2022**, *13*, 129.
19. McEntire, B.; Bock, R.; Rahaman, M.; Bal, B.S.; Webster, T.; Pezzotti, G. Anti-Infective and Osteointegration Characteristics of Silicon Nitride Spinal Fusion Implants. *Bone Jt. J.***2016**, *98*, 32.
20. Calvert, G.C.; Huffmon, G.V.B.I.; Rambo, W.M.J.; Smith, M.W.; McEntire, B.J.; Bal, B.S. Clinical Outcomes for Lumbar Fusion using Silicon Nitride versus other Biomaterials. *J. Spine Surg.***2020**, *6*, 33–48, doi:https://doi.org/10.21037/jss.2019.12.11.
21. Pezzotti, G.; Oba, N.; Zhu, W.; Marin, E.; Rondinella, A.; Boschetto, F.; McEntire, B.J.; Yamamoto, K.; Bal, B.S. Human Osteoblasts Grow Transitional Si/N Apatite in Quickly Osteointegrated Si_3N_4 Cervical Insert. *Acta Biomater.***2017**, *64*, 411–420.
22. Rambo, W.M. Treatment of Lumbar Discitis using Silicon Nitride Spinal Spacers: A Case Series and Literature Review. *Int. J. Surg. Case Rep.***2018**, *43*, 61–68, doi:https://doi.org/10.1016/j.ijscr.2018.02.009.
23. Heimann, R.B. Silicon Nitride Ceramics: Structure, Synthesis, Properties, and Biomedical Applications. *Materials (Basel).***2023**, *16*, 1–16, doi:10.3390/ma16145142.
24. Liu, Z.; Wang, R.; Liu, W.; Liu, Y.; Feng, X.; Zhao, F.; Chen, P.; Shao, L.; Rong, M. Recent Advances in the Application and Biological Mechanism of Silicon Nitride Osteogenic Properties: A Review. *Biomater. Sci.***2023**, Advance Article, doi:https://doi.org/10.1039/D3BM00877K.
25. Yorifuji, M.; Affatato, S.; Tateiwa, T.; Takahashi, Y.; Shishido, T.; Marin, E.; Zanocco, M.; Zhu, W.; Pezzotti, G.; Yamamoto, K. Wear Simulation of Ceramic-on-Crosslinked Polyethylene Hip Prostheses: A New Non-Oxide Silicon Nitride versus the Gold Standard Composite Oxide Ceramic Femoral Heads. *Materials (Basel).***2020**, *13*, 1–11, doi:10.3390/ma13132917.
26. Pezzotti, G. Bioceramics for Hip Joints: The Physical Chemistry Viewpoint. *Materials (Basel).***2014**, *7*, 4367–4410, doi:10.3390/ma7064367.
27. Pezzotti, G.; Zhu, W.; Sugano, N.; Marin, E.; Yamamoto, K.; Nishiike, N.; Hori, T.; McEntire, B.J.; Bock, R.; Bal, B.S. Oxide Ceramic Femoral Heads Contribute to Polyethylene Liner Degradation. *Proc. Annu. Meet. Orthop. Res. Soc.***2018**, 0275.
28. Poliakov, A.; Pakhaliuk, V.; Popov, V.L. Current Trends in Improving of Artificial Joints Design and Technologies for Their Arthroplasty. *Front. Mech. Eng.***2020**, *6*, 2020.0004, doi: https://doi.org/10.3389/fmech.2020.00004.
29. Pezzotti, G.; Yamamoto, K. Artificial Hip Joints: The Biomaterials Challenge. *J. Mech. Behav. Biomed. Mater.***2014**, *31*, 3–20, doi:https://doi.org/10.1016/j.jmbbm.2013.06.001.
30. Krajangta, N.; Sarinnaphakorn, L.; Didron, P.P.; Wasanapiarnpong, T. Development of Silicon Nitride Ceramic for CAD/CAM Restoration. *Dent. Mater. J.***2020**, *3*, 1–6, doi: https://doi.org/10.4012/dmj.2019-119.
31. Mohammadi, H.; Beddu, S.; Petrů, M.; Sepantafar, M.; Ebadi, M.; Yap, B.K.; Bang, L.T.; Yong, T.C.; Ramesh, S.; Rahimian Koloor, S.S.; et al. Advances in Silicon Nitride Ceramic Biomaterials for Dental Applications—A Review. *J. Mater. Res. Technol.***2024**, *28*, 2778–2791, doi:https://doi.org/10.1016/j.jmrt.2023.12.186.
32. Shi, H.Y.; Pang, R.; Yang, J.; Fan, D.; Cai, H.; Jiang, H.B.; Han, J.; Lee, E.-S.; Sun, Y. Overview of Several Typical Ceramic Materials for Restorative Dentistry. *Biomed Res. Int.***2022**, *2022*, 8451445, doi:https://doi.org/10.1155/2022/8451445.

33. Zanocco, M.; Boschetto, F.; Zhu, W.; Marin, E.; McEntire, B.J.; Bal, B.S.; Adachi, T.; Yamamoto, T.; Kanamura, N.; Ohgitani, E.; et al. 3D-Additive Deposition of an Antibacterial and Osteogenic Silicon Nitride Coating on Orthopaedic Titanium Substrate. *J. Mech. Behav. Biomed. Mater.***2020**, *103*, 103557, doi:https://doi.org/10.1016/j.jmbbm.2019.103557.
34. Pezzotti, G.; Asai, T.; Adachi, T.; Ogitani, E.; Yamamoto, T.; Kanamura, N.; Boschetto, F.; Zhu, W.; Zanocco, M.; Marin, E.; et al. Antifungal Activity of Polymethyl Methacrylate/Si_3N_4 Composites against Candida Albicans. *Acta Biomater.***2021**, *126*, 259–276, doi:https://doi.org/10.1016/j.actbio.2021.03.023.
35. Anderson, M.C.; Olsen, R. Bone Ingrowth into Porous Silicon Nitride. *J. Biomed. Mater. Res.***2010**, *92A*, 1598–1605, doi:https://doi.org/10.1002/jbm.a.32498.
36. Vidakis, N.; Petousis, M.; Michailidis, N.; Papadakis, V.; Mountakis, N.; Argyros, A.; Dimitriou, E.; Charou, C.; Moutsopoulou, A. Polylactic Acid/Silicon Nitride Biodegradable and Biomedical Nanocomposites with Optimized Rheological and Thermomechanical Response for Material Extrusion Additive Manufacturing. *Biomed. Eng. Adv.***2023**, *6*, 100103, doi:https://doi.org/10.1016/j.bea.2023.100103.
37. Walkosz, W. Silicon Nitride Ceramics. In *Atomic Scale Characterization and First-Principles Studies of Si_3N_4 Interfaces*; Walkosz, W., Ed.; Springer New York: New York, NY, 2011; pp. 1–10 ISBN 978-1-4419-7817-2.
38. Guo, Z.; Wang, F.; Wang, L.; Tu, K.; Jiang, C.; Xi, Y.; Hong, W.; Xu, Q.; Wang, X.; Yang, B.; et al. A Flexible Neural Implant with Ultrathin Substrate for Low-Invasive Brain–Computer Interface Applications. *Microsystems Nanoeng.***2022**, *8*, 1–12, doi:https://doi.org/10.1038/s41378-022-00464-1.
39. Neumann, A.; Unkel, C.; Werry, C.; Herborn, C.U.; Maier, H.R.; Ragoss, C.; Jahnke, K. Prototype of a Silicon Nitride Ceramic-Based Miniplate Osteofixation System for the Midface. *Otolaryngol. Neck Surg.***2006**, *134*, 923–30, doi:https://doi.org/10.1016/j.otohns.2006.01.022.
40. Huynh, M.A.; Nguyen, T.-H.; Nguyen, T.-K.; Nguyen, N.-T. Convergence of Implantable Bioelectronics and Brain–Computer Interfaces. *ACS Appl. Electron. Mater.***2023**, *5*, 5777–5793, doi:https://doi.org/10.1021/acsaelm.3c00879.
41. Shen, K.; Maharbiz, M.M. Ceramic Packaging in nNeural Implants. *J. Neural Eng.***2021**, *18*, 25002, doi:https://doi.org/10.1088/1741-2552/abd683.
42. Mescola, A.; Canale, C.; Prato, M.; Diaspro, A.; Berdondini, L.; Maccione, A.; Dante, S. Specific Neuron Placement on Gold and Silicon Nitride-Patterned Substrates through a Two-Step Functionalization Method. *Langmuir***2016**, *32*, 6319–6327, doi:https://doi.org/10.1021/acs.langmuir.6b01352.
43. Ahn, S.-H.; Jeong, J.; Kim, S.J. Emerging Encapsulation Technologies for Long-Term Reliability of Microfabricated Implantable Devices. *Micromachines* **2019**, *10*.
44. Wu, H.; Liu, T.; Xu, Z.; Qian, J.; Shen, X.; Li, Y.; Pan, Y.; Wang, D.; Zheng, K.; Boccaccini, A.R.; et al. Enhanced Bacteriostatic Activity, Osteogenesis and Osseointegration of Silicon Nitride/Polyetherketoneketone Composites with Femtosecond Laser Induced Micro/Nano Structural Surface. *Appl. Mater. Today***2020**, *18*, 100523, doi:https://doi.org/10.1016/j.apmt.2019.100523.
45. Marin, E.; Boschetto, F.; Zanocco, M.; Honma, T.; Zhu, W.; Pezzotti, G. Explorative Study on the Antibacterial Effects of 3D-Printed PMMA/Nitrides Composites. *Mater. Des.***2021**, *206*, 109788, doi:https://doi.org/10.1016/j.matdes.2021.109788.
46. Kompella, U.B.; Amrite, A.C.; Pacha Ravi, R.; Durazo, S.A. Nanomedicines for Back of the Eye Drug Delivery, Gene Delivery, and Imaging. *Prog. Retin. Eye Res.***2013**, *36*, 172–198, doi:https://doi.org/10.1016/j.preteyeres.2013.04.001.
47. Praharaj, R.; Swain, S.; Rautray, T.R. Bioceramics for Antibacterial and Antiviral Applications. *Adv. Ceram. Coatings Biomed. Appl.***2023**, 347.
48. Ma, P.; Yang, C.-Y.; Li, C.; Hu, P.; Yang, F.; Lu, J.; Huang, Y.-Y.; Wu, H.; Wu, Q.; Pan, Y.; et al. Blow-Spun Si_3N_4-Incorporated Nanofibrous Dressing with Antibacterial,

Anti-Inflammatory, and Angiogenic Activities for Chronic Wound Treatment. *Adv. Fiber Mater.***2024**, 1–18, doi:https://doi.org/10.1007/s42765-023-00361-w.
49. Zhang, X.Y.; Huo, W.L.; Qi, F.; Qu, Y.N.; Xu, J.; Gan, K.; Ma, N.; Yang, J.L. Ultralight Silicon Nitride Ceramic Foams from Foams Stabilized by Partially Hydrophobic Particles. *J. Am. Ceram. Soc.***2016**, *99*, 2920–2926, doi:https://doi.org/10.1111/jace.14320.
50. Karmacharya, M.; Kumar, S.; Gulenko, O.; Cho, Y.-K. Advances in Facemasks during the COVID-19 Pandemic Era. *ACS Appl. Bio Mater.***2021**, *4*, 3891–3908, doi:https://doi.org/10.1021/acsabm.0c01329.
51. Simpson, S.; McMinn, C.; Mondfrans, S.M. Van; Hendry, J.; Ronayne, S.; Dewhurst, S.; Feng, C.; Bal, B.S.; Bock, R.M.; McEntire, B.J. A Novel Antipathogenic Agent for Nonwoven Fabric. *Biomed. Mater. Devices.* **2023**, *1*, 469–482, doi:https://doi.org/10.1007/s44174-022-00001-8.
52. Wright, E.; Miller, J.J.; Csordas, M.; Gosselin, A.R.; Carter, J.A.; McGrath, J.L.; Latulippe, D. R.; Roussie, J.A. Development of Isoporous Microslit Silicon Nitride Membranes for Sterile Filtration Applications. *Biotechnol. Bioeng.***2020**, *117*, 879–885, doi:https://doi.org/10.1002/bit.27240.
53. Hosseini, M.; Behzadinasab, S.; Benmamoun, Z.; Ducker, W.A. The Viability of SARS-CoV-2 on Solid Surfaces. *Curr. Opin. Colloid Interface Sci.***2021**, *55*, doi:https://doi.org/10.1016/j.cocis.2021.101481.
54. Cutrone, A.; Micera, S. Implantable Neural Interfaces and Wearable Tactile Systems for Bidirectional Neuroprosthetics Systems. *Adv. Healthc. Mater.***2019**, *8*, 1801345, doi:https://doi.org/10.1002/adhm.201801345.
55. Maldonado Vázquez, J.M. Interferometric Biosensors for Rapid Identification of Nosocomial Infections, Universitat Autònoma de Barcelona, **2017**.
56. Bryan, M.R.; Butt, J.N.; Bucukovski, J.; Miller, B.L. Biosensing with Silicon Nitride Microring Resonators Integrated with an On-Chip Filter Bank Spectrometer. *ACS Sensors***2023**, *8*, 739–747, doi:https://doi.org/10.1021/acssensors.2c02276.
57. Zhang, H.; Kim, E.S. Micromachined Acoustic Resonant Mass Sensor. *J. Microelectromechanical Syst.***2005**, *14*, 699–706, doi:https://doi.org/10.1109/JMEMS.2005.845405.
58. Choi, W.; Jeon, E.-S.; Chun, K.-Y.; Kim, Y.-R.; Park, K.-B.; Kim, K.-B.; Han, C.-S. A Low-Noise Silicon Nitride Nanopore Device on a Polymer Substrate. *PLoS One***2018**, *13*, e0200831.
59. Tan, S.; Wang, L.; Yu, J.; Hou, C.; Jiang, R.; Li, Y.; Liu, Q. DNA-Functionalized Silicon Nitride Nanopores for Sequence-Specific Recognition of DNA Biosensor. *Nanoscale Res. Lett.***2015**, *10*, 205, doi:https://doi.org/10.1186/s11671-015-0909-0.
60. Petralia, S.; Cosentino, T.; Sinatra, F.; Favetta, M.; Fiorenza, P.; Bongiorno, C.; Sciuto, E.L.; Conoci, S.; Libertino, S. Silicon Nitride Surfaces as Active Substrate for Electrical DNA Biosensors. *Sensors Actuators B Chem.***2017**, *252*, 492–502, doi:https://doi.org/10.1016/j.snb.2017.06.023.
61. Niedzwiecki, D.J.; Chou, Y.-C.; Xia, Z.; Thei, F.; Drndić, M. Detection of Single Analyte and Environmental Samples with Silicon Nitride Nanopores: Antarctic Dirt Particulates and DNA in Artificial Seawater. *Rev. Sci. Instrum.***2020**, *91*, 31301, doi:https://doi.org/10.1063/1.5138210.
62. AU - Piskunen, P.; AU - Shen, B.; AU - Julin, S.; AU - Ijäs, H.; AU - Toppari, J.J.; AU - Kostiainen, M.A.; AU - Linko, V. DNA Origami-Mediated Substrate Nanopatterning of Inorganic Structures for Sensing Applications. *JoVE***2019**, e60313, doi:doi:10.3791/60313.
63. Pezzotti, G.; Fujita, Y.; Boschetto, F.; Zhu, W.; Marin, E.; Vandelle, E.; McEntire, B.J.; Bal, B.S.; Giarola, M.; Makimura, K.; *et al.* Activity and Mechanism of Action of the Bioceramic Silicon Nitride as an Environmentally Friendly Alternative for the Control of the Grapevine Downy Mildew Pathogen Plasmopara viticola. *Front. Microbiol.***2020**, *11*, 1–17, doi:https://doi.org/10.3389/fmicb.2020.610211.

64. Caubet, M.; Cornu, S.; Saby, N.P.A.; Meunier, J.-D. Agriculture Increases the Bioavailability of Silicon, a Beneficial Element for Crop, in Temperate Soils. *Sci. Rep.***2020**, *10*, 19999, doi: https://doi.org/10.1038/s41598-020-77059-1.
65. Xu, D.; Gao, T.; Fang, X.; Bu, H.; Li, Q.; Wang, X.; Zhang, R. Silicon Addition Improves Plant Productivity and Soil Nutrient Availability Without Changing the Grass:Legume Ratio Response to N Fertilization. *Sci. Rep.***2020**, *10*, 10295, doi:https://doi.org/10.1038/s41598-020-67333-7.
66. Dutra, A.F.; Leite, M.R.L.; Melo, C.C. de F.; Amaral, D.S.; da Silva, J.L.F.; Prado, R. de M.; Piccolo, M. de C.; Miranda, R. de S.; da Silva Júnior, G.B.; Sousa, T.K. dos S.A.; *et al.* Soil and Foliar Si Fertilization Alters Elemental Stoichiometry and Increases Yield of Sugarcane Cultivars. *Sci. Rep.***2023**, *13*, 1–10, doi:https://doi.org/10.1038/s41598-023-43351-z.
67. Sun, S.; Yang, Z.; Song, Z.; Wang, N.; Guo, N.; Niu, J.; Liu, A.; Bai, B.; Ahammed, G.J.; Chen, S. Silicon Enhances Plant Resistance to Fusarium Wilt by Promoting Antioxidant Potential and Photosynthetic Capacity in Cucumber (Cucumis sativus L.). *Front. Plant Sci.***2022**, *13*, 1–16, doi:https://doi.org/10.3389/fpls.2022.1011859.
68. Sakr, N. The Role of Silicon (Si) in Increasing Plant Resistance Against Fungal Diseases. *Hell. Plant Prot. J.***2016**, *9*, 1–15, doi:https://doi.org/10.1515/hppj-2016-0001.
69. Wang, M.; Gao, L.; Dong, S.; Sun, Y.; Shen, Q.; Guo, S. Role of Silicon on Plant–Pathogen Interactions. *Front. Plant Sci.***2017**, *8*, 1–14, doi:https://doi.org/10.3389/fpls.2017.00701.
70. Ahammed, G.J.; Yang, Y. Mechanisms of Silicon-Induced Fungal Disease Resistance in Plants. *Plant Physiol. Biochem.***2021**, *165*, 200–206, doi:https://doi.org/10.1016/j.plaphy.2021.05.031.
71. Gojon, A. Nitrogen Nutrition in Plants: Rapid Progress and New Challenges. *J. Exp. Bot.***2017**, *68*, 2457–2462, doi:https://doi.org/10.1093/jxb/erx171.
72. Ye, J.Y.; Tian, W.H.; Jin, C.W. Nitrogen in Plants: From Nutrition to the Modulation of Abiotic Stress Adaptation. *Stress Biol.***2022**, *2*, 1–14, doi:https://doi.org/10.1007/s44154-021-00030-1.
73. El amrani, A.; Menous, I.; Mahiou, L.; Tadjine, R.; Touati, A.; Lefgoum, A. Silicon Nitride Film for Solar Cells. *Renew. Energy***2008**, *33*, 2289–2293, doi:https://doi.org/10.1016/j.renene.2007.12.015.
74. Lipiński, M. Silicon Nitride for Photovoltaic Application. *Arch. Mater. Sci. Eng.***2010**, *46*, 69–87.
75. Ashour, E.S.M.; Sulaiman, M.Y.; Amin, N.; Ibrahim, Z. Silicon Nitride Passivation of Silicon Nanowires Solar Cell. *J. Phys. Conf. Ser.***2013**, *431*, doi:https://doi.org/10.1088/1742-6596/431/1/012021.
76. Suzuki, T.; Masuda, A.; Ohdaira, K. Effect of a Silicon Nitride Film on the Potential-Induced Degradation of n-Type Front-Emitter Crystalline Silicon Photovoltaic Modules. *Jpn. J. Appl. Phys.***2020**, *59*, 104002, doi:https://doi.org/10.35848/1347-4065/abb39e.
77. Iwahashi, T.; Morishima, M.; Fujibayashi, T.; Yang, R.; Lin, J.; Matsunaga, D. Silicon Nitride Anti-Reflection Coating on the Glass and Transparent Conductive Oxide Interface for Thin Film Solar Cells and Modules. *J. Appl. Phys.***2015**, *118*, 145302, doi:https://doi.org/10.1063/1.4932639.
78. Nagel, H.; Aberle, A.G.; Hezel, R. Optimised Antireflection Coatings for Planar Silicon Solar Cells Using Remote PECVD Silicon Nitride and Porous Silicon Dioxide. *Prog. Photovoltaics Res. Appl.***1999**, *7*, 245–260, doi:https://doi.org/10.1002/(SICI)1099-159X(199907/08)7:4<245::AID-PIP255>3.0.CO;2-3.
79. Ghosh, H.; Mitra, S.; Siddiqui, M.S.; Saxena, A.K.; Chaudhuri, P.; Saha, H.; Banerjee, C. Back Scattering Involving Embedded Silicon Nitride (SiN) Nanoparticles for c-Si Solar Cells. *Opt. Commun.***2018**, *413*, 63–72, doi:https://doi.org/10.1016/j.optcom.2017.12.028.
80. Basch, A.; Beck, F.; Söderström, T.; Varlamov, S.; Catchpole, K.R. Enhanced Light Trapping in Solar Cells Using Snow Globe Coating. *Prog. Photovoltaics Res. Appl.***2012**, *20*, 837–842, doi:https://doi.org/10.1002/pip.2240.

81. Zou, J.; Yu, Y.; Ye, M.; Liu, L.; Deng, S.; Zhang, X. Ultra Efficient Silicon Nitride Grating Coupler with Bottom Grating Reflector. *Opt. Express***2015**, *23*, 26305, doi:https://doi.org/10.1364/oe.23.026305.
82. Iyengar, V. V; Nayak, B.K.; Gupta, M.C. Optical Properties of Silicon Light Trapping Structures for Photovoltaics. *Sol. Energy Mater. Sol. Cells***2010**, *94*, 2251–2257, doi:https://doi.org/10.1016/j.solmat.2010.07.020.
83. Hegedüs, N.; Balázsi, K.; Balázsi, C. Silicon Nitride and Hydrogenated Silicon Nitride Thin Films: A Review of Fabrication Methods and Applications. *Materials (Basel).***2021**, *14*, 5658, doi:10.3390/ma14195658.
84. Hezel, R. *Si Silicon: Silicon Nitride in Microelectronics and Solar Cells*; Springer Science & Business Media, 2013; ISBN 3662099012.
85. Adams, A.C. Silicon Nitride and Other Insulator Films. In *Plasma Deposited Thin Films*; Mort, J., Jansen, F., Eds.; CRC Press: Boca Raton, FL USA, 1986; pp. 129–159.
86. Yen, B.K.; White, R.L.; Waltman, R.J.; Dai, Q.; Miller, D.C.; Kellock, A.J.; Marchon, B.; Kasai, P.H.; Toney, M.F.; York, B.R.; *et al.* Microstructure and Properties of Ultrathin Amorphous Silicon Nitride Protective Coating. *J. Vac. Sci. Technol. A* **2003**, *21*, 1895–1904, doi:https://doi.org/10.1116/1.1615974.
87. Ulrich, R.K.; Phillips, A.J.; Yi, D.H.; Brown, W.D.; Ang, S.S. Mechanical Stability of PECVD Silicon Nitride Protective Films Over Bondwires, Bonds and Bondpads During Thermal Stress. *Microelectron. Reliab.***1991**, *31*, 1237–1249, doi:https://doi.org/10.1016/0026-2714(91)90314-W.
88. Kaloyeros, A.E.; Pan, Y.; Goff, J.; Arkles, B. Review—Silicon Nitride and Silicon Nitride-Rich Thin Film Technologies: State-of-the-Art Processing Technologies, Properties, and Applications. *ECS J. Solid State Sci. Technol.***2020**, *9*, 063006, doi:https://doi.org/10.1149/2162-8777/aba447.
89. Kubacki, R.M. Low Temperature Plasma Deposition of Silicon Nitride to Produce Ultra-Reliable, High Performance, Low Cost Sealed Chip-on-Board (SCOB) Assemblies. *IEEE Trans. Components, Packag. Manuf. Technol. Part A***1995**, *18*, 471–477, doi:10.1109/95.465139.
90. Kaloyeros, A.E.; Jové, F.A.; Goff, J.; Arkles, B. Review—Silicon Nitride and Silicon Nitride-Rich Thin Film Technologies: Trends in Deposition Techniques and Related Applications. *ECS J. Solid State Sci. Technol.***2017**, *6*, P691–P714, doi:https://doi.org/10.1149/2.0011710jss.
91. Roch, I.; Buchaillot, L.; Wallart, X.; Collard, D. Silicon Nitride as an Effective Protection Against Oxidation of a TiNi Thin Film in High Temperature Oxidizing Air Environment at Atmospheric Pressure. *J. Vac. Sci. Technol. B Microelectron. Nanom. Struct. Process. Meas. Phenom.***2001**, *19*, 305–307, doi:https://doi.org/10.1116/1.1338552.
92. Senichev, A.; Peana, S.; Martin, Z.O.; Yesilyurt, O.; Sychev, D.; Lagutchev, A.S.; Boltasseva, A.; Shalaev, V.M. Silicon Nitride Waveguides with Intrinsic Single-Photon Emitters for Integrated Quantum Photonics. *ACS Photonics***2022**, *9*, 3357–3365, doi:https://doi.org/10.1021/acsphotonics.2c00750.
93. Fan, Y.-R.; Lyu, C.; Yuan, C.-Z.; Deng, G.-W.; Zhou, Z.-Y.; Geng, Y.; Song, H.-Z.; Wang, Y.; Zhang, Y.-F.; Jin, R.-B.; *et al.* Multi-Wavelength Quantum Light Sources on Silicon Nitride Micro-Ring Chip. *Laser Photon. Rev.***2023**, *17*, 2300172, doi:https://doi.org/10.1002/lpor.202300172.
94. Sharma, T.; Wang, J.; Kaushik, B.K.; Cheng, Z.; Kumar, R.; Wei, Z.; Li, X. Review of Recent Progress on Silicon Nitride-Based Photonic Integrated Circuits. *IEEE Access***2020**, *8*, 195436–195446, doi:https://doi.org/10.1109/ACCESS.2020.3032186.
95. Fink, J.M.; Kalaee, M.; Pitanti, A.; Norte, R.; Heinzle, L.; Davanço, M.; Srinivasan, K.; Painter, O. Quantum Electromechanics on Silicon Nitride Nanomembranes. *Nat. Commun.***2016**, *7*, 1–10, doi:https://doi.org/10.1038/ncomms12396.

96. Senichev, A.; Martin, Z.O.; Peana, S.; Sychev, D.; Xu, X.; Lagutchev, A.S.; Boltasseva, A.; Shalaev, V.M. Room-Temperature Single-Photon Emitters in Silicon Nitride. *Sci. Adv.***2021**, *7*, 1–7, doi:https://doi.org/10.1126/sciadv.abj0627.
97. Zhao, Y.; Okawachi, Y.; Jang, J.K.; Ji, X.; Lipson, M.; Gaeta, A.L. Near-Degenerate Quadrature-Squeezed Vacuum Generation on a Silicon-Nitride Chip. *Phys. Rev. Lett.***2020**, *124*, 193601, doi:https://doi.org/10.1103/PhysRevLett.124.193601.
98. Elshaari, A.W.; Pernice, W.; Srinivasan, K.; Benson, O.; Zwiller, V. Hybrid Integrated Quantum Photonic Circuits. *Nat. Photonics***2020**, *14*, 285–298, doi:https://doi.org/10.1038/s41566-020-0609-x.
99. Blumenthal, D.J.; Heideman, R.; Geuzebroek, D.; Leinse, A.; Roeloffzen, C. Silicon Nitride in Silicon Photonics. *Proc. IEEE***2018**, *106*, 2209–2231, doi:10.1109/JPROC.2018.2861576.
100. Xiang, C.; Jin, W.; Bowers, J.E. Silicon Nitride Passive and Active Photonic Integrated Circuits: Trends and Prospects. *Photonics Res.***2022**, *10*, A82, doi:https://doi.org/10.1364/prj.452936.
101. Gritsenko, V.A.; Perevalov, T. V; Orlov, O.M.; Krasnikov, G.Y. Nature of Traps Responsible for the Memory Effect in Silicon Nitride. *Appl. Phys. Lett.***2016**, *109*, 62904, doi:https://doi.org/10.1063/1.4959830.

Index

B. S. Bal et al. (eds.), *Silicon Nitride Bioceramics*,
https://doi.org/10.1007/978-3-031-67047-3

GPSR Compliance
The European Union's (EU) General Product Safety Regulation (GPSR) is a set of rules that requires consumer products to be safe and our obligations to ensure this.

If you have any concerns about our products, you can contact us on

ProductSafety@springernature.com

In case Publisher is established outside the EU, the EU authorized representative is:

Springer Nature Customer Service Center GmbH
Europaplatz 3
69115 Heidelberg, Germany

www.ingramcontent.com/pod-product-compliance
Lightning Source LLC
Chambersburg PA
CBHW070014020626
46050CB00003B/35
9783031670466